SPECTRAL THEORY OF DIFFERENTIAL OPERATORS

Self-Adjoint Differential Operators

SPECTRAL THEORY OF DIFFERENTIAL OPERATORS

Self-Adjoint Differential Operators

V. A. Il'in
Moscow State University
Moscow, Russia

CONSULTANTS BUREAU • NEW YORK, LONDON, AND MOSCOW

Library of Congress Cataloging-in-Publication Data

Il'in, V. A.
[Spektral'naía teoriía differentsial'nykh operatorov. English]
Spectral theory of differential operators : self-adjoint differential operators / V.A. Il'in.
p. cm.
Includes bibliographical references and index.
ISBN 0-306-11037-7
1. Selfadjoint operators. I. Title.
QA329.2.I4513 1995
515'.7242--dc20 95-36856
CIP

ISBN 0-306-11037-7

A Division of Plenum Publishing Corporation
233 Spring Street, New York, N.Y. 10013

10 9 8 7 6 5 4 3 2 1

Preface

This monograph presents a systematized exposé of the major results obtained by the author in the field of the spectral theory of self-adjoint elliptic operators.

Our main goal was to demonstrate the efficiency of a number of novel techniques, based mostly on the mean-value theorems for regular solutions of elliptic equations with a spectral parameter.

This monograph consists of four chapters and two appendices.

Chapter 1 deals with a systematic study of the so-called fundamental systems of functions (FSF) of the Laplace operator. The concept of FSF (first introduced by the author in 1967–1968) makes it possible, in studying spectral decompositions, to avoid specifying the boundary conditions in any form and to develop a technique comprising all self-adjoint nonnegative extensions of the Laplace operator (including self-adjoint extensions in a broader domain) possessing a purely point spectrum (admitting, however, an infinite multiplicity for the eigenvalues and a dense set of limit points — a very real possibility, as we shall see later).

In Section 1.1 we establish an estimate exact to within an order of magnitude, both from below and from above, for a certain "aggregate" composed of the sum of the squares of the fundamental functions constituting an arbitrary FSF, and provide a detailed analysis of the spectrum of an arbitrary FSF. We show that, for an arbitrarily specified countable set of real nonnegative numbers in an N-dimensional domain ($N \geq 2$), there exists a FSF of the Laplace operator such that for this system this countable set is a subset of eigenvalues. We note that the eigenvalues that remain after eliminating all the eigenvalues of the countable set are not distributed more densely than the eigenvalues of the first boundary-value problem for the Laplace operator.

This fact allows us to draw a number of important conclusions concerning the growth of the leading term with respect to λ in the asymptotic behavior of the number $n(\lambda)$ of eigenvalues not exceeding λ. These conclusions do not fit in with the conventional concepts stemming from the works of H. Weil and T. Carleman.

In Section 1.2, we construct the so-called fractional kernels for an arbitrary FSF of the Laplace operator. These kernels are the kernels of integral operators corresponding to arbitrary real positive powers of the integral operator which is generated by the respective Green's function for the equation $-\Delta u + u = 0$.

Making use of our technique, we show that, for any interior point x and for any real $\alpha > 0$, there exists a fractional kernel $T_\alpha(x, y)$ representable in the form

$$T_\alpha(x, y) = \overset{\circ}{T}_\alpha(x, y) + \chi_\alpha(x, y),$$

where $\overset{\circ}{T}_\alpha(x, y)$ is the so-called Bessel–MacDonald kernel, which is expressed through the cylindrical MacDonald function $K_\nu(r)$ as

$$\overset{\circ}{T}_\alpha(x, y) = 2^{1-\alpha}[(2\pi)^{N/2}\Gamma(\alpha)]^{-1}|x - y|^{\alpha - N/2} K_{N/2-\alpha}(|x - y|),$$

and $\chi_\alpha(x, y)$ is a function possessing continuous partial derivatives of an arbitrarily high order with respect to the coordinates of points x and y in an open domain.

In Section 1.3, we apply our technique to estimate, in the L_2 metric, the remainder term of a spectral function corresponding to an arbitrary FSF of the Laplace operator. Next, in terms of the estimate obtained, we derive a number of corollaries which, in particular, enable us to establish in Section 1.4, for a compactly supported function, the definitive conditions for uniform convergence of the Fourier series in an arbitrary FSF of the Laplace operator in the Sobolev–Liouville L_p^α classes.

For an arbitrary N-dimensional domain and an arbitrary FSF of the Laplace operator, these conditions for uniform convergence, definitive in terms of the L_p^α classes, take the form of three inequalities:

$$\alpha \geq \frac{N-1}{2}, \qquad \alpha p > N, \qquad p \geq 1. \tag{$*$}$$

The next major result in Section 1.4 is the proof (for each individual FSF of the Laplace operator) of the negative-type theorem stating that the Fourier series in this FSF lacks the property of localization for a compactly supported function in the Sobolev–Liouville class L_p^α with any fixed $\alpha < (N-1)/2$, $p \geq 1$.

We also show in Section 1.4 that for an arbitrary compactly supported function, the inequality $\alpha \geq (N-1)/2$ is the definitive condition (in terms of the Sobolev–Liouville classes L_p^α) ensuring the localization of its Fourier series in an arbitrary FSF of the Laplace operator.

In Section 1.5, we discuss the possible generalizations of the theory we have developed.

Chapter 2 is dedicated to the study of arbitrary self-adjoint nonnegative extensions of the Laplace operator whose spectrum is not necessarily a point spectrum.

We show that our technique, with allowance for the familiar Gårding–Browder–Mautner theorem on the ordered spectral representation of the space L_2 with respect to an arbitrary self-adjoint extension of the elliptic operator, can be extended to the case of an arbitrary self-adjoint nonnegative extension of the Laplace operator with any type (point, continuous, or mixed) of spectrum.

In Chapter 2 we also show that our method enables one to establish exact conditions for uniform convergence of not only the spectral decompositions themselves, but also of the so-called *Riesz means* of these spectral decompositions. The transition to the Riesz means is advantageous in the sense that it enables one to relax the constraint of smoothness imposed on an expandable function to ensure its uniform convergence.

Finally, another important topic covered in Chapter 2 is the specification of conditions for uniform convergence of spectral decompositions and their Riesz means definitive not only in terms of the Sobolev–Liouville classes L_p^α, but also in terms of each of the Nikol'skii H_p^α, Besov $B_{p,\theta}^\alpha$, and Zygmund–Hölder C_α classes.

In Sections 2.2–2.5 we establish that the three inequalities

$$\alpha \geq \frac{N-1}{2} - s, \qquad \alpha p > N, \qquad p \geq 1$$

constitute the condition for a uniform (on any compact set) convergence of the Riesz means of order s, definitive in terms of each of the four classes

$$L_p^\alpha, \quad H_p^\alpha, \quad B_{p,\theta}^\alpha \qquad (\text{for any} \quad \theta \geq 1) \qquad C^\alpha,$$

for an arbitrary N-dimensional domain, for any s on the half-interval $0 \leq s < (N-1)/2$, and for any function compactly supported in this domain. For the case of the Zygmund–Hölder class C^α, only the first of the three inequalities applies. Ultimately, the inequality

$$\alpha \geq \frac{N-1}{2} - s$$

is the definitive condition, in terms of each of the four classes

$$L_2^\alpha, \quad H_2^\alpha, \quad B_{2,\theta}^\alpha \qquad (\text{for any} \quad \theta \geq 1) \qquad C^\alpha,$$

which ensures the localization of the Riesz means of order s of spectral decompositions.

Of prime importance is here the negative-type theorem, which we prove in Section 2.4. For each individual self-adjoint nonnegative extension of the Laplace operator in an arbitrary N-dimensional domain and for any s in the half-interval $0 \leq s < (N-1)/2$, this theorem shows that a compactly supported function belonging to the Zygmund–Hölder class C^α with an arbitrarily fixed $\alpha < (N-1)/2 - s$ (and, consequently, belonging to each of the classes L_p^α, H_p^α, and $B_{p,\theta}^\alpha$ with arbitrarily fixed $\alpha < (N-1)/2 - s$, $p \geq 1$, $\theta \geq 1$) lacks the property of localization of the Riesz means of order s for its spectral decompositions.

We wish to emphasize that the results, obtained in Sections 2.2–2.5, are new ones (and definitive in each of the four classes mentioned) also for the Riesz means of expansions into an N-fold trigonometric Fourier series (with spherical sums).

In Sections 2.6 and 2.7 we show that our method allows one, without resorting to the conventional Carleman technique and to the Tauber theorem formalism, to estimate (in both the L_2 and L_∞ metrics) the remainder term of the Riesz means of any nonnegative order for the spectral function corresponding to an arbitrary self-adjoint nonnegative extension of the Laplace operator. In contrast to other authors who are concerned with this problem, we have obtained an estimate for the remainder term of the Riesz means of a spectral function for which the order of smallness at $s > (N-1)/2$ grows unboundedly as the order s of the Riesz means is allowed to increase.

Chapter 3 is devoted to the problem of the Riesz equisummability of spectral decompositions corresponding to two arbitrary self-adjoint nonnegative extensions of the Laplace operator. This problem is dealt with in both a classical and a generalized sense.

As shown in Chapter 2, if a function $f(x)$, compactly supported in an arbitrary N-dimensional domain G, belongs in this domain to one of the four classes L_2^α, H_2^α, $B_{2,\theta}^\alpha$, or C^α, then for any s in the half-interval $0 \leq s < (N-1)/2$ the condition $\alpha \geq (N-1)/2 - s$ ensures the classical equisummability of the Riesz means of order s, that is, ensures a uniform (on any compact set of domain G) tendency to zero of the difference of the Riesz means of order s for the spectral decompositions corresponding to two arbitrary self-adjoint nonnegative extensions of the Laplace operator in domain G (or in any domain to which this domain is interior).

In Section 3.1 we prove that the above result is a definitive one, that is, for a function $f(x)$, finite in domain G and belonging to the Zygmund–Hölder class (the more so, to any of the classes L_2^α, H_2^α and $B_{2,\theta}^\alpha$), for any fixed positive α subject to $\alpha < (N-1)/2$, by and large, no equisummability (either pointwise or uniform on any compact set of domain G) of the Riesz means of order s of two spectral decompositions takes place.

This result signifies, in particular, that the main theorems of negative type, as established in Chapters 1 and 2, are, in principle, impossible to obtain by analyzing the convergence of the Riesz means of one special decomposition (for example, by expanding in an N-fold Fourier integral) and subsequently making use of the equisummability of the Riesz means of two spectral decompositions.

In Section 3.1 we also establish an estimate, exact to within an order of magnitude and uniform on any compact set of domain G, for the difference of the Riesz means of order $s \geq 0$ of two arbitrary spectral decompositions of an arbitrary function in the class L_2.

In Section 3.2, the Riesz equisummability of two spectral decompositions of an arbitrary function in the class L_2 is studied in a generalized sense, that is, in the sense of convergence to zero of the difference of the Riesz means of the two spectral decompositions almost everywhere in the domain of interest.

We prove that if a function $f(x)$ belongs to the class L_2 in an arbitrary N-dimensional domain G, then the Riesz means of any positive order s of two arbitrary self-adjoint nonnegative extensions of the Laplace operator in this domain are equisummable in G not only in the generalized sense, but also the difference of the said Riesz means of order s and of size λ has almost everywhere in G an order of smallness $o(\lambda^{-s/2})$ growing unboundedly as the order s of the Riesz means is allowed to increase.

It is also shown in Section 3.2 that if a function $f(x)$ belongs, in an arbitrary N-dimensional domain G, to the class L_2 and becomes zero within a certain domain D interior to G, then the Riesz means of order $s > 0$ and of size λ of any self-adjoint nonnegative extension of the Laplace operator in domain G has an order of $o(\lambda^{-s/2})$ almost everywhere in domain D.

In Chapter 4, we apply our technique to the case of a general self-adjoint elliptic operator L of second order with smooth coefficients.

We consider self-adjoint nonnegative extensions of this operator in the space L_2 with a strictly positive weight factor $\rho_0(x)$. The use of this factor makes it possible to consider an elliptic operator of second order on an arbitrary Riemannian manifold, in particular, the Beltrami–Laplace operator on a closed manifold in an arbitrary (not necessarily symmetric or harmonic) Riemannian space.

The starting point of our technique is the mean-value formula for the regular solution of an elliptic equation of second order with spectral parameter and weight factor $\rho_0(x)$ in the special form as established by E. I. Moiseev.

In Section 4.1 we give the Moiseev mean-value formula and some needed results from the theory of ordered spectral representations of the weighted space L_2 and from the theory of fractional powers of the main integral operator. The major result of Section 4.1 is an integral estimate, exact to within an order of magnitude, for the squares of fundamental functions.

In Section 4.2, we prove, for an individual self-adjoint nonnegative extension of an elliptic operator of second order, the main theorem of negative type stating that the Riesz means of order s of the spectral decomposition of a finite function belonging to the Zygmund–Hölder class C^α with an arbitrarily fixed α on the $0 < \alpha < (N-1)/2 - s$ interval lack the property of localization for any s in the half-interval

$$0 \le s < \frac{N-1}{2}.$$

Whence it follows, as in the case of the Laplace operator, that the property of localization of the Riesz means of order s is also lacking in the spectral decomposition of a finite function belonging to any of the classes L_p^α, H_p^α, and $B_{p,\theta}^\alpha$ at any fixed α, p, and θ subject to the conditions

$$0 < \alpha < \frac{N-1}{2} - s, \qquad p \ge 1, \qquad \theta \ge 1.$$

In proving the central theorem of negative type, we first establish a lower bound estimate for the Lebesgue function of the Riesz means of a "small" nonnegative order

s (if this can be accomplished using a formalism similar to that expounded in Chapter 2); next, we estimate, in the L_1 metric, the Riesz means of a spectral function of any order $s < (N-1)/2$.

In Section 4.3, the main theorem of positive type is proved, in which the reasoning, for simplicity, is focused on the spectral decompositions.

Undoubtedly, of independent interest are the lemma on the Fourier images of a finite function in the Nikol'skii class and the mean-value estimation over a geodetical sphere for a function of the Nikol'skii class.

A result central to Chapter 4 is the proof that, for an arbitrary self-adjoint nonnegative extension of the general elliptic operator of second order and for an arbitrary compactly supported function, the above three inequalities (*) are definitive conditions that determine the convergence of a spectral decomposition on any compact set in the classes L_p^α, H_p^α, $B_{p,\theta}^\alpha$ (for any $\theta \geq 1$) and C^α, and the condition $\alpha \geq (N-1)/2$ is a definitive one for the localization of a spectral decomposition in each of the classes L_2^α, H_2^α, $B_{2,\theta}^\alpha$ (for any $\theta \geq 1$) and C^α.

This monograph has two appendices, which deal with situations in which the function to be expanded is not compactly supported.

In Appendix I, conditions that are definitive in the periodic Sobolev–Liouville classes have been established for the uniform convergence of an N-fold trigonometric Fourier series with spherical partial functions for an arbitrary odd $N > 1$ within a closed N-dimensional cube.

In Appendix II, for a function with no compact support, the definitive (in the Sobolev–Liouville classes) conditions of the uniform (on any compact set) convergence of expansions in the eigenfunctions of the first, second, or third homogeneous boundary-value problem have been established for an elliptic operator of second order in an arbitrary bounded N-dimensional domain (for any odd $N > 1$).

Many of the results in this monograph have been presented in a conceptually revised form, not necessarily identical with their original argumentation.

Some of the results presented in this book have not been reported before. Among them, we mention, for instance, the following:

(i) a method for estimating, in the L_∞ metric, the remainder term of the Riesz means of any nonnegative order of a spectral function, without resorting to the conventional Carleman technique and to the Tauber theorem formalism;

(ii) an integral estimate of the squares of fundamental functions for a general elliptic operator of second order;

(iii) conditions, exact in the periodic Sobolev–Liouville classes, for the uniform (within a closed N-dimensional cube) convergence of a multiple trigonometric Fourier series with spherical partial sums (for odd $N > 1$);

(iv) conditions, exact in the Sobolev–Liouville classes, for the uniform (on any compact set) convergence of expansions in the eigenfunctions of, respectively, first, second, and third boundary-value problems for an elliptic operator of second order in a bounded N-dimensional (for odd $N > 1$) domain.

Many of the results in this book emerged from stimulating discussions which the author had, on numerous occasions, with A. N. Tikhonov, A. A. Samarskii, S. M. Nikol'skii, V. P. Maslov, A. V. Bitsadze, L. D. Kudryavtsev, and with his disciples and colleagues Sh. A. Alimov, E. I. Moiseev and I. A. Shishmarev.

The author gratefully acknowledges those discussions, which were of tremendous help to him in writing this book.

The author is also thankful to A. A. Dezin, who read the manuscript and made a number of valuable remarks concerning its contents.

Contents

Chapter 1. Expansion in the Fundamental System of Functions of the Laplace Operator 1

1.1 Fundamental Systems of Functions and Their Properties 2

1.1.1 Concept of Fundamental System of Functions. Mean-Value Formula 2

1.1.2 Exact Estimates for the Sum of the Squares of Eigenfunctions and the Resulting Corollaries 5

1.1.3 The Spectral Structure of an Arbitrary FSF 19

1.1.4 Proof of Lemma 1.1 26

1.2 Fractional Kernels 28

1.2.1 Existence and Representation of Fractional Kernels 29

1.2.2 Proof of Ancillary Lemmas 36

1.3 Estimate for the Remainder Term of a Spectral Function in the Metric L_2 and the Resulting Corollaries 40

1.3.1 Estimate for the Remainder Term of a Spectral Function in the Metric L_2 40

1.3.2 Proof of Lemma 1.4 47

1.3.3 Corollaries to the Theorem on Estimation of Spectral Function in the Metric L_2 53

1.4 Exact Conditions for the Localization and Uniform Convergence of Expansions with Respect to an Arbitrary FSF in the Sobolev–Liouville Classes 68

1.4.1 Conditions for the Uniform Convergence of Fourier Series in an Arbitrary FSF 69

1.4.2 Conditions for Fourier Series Localization with Respect to an Arbitrary FSF 75

1.4.3 Conditions for the Nonlocalization of Fourier Series with Respect to an Arbitrary FSF 76

1.4.4 On the Exactness of the Established Conditions for Uniform Convergence and Localization 78

1.4.5 Conditions for Fourier Series Divergence with Respect to an Arbitrary FSF in the Metric L_p and on a Set of Positive Measure . . . 79
1.5 On the Potential Generalization of the Theory . . . 79

Comments on Chapter 1 **80**

Chapter 2. Spectral Decompositions Corresponding to an Arbitrary Self-Adjoint Nonnegative Extension of the Laplace Operator **83**

2.1 Self-Adjoint Nonnegative Extensions of Elliptic Operators. Ordered Spectral Representations of the Space L_2. Classes of Differentiable Functions of N Variables . . . 84
2.1.1 Self-Adjoint Nonnegative Extensions of Elliptic Operators . . . 84
2.1.2 Ordered Spectral Representations of the Space $L_2(G)$ (with Respect to Extension $\hat{A}$) . . . 87
2.1.3 Classes of Differentiable Functions of N Variables . . . 88
2.2 Formulation and Analysis of Main Results . . . 93
2.2.1 Formulation of the Main Theorems and the Resulting Corollaries . . . 93
2.2.2 Brief Analysis of Results . . . 97
2.3 Certain Properties of the Fundamental Functions of an Arbitrary Ordered Spectral Representation in the Space L_2 . . . 98
2.3.1 The Mean-Value Formula. Expression for the Fourier Image of a Function from the Class of Radial Functions . . . 98
2.3.2 Integral Estimate for the Square of the Fundamental Function 99
2.3.3 Fractional Kernels . . . 103
2.4 Proof of Negative Theorem 2.1 . . . 109
2.4.1 Lower-Bound Estimate for the Lebesgue Functions of the Riesz Means . . . 109
2.4.2 Proof of Ancillary Estimates . . . 120
2.4.3 Lemma on the Relation between the Riesz Means of Integrals $\int_0^\lambda (t+1)^\beta U(t)d\rho(t)$ and $\int_0^\lambda U(t)d\rho(t)$ . . . 126
2.4.4 Lemma on the Unboundedness in L_1 of the Riesz Means of Order $s \geq 0$ of the Spectral Decompositions of Kernels of Order $\alpha < (N-1)/4 - s/2$ . . . 129
2.4.5 Direct Proof of Theorem 2.1 . . . 133
2.5 Proof of Positive Theorem 2.3 . . . 135
2.5.1 Lemmas on the Fourier Images of a Finite Function in the Nikol'skii Class . . . 135
2.5.2 Properties of Derivatives Taken with Respect to the Sphere Radius and to the Sphere Means for Functions of the Nikol'skii Class . . . 148

2.5.3 Main Estimate for the Riesz Means of a Spectral Decomposition . . . 165
2.5.4 Direct Proof of Theorem 2.3 . . . 173
2.6 Estimate for the Remainder Term of the Riesz Means of a Spectral Function in the Metric L_2 . . . 174
2.7 Estimate for the Remainder Term of the Riesz Means of a Spectral Function in the Metric L_∞ . . . 180
2.7.1 Proof of the Main Theorem . . . 180
2.7.2 Proof of Ancillary Lemma 2.12 . . . 189

Comments on Chapter 2 **194**

Chapter 3. On the Riesz Equisummability of Spectral Decompositions in the Classical and the Generalized Sense **197**
3.1 On the Riesz Equisummability of Spectral Decompositions in the Classical Sense . . . 199
3.1.1 Lemma on the Estimation in L_1 of the Difference of the Riesz Means for Spectral Functions . . . 201
3.1.2 Lemma on the Unboundedness of the Difference of the Fractional-Kernel Riesz Means . . . 215
3.1.3 Direct Proof of Theorem 3.1 . . . 219
3.1.4 Uniform and Accurate, to within an Order of Magnitude, Estimate of the Riesz Means for Two Spectral Decompositions . . . 221
3.2 On the Riesz Equisummability of Spectral Decompositions in the Generalized Sense . . . 226
3.2.1 Statement of Results . . . 226
3.2.2 Proof of Ancillary Statements . . . 227
3.2.3 Direct Proof of Theorem 3.4 . . . 245

Comments on Chapter 3 **252**

Chapter 4. Self-Adjoint Nonnegative Extensions of an Elliptic Operator of Second Order **253**
4.1 Ancillary Propositions about Fundamental Functions . . . 254
4.1.1 Ordered Weighted Spectral Representations of the Space L_2 . . . 254
4.1.2 Moiseev Mean-Value Formula . . . 256
4.1.3 Estimation of the Integral of the Squares of Fundamental Functions . . . 259
4.1.4 Fractional Powers of Self-Adjoint Extensions of Elliptic Operators . . . 275
4.2 Theorems of Negative Type . . . 277
4.2.1 Proof of the Central Lemma for the Riesz Means of a Small Nonnegative Order . . . 278

4.2.2 Proof of the Main Inequality (4.2.14) 285
4.2.3 On the Exactness (to within an Order of Magnitude) of the Main Estimate for the Central Lemma 295
4.2.4 Lemma on the Riesz Means of Any Nonnegative Order 297
4.2.5 Direct Proof of Theorem 4.1 301
4.3 Theorems of Positive Type 303
4.3.1 Lemma on the Fourier Images of a Finite Function in the Nikol'skii Class 305
4.3.2 Proof of the Main Local Asymptotic Estimate 313
4.3.3 Proof of Estimate (4.3.53) 325
4.3.4 Direct Proof of the Localization Theorem 333
4.3.5 The Mean Geodetic-Sphere Estimate for a Function of the Nikol'skii Class 335
4.3.6 Proof of the Main Asymptotic Estimate 342
4.3.7 Direct Proof of the Uniform Convergence Theorem 347

Comments on Chapter 4 **348**

Appendix 1. Conditions for the Uniform Convergence of Multiple Trigonometric Fourier Series with Spherical Partial Sums **351**

Appendix 2. Conditions for the Uniform Convergence of Decompositions in Eigenfunctions of the First, Second, and Third Boundary-Value Problems for an Elliptic Operator of Second Order **355**

Epilogue **367**

References **379**

Index **387**

Nomenclature

E^N, N-dimensional Euclidean space.

G, arbitrary domain in E^N.

Ω, arbitrary N-dimensional subdomain of domain G.

D, most commonly, subdomain in Ω or G.

$J_\nu(x)$, Bessel function of the first kind of order ν.

$Y_\nu(x)$, Neyman function, or a cylindrical function of the second kind of order ν.

$K_\nu(x)$, MacDonald function, or a cylindrical function of the third kind of order ν.

$\Gamma(s)$, Euler's gamma function.

$\omega_N = 2(\pi)^N[\Gamma(N/2)]^{-1}$, unit sphere surface area in space E^N.

$\rho(\lambda)$ or $\rho(t)$, spectral measure, generated by an ordered spectral representation of the space L_2 with respect to a particular self-adjoint extension of an elliptic operator.

$E_\lambda(x)$ or $\sigma_\lambda(x, f)$, spectral decomposition of function $f(x)$ at point x.

$E_\lambda^s f(x)$ or $\sigma_\lambda^s(x, f)$, Riesz means of order s of spectral decomposition of function $f(x)$ at point x.

$\theta(x, y, \lambda)$, spectral function of a particular FSF or a particular self-adjoint extension of an elliptic operator.

$\theta^s(x, y, \lambda)$, Riesz means of order s of a spectral function.

$H_p^\alpha(G)$, Nikol'skii class of functions.

$L_p^\alpha(G)$, Sobolev–Liouville class of functions.

$B_{p,\theta}^\alpha(G)$, Besov class of functions.

$C^\alpha(G)$, Zygmund–Hölder class of functions.

$W_p^n(G)$, Sobolev class of functions of integral order n.

Chapter 1

Expansion in the Fundamental System of Functions of the Laplace Operator

In this chapter, we introduce the concept of a fundamental system of functions (FSF) for the simplest elliptic operator, the Laplace operator, defined in an arbitrary (not necessarily bounded) N-dimensional domain. The FSF encompasses the eigenfunction systems of all self-adjoint boundary-value problems for the Laplace operator; for such systems, the spectrum is a pure point spectrum, admitting of an infinite multiplicity and everywhere dense set of limit points for the eigenvalues — quite a realistic situation, as we shall see later.

In Section 1.1, we specify the basic properties of such systems and the properties of their respective set of eigenvalues.

We show that for a certain "aggregate" composed of the sum of the squares of fundamental functions, there exists an estimate exact to within an order of magnitude both from above and from below.

It will also be shown that in an N-dimensional domain ($N \geq 2$), the set of all eigenvalues can include, for a subset, an arbitrarily assigned countable set of nonnegative real numbers.

Section 1.2 is concerned with the construction, for the systems of interest, of the so-called fractional kernels which represent, for a particular boundary-value problem, the kernels of integral operators corresponding to any positive power of an integral operator as generated by Green's function of the respective boundary-value problem.

In Section 1.3, without resorting to the Carleman method and the Tauber theorem techniques, we establish an exact (to within an order of magnitude) estimate in the L_2 metric of the remainder term of the spectral function corresponding to an arbitrary FSF.

In Section 1.4, for compactly supported functions we specify conditions, exact in the Sobolev–Liouville classes, for equiconvergence and localization of the spectral decompositions of an arbitrary FSF of the Laplace operator. These newly established conditions have been shown to be novel (and also exact) even for multiple trigonometric Fourier series with spherical partial sums.

In Section 1.5, we discuss potential generalizations of our theory. We provide evidence that, given the condition established by A. M. Olevsky, the FSF is not necessarily subject to the orthonormalization requirement in the domain of interest or in any other domain exterior to it.

1.1. FUNDAMENTAL SYSTEMS OF FUNCTIONS AND THEIR PROPERTIES

1.1.1. Concept of Fundamental System of Functions. Mean-Value Formula

Numerous studies of the spectral theory of elliptic operators are based, as a rule, on a technique which requires, for its implementation, the specification of boundary conditions — either in an explicit form or by specifying the domain within which an appropriate operator is defined. In order to avoid specifying the boundary conditions in any form and to design a technique that would enable one to study an arbitrary nonnegative self-adjoint extension of the Laplace operator (whose spectrum is a pure point spectrum), we introduce the concept of a fundamental system of functions (FSF) of the Laplace operator. In essence, we introduce the concept of a complete orthonormalized system associated only with a formal differential operator, the Laplace operator.

Let G be an arbitrary (not necessarily bounded) domain in the N-dimensional Euclidean space E^N.

DEFINITION 1. *We refer to a complete system $\{u_n(x)\}$, orthonormalized in an arbitrary N-dimensional domain G, as the fundamental system of functions (FSF) of the Laplace operator in this domain if each function $u_n(x)$ belongs, in the open domain G, to the class $C^{(2)}$ and satisfies the equation*

$$u_n + \lambda_n u_n = 0$$

for a nonnegative number λ_n inside G.

The concept of the fundamental system of functions of the Laplace operator can also be introduced locally, that is, with respect to an arbitrary N-dimensional subdomain Ω of domain G, rather than with respect to the entire N-dimensional domain G within which the system in question is orthonormalized.

DEFINITION 2. *A complete system $u_n(x)$, orthonormalized in an arbitrary N-dimensional domain G, is referred to as the fundamental system of functions (FSF) of* *the Laplace operator in a subdomain* Ω *of domain* G *if each function* $u_n(x)$ *belongs, in the open domain* Ω, *to the class* $C^{(2)}$ *and satisfies the equation*

$$u_n + \lambda_n u_n = 0$$

for a nonnegative number λ_n *inside* Ω.

The concept of an FSF of the Laplace operator (even though in the entire domain G) includes, as a special case, the eigenfunction systems of all classical self-adjoint boundary-value problems for the Laplace operator in an arbitrary domain G (for example, the eigenfunction systems of the first, second, and third boundary-value problems in this domain), an N-fold trigonometric system (in the case where the domain G is an N-dimensional rectangular parallelepiped and the boundary conditions are defined as equal values for the functions $u_n(x)$ themselves or for their normal derivatives on the opposite faces of the parallelepiped), or, in a broader sense, the eigenfunction system of any nonnegative self-adjoint extension of the Laplace operator exhibiting a pure point spectrum in domain G[1)].

We point out that an FSF of the Laplace operator, as any orthonormalized system (either in the entire domain G or in an arbitrary subdomain Ω of the domain G), *is at most countable*, and all of its elements can be numbered.

In what follows we will refer to the separate elements $u_n(x)$ of the FSF in question as the *fundamental functions*, and to their respective numerical values λ_n as the *fundamental values*, or the *eigenvalues*. The collection of all the fundamental values is referred to as the *spectrum* of a particular FSF.

Here and in Section 1.1.2 we focus on the properties of fundamental functions of an arbitrary FSF, and in Section 1.1.3, on the structure of its spectrum.

Since the concept of an FSF of the Laplace operator in the subdomain Ω of domain G is more general than the concept of a FSF over the entire domain G, henceforth we shall formulate the properties of fundamental functions for a FSF of the Laplace operator in an arbitrary subdomain Ω of domain G, admitting, however, the eventual coincidence of Ω and G.

In what follows we agree to use the symbol

$$\int \cdots \int_{\omega} f(x + r\omega)\, d\omega \tag{1.1.1}$$

to denote an integral of the function f over all the angles on the surface of an N-dimensional sphere of radius r centered at point x.

A major characteristic property of the element of an arbitrary FSF is the mean-value formula.

Assume that $u_n(x)$ is the element of an arbitrary FSF of the Laplace operator in any N-dimensional subdomain Ω of an arbitrary N-dimensional domain G, corresponding to the eigenvalue $\lambda_n \neq 0$; x is a point in the subdomain Ω; r is a positive

1) For the definition of an arbitrary nonnegative self-adjoint extension of the Laplace operator, see Chapter 2.

number such that an N-dimensional ball of radius r centered at point x is contained in Ω. Then the following formula holds:

$$\int_{\omega} \cdots \int u_n(x + r\omega)\, d\omega = (2\pi)^{N/2}\, u_n(x) (r\sqrt{\lambda_n})^{(2-N)/2}\, J_{(N-2)/2}(r\sqrt{\lambda_n}), \quad (1.1.2)$$

currently referred to as the mean-value formula[2)].

The derivation of this formula given here is quite simple. To start with, we note that it will suffice to establish the validity of formula (1.1.2) for the values of r which do not reduce the Bessel function $J_{(N-2)/2}(r\sqrt{\lambda_n})$, a cofactor in (1.1.2), to zero. (If so, formula (1.1.2), by its continuity, will hold for the values of r for which the Bessel function in question becomes zero.)

Let r be a value for which the Bessel function in question becomes zero. We denote by the symbol $Y_\nu(t)$ the so-called *Neyman function of order ν in the argument t*[3)]. We apply the second Green formula over an N-dimensional ball Ω_r^x of radius r centered at point x to two functions: to the fundamental function $u_n(y)$ and to the following function:

$$v_n(y) = -\frac{\lambda_n^{\nu/2}}{2^{\nu+2}\pi^\nu} \left[\frac{Y_\nu(|x-y|\sqrt{\lambda_n})}{|x-y|^\nu} - \frac{J_\nu(|x-y|\sqrt{\lambda_n})}{|x-y|^\nu}\, \frac{Y_\nu(r\sqrt{\lambda_n})}{J_\nu(r\sqrt{\lambda_n})} \right].$$

(For convenience, we have put $\nu = (N-2)/2$; the symbol $|x-y|$ refers to the N-dimensional distance between points x and y.)

The second Green formula having been applied, the two volume integrals over the ball Ω_r^x cancel each other, one of the integrals over the surface of this ball goes to zero, and the other integral over the ball surface and the "outlier" at point x, with allowance made for the value of the Wronskian of the Bessel function,

$$\left[J_\nu(r\sqrt{\lambda_n})\, \frac{dY_\nu(r\sqrt{\lambda_n})}{dr} - Y_\nu(r\sqrt{\lambda_n})\, \frac{dJ_\nu(r\sqrt{\lambda_n})}{dr} \right] = \frac{2}{\pi r},$$

yields formula (1.1.2).

We say that a function f *belongs in domain Ω to the class of radial functions* if this function is dependent only on the distance $r = |x-y|$ between a variable point y in domain Ω and a fixed point x in the same domain, and if the f is distinct from zero only within a domain interior to the ball Ω.

To conclude this section, we give an expression for the Fourier coefficient of an arbitrary function f belonging to the class of radial functions in domain Ω and expanded in an arbitrary FSF of the Laplace operator in the subdomain Ω of domain G. Since the function f is distinct from zero only within a ball of radius R centered at

2) The symbol $J_\nu(t)$ denotes a Bessel function of the first kind of order ν in the argument t. For $\lambda_n = 0$, formula (1.1.2) changes over into the familiar mean-value formula for a harmonic function.

3) For the definition and properties of the Neyman function (also referred to as a cylindrical function of the second kind), see Bateman and Erdélyi [9, p.12].

point x, completely interior to Ω, and is dependent solely on the distance $r = |x - y|$, for the desired Fourier coefficient f_n written in a spherical framework with the origin at point x we obtain

$$f_n = \int_0^R f(r) \left(\int \underset{\omega}{\cdots} \int u_n(x + r\omega)\, d\omega \right) r^{N-1}\, dr.$$

A comparison of this equality with the mean-value formula (1.1.2) leads to the following expression for the desired Fourier coefficient:

$$f_n = (f, u_n) = (2\pi)^{N/2} u_n(x)\, \lambda_n^{(2-N)/4} \int_0^R f(r)\, r^{N/2} J_{(N-2)/2}(r\sqrt{\lambda_n}) dr. \qquad (1.1.3)$$

1.1.2. Exact Estimates for the Sum of the Squares of Eigenfunctions and the Resulting Corollaries

In contrast to the ordinary differential operator, there exists no asymptotic representation of separate eigenfunctions for the elliptic operator (the Laplace operator included) in an N-dimensional domain ($N \geq 2$) even for the simplest boundary-value problems for large values of the spectral parameter λ_n[4]. This notwithstanding, there is a way to establish, for an arbitrary FSF of the Laplace operator in an N-dimensional subdomain Ω of an arbitrary N-dimensional domain G, an estimate, both upper- and lower-bound, exact to within an order of magnitude of a certain "aggregate" composed of the squares of eigenfunctions.

THEOREM 1.1. *If $\{u_n(x)\}$ is an arbitrary FSF of the Laplace operator in an arbitrary N-dimensional subdomain Ω of an arbitrary N-dimensional domain G, then the estimate*

$$\sum_{\mu \leq \sqrt{\lambda_n} \leq \mu+1} u_n^2(x) \leq C(\mu + 1)^{N-1} \qquad (1.1.4)$$

holds for any $\mu \geq 0$ uniform with respect to x in each subdomain Ω' strictly interior to domain Ω.

PROOF: (1) To begin with, we wish to prove the validity of estimate (1.1.4) for the case where μ is greater than a certain positive number μ_0 whose choice will be be specified later. We fix an arbitrary subdomain Ω' strictly interior to domain Ω and denote by R a positive number smaller than $\pi/2$ and smaller than the distance

---

[4] For example, for the Laplace operator within a square $[0 \leq x_1 \leq \pi] \times [0 \leq x_2 \leq \pi]$ with a uniform boundary condition of the first kind, two eigenfunctions $u_{mn}(x_1, x_2) = \sin(mx_1)\sin(nx_2)$ and $u_{nm}(x_1, x_2) = \sin(nx_1)\sin(mx_2)$ correspond to each eigenvalue $\lambda_{mn} = m^2 + n^2$ on the square boundary; these eigenfunctions behave in a markedly distinct manner with the eigenvalue $m^2 + n^2$ increasing when the integer m is large and the integer n is small.

between Ω' and the boundary of domain Ω. Assuming x to be a fixed point inside domain Ω', let us consider a function of argument $r = |x - y|$:

$$v(r) = \begin{cases} \mu^{N/2}(2\pi)^{-N/2} r^{(2-N)/2} J_{(N-2)/2}(\mu r) & \text{for} \quad R/2 < r < R, \\ 0 & \text{for other} \quad r. \end{cases} \tag{1.1.5}$$

Since function (1.1.5) belongs in domain Ω to the class of radial functions, for its Fourier coefficient v_n in system $\{u_n(y)\}$ a relation of the form (1.1.3) holds:

$$\begin{aligned} v_n &= (v, u_n) = (2\pi)^{N/2} u_n(x) \lambda_n^{(2-N)/4} \int_0^R v(r) r^{N/2} J_{(N-2)/2}(r\sqrt{\lambda_n})\, dr \\ &= \mu^{N/2} \lambda_n^{(2-N)/4} u_n(x) \int_{R/2}^R J_{(N-2)/2}(\mu r) J_{(N-2)/2}(r\sqrt{\lambda_n}) r\, dr. \end{aligned} \tag{1.1.6}$$

We wish to prove, using formula (1.1.6), that if the positive number μ_0 is large enough, then for $\mu \geq \mu_0$ for all n's such that $\mu \leq \sqrt{\lambda_n} \leq \mu + 1$, there is a constant $\alpha > 0$ such that

$$|v_n| \geq \alpha\, |u_n(x)|. \tag{1.1.7}$$

Inequality (1.1.7) having been proved, we write the Bessel inequality for function (1.1.5) in the orthonormal system $\{u_n(y)\}$ and leave on the left-hand side of this inequality only the n-labeled summands satisfying the condition $\mu \leq \sqrt{\lambda_n} \leq \mu + 1$ to obtain[5)]

$$\sum_{\mu \leq \sqrt{\lambda_n} \leq \mu+1} u_n^2(x) \leq \frac{1}{\alpha^2} \int_\Omega v^2(|x-y|) dy = \frac{\mu^N \omega_N}{(2\pi)^N \alpha^2} \int_{R/2}^R r\, J^2_{(N-2)/2}(\mu r)\, dr = O(\mu^{N-1}),$$

which completes the derivation of estimate (1.1.4) for this particular case.

Thus, to prove estimate (1.1.4) for the first case of interest, it suffices to prove that, given a certain $\mu_0 > 0$, there is a constant $\alpha > 0$ such that for all n satisfying the condition $\mu_0 \leq \mu \leq \sqrt{\lambda_n} \leq \mu + 1$, inequality (1.1.7) remains valid.

We observe that if $\mu_0 > 0$ is large enough, then, given $\mu_0 \leq \mu \leq \sqrt{\lambda_n} \leq \mu + 1$, the following estimates hold:

$$\frac{\mu}{\sqrt{\lambda_n}} = 1 + O\left(\frac{1}{\mu}\right), \quad \left(\frac{\mu}{\sqrt{\lambda_n}}\right)^{(N-2)/2} = 1 + O\left(\frac{1}{\mu}\right), \quad \frac{1}{\lambda_n^{1/4}} = \frac{1}{\sqrt{\mu}}\left[1 + O\left(\frac{1}{\mu}\right)\right]. \tag{1.1.8}$$

[5)] Here we make use of the estimate $|J_\nu(x)| \leq C(\nu)x^{-1/2}$, which holds for all $x \geq 0$ and $\nu \geq -1/2$ and which immediately follows the representation of $J_\nu(x)$ in the form of a series for $|x| \leq 1$ and the asymptotic representation for $|x| > 1$ (see Bateman and Erdélyi [9, pp. 12 and 98]). Here ω_N refers to the surface area of an N-dimensional sphere of unit radius equal to $2(\sqrt{\pi})^N[\Gamma(N/2)]^{-1}$, where $\Gamma(s)$ is the Euler Γ-function.

The second estimate (1.1.8) and relation (1.1.6) allow us to assert that, to prove the validity of inequality (1.1.7) for all the n's for which $\mu_0 \le \mu \le \sqrt{\lambda_n} \le \mu+1$, it suffices to make sure that there exists a constant $\beta > 0$ such that, for all the n's specified, the following inequality holds:

$$\mu \int_{R/2}^{R} J_{(N-2)/2}(\mu r)\, J_{(N-2)/2}(r\sqrt{\lambda_n})r\, dr > \beta. \tag{1.1.9}$$

We make use of the known asymptotic formula for the Bessel function $J_\nu(x)$ holding for the values of argument x in excess of unity[6]:

$$J_\nu(x) = \sqrt{\frac{2}{\pi x}} \cos\left(x - \nu\frac{\pi}{2} - \frac{\pi}{4}\right) + O\left(\frac{1}{x^{3/2}}\right). \tag{1.1.10}$$

Now we fix the positive number μ_0 large enough for both estimates (1.1.8) and the inequality $\mu_0 \ge 2/R$ to hold true. Then the arguments of both Bessel functions under the integral sign of the left-hand side of (1.1.9) are in excess of unity, and, making use of the asymptotic formula (1.1.10) for the Bessel functions in question, we arrive at the following asymptotic representation[7]:

$$\begin{aligned} \mu J_{(N-2)/2}(\mu r)\, J_{(N-2)/2}(r\sqrt{\lambda_n})r = \frac{\sqrt{\mu}}{\pi\lambda_n^{1/4}} \Big\{\sin\Big[r\Big(\mu+\sqrt{\lambda_n}\Big) - (N-1)\frac{\pi}{2}\Big] \\ + \cos\Big[r\Big(\mu-\sqrt{\lambda_n}\Big)\Big]\Big\} + \frac{1}{r}O\Big(\frac{1}{\mu}\Big). \end{aligned} \tag{1.1.11}$$

Since, by virtue of the third of relations (1.1.8), the representation $\sqrt{\mu}/\lambda_n^{1/4} = 1+O(1/\mu)$ holds with reference to equality (1.1.11) in order to prove inequality (1.1.9) for the n's of interest, it suffices to verify the following three estimates:

$$\frac{1}{\pi}\int_{R/2}^{R} \sin\Big[r\Big(\mu+\sqrt{\lambda_n}\Big) - (N-1)\frac{\pi}{2}\Big]\, dr = O\Big(\frac{1}{\mu}\Big), \tag{1.1.12}$$

$$\int_{R/2}^{R} \frac{1}{r}O\Big(\frac{1}{\mu}\Big)\, dr = O\Big(\frac{1}{\mu}\Big), \tag{1.1.13}$$

$$\frac{1}{\pi}\int_{R/2}^{R} \cos\Big[r\big(\mu-\sqrt{\lambda_n}\big)\Big)\, dr \ge \gamma > 0. \tag{1.1.14}$$

---

[6] See, for example, Bateman and Erdélyi [9, p. 98].

[7] In doing so, also we make use of the third relation in (1.1.8).

The validity of estimates (1.1.12) and (1.1.13) is checked via simple integration. To prove inequality (1.1.14), we note that the quantity on the left-hand side of this inequality is exactly equal to

$$\frac{1}{\pi}\left\{R\frac{\sin\left[R(\mu-\sqrt{\lambda_n})\right]}{R(\mu-\sqrt{\lambda_n})}-\frac{R}{2}\frac{\sin\left[\frac{R}{2}(\mu-\sqrt{\lambda_n})\right]}{\frac{R}{2}(\mu-\sqrt{\lambda_n})}\right\}. \tag{1.1.15}$$

By condition, $R<\pi/2$. Along with this, the inequality $0\leq\sqrt{\lambda_n}-\mu\leq 1$ holds for all the n's in question. For these reasons, the arguments of both sine functions in (1.1.15) do not exceed $\pi/2$ in modulus. However, given $|\rho|\leq\pi/2$, the inequality

$$1\geq\frac{\sin\rho}{\rho}\geq\frac{2}{\pi} \tag{1.1.16}$$

holds[8)].

Inequality (1.1.16) allows us to assert that the value of (1.1.15) is, in any case, not smaller than that of

$$\frac{1}{\pi}\left\{R\frac{2}{\pi}-\frac{R}{2}\right\}=\frac{R(4-\pi)}{2\pi^2}=\gamma>0.$$

This proves inequality (1.1.14). Thus, the proof of estimate (1.1.4) for the case where μ exceeds a prefixed sufficiently large number $\mu_0>0$ has been completed.

(2) To complete the proof of Theorem 1.1, it suffices to establish estimate (1.1.4) for the case in which μ satisfies the inequality $0\leq\mu\leq\mu_0$.

Note that in this case, in order to prove estimate (1.1.4), it suffices to establish the estimate

$$\sum_{\sqrt{\lambda_n}\leq\mu_0}u_n^2(x)=O(\mu_0^N), \tag{1.1.17}$$

uniform with respect to x in any subdomain Ω' strictly interior to domain Ω.

By analogy with the above, we fix an arbitrary subdomain Ω' strictly interior to domain Ω, an arbitrary point x in Ω', and a positive number R smaller than the distance between Ω' and the boundary of domain Ω, and consider the following function of argument $r=|x-y|$:

$$w(r)=\begin{cases}\Gamma\left(\dfrac{N}{2}+1\right)\pi^{-N/2}R^{-N} & \text{for}\quad r<R,\\ 0 & \text{for}\quad r>R.\end{cases} \tag{1.1.18}$$

Making use of an equality of the form (1.1.3) to calculate the Fourier coefficient w_n of function (1.1.18) in system $\{u_n(y)\}$, we obtain, with allowance made

---

8) Suffice it to note that the function $(\sin\rho)/\rho$ decreases in the segment $0\leq\rho\leq 1$, since everywhere in this segment its derivative $\rho^{-2}\cos\rho\,(\rho-\tan\rho)$ is nonpositive, keeping in mind that the function $(\sin\rho)/\rho$ is equal to 1 at $\rho=0$ and to $2/\pi$ at $\rho=\pi/2$.

for $\int \rho^{\nu} J_{\nu-1}(\rho)d\rho = \rho^{\nu} J_{\nu}(\rho)$[9],

$$w_n = (2\pi)^{N/2}\, u_n(x)\, \lambda_n^{(2-N)/4}\, \Gamma\left(\frac{N}{2}+1\right)\, \pi^{-N/2}\, R^{-N} \int_0^R r^{N/2}\, J_{(N-2)/2}\left(r\sqrt{\lambda_n}\right) dr$$

$$= \Gamma\left(\frac{N}{2}+1\right)\, 2^{N/2}\, \frac{J_{N/2}(R\sqrt{\lambda_n})}{(R\sqrt{\lambda_n})^{N/2}}\, u_n(x).$$

Thus, Parseval's relation for function (1.1.18) in the complete orthonormalized system $\{u_n(y)\}$ takes the form

$$\sum_{n=1}^{\infty} u_n^2(x) \left\{\Gamma\left(\frac{N}{2}+1\right)\, 2^{N/2}\, \frac{J_{N/2}(R\sqrt{\lambda_n})}{(R\sqrt{\lambda_n})^{N/2}}\right\}^2$$
$$= w_N \int_0^R \frac{\Gamma^2\left(\frac{N}{2}+1\right)}{\pi^N\, R^{2N}}\, r^{N-1} dr = \frac{\Gamma\left(\frac{N}{2}+1\right)}{\pi^{N/2}\, R^N}. \tag{1.1.19}$$

Since it suffices to specify estimate (1.1.17) for large μ_0 only, we put $R = 1/(2\mu_0)$ in (1.1.19). Thereby, for all the n's for which $\sqrt{\lambda_n} \leq \mu_0$, the argument $R\sqrt{\lambda_n}$ of the Bessel function on the left-hand side of (1.1.19) will not exceed 1/2. It follows immediately from the power-series representation of the Bessel function that for all the n's the term enclosed within braces on the left-hand side of (1.1.19) will exceed in value a certain positive number α_0. Therefore, given $R = 1/(2\mu_0)$, we obtain from (1.1.19)

$$\alpha_0^2 \sum_{\sqrt{\lambda_n} \leq \mu} u_n^2(x) \leq 2^N\, \frac{\Gamma\left(\frac{N}{2}+1\right)}{\pi^{N/2}}\, \mu_0^N.$$

The derivation of estimate (1.1.17) is thus complete, which proves Theorem 1.1.

REMARK 1. If the points of condensation of the fundamental values $\{\lambda_n\}$ fall within the segment $[\mu^2, (\mu+1)^2]$, then the series on the left-hand side of (1.1.4) is convergent (in each subdomain Ω' strictly interior to domain Ω). Such a case, as we shall see in Section 3, is quite realistic.

REMARK 2. *Theorem 1.1 may be reformulated in the following way: if $\{u_n(x)\}$ is an arbitrary FSF of the Laplace operator in any N-dimensional subdomain Ω of an arbitrary N-dimensional domain G, then the estimate*

$$\sum_{|\sqrt{\lambda_n}-\mu|\leq 1} u_n^2(x) = O(\mu^{N-1}) \tag{1.1.20}$$

[9] See, for example, Bateman and Erdélyi [9, p. 20, formula (52)].

holds for any $\mu \geq 1$ uniformly with respect to x in each subdomain Ω' strictly interior to domain Ω.

(Suffice it to note that the summation in (1.1.20) is made over the values of n for which the term $\sqrt{\lambda_n}$ belongs to the segment $[\mu - 1, \mu + 1]$; since this segment is the sum of two segments $[\mu - 1, \mu]$ and $[\mu, \mu + 1]$, for either of these estimate (1.1.4) holds for $\mu \geq 1$.)

Several simple corollaries follow from Theorem 1.1.

COROLLARY 1. *If $\{u_n(x)\}$ is an arbitrary FSF of the Laplace operator in any N-dimensional subdomain Ω of an arbitrary N-dimensional domain G, then the estimate*

$$\sum_{|\sqrt{\lambda_n}-\mu|\leq\rho} u_n^2(x) = \rho\, O(\mu^{N-1}) \tag{1.1.21}$$

holds for any $\mu \geq 1$ and any ρ in the segment $1 \leq \rho \leq \mu$ uniformly with respect to x in each subdomain Ω' strictly interior to domain Ω.

To prove this corollary, it suffices to represent the segment $[\mu - \rho, \mu + \rho]$, within whose confines $\sqrt{\lambda_n}$ is varied in (1.1.21), as the sum of no more[10] than $[2\rho] + 1$ segments with a length not in excess of unity, and to specify an estimate of the form (1.1.4) for each of the segments in question.

COROLLARY 2. *If $\{u_n(x)\}$ is an arbitrary FSF of the Laplace operator in any N-dimensional subdomain Ω of an arbitrary N-dimensional domain G, then the estimates*

$$\sum_{1\leq\lambda_n\leq\lambda} u_n^2(x)\,\lambda_n^{\delta-N/2} = O(\lambda^{\delta}), \tag{1.1.22}$$

$$\sum_{\lambda_n\geq\lambda} u_n^2(x)\,\lambda_n^{-\delta-N/2} = O(\lambda^{-\delta}) \tag{1.1.23}$$

hold for any $\delta > 0$ and for all $\lambda \geq 1$ uniformly with respect to x in each subdomain Ω' strictly interior to domain Ω.

In particular, for any $\delta > 0$ one may assert the uniform-in-Ω' boundedness of the series sum

$$\sum_{\lambda_n\geq 1} u_n^2(x)\,\lambda_n^{-\delta-N/2}. \tag{1.1.24}$$

PROOF: To prove estimate (1.1.22), we denote by the symbol $[\sqrt{\lambda}]$ the integral part of the number $\sqrt{\lambda}$ and note that the left-hand side of (1.1.22) is in any case not larger than the sum

---

[10] The symbol $[2\rho]$ signifies the integral part of a number 2ρ.

$$\sum_{k=1}^{[\sqrt{\lambda}]}\left[\sum_{k\le\sqrt{\lambda_n}\le k+1} u_n^2(x)\,\lambda_n^{\delta-N/2}\right]. \tag{1.1.25}$$

We note further that for a fixed k, the value of $\lambda_n^{\delta-N/2}$ under the sign of the internal sum (enclosed in square brackets) in (1.1.25) in any case does not exceed the number $C_0\,k^{2\delta-N}$, where $C_0 = 1$ for $\delta < N/2$ and $C_0 = 2^{2\delta-N}$ for $\delta > N/2$. This follows from the fact that for all $k \le \sqrt{\lambda_n} \le k+1$,

$$\lambda_n^{\delta-N/2} \le \begin{cases} k^{2\delta-N} & \text{for} \quad \delta \le N/2, \\ (k+1)^{2\delta-N} \le (2k)^{2\delta-N} & \text{for} \quad \delta > N/2. \end{cases}$$

Thus, the sum (1.1.25) and, consequently, the left-hand side of (1.1.22) are majorized by the sum

$$C_0\sum_{k=1}^{[\sqrt{\lambda}]} k^{2\delta-N}\left[\sum_{k\le\sqrt{\lambda_n}\le k+1} u_n^2(x)\right]. \tag{1.1.26}$$

Invoking estimate (1.1.4) for the sum in the square brackets in (1.1.26), we obtain that the left-hand side of (1.1.22) is majorized by the value of

$$C_0C\sum_{k=1}^{[\sqrt{\lambda}]} k^{2\delta-N}(k+1)^{N-1} \le C_0C\sum_{k=1}^{[\sqrt{\lambda}]} k^{2\delta-N}(2k)^{N-1}$$

$$= C_0C\,2^{N-1}\sum_{k=1}^{[\sqrt{\lambda}]} k^{2\delta-1} = O([\sqrt{\lambda}]^{2\delta}) = O(\lambda^{\delta}).$$

Thereby estimate (1.1.22) is proved.

To prove estimate (1.1.23), we majorize the left-hand side of (1.1.23) by the sum

$$\sum_{k=0}^{\infty}\left[\sum_{\sqrt{\lambda}+k\le\sqrt{\lambda_n}\le\sqrt{\lambda}+k+1} u_n^2(x)\,\lambda_n^{-\delta-N/2}\right]. \tag{1.1.27}$$

Since for all $\sqrt{\lambda}+k \le \sqrt{\lambda_n} \le \sqrt{\lambda}+k+1$ inequality $\lambda_n^{-\delta-N/2} \le (\sqrt{\lambda}+k)^{-2\delta-N}$ holds, the sum (1.1.27), in turn, can be majorized by the sum

$$\sum_{k=0}^{\infty}(\sqrt{\lambda}+k)^{-2\delta-N}\left[\sum_{\sqrt{\lambda}+k\le\sqrt{\lambda_n}\le\sqrt{\lambda}+k+1} u_n^2(x)\right]. \tag{1.1.28}$$

Invoking estimate (1.1.4) for the sum in the square brackets in (1.1.28), we obtain that the left-hand side of (1.1.23) is majorized by the value of

$$C\sum_{k=0}^{\infty}(\sqrt{\lambda}+k)^{-2\delta-N}(\sqrt{\lambda}+k+1)^{N-1}$$
$$\leq C\sum_{k=0}^{\infty}(\sqrt{\lambda}+k)^{-2\delta-N}\left[2(\sqrt{\lambda}+k)\right]^{N-1} = C\,2^{N-1}\sum_{k=0}^{\infty}(\sqrt{\lambda}+k)^{-2\delta-1}$$
$$\leq C\,2^{N-1}\left\{\lambda^{-\delta-1/2}+\int_0^{\infty}(\sqrt{\lambda}+x)^{-2\delta-1}dx\right\}\leq C\,2^{N-1}\lambda^{-\delta}\left[1+\tfrac{1}{2\delta}\right].$$

This completes the proof of estimate (1.1.23).

REMARK 3. The fact that estimate (1.1.4) is exact to within an order of magnitude in the class of all FSF of the Laplace operator is readily perceived if we consider, for an FSF, the system of eigenfunctions of the Laplace operator in an N-dimensional ball of radius R with a homogeneous boundary condition of the first or second kind on the boundary of this ball. It is easily verified that for the eigenfunction system in question not only the sum in the left-hand side of (1.1.4), but also the square of the unique radially symmetric eigenfunction, has at the center of the ball an order of magnitude specified by the right-hand side of (1.1.4)[11].

---

[11] The normalized spherically symmetric eigenfunctions of an N-dimensional ball of radius R take the following form:

(1) for the case of the first boundary-value problem $u(R) = 0$:

$$u_n(r) = \frac{\sqrt{2}}{R\sqrt{\omega_N}}\left[J_{(N-4)/2}(\mu_n)\right]^{-1} J_{(N-2)/2}\left(\frac{r}{R}\mu_n\right) r^{(2-N)/2}, \qquad n = 1, 2, \ldots,$$

where μ_n are the zeros for the Bessel function $J_{(N-2)/2}(x)$;

(2) for the case of the second boundary-value problem $u'(R) = 0$:

$$u_n(r) = \frac{\sqrt{2}}{R\sqrt{\omega_N}}\left[J_{(N-2)/2}(\mu_n)\right]^{-1} J_{(N-2)/2}\left(\frac{r}{R}\mu_n\right) r^{(2-N)/2}, \qquad n = 1, 2, \ldots,$$

where μ_n are the zeros for the Bessel function $J_{n/2}(x)$.

It follows from the asymptotic behavior of the Bessel function at large values of the argument and from the definition of zeros, that for the first boundary-value problem

$$J_{(N-4)/2}(\mu_n) = (-1)^n\sqrt{2/\pi}\,\frac{1}{\sqrt{\mu_n}} + o(1/\sqrt{\mu_n}),$$

and for the second boundary-value problem

$$J_{(N-2)/2}(\mu_n) = (-1)^n\sqrt{2/\pi}\,\frac{1}{\sqrt{\mu_n}} + o(1/\sqrt{\mu_n}).$$

Therefore, for either boundary-value problem the values at the center of the ball are $u_n(0) = C_n\mu_n^{(N-1)/2}$, where $\sqrt{\lambda_n} = \mu_n/R$,

$$\lim_{n\to\infty}|C_n| = \frac{1}{2}\sqrt{\pi/\omega_N}\,(2R)^{-N/2}\left[\Gamma(N/2)\right]^{-1}.$$

Now we establish a stronger result to show that *estimate (1.1.4) is exact to within an order of magnitude for each individual FSF of the Laplace operator.*

THEOREM 1.2. *If $\{u_n(x)\}$ is an arbitrary FSF of the Laplace operator in an N-dimensional subdomain Ω of an arbitrary N-dimensional domain G, then there exist positive numbers T and α such that for each $\mu \geq 0$ uniformly with respect to x within each subdomain Ω' strictly interior to Ω the following lower-bound estimate holds:*

$$\sum_{\mu \leq \sqrt{\lambda_n} \leq \mu+T} u_x^2(x) \geq \alpha(\mu+1)^{N-1}. \tag{1.1.29}$$

REMARK 4. A comparison of estimates (1.1.4) and (1.1.29) shows that both estimates are exact to within an order of magnitude.

PROOF OF THEOREM 1.2. We note that it is sufficient to prove the following proposition: If $\{u_n(x)\}$ is an arbitrary FSF of the Laplace operator in any N-dimensional subdomain Ω of an arbitrary N-dimensional domain G, then there exist positive numbers M, $\mu_0 \geq 2M$, and α such that for each $\mu \geq \mu_0$ uniformly with respect to x in each subdomain Ω' strictly interior to domain Ω the following lower-bound estimate holds:

$$\sum_{|\sqrt{\lambda_n}-\mu| \leq M} u_n^2(x) \geq \alpha\, \mu^{N-1}. \tag{1.1.30}$$

Indeed, the formulated proposition having been proved, we obtain (setting $T = 2\mu_0$ and accepting without loss of generality that $\mu_0 > 1$) that all the $\sqrt{\lambda_n}$ involved in the sum on the left-hand side of (1.1.29) satisfy the condition $|\sqrt{\lambda_n}-(\mu+T/2)| \leq T/2$, in which $\mu + T/2 \geq \mu_0$, $T/2 = \mu_0 > 2M > M$; this (by virtue of the proposition formulated) ensures the boundness from below of the sum in (1.1.29) by a value of $\alpha(\mu + T/2)^{N-1}$, which (by virtue of the fact that $T/2 = \mu_0 \geq 1$) obviously exceeds the value $\alpha(\mu+1)^{N-1}$.

Thus, all we have to do is to prove the formulated proposition, which we now do.

By analogy with the above, we fix a subdomain Ω' strictly interior to domain Ω, an arbitrary point x in Ω', and a positive number R, which is in any case smaller than the distance between Ω' and the boundary of domain Ω. (An exact choice of this number will be specified below.) Let $v(r)$ be a function as defined by relation (1.1.5), and v_n the Fourier coefficient of this function as defined by relation (1.1.6). If M is an arbitrary positive number, then Parseval's relation for function (1.1.5) can be written in the form

$$\begin{aligned} \int_G v^2(|x-y|)dy &= \frac{\mu^N \omega_N}{(2\pi)^N} \int_{R/2}^{R} r\, J^2_{(N-2)/2}(\mu r)dr \\ &= \sum_{|\sqrt{\lambda_n}-\mu| \leq M} v_n^2 + \sum_{|\sqrt{\lambda_n}-\mu| > M} v_n^2. \end{aligned} \tag{1.1.31}$$

It follows from the asymptotic formula (1.1.10) that

$$J^2_{(N-2)/2}(\mu r) = \frac{2}{\pi\mu r}\cos^2\left(\mu r - \frac{\pi N}{4} + \frac{\pi}{4}\right) + O\left(\frac{1}{\mu^2 r^2}\right) \tag{1.1.32}$$

for all r in the segment $R/2 \le r \le R$ and for all sufficiently large μ (for $\mu > 2/R$). Relation (1.1.32) implies that for all $\mu > 2/R$, the equality

$$\int_G v^2(|x-y|)dy = \frac{2\omega_N}{\pi(2\pi)^N}\,\mu^{N-1}\left[\int_{R/2}^{R} \cos^2\left(\mu r - \frac{\pi N}{4} + \frac{\pi}{4}\right) dr + O\left(\frac{1}{\mu}\right)\right] \tag{1.1.33}$$

holds.

Since, by virtue of the equality

$$\cos^2\left(\mu r - \frac{\pi N}{4} + \frac{\pi}{4}\right) = \frac{1}{2} + \frac{1}{2}\cos\left(2\mu r - \frac{\pi N}{2} + \frac{\pi}{2}\right),$$

the estimate

$$\int_{R/2}^{R} \cos^2\left(\mu r - \frac{\pi N}{4} + \frac{\pi}{4}\right) dr = \frac{R}{4} + O\left(\frac{1}{\mu}\right)$$

holds, we obtain from (1.1.33) that

$$\int_G v^2(|x-y|)dy = \frac{\omega_N}{(2\pi)^{N+1}}\,\mu^{N-1}\left[R + O\left(\frac{1}{\mu}\right)\right]. \tag{1.1.34}$$

It follows from (1.1.34) that there exists a positive constant β such that, given any fixed R and all sufficiently large μ, the following inequality holds[12)]:

$$\int_G v^2(|x-y|)dy > \beta\, R\, \mu^{N-1}. \tag{1.1.35}$$

Now ensure that, given any arbitrary fixed number $M > 0$ and any $\mu > 2M$ uniformly with respect to x in each subdomain Ω' strictly interior to domain Ω, the following inequality holds:

$$\sum_{|\sqrt{\lambda_n}-\mu| \le M} v_n^2 \le \gamma\, R^2 \sum_{|\sqrt{\lambda_n}-\mu|\le M} u_n^2(x), \tag{1.1.36}$$

where γ denotes a positive constant.

---

12) For example, one may take $\beta = \frac{\omega_N}{2(2\pi)^{N+1}}$.

In point of fact, starting from the representation (1.1.6) for the Fourier coefficient v_n and making use of the estimate

$$|J_\nu(x)| \leq \frac{C}{\sqrt{x}} \tag{1.1.37}$$

for Bessel functions valid for each $\nu \geq -1/2$ and for all $x > 0$, we arrive at the inequality

$$v_n^2 \leq C^2 \left(\frac{\mu}{\sqrt{\lambda_n}}\right)^{N-1} \frac{R}{4}\, u_n^2(x), \tag{1.1.38}$$

where C is the constant in the estimate (1.1.37).

Next, given $|\lambda_n - \mu| \leq M$ and provided that $\mu \geq 2M$, the inequality $\mu/2 \leq \sqrt{\lambda_n}$ holds implying that $(\mu/\sqrt{\lambda_n})^{N-1} \leq 2^{N-1}$. Invoking (1.1.38), the latter inequality lends support to the validity of inequality (1.1.36) with a constant $\gamma = C^2 2^{N-3}$.

A comparison of Parseval's relation (1.1.31) with the inequalities (1.1.35) and (1.1.36) verified above brings us to the conclusion that, in order to prove estimate (1.1.30), *it is sufficient to establish that one can fix $M > 0$ so large and $R > 0$ so small that the inequality*

$$\sum_{|\sqrt{\lambda_n}-\mu|>M} v_n^2 < \frac{\beta}{2} R\mu^{N-1} \tag{1.1.39}$$

holds for all sufficiently large μ uniformly with respect to x in Ω' with the same constant β as that in inequality (1.1.35).

In order to prove inequality (1.1.39), we divide the sum on the left-hand side of (1.1.39) into the following three sums:

$$S_1 = \sum_{\sqrt{\lambda_n}\leq 1} v_n^2, \tag{1.1.40}$$

$$S_2 = \sum_{1<\sqrt{\lambda_n}<\mu/2} v_n^2 + \sum_{\sqrt{\lambda_n}>3\mu/2} v_n^2, \tag{1.1.41}$$

$$S_3 = \sum_{\mu/2\geq|\sqrt{\lambda_n}-\mu|>M} v_n^2. \tag{1.1.42}$$

It suffices to prove that one can fix the number $M > 0$ so large and the number $R > 0$ so small that for all sufficiently large μ any of the sums S_1, S_2, and S_3 for all x in Ω' do not exceed the number $\frac{\beta}{6}R\mu^{N-1}$, where β is the same constant as that in inequality (1.1.35).

In estimating the sum (1.1.40), we rely on the representation (1.1.6) for the Fourier coefficient v_n. Making use in representation (1.1.6) of the following well-known estimates for the Bessel functions[13]:

$$\frac{\left|J_{(N-2)/2}\left(r\sqrt{\lambda_n}\right)\right|}{\left(r\sqrt{\lambda_n}\right)^{(N-2)/2}} \leq C_1, \qquad \left|J_{(N-2)/2}(\mu r)\right| \leq \frac{C_2}{\sqrt{\mu r}},$$

[13] See, for example, Bateman and Erdélyi [9, pp. 12 and 98].

we obtain for all points x in subdomain Ω'

$$v_n^2 = C_1^2 C_2^2 \mu^{N-1} u_n^2(x) \Big(\int_{R/2}^{R} r^{(N-1)/2} dr \Big)^2 \le C_3 \mu^{N-1} R^{N+1} u_n^2(x), \tag{1.1.43}$$

where

$$C_3 = \frac{4\left(2^{(N+1)/2} - 1\right)^2}{(N+1)^2 2^{N+1}} C_1^2 C_2^2.$$

Invoking estimate (1.1.17) taken at $\mu_0 = 1$ and denoting by C_4 a constant that sets a limit to the growth of the O-terms in estimate (1.1.17), we obtain, through the use of this estimate and inequality (1.1.43),

$$S_1 = \sum_{\sqrt{\lambda_n} \le 1} v_n^2 \le C_3 C_4 \mu^{N-1} R^{N+1}. \tag{1.1.44}$$

We now fix the positive number R so small as to satisfy the inequality

$$R^N \le \beta [6 C_3 C_4]^{-1}, \tag{1.1.45}$$

where β is a constant the same as that in inequality (1.1.35).

A comparison of inequalities (1.1.44) and (1.1.45) reveals that

$$S_1 \le \frac{\beta}{6} R \mu^{N-1}, \tag{1.1.46}$$

which provides the desired estimate for S_1. Henceforth, we assume R to be fixed in accordance with inequality (1.1.45).

To estimate the sums (1.1.41) and (1.1.42), we shall need the following auxiliary assertion.

LEMMA 1.1. *Given any two numbers $\sqrt{\lambda_n}$ and μ positive and different from each other, for the quantity*

$$I_n^\mu(R) = \mu^{1/2} \lambda_n^{1/4} \int_{R/2}^{R} r\, J_{(N-2)/2}(\mu r)\, J_{(N-2)/2}(r\sqrt{\lambda_n}) dr \tag{1.1.47}$$

the following estimate is valid:

$$I_n^\mu(R) = O(|\sqrt{\lambda_n} - \mu|^{-1}), \tag{1.1.48}$$

in which the constant that sets a limit to the growth of the O-terms is independent of R.

Without digressing from our line of reasoning in proving Theorem 1.2, we refer the reader to Section 1.1.4 (*vide infra*) for the proof of Lemma 1.1.

To estimate the sum (1.1.41), we note that for all the indices n involved in this sum, the following inequalities hold:

$$|\sqrt{\lambda_n} - \mu| > \frac{1}{2}\mu, \qquad |\sqrt{\lambda_n} - \mu| > \frac{1}{3}\sqrt{\lambda_n}.$$

From these inequalities we infer that for all the indices n encountered in the sum (1.1.41) the estimate

$$|\sqrt{\lambda_n} - \mu|^{-1} = O\left(\lambda_n^{-3/8}\mu^{-1/4}\right)$$

is valid. This estimate, combined with (1.1.48), yields

$$I_n^{\mu}(R) = O\left(\lambda_n^{-3/8}\mu^{-1/4}\right). \tag{1.1.49}$$

Estimate (1.1.49), in combination with the expression (1.1.6) for the Fourier coefficient, leads us to the relation

$$v_n = u_n(x)\left(\frac{\mu}{\sqrt{\lambda_n}}\right)^{(N-1)/2} I_n^{\mu}(R) = u_n(x)\,O\left(\mu^{N/2-3/4}\lambda_n^{-(N/4+1/8)}\right). \tag{1.1.50}$$

It follows from this relation that

$$S_2 = \sum_{1\le\sqrt{\lambda_n}<\mu/2} v_n^2 + \sum_{\sqrt{\lambda_n}>\frac{3}{2}\mu} v_n^2 \le \mu^{N-3/2} C_5 \sum_{\sqrt{\lambda_n}\ge 1} u_n^2(x)\,\lambda_n^{-N/2-1/4}, \tag{1.1.51}$$

where C_5 denotes a constant in the estimate of the O-terms in relation (1.1.50).

Taking into account that the sum of the series (1.1.24) at $\delta = 1/4$ is bounded, in each subdomain Ω' strictly interior to domain Ω, by a certain Ω'-dependent constant C_6, we obtain from (1.1.51) that

$$S_2 \le \mu^{N-1} C_5 C_6 \frac{1}{\sqrt{\mu}}. \tag{1.1.52}$$

We fix now μ_0 so as to satisfy the inequality

$$\frac{1}{\sqrt{\mu_0}} \le \frac{\beta R}{6 C_5 C_6}, \tag{1.1.53}$$

where β is the same constant as that in inequality (1.1.35), and R is the number that has been fixed above in conformity with inequality (1.1.45).

We deduce from (1.1.52) and (1.1.53) that, for any $\mu \ge \mu_0$ and for all points in the subdomain Ω', the following inequality holds:

$$S_2 \le \frac{\beta}{6} R \mu^{N-1}. \tag{1.1.54}$$

Thus, the desired estimate for the sum (1.1.41) has been obtained.

To complete the proof of Theorem 1.2, all we have to do is obtain a similar estimate for the sum (1.1.42).

It suffices to prove that the number M relevant to this sum can be fixed so large that, given the number $R > 0$ fixed above (in accordance with inequality (1.1.45)) and all μ, the sum (1.1.42) for all x in subdomain Ω' does not exceed the value $\frac{\beta}{6}\, R\mu^{N-1}$.

We consider the sequence of segments $[1,2]$, $[2,4]$, ... , $[2^{k-1}, 2^k]$, ... and denote by p the *least* of the indices k for which $2^k \geq \mu/2$. We show that, for $R > 0$ as fixed above and for all μ, the index m can be fixed so large that the sum (1.1.42) with $M = 2^{m-1}$ for all x in subdomain Ω' does not exceed the value $\frac{\beta}{6}\, R\mu^{N-1}$.

We note that

$$S_3 = \sum_{M<|\sqrt{\lambda_n}-\mu|\leq\mu/2} v_n^2 \leq \sum_{k=m}^{p} \left[\sum_{2^{k-1}<|\sqrt{\lambda_n}-\mu|\leq 2^k} v_n^2 \right]. \tag{1.1.55}$$

To estimate v_n^2 on the right-hand side of (1.1.55), we make use of expression (1.1.50) for v_n and recall that for all the indices n relevant to (1.1.55), the relation $\mu/\sqrt{\lambda_n} \leq 2$ holds. From this relation and from (1.1.50), we obtain

$$v_n^2 \leq u_n^2(x)\, 2^{N-1}\, [I_n^\mu(R)]^2\,. \tag{1.1.56}$$

Comparing (1.1.56) and (1.1.48), we obtain that there is a constant C_7 such that

$$v_n^2 \leq C_7\, u_n^2(x)\, |\sqrt{\lambda_n} - \mu|^{-2}. \tag{1.1.57}$$

The substitution of (1.1.57) into the right-hand side of (1.1.55) yields the estimate

$$\begin{aligned} S_3 &\leq C_7 \sum_{k=m}^{p} \left[\sum_{2^{k-1}\leq|\sqrt{\lambda_n}-\mu|\leq 2^k} u_n^2(x)\, |\sqrt{\lambda_n} - \mu|^{-2} \right] \\ &\leq C_7 \sum_{k=m}^{p} \frac{1}{4^{k-1}} \left\{ \sum_{|\sqrt{\lambda_n}-\mu|\leq 2^k} u_n^2(x) \right\}. \end{aligned} \tag{1.1.58}$$

(We have removed the term $|\sqrt{\lambda_n} - \mu|^{-2}$ (its greatest value being $1/4^{k-1}$) from the inner square-bracketed sum and have replaced, in the remaining expression, summation over n (satisfying the inequalities $2^{k-1} \leq |\sqrt{\lambda_n} - \mu| \leq 2^k$) by summation of the values of n satisfying only one inequality $|\sqrt{\lambda_n} - \mu| \leq 2^k$.)

Now, to estimate the sum enclosed in the braces in (1.1.58), we make use of estimate (1.1.21) (*vide supra* Corollary 1 to Theorem 1.1) and take it at $\rho = 2^k$. We obtain, as a result, that for each index k uniformly with respect to x in subdomain Ω', each term enclosed in the braces on the right-hand side of (1.1.58) is equal to

$2^k\, O(\mu^{N-1})$. Denoting by C_8 the constant in the O-term just now written, we obtain that

$$S_3 \le 4C_7\, C_8\, \mu^{N-1} \sum_{k=m}^{p} \frac{1}{2^k} \le 4C_7\, C_8\, \mu^{N-1}\, \frac{1}{2^{m-1}}. \tag{1.1.59}$$

It remains to fix m large enough so as to satisfy the inequality

$$\frac{1}{M} = \frac{1}{2^{m-1}} \le \frac{\beta\, R}{24\, C_7\, C_8}, \tag{1.1.60}$$

with $\beta > 0$ and $R > 0$ fixed in accordance with inequalities (1.1.35) and (1.1.45).

We deduce from (1.1.59) and (1.1.60) that the sum (1.1.42) with the values of $M = 2^{m-1}$ as specified by condition (1.1.60) satisfies the inequality

$$S_3 \le \frac{\beta}{6}\, R\, \mu^{N-1} \tag{1.1.61}$$

for all μ, for $R > 0$ as fixed above, and for all x in subdomain Ω'.

It follows from inequalities (1.1.46), (1.1.54), and (1.1.61) that, given $M > 0$ and $R > 0$ as fixed above and all $\mu > \mu_0$ uniformly with respect to x in subdomain Ω', the inequality (1.1.39) holds, which (by virtue of the aforesaid) completes the proof of Theorem 1.2.

Theorems 1.1 and 1.2 can be grouped together into a general statement.

THEOREM 1.3. *If $\{u_n(x)\}$ is an arbitrary FSF of the Laplace operator in any N-dimensional subdomain Ω of an arbitrary N-dimensional domain G, then there exist positive numbers T, α, and β such that for each $\mu \ge 0$ uniformly with respect to x in each subdomain Ω' strictly interior to domain Ω the following two-sided estimate holds:*

$$\alpha(\mu+1)^{N-1} \le \sum_{\mu \le \sqrt{\lambda_n} \le \mu+T} u_n^2 \le \beta\,(\mu+1)^{N-1}. \tag{1.1.62}$$

1.1.3. The Spectral Structure of an Arbitrary FSF

The works of many famous mathematicians (the pioneering works of Weyl [78] and Carleman [13], the later works by Agmon [1] and Mizohata [61], and the relatively recent work of Seeley [71], Ivrii [50], and other authors) were concerned with the asymptotic distribution of eigenvalues in the simplest boundary-value problems for elliptic operators. At present, the spectrum of the first boundary-value problem for the Laplace operator in an arbitrary bounded N-dimensional domain G has been studied in great detail. If we denote by $n(\lambda)$ the number of eigenvalues of the afore said problem not greater than λ (for any $\lambda > 0$), that is, we set

$$n(\lambda) = \sum_{\lambda_n \le \lambda} 1,$$

then for $n(\lambda)$ at large λ the following asymptotic formula is valid:

$$n(\lambda) = A\,\lambda^{N/2} + O(\lambda^{(N-1)/2}), \tag{1.1.63}$$

in which the constant A is equal to $(4\pi)^{-N/2} \cdot \text{mes}\, G$ (where mes G denotes the measure of domain G).

We note, however, that, only a few years ago, formula (1.1.63) as it is written here was established for the first boundary-value problem eigenvalues in an N-dimensional rectangular parallelepiped and in certain exotic domains G, whereas for other domains the remainder term $O(\lambda^{(N-1)/2})$ was established in the rougher form of $O(\lambda^{(N-1)/2} \ln \lambda)$. It was only in the above-mentioned work of Seeley that formula (1.1.63) was established for the eigenvalues of the first boundary-value problem in an arbitrary bounded N-dimensional domain G.

A trivial consequence of the asymptotic formula (1.1.63) is that for the eigenvalues of the first boundary-value problem in an arbitrary bounded N-dimensional domain G there exist positive constants T, α, and β such that for all $\mu \geq 0$ the two-sided estimate

$$\alpha\,(\mu+1)^{N-1} \leq \sum_{\mu \leq \sqrt{\lambda_n} \leq \mu+T} 1 \leq \beta\,(\mu+1)^{N-1} \tag{1.1.64}$$

holds.

Thus, the eigenvalues λ_n of the first boundary-value problem for the Laplace operator are distributed in such a way that, given any $\mu \geq 0$, a segment $[\mu, \mu+T]$ of length T accommodates not fewer than $\alpha\,(\mu+1)^{N-1}$ and not more than $\beta\,(\mu+1)^{N-1}$ roots $\sqrt{\lambda_n}$ for the eigenvalues λ_n.

We now show that the spectrum of an arbitrary FSF of the Laplace operator in any N-dimensional subdomain Ω of an arbitrary N-dimensional domain G is not less closely spaced than the spectrum of the first boundary-value problem that we have considered above. To be more exact, the following assertion will be proved.

THEOREM 1.4. *If $\{\lambda_n\}$ is the spectrum of an arbitrary FSF of the Laplace operator in any N-dimensional subdomain Ω of an arbitrary N-dimensional domain G, then there exist positive numbers T and α_0 such that for each $\mu \geq 0$ the following lower-bound estimate holds:*

$$\sum_{\mu \leq \sqrt{\lambda_n} \leq \mu+T} 1 \geq \alpha_0\,(\mu+1)^{N-1}. \tag{1.1.65}$$

PROOF: We fix a certain subdomain Ω' of domain Ω with a measure $\text{mes}\,\Omega'$ finite and distinct from zero. Making use of the fact that the norms of all the elements $u_n(x)$ in the domain G are equal to unity and invoking estimate (1.1.2) (*vide supra*

Theorem 1.2), we obtain

$$\sum_{\mu\le\sqrt{\lambda_n}\le\mu+T} 1 = \sum_{\mu\le\sqrt{\lambda_n}\le\mu+T} \int_G u_n^2 dx \ge \sum_{\mu\le\sqrt{\lambda_n}\le\mu+T} \int_{\Omega'} u_n^2 dx$$

$$= \int_{\Omega'} \left[\sum_{\mu\le\sqrt{\lambda_n}\le\mu+T} u_n^2 \right] dx \ge \alpha\,(\mu+1)^{N-1}\cdot(\operatorname{mes}\Omega').$$

Having set $\alpha_0 = \alpha\cdot(\operatorname{mes}\Omega')$, we arrive at estimate (1.1.65).

The proof of Theorem 1.4 is complete.

COROLLARY 1. *If the symbol $n(\lambda)$ denotes the number of eigenvalues λ_n not in excess of λ for an arbitrary FSF of the Laplace operator in any N-dimensional subdomain Ω of an arbitrary N-dimensional domain Ω, then there exists a positive constant α_1 such that for all sufficiently large $\lambda > 0$ the following lower-bound estimate holds:*

$$n(\lambda) \ge \alpha_1\,\lambda^{N/2}. \tag{1.1.66}$$

COROLLARY 2. *If the eigenvalues of an arbitrary FSF of the Laplace operator in any N-dimensional subdomain Ω of an arbitrary N-dimensional domain G lack end points of condensation, then, with the eigenvalues numbered in nondecreasing order, one may assert the validity of the estimate*

$$\sqrt{\lambda_{n+1}} - \sqrt{\lambda_n} = O(1).$$

Naturally, the following question arises: For an arbitrary FSF of the Laplace operator, does there exist any upper estimate either for the number $n(\lambda)$ of the eigenvalues not in excess of λ, or for the sum on the left-hand side of (1.1.65)? Or, in general terms, can a FSF of the Laplace operator possess eigenvalue end points of condensation?

We now proceed with the elucidation of these issues to show that the spectrum of an arbitrary FSF of the Laplace operator (even over the entire domain G) displays a more complex structure than that of the spectrum of those boundary-value problems that were the objects of study in the previously mentioned works of Weyl, Carleman, Agmon, Mizohata, Seeley, and Ivrii.

To begin, we prove the following proposition.

THEOREM 1.5. *Let $\{\lambda_n^0\}$ be an arbitrarily assigned sequence of numbers satisfying the condition $0 \le \lambda_n^0 \le n$. Then there exists a FSF of the Laplace operator in the square $Q = [0 \le x \le \pi] \times [0 \le y \le \pi]$ for which the sequence $\{\lambda_n^0\}$ is a sequence of eigenvalues. The more so, with the initial sequence $\{\lambda_n^0\}$ eliminated from the set of all eigenvalues, the remaining eigenvalues obey the same asymptotic distribution law as that for the first boundary-value problem eigenvalues in the square Q.*

PROOF: We consider, for each number $n = 1, 2, 3, \ldots$ in an interval $0 \le x \le \pi$, the eigenvalue problem for the simplest ordinary differential operator:

$$\begin{cases} X''(x) + \mu X(x) = 0 & \text{for } \quad 0 < x < \pi, \\ X(0) = 0, \qquad X'(\pi) - h_n X(\pi) = 0, \end{cases} \tag{1.1.67}$$

where $h_n = \sqrt{n^2 - \lambda_n^0}\coth\left(\pi\sqrt{n^2 - \lambda_n^0}\right)$. It is a well-known fact that for each number n problem (1.1.67) with a uniform boundary condition of the first kind at the left-hand end $x = 0$ and with a uniform boundary condition of the third kind at the right-hand end $x = \pi$ has a complete orthonormal system of eigenfunctions[14]. Since the function

$$X(x) = \begin{cases} \sinh(x\sqrt{-\mu}) & \text{for } \quad \mu < 0, \\ x & \text{for } \quad \mu = 0, \\ \sin(x\sqrt{\mu}) & \text{for } \quad \mu > 0 \end{cases}$$

satisfies the equation $X'' + \mu X = 0$ and the boundary condition at the left-hand end $X(0) = 0$, we obtain, from the boundary condition at the right-hand end $X'(\pi) - h_n X(\pi) = 0$, an equation for determining the eigenvalues of problem (1.1.67). At $\mu < 0$, the boundary condition $X'(\pi) - h_n X(\pi) = 0$ leads to the equation $\sqrt{-\mu}\coth(\pi\sqrt{-\mu}) = h_n \sinh(\pi\sqrt{-\mu})$, which reduces to the following form:

$$\sqrt{-\mu}\coth(\pi\sqrt{-\mu}) = \sqrt{n^2 - \lambda_n^0}\coth(\pi\sqrt{n^2 - \lambda_n^0}).$$

The latter equation, by virtue of the function $x\coth(\pi x)$ increasing on the ray $x \ge 0$, has a single root, the negative eigenvalue $\mu_n^0 = \lambda_n^0 - n^2 < 0$.

The value of $\mu = 0$ is not an eigenvalue to problem (1.1.67), since the function $X(x) = x$ satisfying the equation $X'' + \mu X = 0$ and the boundary condition $X(0) = 0$ at $\mu = 0$ in an obvious way does not satisfy the second boundary condition $X'(\pi) - h_n X(\pi) = 0$, which is reducible to the equality $1 = h_n \pi$. The latter equality is infeasible, since $h_n > 1/\pi$ for all the numbers n[15].

At $\mu > 0$, the boundary condition $X'(\pi) - h_n X(\pi) = 0$ leads to a transcendental equation $\sqrt{\mu}\,\cos(\pi\sqrt{\mu}) = h_n \sin(\pi\sqrt{\mu})$, which can be rewritten in the form $\sqrt{\mu}\coth(\pi\sqrt{\mu}) = h_n$. The latter equation at any fixed number n has a countable set of positive roots μ_n^k $(k = 1, 2, \ldots)$ which are positive eigenvalues to problem (1.1.67) numbered in increasing order. This stems from the fact that, given any $k = 1, 2, \ldots$, the function $\coth(\pi\sqrt{\mu})$ runs through all its values from $-\infty$ to ∞ with μ varying within the limits of $k^2 < \mu < (k+1)^2$. Thus, the k-numbered eigenvalues μ_n^k satisfy, in an obvious way, the inequalities

$$k^2 \le \mu_n^k \le (k+1)^2. \tag{1.1.68}$$

[14] This follows from the theory of integral equations and from the symmetry of the Green's function of this particular problem, which has been dealt with in a book by Smirnov [73, p. 529].

[15] This stems from the fact that h_n is equal to the value of the function $x\coth(\pi x)$ at the point $x = \sqrt{n^2 - \lambda_n^0} > 0$. Since the function $x\coth(\pi x)$ increases on the ray $x \ge 0$ and goes to $1/\pi$ at the point $x = 0$, all its values at $x > 0$ are in excess of $1/\pi$.

To summarize, we can say that for each $n = 1, 2, \ldots$ problem (1.1.67) has one negative eigenvalue $\mu_n^0 = \lambda_n^0 - n^2$ and a countable set of positive eigenvalues μ_n^k $(k = 1, 2, \ldots)$ satisfying inequalities (1.1.68). These eigenvalues are associated with the eigenfunctions

$$X_n^0(x) = N_n^0 \sinh(x\sqrt{-\mu_n^0}), \quad X_n^k(x) = N_n^k \sin(x\sqrt{\mu_n^k}), \quad (k = 1, 2, \ldots). \quad (1.1.69)$$

In the expressions for the eigenfunctions (1.1.69), the symbols N_n^k denote normalization factors chosen in such a manner as to make the norm of each function $X_n^k(x)$ equal to unity. As has been noted above, the system of all eigenfunctions (1.1.69) is (for each n) complete and orthonormal.

Since the system $\left\{(\sqrt{\frac{\pi}{2}})^{-1} \sin(ny)\right\}$, $n = 1, 2, \ldots$, is known to be a complete system orthonormal on the interval $0 \leq y \leq \pi$, the system

$$u_n^k(x, y) = X_n^k(x) \left(\sqrt{\frac{\pi}{2}}\right)^{-1} \sin(ny), \qquad n = 1, 2, \ldots, \quad k = 0, 1, \ldots, \quad (1.1.70)$$

is complete and orthonormal on the square $Q = [0 \leq x \leq \pi] \times [0 \leq y \leq \pi]$, each function $u_n^k(x, y)$ for the nonnegative number $\lambda_n^k = \mu_n^k + n^2$ being a solution of the equation

$$\left(\frac{\partial^2}{\partial x^2} + \frac{\partial^2}{\partial y^2}\right) u_n^k(x, y) + \lambda_n^k\, u_n^k(x, y) = 0$$

on the square Q.

Thus, the system (1.1.70) is FSF of the Laplace operator in the square Q, with the eigenvalues $\{\lambda_n^k\}$ of this system containing an arbitrarily assigned sequence $\{\lambda_n^0\}$ as a subsequence.

If we eliminate the initially assigned sequence $\{\lambda_n^0\}$, $n = 1, 2, \ldots$, from the set of eigenvalues $\{\lambda_n^k\}$, $n = 1, 2, \ldots$, $k = 0, 1, \ldots$, then the remaining fundamental values $\{\lambda_n^k\}$ for $n = 1, 2, \ldots$, $k = 1, 2, \ldots$, will satisfy, by virtue of condition (1.1.68), the inequalities

$$n^2 + k^2 \leq \lambda_n^k \leq n^2 + (k+1)^2. \quad (1.1.71)$$

Since the eigenvalues of the first boundary-value problem for the Laplace operator in the square Q form a set

$$\widehat{\lambda}_n^k = n^2 + k^2, \qquad n = 1, 2, \ldots, \quad k = 1, 2, \ldots,$$

it follows from inequalities (1.1.71) that the fundamental values that remain after the initially assigned sequence is eliminated obey the same asymptotic distribution law as that for the eigenvalues of the first boundary-value problem for the Laplace operator in the square Q.

The proof of Theorem 1.5 is thus complete.

REMARK 1. We now specify the boundary conditions which are satisfied, at the boundary of square Q, by the eigenfunctions (1.1.70) that we have constructed.

It stands to reason that $u_n^k(x,0) = u_n^k(x,\pi) = u_n^0(0,y) = 0$; $\frac{\partial u}{\partial y}(\pi,y) - H\,u(\pi,y) = 0$, where H is a self-adjoint operator which brings a function $f(y)$, expandable in the system $\left\{\left(\sqrt{\frac{\pi}{2}}\right)^{-1}\sin(ny)\right\}$, of the form $f(y) \sim \sum b_n \left(\sqrt{\frac{\pi}{2}}\right)^{-1}\sin(ny)$ into correspondence with the function $Hf(y)$, expandable in the same system, of the form $Hf(y) \sim \sum h_n\, b_n \left(\sqrt{\frac{\pi}{2}}\right)^{-1}\sin(ny)$. The operator H can be expressed explicity via the integral operator.

The following three remarks outline routes to a generalization of Theorem 1.5.

REMARK 2. In Theorem 1.5, the condition $0 \leq \lambda_n^0 \leq n$ may be ignored. This having been done, the method we used in proving Theorem 1.5 allows us to construct a complete orthonormal system of functions of the Laplace operator in the square Q; however, not all of the eigenvalues $\lambda_n^k = \mu_n^k + n^2$, $n = 1,2,\dots$, $k = 1,2,\dots$, will be positive.

REMARK 3. Under the conditions of Theorem 1.5, the square Q may be replaced by a circle K. If we take an arbitrary sequence of positive numbers $\{\lambda_n^0\}$ and consider, for each $n = 1,2,\dots$, the respective problem for the Bessel function of order n instead of problem (1.1.67), we can construct a FSF of the Laplace operator in a circle K; for this FSF, the preassigned sequence $\{\lambda_n^0\}$ will be a subsequence of the fundamental values.

REMARK 4. In a similar manner, based on an arbitrary preassigned sequence of nonnegative numbers $\{\lambda_n^0\}$, we can construct a FSF of the Laplace operator in an N-dimensional cylinder defined as the product of an interval $[0 \leq x_1 \leq \pi]$ and an $(N-1)$-dimensional bounded domain $\widehat{G}$; for this FSF, the sequence $\{\lambda_n^0\}$ is a subsequence of the fundamental values. (It suffices, in the reasoning relevant to Theorem 1.5, to replace n^2 and $(\sqrt{\pi/2})^{-1}\sin(ny)$, respectively, by the eigenvalues $\widetilde{\lambda}_n$ and eigenfunctions $v_n(x_1, x_2, \dots, x_{N-1})$ of domain $\widetilde{G}$.) The fundamental values that remain after the elimination of $\{\lambda_n^0\}$ obey the same distribution law as the eigenvalues of the first boundary-value problem for the Laplace operator in an N-dimensional cylinder.

The theorem we have proved provides a number of far-reaching corollaries that give a deeper insight into the possible spectral structure of an arbitrary FSF of the Laplace operator in a bounded N-dimensional ($N \geq 2$) domain.

COROLLARY 1. *There exists, in a bounded N-dimensional ($N \geq 2$) domain, a FSF of the Laplace operator such that for its fundamental values any of the following four possibilities is implemented:*

(1) there are fundamental values of infinite multiplicity;

(2) the fundamental values have arbitrary isolated points of condensation;

(3) the fundamental values possess points of condensation extending throughout the entire ray $\lambda \geq 0$;

(4) the fundamental values possess points of condensation extending throughout any closed set of the ray $\lambda \geq 0$.

In fact, any preassigned sequence $\{\lambda_n^0\}$ can be chosen in such a manner that it possesses:

(1) elements each encountered an infinite number of times;

(2) preassigned isolated limit points;

(3) limit points extending throughout the entire ray $\lambda \geq 0$[16];

(4) limit points extending throughout a preassigned closed subset of the ray $\lambda \geq 0$.

COROLLARY 2. *Whatever the preassigned function $f(\lambda)$, continuous on the ray $\lambda \geq 0$ and growing as $\lambda \to \infty$ faster than $\lambda^{N/2}$, there exists, for a bounded N-dimensional ($N \geq 2$) domain, a FSF of the Laplace operator such that its set of fundamental values has no end points of condensation, and for this set the function $f(\lambda)$ is the main term in the asymptotic behavior of the number $n(\lambda)$ of fundamental values not exceeding λ, with the representation*

$$n(\lambda) = f(\lambda) + O(\lambda^{N/2}) \tag{1.1.72}$$

holding for $n(\lambda)$.

Corollary 2 disproves the repeatedly entertained view that the main term in the asymptotic behavior of the number $n(\lambda)$ of eigenvalues not exceeding λ must not follow other than a power growth.

To begin the proof of Corollary 2, we denote by $f^{-1}(t)$ the function inverse to $f(\lambda)$. It is clear that $f^{-1}(t)$ is likewise an increasing function and is continuous on the ray $t \geq 0$. We set $\lambda_n^0 = f^{-1}(n)$, $n = 1, 2, \ldots$. We denote by $[f(\lambda)]$ the integral part of the number $f(\lambda)$; it is clear[17] that

$$\sum_{\lambda_n^0 \leq \lambda} 1 = [f(\lambda)] = f(\lambda) + O(1). \tag{1.1.73}$$

On the other hand, denoting by $\{\lambda_n'\}$ the fundamental values that remain after the sequence $\{\lambda_n^0\}$ has been eliminated, we obtain

$$\sum_{\lambda_n' \leq \lambda} 1 = O(\lambda^{N/2}). \tag{1.1.74}$$

Thus, representation (1.1.72) follows from (1.1.73) and (1.1.74).

Despite the fact that the spectrum of an arbitrary FSF of the Laplace operator, as we have shown above, may exhibit a complex and nonclassical behavior, we will

[16] It suffices to take for $\{\lambda_n^0\}$ the sequence of all rational positive numbers.

[17] Indeed, if $\lambda_{n_0}^0$ is the largest of the values λ_n^0 satisfying the condition $\lambda_n^0 \leq \lambda$, then we obtain from the condition $\lambda_{n_0}^0 \leq \lambda < \lambda_{n_0+1}^0$ that $n_0 = f(\lambda_{n_0}^0) \leq f(\lambda) < f(\lambda_{n_0+1}^0) = n_0 + 1$. Consequently,

$$\sum_{\lambda_n^0 \leq \lambda} 1 = n_0 = [f(\lambda)] = f(\lambda) + O(1).$$

design in the subsequent sections of this chapter a technique enabling us to establish in a definite sense the final conditions for uniform convergence and localization of a Fourier series in an arbitrary FSF of the Laplace operator; these conditions are novel (and likewise final ones in a definite sense) even for the thoroughly studied multiple trigonometric Fourier series (with spherical partial sums).

1.1.4. Proof of Lemma 1.1

Making use of the recurrence relations for Bessel functions,[18)]

$$\int r^{\nu} J_{\nu-1}(\mu r)dr = \frac{1}{\mu} r^{\nu} J_{\nu}(\mu r), \quad \frac{d}{dr}\left[\frac{J_{\nu}(r\sqrt{\lambda_n})}{r^{\nu}}\right] = -\sqrt{\lambda_n}\,\frac{J_{\nu+1}(r\sqrt{\lambda_n})}{r^{\nu}},$$

we integrate the integral (1.1.47) by parts two times to obtain

$$\begin{aligned} I_n^{\mu}(R) = {} & \frac{\lambda_n^{1/4}}{\mu^{1/2}}\left[r\, J_{N/2}(\mu r)\, J_{N/2-1}(r\sqrt{\lambda_n})\right]\Big|_{r=R/2}^{r=R} \\ & + \frac{\lambda_n^{3/4}}{\mu^{3/2}}\left[r\, J_{N/2+1}(\mu r)\, J_{N/2}(r\sqrt{\lambda_n})\right]\Big|_{r=R/2}^{r=R} \\ & + \frac{\lambda_n^{5/4}}{\mu^{3/2}}\int\limits_{R/2}^{R} r\, J_{N/2+1}(\mu r)\, J_{N/2+1}(r\sqrt{\lambda_n})dr. \end{aligned} \tag{1.1.75}$$

Further, we make use of the recurrence relations[19)]

$$J_{N/2+1}(\mu r) = -J_{N/2-1}(\mu r) + \frac{N}{\mu r} J_{N/2}(\mu r),$$

$$J_{N/2+1}(r\sqrt{\lambda_n}) = -J_{N/2-1}(r\sqrt{\lambda_n}) + \frac{N}{r\sqrt{\lambda_n}} J_{N/2}(r\sqrt{\lambda_n}).$$

for the terms under the integration sign on the right-hand side of (1.1.75). The resultant relation (1.1.75) takes the following form:

---

18) See, for example, Bateman and Erdélyi [9, p. 20, formulas (52) and (53)].
19) See, for example, Bateman and Erdélyi [9, p. 20, formula (56)].

$$
\begin{aligned}
I_n^\mu(R)\left(1-\frac{\lambda_n}{\mu^2}\right) &= \frac{\lambda_n^{1/4}}{\mu^{1/2}}\left[r\,J_{N/2}(\mu r)\,J_{N/2-1}(r\sqrt{\lambda_n})\right]\Big|_{r=R/2}^{r=R} \\
&+\frac{\lambda_n^{3/4}}{\mu^{3/2}}\left[r\,J_{N/2+1}(\mu r)\,J_{N/2}(r\sqrt{\lambda_n})\right]\Big|_{r=R/2}^{r=R} \\
&-N\frac{\lambda_n^{3/4}}{\mu^{3/2}}\int\limits_{R/2}^{R} J_{N/2-1}(\mu r)\,J_{N/2}(r\sqrt{\lambda_n})dr \\
&-N\frac{\lambda_n^{5/4}}{\mu^{3/2}}\int\limits_{R/2}^{R} J_{N/2}(\mu r)\,J_{N/2-1}(r\sqrt{\lambda_n})dr \\
&+N\frac{\lambda_n^{3/4}}{\mu^{5/2}}\int\limits_{R/2}^{R} J_{N/2}(\mu r)\,J_{N/2}(r\sqrt{\lambda_n})\frac{dr}{r}.
\end{aligned} \tag{1.1.76}
$$

In our subsequent reasoning we assume that $\sqrt{\lambda_n} < \mu$. Majorizing the moduli of Bessel functions under the integration sign of the last integral on the right-hand side of (1.1.76) with unity, and the moduli of the remaining Bessel functions on the right-hand side of (1.1.76) by the estimate $|J_\nu(x)| \le C(\nu)x^{-1/2}$, we obtain from (1.1.76) the estimate

$$I_n^\mu(R)\left(1-\frac{\lambda_n}{\mu^2}\right) = O\left(\frac{1}{\mu}\right), \tag{1.1.77}$$

in which the estimate that sets a limit to the growth of the O-terms is independent of R. Taking into account that $\sqrt{\lambda_n} < \mu$, we obtain an estimate from (1.1.77),

$$I_n^\mu(R) = O((\mu-\sqrt{\lambda_n})^{-1}). \tag{1.1.78}$$

For the case $\sqrt{\lambda_n} > \mu$, having interchanged the roles of μ and $\sqrt{\lambda_n}$ in our above reasoning, we arrive at the estimate

$$I_n^\mu(R)\left(1-\frac{\mu^2}{\lambda_n}\right) = O\left(\frac{1}{\sqrt{\lambda_n}}\right)$$

in place of (1.1.77), from which we obtain, recalling that $\sqrt{\lambda_n} > \mu$,

$$I_n^\mu(R) = O((\sqrt{\lambda_n}-\mu)^{-1}). \tag{1.1.79}$$

The validity of estimate (1.1.48) for any $\sqrt{\lambda_n} \neq \mu$ follows from (1.1.78) and (1.1.79): This completes the proof of Lemma 1.1.

1.2. FRACTIONAL KERNELS

In this section we will study the expansion of functions in an arbitrary FSF $\{u_n(y)\}$ of the Laplace operator in any N-dimensional subdomain Ω of an arbitrary N-dimensional domain G. The functions of interest in domain Ω belong to the class of radial functions exhibiting a power singularity of the type $|x-y|^{2\alpha-N}$ $(\alpha > 0)$ at a single fixed point x in the open domain Ω. To be more precise, we are concerned with the expansion of functions that display, at a fixed point x, the same singularity typical of the so-called *Bessel–MacDonald kernel*[20)]

$$|x-y|^{\alpha-N/2}K_{N/2-\alpha}(|x-y|) \tag{1.2.1}$$

at this point.

We will dwell upon the specificity of the Fourier coefficient of an arbitrary function belonging in domain Ω to the class of radial functions displaying, at a fixed point x, a singularity of the type (1.2.1) and satisfying everywhere, except for this point, one or another condition of smoothness. Starting from a special form of the Fourier coefficient for this function, we will establish the existence and derive an explicit representation of the so-called fractional kernels, that is, functions $T_\alpha(x,y)$ which, for any fixed real number $\alpha > 0$ and any fixed point x in the open domain Ω, possess a bilinear series of the form

$$\sum_{n=1}^{\infty} u_n(x)u_n(y)(1+\lambda_n)^{-\alpha}$$

for the Fourier series in the FSF $\{u_n(y)\}$ under study.

The term "fractional kernel" is used on the grounds that the function $T_\alpha(x,y)$, for the case where Ω coincides with the entire domain G, serves as the kernel of the homogeneous integral equation

$$u_n(x) = (1+\lambda_n)^{\alpha}\int_G T_\alpha(x,y)u_n(y)dy.$$

Viewed from the standpoint of linear operator theory in the case $\Omega \equiv G$, the fractional kernels $T_\alpha(x,y)$ are kernels of the integral operators A^α which serve as positive real powers of order α of the integral operator

$$\mathbf{A}f = \int_G T_1(x,y)f(y)dy,$$

20) The symbol $K_\nu(r)$ denotes the so-called MacDonald function of order ν or, as it is also referred to, cylindrical functions of the third kind of order ν exponentially decaying as $r \to \infty$ (see, for example, Bateman and Erdélyi [9, pp. 13 and 17].

generated by the respective Green's functions $T_1(x,y)$ for the Helmholtz operator $\Delta u - u$ in domain G.

The fractional kernels, apart from being of interest in their own right, serve, as we shall see below in Sections 1.3 and 1.4, as a major tool in our method of studying the problems of uniform convergence and localization of the Fourier series in an arbitrary FSF of the Laplace operator. These kernels are quite expedient in establishing the embedding theorems of so-called Liouville classes that generalize the Sobolev classes W_p^ℓ of differentiable functions of N variables onto the fractional differentiability exponents.

1.2.1. Existence and Representation of Fractional Kernels

Assume that $\{u_n(y)\}$ is an arbitrary FSF of the Laplace operator in any N-dimensional subdomain Ω of an arbitrary N-dimensional domain G. We fix an arbitrary point x of the open domain Ω, and our objective now is to establish, for any real number $\alpha > 0$, a function $T_\alpha(x,y)$ such that its nth Fourier coefficient in the expansion in the system $\{u_n(y)\}$ possesses the form $u_n(x)(1+\lambda_n)^{-\alpha}$. We call this function (providing it exists) a *kernel of order* α (for a particular FSF) and denote it by the symbol $T_\alpha(x,y)$. The kernel $T_1(x,y)$ is thus presumed to play the role of Green's function for the operator $\Delta u - u$.

THEOREM 1.6. *For any real* $\alpha > 0$, *there exists a kernel* $T_\alpha(x,y)$ *representable in the form*

$$T_\alpha(x,y) = A_N^\alpha |x-y|^{\alpha - N/2} K_{N/2-\alpha}(|x-y|) + \chi_\alpha(x,y), \tag{1.2.2}$$

where A_N^α *is a constant defined by the equality*

$$A_N^\alpha = 2^{1-\alpha}\left[(2\pi)^{N/2}\Gamma(\alpha)\right]^{-1}, \tag{1.2.3}$$

the symbol $K_\nu(r)$ *denotes the MacDonald function of order* ν *of the argument* r, *and* $\chi_\alpha(x,y)$ *is a function possessing, in the open domain* Ω, *continuous partial derivatives with respect to the coordinates of points* x *and* y *of any high order.*

PROOF: We fix an arbitrary point x of the open domain Ω and a positive number R smaller than the mid-distance between x and the boundary of Ω. We consider, for any fixed index m, two functions of the distance $r = |x-y|$ between a variable point y of domain G and a prefixed point x:

$$v_\alpha(r) = \begin{cases} A_N^\alpha\, r^{\alpha-N/2} K_{N/2-\alpha}(r) & \text{for } r \le R, \\ 0 & \text{for } r > R, \end{cases} \tag{1.2.4}$$

$$w_\alpha(r) = \begin{cases} \sum_{k=0}^{m} a_k\, r^{2k} & \text{for } r \le R, \\ 0 & \text{for } r > R. \end{cases} \tag{1.2.5}$$

Here A_N^α is a constant defined by the equality (1.2.3), $a_0, a_1, \dots, a_m$ are constants chosen in such a manner that at $r = R - 0$ the difference $v_\alpha(r) - w_\alpha(r)$ and all the derivatives of this difference with respect to r of order m inclusive are equal to zero. It is to be ascertained now that the last condition provides for an unambiguous choice of the constants $a_0, a_1, a_2, \dots, a_m$.

Along with the operation of ordinary differentiation with respect to the variable r, we introduce an operation of the so-called *Bessel differentiation* with respect to the variable r, denoted by the symbol D and defined by the equality

$$Df(r) = \frac{1}{r} f'(r).$$

The symbol D^k will be used to mean the result of the Bessel differentation operation D applied k times. The conditions for choosing the coefficients $a_0, a_1, a_2, \dots, a_m$ are specified by the equalities[21)]

$$f(R) = 0, \quad f'(R) = 0, \quad f^{(2)}(R) = 0, \quad \dots, \quad f^{(m)}(R) = 0, \tag{1.2.6}$$

for the function $f(r) = v_\alpha(r) - w_\alpha(r)$. Since $R > 0$, it is evident that equalities (1.2.6) are equivalent to the equalities

$$f(R) = 0, \quad Df(R) = 0, \quad D^2 f(R) = 0, \quad \dots, \quad D^m f(R) = 0. \tag{1.2.7}$$

Taking into account functions (1.2.5) and (1.2.4), equalities (1.2.7) lead to the following quadratic system of linear equations for determining the coefficients $a_0, a_1, a_2, \dots, a_m$:

$$\left\{\begin{array}{lcll}
a_0 + a_1 R^2 + a_2 R^4 + a_3 R^6 + & \dots & +a_m R^{2m} & = v_\alpha(R), \\
2a_1 + 4a_2 R^2 + 6a_3 R^4 + & \dots & +2m a_m R^{2m-2} & = Dv_\alpha(R), \\
4 \cdot 2a_2 + 6 \cdot 4 \cdot a_3 R^2 + & \dots & +2m(2m-2) a_m R^{2m-4} & = D^2 v_\alpha(R), \\
 & \vdots & & \vdots \\
 & & 2m(2m-2)\dots 4 \cdot 2 a_m R^{2m-4} & = D^m v_\alpha(R).
\end{array}\right.$$

Since the determinant of this system is $2^m \cdot 4^{m-1} \cdot 6^{m-2} \dots (2m-2)^2\, 2m$ and is distinct from zero, there exists a unique solution to this system, and this solution is defined by Kramer's formulas.

In what follows, the parcticular form of the coefficients $a_0, a_1, a_2, \dots, a_m$ is not essential, and, we dispense with writing out these coefficients *in extenso*.

By virtue of equality (1.1.3), the Fourier coefficient $f_n = (f, u_n(y))$ of the function $f(r) = v_\alpha(r) - w_\alpha(r)$ takes the form

$$f_n = (2\pi)^{N/2} \lambda_n^{(2-N)/4} u_n(x) \int_0^R [v_\alpha(r) - w_\alpha(r)] r^{N/2} J_{(N-2)/2}\left(r\sqrt{\lambda_n}\right) dr. \tag{1.2.8}$$

21) From now on, the derivatives at the point $r = R$ refer to the left-hand derivatives at this point.

We subject the integral on the right-hand side of (1.2.8) to integration $(m+1)$ times by parts, making use of the recurrence relation $\int r^{\nu} J_{\nu-1}\left(r\sqrt{\lambda_n}\right) dr = \lambda_n^{-1/2} r^{\nu} J_{\nu}(r\sqrt{\lambda_n})$ and the Bessel differentiation operation D as defined above.

The result obtained is

$$f_n = (2\pi)^{N/2} u_n(x)$$

$$\times \left\{ \sum_{k=1}^{m+1} (-1)^{k+1} \lambda_n^{-N/4-(k-1)/2} \left[D^{k-1} f(r) r^{N/2+k-1} J_{N/2+k-1}\left(r\sqrt{\lambda_n}\right)\right] \Big|_{r=0}^{r=R} \right.$$

$$\left. +(-1)^{m+1} \lambda_n^{-N/4-m/2} \int_0^R D^{m+1}[v_\alpha(r) - w_\alpha(r)] r^{N/2+m+1} J_{N/2+m}\left(r\sqrt{\lambda_n}\right) dr \right\}. \tag{1.2.9}$$

We note further that by virtue of the recurrence relation[22)]

$$\frac{d}{dr}\left[r^{-\nu} K_{\nu}(r)\right] = -r^{\nu} K_{\nu+1}(r)$$

the relation

$$D^k [v_\alpha(r)] = 2^{1-\alpha} \left[(2\pi)^{N/2}\Gamma(\alpha)\right]^{-1} (-1)^k r^{-(N/2-\alpha+k)} K_{N/2-\alpha+k}(r) \tag{1.2.10}$$

holds. It is to be taken into account that the operation D, applied at any time, lowers the power of the polynomial $w_\alpha(r)$ by two and that

$$\left\{ \begin{aligned} & D^m w_\alpha(r) = D^m w_\alpha(R) = D^m v_\alpha(R) = \\ & \quad = 2^{1-\alpha}\left[(2\pi)^{N/2}\Gamma(\alpha)\right]^{-1} (-1)^m R^{-(N/2-\alpha+m)} K_{N/2-\alpha+m}(R), \\ & D^{m+1} w_\alpha(R) \equiv 0. \end{aligned} \right. \tag{1.2.11}$$

We observe now that all the substitutions in relation (1.2.9) reduce to zero. Indeed, the substitutions at $r = R$ become zero by virtue of relations (1.2.7), and the substitutions at $r = 0$ become zero by virtue of the boundedness of all Bessel derivatives $D^{k-1} w_\alpha(r)$ for $k = 1, 2, \ldots, m+1$, by virtue of relation (1.2.10), and by virtue of the behavior of the cylindrical functions

$$J_{N/2+k-1}\left(r\sqrt{\lambda_n}\right), \qquad K_{N/2-\alpha+k-1}(r)$$

as $r \to 0$.

---

22) See, for example, Bateman and Erdélyi [9, p. 91].

The reduction to zero of all the substitutions in (1.2.9), the relation (1.2.10), and the second relation (1.2.11) enable us to rewrite relation (1.2.9) in the following form:

$$f_n = \lambda_n^{-N/4-m/2} 2^{1-\alpha}[\Gamma(\alpha)]^{-1} u_n(x) \int_0^R r^\alpha J_{N/2+m}\left(r\sqrt{\lambda_n}\right) K_{N/2-\alpha+m+1}(r)dr. \tag{1.2.12}$$

Let us establish another expression for the Fourier coefficient f_n; to this effect, we perform the transformation $\int_0^R = \int_0^\infty - \int_R^\infty$ and make use of the familiar expression for the integral[23)]

$$\lambda_n^{-N/4-m/2} 2^{1-\alpha}[\Gamma(\alpha)]^{-1} \int_0^\infty r^\alpha J_{N/2+m}\left(r\sqrt{\lambda_n}\right) K_{N/2-\alpha+m+1}(r)dr = (1+\lambda_n)^{-\alpha}. \tag{1.2.13}$$

With the aid of (1.2.12) and (1.2.13) we obtain the following relationship for the Fourier coefficient:

$$\begin{aligned} f_n &= u_n(x)(1+\lambda_n)^{-\alpha} \\ &- u_n(x)\lambda_n^{-N/4-m/2} 2^{1-\alpha}[\Gamma(\alpha)]^{-1} \int_R^\infty r^\alpha J_{N/2+m}\left(r\sqrt{\lambda_n}\right) K_{N/2-\alpha+m+1}(r)dr. \end{aligned} \tag{1.2.14}$$

Thus, we conclude that the Fourier coefficient of the function f of interest, which has at the point $y = x$ a singularity of the type $A_N^\alpha |x-y|^{\alpha-N/2} K_{N/2-\alpha}(|x-y|)$ and possesses everywhere, except for this point, continuous partial derivatives of a preassigned order up to m, is representable in the form

$$f_n = u_n(x)(1+\lambda_n)^{-\alpha} + \overset{m}{\gamma}_n u_n(x), \tag{1.2.15}$$

where for the quantity $\overset{m}{\gamma}_n$, independent of x and y, either of the following two representations holds:

$$\overset{m}{\gamma}_n = -\lambda_n^{-N/4-m/2} 2^{1-\alpha}[\Gamma(\alpha)]^{-1} \int_R^\infty r^\alpha J_{N/2+m}\left(r\sqrt{\lambda_n}\right) K_{N/2+m+1-\alpha}(r)dr, \tag{1.2.16}$$

$$\begin{aligned} \overset{m}{\gamma}_n &= -(1+\lambda_n)^{-\alpha} \\ &+ \lambda_n^{-N/4-m/2} 2^{1-\alpha}[\Gamma(\alpha)]^{-1} \int_0^R r^\alpha J_{N/2+m}\left(r\sqrt{\lambda_n}\right) K_{N/2+m+1-\alpha}(r)dr. \end{aligned} \tag{1.2.17}$$

---

23) See, for example, Bateman and Erdélyi [9, p. 108, formula (39)].

Note that the function f, which is the difference of functions (1.2.4) and (1.2.5), can be represented everywhere in domain G in the form

$$f = A_N^\alpha |x-y|^{\alpha-N/2} K_{N/2-\alpha}(|x-y|) + \varphi_\alpha(|x-y|), \tag{1.2.18}$$

where the function $\varphi_\alpha(|x-y|)$ is

$$\varphi_\alpha(|x-y|) = \begin{cases} -w_\alpha(|x-y|) & \text{for } |x-y| \le R, \\ -A_N^\alpha |x-y|^{\alpha-N/2} K_{N/2-\alpha}(|x-y|) & \text{for } |x-y| > R. \end{cases} \tag{1.2.19}$$

The available choice of the $w_\alpha(r)$ functional coefficients a_0, a_1, $a_2, \dots$, a_m and the representation (1.2.19) imply that the function $\varphi_\alpha(|x-y|)$ in domain Ω possesses continuous partial derivatives with respect to the coordinates of both point x and point y up to order m inclusive[24)].

It follows from relations (1.2.15) and (1.2.18) that a function $T_\alpha(x,y)$ which has the quantity $u_n(x)(1+\lambda_n)^{-\alpha}$ for its Fourier coefficient in the system $\{u_n(y)\}$ is represented in the form

$$T_\alpha(x,y) = A_N^\alpha |x-y|^{\alpha-N/2} K_{N/2-\alpha}(|x-y|) + \varphi_\alpha(|x-y|) - \Psi_\alpha(x,y), \tag{1.2.20}$$

where $\varphi_\alpha(|x-y|)$ is a function defined by relation (1.2.19), and $\Psi_\alpha(x,y)$ is a function whose Fourier series in the system $\{u_n(y)\}$ is a bilinear series of the form

$$\Psi_\alpha(x,y) = \sum_{n=1}^{\infty} \overset{m}{\gamma}_n u_n(x) u_n(y). \tag{1.2.21}$$

We have put an equality sign in relation (1.2.21) and now wish to prove that not only the series on the right-hand side of (1.2.21), but also those series that are derived from it via term-by-term differentiation with respect to the coordinates of the points x and y to an order infinitely increasing with m, are uniformly convergent on any compact set of domain Ω.

This will complete the proof of Theorem 1.6, since the function $\chi_\alpha(x,y)$ involved in the condition of the theorem can be taken as $\chi_\alpha(x,y) = \varphi_\alpha(|x-y|) - \Psi_\alpha(x,y)$, and both functions $\varphi_\alpha(|x-y|)$ and $\Psi_\alpha(x,y)$, by virtue of what was said above, will be shown to possess, in the open region Ω, continuous partial derivatives with respect to the coordinates of points x and y of an order infinitely increasing with m.

We note, for one thing, from relation (1.2.16) and from estimates of the functions $J_\nu(x)$ and $K_\nu(x)$ for values of the argument in excess of $R > 0$ that there exists a

---

[24)] Indeed, the function $-w_\alpha(|x-y|)$ possesses at $|x-y| \le R$ continuous partial derivatives of arbitrarily high order with respect to the coordinates of the points x and y; the function $-A_N^\alpha |x-y|^{\alpha-N/2} K_{N/2-\alpha}(|x-y|)$ possesses at $|x-y| \ge R$ continuous derivatives of arbitrarily high order with respect to the coordinates of the points x and y; by virtue of relations (1.2.6), partial derivatives of order up to m of the function $-w_\alpha(|x-y|)$ at $|x-y| = R-0$ coincide with the respective partial derivatives of the function $-A_N^\alpha |x-y|^{\alpha-N/2} K_{N/2-\alpha}(|x-y|)$.

constant $C(R,\alpha,m,R)$ such that for all $\lambda_n \geq 1$[25] we have

$$|\overset{m}{\gamma}_n| \leq C(R,\alpha,m,N)(1+\lambda_n)^{-N/4-m/2}. \tag{1.2.22}$$

We note further from relation (1.2.17) and from the estimates $|J_\nu(x)| \leq C_1(\nu)x^\nu$ and $x^\nu K_\nu(x) < C_2(\nu)$ (for all $x \geq 0$, $\nu \geq 0$) that for $m > \alpha - N/2 - 1$ and for all λ_n there is

$$|\overset{m}{\gamma}_n| \leq C_1(R,\alpha,m,N). \tag{1.2.23}$$

As has been mentioned above, it suffices to prove that, for arbitrary fixed non-negative integers p and q and for all sufficiently large m, the series

$$\sum_{n=1}^{\infty} \overset{m}{\gamma}_n u_n^{(p)}(x) u_n^{(q)}(y) \tag{1.2.24}$$

converges uniformly with respect to x and y both belonging to an arbitrary compact set of domain Ω; here the symbol $u_n^{(p)}(x)$ refers to any partial derivative of the function $u_n(x)$ of order p, and $u_n^{(q)}(y)$ refers to a partial derivative of the function $u_n(y)$ of order q.

It suffices to prove, by virtue of the Cauchy–Buniakowski inequality, that for an arbitrary fixed $p \geq 0$ and for all sufficiently large m, the series

$$\sum_{n=1}^{\infty} \left|\overset{m}{\gamma}_n\right| [u_n^{(p)}(x)]^2 \tag{1.2.25}$$

converges uniformly with respect to x on any compact set of domain Ω and that for an arbitrary fixed $q \geq 0$ and for all sufficiently large m, the series

$$\sum_{n=1}^{\infty} \left|\overset{m}{\gamma}_n\right| [u_n^{(q)}(y)]^2 \tag{1.2.26}$$

converges uniformly with respect to y on any compact set of domain Ω.

We confine ourselves to a proof of the uniform convergence of the series (1.2.25), since the uniform convergence of the series (1.2.26) can be proved in much the same manner.

Let μ_0 be a fixed sufficiently large number (its choice will be discussed in some detail below).

Assume that

$$\overset{m\prime}{\gamma}_n = \begin{cases} \overset{m}{\gamma}_n & \text{for} \quad \lambda_n \geq \mu_0, \\ 0 & \text{for} \quad \lambda_n < \mu_0, \end{cases} \qquad \overset{m\prime\prime}{\gamma}_n = \begin{cases} 0 & \text{for} \quad \lambda_n \geq \mu_0, \\ \overset{m}{\gamma}_n & \text{for} \quad \lambda_n < 0. \end{cases}$$

[25] If we fix an arbitrary compact set K of domain Ω and take for x an arbitrary point of the compact set K and for R a number less than half the distance between the compact set K and the boundary of Ω, then the constant C for all points x of the compact set K is dependent only on α, m, and N.

Then the series (1.2.25) represents the sum of two series

$$\sum_{n=1}^{\infty}\left|\overset{m}{\gamma}{}'_n\right|[u_n^{(p)}(x)]^2 \quad \text{and} \quad \sum_{n=1}^{\infty}|\overset{m}{\gamma}{}''_n|[u_n^{(p)}(x)]^2, \tag{1.2.27}$$

and all we have to do is to prove that each of the series (1.2.27) at an arbitrary fixed $p \geq 0$ and any sufficiently large m converges uniformly with respect to x on any compact set of domain Ω.

To begin, we prove that the former of the series (1.2.27) converges, given any fixed $p \geq 0$ and any sufficiently large m, uniformly with respect to x on any compact set of domain Ω.

The following lemma lends support to this statement.

LEMMA 1.2. *Given any $\delta > 0$ and fixed sufficiently large $\mu_0 > 0$, the series*

$$\sum_{n=1}^{\infty}[u_n^{(p)}(x)]^2\beta_n^2, \tag{1.2.28}$$

where

$$\beta_n = \begin{cases} \frac{1}{2}(1+\lambda_n)^{-N/4-p/2-\delta/4} & \text{for} \quad \lambda_n \geq \mu_0, \\ 0 & \text{for} \quad \lambda_n < \mu_0, \end{cases} \tag{1.2.29}$$

converges uniformly with respect to x on an arbitrary compact set of domain Ω.

Indeed, the Lemma 1.2 having been proved, we obtain, invoking the definition of $\overset{m}{\gamma}{}'_n$ and estimate (1.2.22), that the former of the series (1.2.27) at any $m > N/2 + 2p$ is majorized uniformly (on a fixed compact set) to within a constant factor $\frac{1}{4}C(R, \alpha, m, N)$ by the convergent series (1.2.28) in which the β_n is defined via equality (1.2.29) for $\delta = m - N/2 - 2p > 0$.

In anticipation of a proof of Lemma 1.2 (*vide infra*, Section 2), we now show that the latter of the series (1.2.27) converges, for an arbitrary fixed $p \geq 0$ and for all sufficiently large m uniformly with respect to x on any compact set of domain Ω. To this effect, it suffices to prove the following lemma.

LEMMA 1.3. *For any nonnegative integer p, the series*

$$\sum_{n=1}^{\infty}[u_n^{(p)}(x)]^2\delta_n^2, \tag{1.2.30}$$

in which for fixed λ_0 (vide supra, Lemma 1.2) the number δ_n takes the form

$$\delta_n = \begin{cases} 3/4 & \text{for} \quad \lambda_n < \mu_0, \\ 0 & \text{for} \quad \lambda_n \geq \mu_0, \end{cases}$$

converges uniformly with respect to x on an arbitrary compact set of domain Ω.

In fact, Lemma 1.3 having been proved, we obtain by the definition of $\overset{m}{\gamma}{}''_n$ and the estimate (1.2.23) that the latter of the series (1.2.27) is majorized, within a constant factor on any fixed compact set of domain Ω, by the series (1.2.30) uniformly convergent on this compact set.

Thus, to complete the proof of Theorem 1.6, we have to prove the previously formulated Lemmas 1.2 and 1.3.

1.2.2. Proof of Ancillary Lemmas

PROOF OF LEMMA 1.2. We fix an arbitrary $\delta > 0$, an arbitrary compact set K of domain Ω, and an arbitrary number $R > 0$ smaller than the mid-distance between the compact set K and the boundary of domain Ω; given an arbitrary point x in the compact set K, we consider a function f which is the difference of functions (1.2.4) and (1.2.5) with $\alpha = N/4 + p/2 + \delta/4$ and with m greater in value than $p + \delta$. By virtue of the relations (1.2.15) and (1.2.22), we obtain for all the indices n for which $\lambda_n \geq 1$

$$f_n = u_n(x)\left\{(1+\lambda_n)^{-(N/4+p/2+\delta/4)} + O\left[(1+\lambda_n)^{-(N/4+m/2)}\right]\right\}, \tag{1.2.31}$$

where the quantity O is independent of x.

We remark now that by construction the function f and all its partial derivatives with respect to the coordinates of the point x up to order m inclusive everywhere, except for the points $y = x$, are continuous in the collection of points x, y for $x \in K$, $y \in G$.

We note further that the function f exhibits a specific feature such that, given $y = x$, any partial derivative $f^{(p)}$ of order p at any $\delta > 0$ admits of an integrable square, that is, for any x in the compact set K, there exists an integral

$$\int_G \left[f^{(p)}\right]^2 dy. \tag{1.2.32}$$

In view of these assumptions, the Fourier coefficient $f_n^{(p)}$ of any partial derivative $f^{(p)}$ with respect to the coordinates of the point x of order p of the function f is derived via differentiating the Fourier coefficient (1.2.31) of the function f, that is, it takes the form

$$f_n^{(p)} = u_n^{(p)}(x)\left\{(1+\lambda_n)^{-(N/4+p/2+\delta/4)} + O\left[(1+\lambda_n)^{-(N/4+m/2)}\right]\right\}. \tag{1.2.33}$$

Since $N/4 + m/2 = N/4 + p/2 + \delta/2 > N/4 + p/2 + \delta/4$, the μ_0 can be fixed so large that for all $\lambda_n \geq \mu_0$ the inequality

$$\left|O\left[(1+\lambda_n)^{-(N/4+m/2)}\right]\right| \leq \frac{1}{2}(1+\lambda_n)^{-(N/4+p/2+\delta/4)} \tag{1.2.34}$$

holds.

We obtain from (1.2.33) and (1.2.34) the inequality

$$|f_n^{(p)}| \geq \frac{1}{2}|u_n^{(p)}(x)|(1+\lambda_n)^{-(N/4+p/2+\delta/4)}, \tag{1.2.35}$$

which holds for all $\lambda_n \geq \mu_0$.

We observe now that the existence of the integral (1.2.32) and the completeness and orthonormality of the system $\{u_n(y)\}$ imply, for any x in the compact set K, the validity of Parseval's equality

$$\sum_{n=1}^{\infty}[f_n^{(p)}]^2 = \int_G [f^{(p)}]^2 dy, \tag{1.2.36}$$

by virtue of which the series of nonnegative terms on the left-hand side of (1.2.36) converges to the sum on the right-hand side of (1.2.36) at each point x of the compact set K. The more so, since each term of the series $[f_n^{(p)}]^2$ and its sum on the right-hand side of (1.2.36) are continuous functions of the point x on the compact set K, then in accordance with Dini's theorem, the series on the left-hand side of (1.2.36) converges uniformly with respect to x on the compact set K.

Finally, by the definition (1.2.29) of the quantities β_n and by the inequality (1.2.35) one concludes that series (1.2.28) everywhere on the compact set K is majorized by the series on the left-hand side of (1.2.28).

The uniform convergence of series (1.2.28) on a compact set K is proved. This completes the proof of Lemma 1.2.

PROOF OF LEMMA 1.3. We fix an arbitrary integer $p \geq 0$, an arbitrary compact set of domain Ω, and an arbitrary number $R > 0$ smaller than the mid-distance between the compact set K and the boundary of domain Ω; given for x an arbitrary point of the compact set K, we consider the following function of the distance $r = |x - y|$ between x and a variable point y:

$$f(r) = \begin{cases} \dfrac{\Gamma(N/2+p+2)}{(p+1)!\pi^{N/2}R^N}\left(1-\dfrac{r^2}{R^2}\right)^{p+1} & \text{for} \quad r \leq R, \\ 0 & \text{for} \quad r > R. \end{cases} \tag{1.2.37}$$

We now ensure that the Fourier coefficient f_n of the function (1.2.37) in the system $\{u_n(y)\}$ takes the form

$$f_n = 2^{N/2+p+1}\Gamma(N/2+p+2)\left(R\sqrt{\lambda_n}\right)^{-(N/2+p+1)} J_{N/2+p+1}\left(R\sqrt{\lambda_n}\right) u_n(x). \tag{1.2.38}$$

Indeed, making use of representation (1.1.3) for the Fourier coefficient f_n, we integrate the integral in (1.1.3) $(p+1)$ times by parts, based on the recurrence formula

$\int r^\nu J_{\nu-1}(r\sqrt{\lambda_n})dr = \lambda_n^{-1/2} r^\nu J_\nu(r\sqrt{\lambda_n})$ and on the use of the Bessel derivative concept $Df(r) = r^{-1}\partial f(r)/\partial r$, to obtain the following representation for the Fourier coefficient f_n:

$$f_n = (2\pi)^{N/2} u_n(x)\Bigg\{\sum_{k=1}^{p+1}(-1)^{k+1}\lambda_n^{-N/4-(k-1)/2} \times \left[r^{N/2+k-1} J_{N/2+k-1}\left(r\sqrt{\lambda_n}\right) D^{k-1} f(r)\right]\Big|_{r=0}^{r=R} + (-1)^{p+1}\lambda_n^{-N/4-p/2}\int_0^R r^{N/2+p+1} J_{N/2+p}\left(r\sqrt{\lambda_n}\right) D^{p+1} f(r)dr\Bigg\}. \tag{1.2.39}$$

Since for any $k \le p+1$ one has

$$D^k\left[\left(1-\frac{r^2}{R^2}\right)^{p+1}\right] = (-1)^k\, 2^k R^{-2k}(p+1)p(p-1)\dots(p-k+2)\left(1-\frac{r^2}{R^2}\right)^{p+1-k}$$

and, in particular,

$$D^{p+1}\left[\left(1-\frac{r^2}{R^2}\right)^{p+1}\right] = (-1)^{p+1}\, 2^{p+1}(p+1)!R^{-2p-2},$$

all the substitutions in (1.2.39) become zero, and we obtain that

$$f_n = 2^{N/2+p+1} u_n(x)\Gamma\left(N/2+p+2\right) \times R^{-(N+2p+2)}\lambda_n^{-(N/4+p/2)}\int_0^R r^{N/2+p+1} J_{N/2+p}(r\sqrt{\lambda_n})dr$$

$$= 2^{N/2+p+1}\Gamma\left(N/2+p+2\right)(R\sqrt{\lambda_n})^{-(N/2+p+1)} J_{N/2+p+1}(R\sqrt{\lambda_n})u_n(x).$$

This lends support to relation (1.2.38).

Note that, by construction, the function (1.2.37) and all its partial derivatives with respect to the coordinates of the point x up to order p inclusive are continuous in the collection of points x, y for $x \in K$, $y \in G$. Under these assumptions, the Fourier coefficient $f_n^{(p)}$ of any partial derivative $f^{(p)}$ with respect to the coordinates of the point x of order p of the function (1.2.37) is derived via appropriate differentiation of the Fourier coefficient (1.2.38), that is, it takes the form

$$f_n^{(p)} = 2^{N/2+p+1}\Gamma\left(N/2+p+2\right)(R\sqrt{\lambda_n})^{-(N/2+p+1)} J_{N/2+p+1}(R\sqrt{\lambda_n})u_n^{(p)}(x). \tag{1.2.40}$$

One infers, based on the representation of the Bessel function $J_\nu(x)$ in the form of a power series and on the property of an alternating series with its terms monotonically and absolutely decreasing, that for all x within the segment $0 \leq x \leq 1$ and for all $\nu \geq 0$, the inequality

$$2^\nu \Gamma(\nu+1)\frac{J_\nu(x)}{x^\nu} > 1 - \frac{1}{4(\nu+1)} \geq \frac{3}{4} \tag{1.2.41}$$

holds.

Let $\mu_0 > 1$ be a fixed number as specified by Lemma 1.2. Given R smaller than $1/\mu_0$, we find that for all indices n such that $\lambda_n \leq \mu_0$, the argument $R\sqrt{\lambda_n}$ of the Bessel function in (1.2.40) belongs to the segment $[0,1]$. If so, a comparison of representation (1.2.40) and inequality (1.2.41) reveals that for an arbitrary fixed $p \geq 0$ and for all indices n such that $\lambda_n < \mu_0$, the inequality

$$|f_n^{(p)}| \geq \frac{3}{4}|u_n^{(p)}(x)| \tag{1.2.42}$$

remains valid.

Writing Parseval's equality in the system $\{u_n(y)\}$ for the function $f^{(p)}$ continuous for $x \in K$, $y \in G$,

$$\sum_{n=1}^{\infty}[f_n^{(p)}]^2 = \int_G [f^{(p)}]^2 dy, \tag{1.2.43}$$

we shall see that the series of nonnegative terms on the left-hand side of (1.2.43) is convergent at each point x of the compact set K. The more so, by virtue of Dini's theorem, the convergence of this series remains uniform with respect to x on the compact set K. Note that the series (1.2.30) for all x in the compact set K is majorized by the series on the left-hand side of (1.2.43) by virtue of inequality (1.2.42) and by the definition of δ_n as specified in the formulation of Lemma 1.3.

This completes the proof of Lemma 1.3.

REMARK. The above reasonings imply that for any $\alpha > N/4$ the integral

$$\int_G T_\alpha^2(x,y)dy \tag{1.2.44}$$

exists and is a continuous function of the point x for all x on an arbitrary compact set of subdomain Ω.

1.3. ESTIMATE FOR THE REMAINDER TERM OF A SPECTRAL FUNCTION IN THE METRIC L_2 AND THE RESULTING COROLLARIES

1.3.1. Estimate for the Remainder Term of a Spectral Function in the Metric L_2

Let $\{u_n(x)\}$ be an arbitrary FSF of the Laplace operator in any N-dimensional subdomain Ω of an arbitrary N-dimensional domain G.

We fix an arbitrary real $\lambda > 0$ and consider the sum

$$\theta(x, y, \lambda) = \sum_{\lambda_n < \lambda} u_n(x)u_n(y) \tag{1.3.1}$$

for any point x interior to domain Ω and for any point y of domain G. The sum (1.3.1) for our fixed $\lambda > 0$ can contain either a finite or an infinite number of summands. If the fundamental values $\{\lambda_n\}$ possess limit points in the segment $[0, \lambda]$, then the sum (1.3.1) contains an infinite number of summands and reduces to a series which, by virtue of the Cauchy–Buniakowski inequality and estimate (1.1.17), is absolutely convergent for any points x and y interior to domain Ω.

If $f(y)$ is an arbitrary summable function compactly supported in domain Ω and continued with zero beyond the confines of domain Ω, then the quantity

$$\sigma_\lambda(x, f) = \int \theta(x, y, \lambda) f(y)dy = E_\lambda f(x) = \sum_{\lambda_n < \lambda} u_n(x) \int u_n(y) f(y)dy \tag{1.3.2}$$

may in a natural manner be called the *spectral decomposition* of the function f in the FSF $\{u_n(x)\}$ considered.

The function $\theta(x, y, \lambda)$ itself, whose convolution with $f(y)$ yields the spectral decomposition (1.3.2), is naturally referred to as a *spectral function* corresponding to the FSF $\{u_n(x)\}$ under study.

In this section we establish an asymptotic estimate of the spectral function $\theta(x, y, \lambda)$ at large values of λ.

We fix an arbitrary compact set K of the domain Ω, an arbitrary point x of the compact set K, and an arbitrary positive number R smaller than the distance between the compact set K and the boundary of Ω; we are concerned with the following function of the distance $r = |x - y|$ between a fixed point x and a variable point y:

$$\overset{\lambda}{v}(r) = \begin{cases} \lambda^{N/4}(2\pi r)^{-N/2} J_{N/2}(r\sqrt{\lambda}) & \text{for} \quad r \le R, \\ 0 & \text{for} \quad r > R. \end{cases} \tag{1.3.3}$$

We wish to prove that the function (1.3.3) is the main term in the asymptotic estimate of the spectral function $\theta(x, y, \lambda)$ for large λ. To be more exact, we will prove the following statement.

THEOREM 1.8. *For the spectral function $\theta(x, y, \lambda)$ corresponding to an arbitrary FSF $\{u_n(x)\}$ of the Laplace operator in any N-dimensional subdomain Ω of an arbitrary N-dimensional domain G, the following representation is valid:*

$$\theta(x, y, \lambda) = \overset{\lambda}{v}(|x-y|) + \widehat{\theta}(x, y, \lambda), \tag{1.3.4}$$

where $\overset{\lambda}{v}(r)$ is the function (1.3.3), and $\widehat{\theta}(x, y, \lambda)$ is the remainder term for which the estimate below, exact to within an order of magnitude,

$$\left\|\widehat{\theta}(x, y, \lambda)\right\|_{L_2(G)} \geq CR^{-1}\lambda^{(N-1)/4}, \tag{1.3.5}$$

holds for all large enough $\lambda > 0$ uniformly with respect to x on an arbitrary compact set K of the domain Ω (the L_2-norm in this estimate is taken in the coordinates of the point y).

REMARK. It follows from estimate (1.3.5) that, given any $\lambda \geq 0$, the estimate

$$\|\widehat{\theta}(x, y, \lambda)\|_{L_2(G)} \leq CR^{-1}(1+\lambda)^{(N-1)/4}, \tag{1.3.5'}$$

also uniform with respect to x on an arbitrary compact set K of subdomain Ω, holds for the remainder term $\widehat{\theta}(x, y, \lambda)$.

PROOF: We fix an arbitrary compact set of domain Ω, an arbitrary point x of compact K, and an arbitrary positive number R smaller than the distance between compact K and the boundary of domain Ω; further, we consider the function (1.3.3) in which $r = |x - y|$ refers to the distance between a fixed point x and a variable point y.

In accordance with equality (1.1.3), the Fourier coefficient $\overset{\lambda}{v}_n(x)$ of this function in its expansion in the system $\{u_n(y)\}$ is

$$\begin{aligned} \overset{\lambda}{v}_n(x) &= (2\pi)^{N/2}u_n(x)\lambda_n^{(2-N)/4}\int_0^R \overset{\lambda}{v}(r)r^{N/2}J_{(N-2)/2}\left(r\sqrt{\lambda_n}\right)dr \\ &= \lambda^{N/4}\lambda_n^{(2-N)/4}u_n(x)\int_0^R J_{N/2}\left(r\sqrt{\lambda}\right)J_{(N-2)/2}\left(r\sqrt{\lambda_n}\right)dr. \end{aligned} \tag{1.3.6}$$

In the latter integral in (1.3.6) we perform a transformation of the form $\int_0^R = \int_0^\infty - \int_R^\infty$ and make use of the familiar value for this integral[26)]

$$\lambda^{N/4}\lambda_n^{(2-N)/4}\int_0^\infty J_{N/2}\left(r\sqrt{\lambda}\right)J_{(N-2)/2}\left(r\sqrt{\lambda_n}\right)dr = \begin{cases} 1 & \text{for} \quad \lambda_n < \lambda, \\ 1/2 & \text{for} \quad \lambda_n = \lambda, \\ 0 & \text{for} \quad \lambda_n > \lambda. \end{cases} \tag{1.3.7}$$

26) See, for example, Bateman and Erdélyi [9, p. 107, formula (34)].

Making use of the notations

$$\delta_n(\lambda) = \begin{cases} 1 & \text{for } \lambda_n < \lambda, \\ 1/2 & \text{for } \lambda_n = \lambda, \\ 0 & \text{for } \lambda_n > \lambda, \end{cases}$$

$$\overset{\lambda}{I}_n(R) = \int_R^{\infty} J_{N/2}(r\sqrt{\lambda}) J_{(N-2)/2}(r\sqrt{\lambda_n})dr, \tag{1.3.8}$$

we obtain from (1.3.6) with the aid of (1.3.7)

$$\overset{\lambda}{v}_n(x) = \delta_n(\lambda)u_n(x) - \lambda^{N/4}\lambda_n^{(2-N)/4}u_n(x)\overset{\lambda}{I}_n(R). \tag{1.3.9}$$

We multiply both sides of (1.3.9) by the fundamental function $u_n(y)$ and sum up the resultant equality over all the numbers n from 1 to infinity. The formally obtained equality is thus

$$\begin{aligned} \sum_{n=1}^{\infty} \overset{\lambda}{v}_n(x)u_n(y) = \sum_{\lambda_n<\lambda} u_n(x)u_n(y) + \frac{1}{2} \sum_{\lambda_n=\lambda} u_n(x)u_n(y) \\ -\lambda^{N/4} \sum_{n=1}^{\infty} \lambda_n^{(2-N)/4} \overset{\lambda}{I}_n(R)u_n(x)u_n(y). \end{aligned} \tag{1.3.10}$$

We now ensure that for any fixed $\lambda > 0$ and for any fixed point x of the compact set K all the series (1.3.10) are convergent at least in the metric $L_2(G)$ (with respect to the coordinates of the point y), so that the equality (1.3.10) may in any event be regarded as an equality of the elements of the space $L_2(G)$. Indeed, given any fixed $\lambda > 0$ and any fixed point x of the compact set K, the function (1.3.3) as a function of the point y belongs to the class $L_2(G)$, and therefore the Fourier series on the left-hand side of (1.3.10) of this function with respect to the complete orthonormal system $\{u_n(y)\}$ converges to this function in the metric $L_2(G)$.

As regards the first and second series on the right-hand side of (1.3.10), by virtue of Parseval's equality for the complete orthonormal system $\{u_n(y)\}$, these series (given any fixed $\lambda > 0$ and $x \in K$) converge in the metric $L_2(G)$ with respect to y, respectively, to

$$\left[\sum_{\lambda_n<\lambda} u_n^2(x)\right]^{1/2} = \left[\int_G \theta^2(x, y, \lambda)dy\right]^{1/2} \quad \text{and} \quad \frac{1}{2}\left[\sum_{\lambda_n=\lambda} u_n^2(x)\right]^{1/2}.$$

Hence it follows that the last (third) series on the right-hand side of (1.3.10) converges for any fixed $\lambda > 0$ and for any fixed point x of the compact set K in the metric $L_2(G)$ with respect to the coordinates of the point y.

Thus, we have established the validity of equality (1.3.10) (for any fixed $\lambda > 0$ and for any fixed point x of the compact set K) viewed as an equality of the $L_2(G)$ elements. By definition (1.3.1) of the spectral function and by virtue of the convergence of the Fourier series of the function (1.3.3) to this function in the metric $L_2(G)$, we can rewrite equality (1.3.10) as

$$\theta(x, y, \lambda) = \overset{\lambda}{v}(|x-y|) + \widehat{\theta}(x, y, \lambda), \tag{1.3.11}$$

where $\overset{\lambda}{v}(r)$ is the function (1.3.3), and $\widehat{\theta}(x, y, \lambda)$ represents the sum for the series

$$\widehat{\theta}(x, y, \lambda) = -\frac{1}{2} \sum_{\lambda_n = \lambda} u_n(x) u_n(y) + \lambda^{N/4} \sum_{n=1}^{\infty} \lambda_n^{(2-N)/4} \overset{\lambda}{I}_n(R) u_n(x) u_n(y) \tag{1.3.12}$$

convergent in $L_2(G)$ with respect to the coordinates of the point y (for fixed $\lambda > 0$ and $x \in K$).

To prove Theorem 1.8, we must ensure that, for all sufficiently large $\lambda > 0$, uniformly with respect to x on an arbitrary fixed compact set K, the following estimate holds:

$$\|\widehat{\theta}(x, y, \lambda)\|_{L_2(G)} \leq C R^{-1} \lambda^{(N-1)/4}, \tag{1.3.13}$$

in which the L_2-norm is specified in the coordinates of the point y.

To prove the estimate (1.3.13), it will suffice, by virtue of relation (1.3.12) and Parseval's equality, to establish that the two equalities

$$\left\| \sum_{\lambda_n = \lambda} u_n(x) u_n(y) \right\|_{L_2(G)}^2 = \sum_{\lambda_n = \lambda} u_n^2(x) \leq C_1 \lambda^{(N-1)/2}, \tag{1.3.14}$$

$$\left\| \sum_{n=1}^{\infty} \lambda^{N/4} \lambda_n^{(2-N)/4} \overset{\lambda}{I}_n(R) u_n(x) u_n(y) \right\|_{L_2(G)}^2 = \lambda^{N/2} \sum_{n=1}^{\infty} \lambda_n^{(2-N)/2} u_n^2(x) \left[\overset{\lambda}{I}_n(R) \right]^2 \leq C_2 R^{-2} \lambda^{(N-1)/2} \tag{1.3.15}$$

hold, uniform with respect to x on the fixed compact set K.

The validity of inequality (1.3.14) follows immediately from Theorem 1.1 (*vide supra*, Section 1.1.2) or, to be more exact, estimate (1.1.4) taken at $\mu = \sqrt{\lambda}$.

To prove the validity of inequality (1.3.15), it suffices to establish the following six estimates:

$$\lambda^{N/2} \sum_{\sqrt{\lambda_n} \leq 1} \lambda_n^{(2-N)/2} u_n^2(x) \left[\overset{\lambda}{I}_n(R) \right]^2 \leq C_3 R^{-2} \lambda^{(N-1)/2}, \tag{1.3.16}$$

$$\lambda^{N/2} \sum_{1<\sqrt{\lambda_n}\leq\sqrt{\lambda}/2} \lambda_n^{(2-N)/2} u_n^2(x) \left[\overset{\lambda}{I}_n(R)\right]^2 \leq C_4 R^{-2} \lambda^{(N-1)/2}, \tag{1.3.17}$$

$$\lambda^{N/2} \sum_{\sqrt{\lambda}/2\leq\sqrt{\lambda_n}\leq\sqrt{\lambda}-1} \lambda_n^{(2-N)/2} u_n^2(x) \left[\overset{\lambda}{I}_n(R)\right]^2 \leq C_5 R^{-2} \lambda^{(N-1)/2}, \tag{1.3.18}$$

$$\lambda^{N/2} \sum_{|\sqrt{\lambda_n}-\sqrt{\lambda}|<1} \lambda_n^{(2-N)/2} u_n^2(x) \left[\overset{\lambda}{I}_n(R)\right]^2 \leq C_6 R^{-2} \lambda^{(N-1)/2}, \tag{1.3.19}$$

$$\lambda^{N/2} \sum_{\sqrt{\lambda}+1\leq\sqrt{\lambda_n}\leq 3\sqrt{\lambda}/2} \lambda_n^{(2-N)/2} u_n^2(x) \left[\overset{\lambda}{I}_n(R)\right]^2 \leq C_7 R^{-2} \lambda^{(N-1)/2}, \tag{1.3.20}$$

$$\lambda^{N/2} \sum_{\sqrt{\lambda_n}\geq 3\sqrt{\lambda}/2} \lambda_n^{(2-N)/2} u_n^2(x) \left[\overset{\lambda}{I}_n(R)\right]^2 \leq C_8 R^{-2} \lambda^{(N-1)/2}, \tag{1.3.21}$$

which are uniform with respect to x on an arbitrary compact set K.

The following lemma is essential for establishing the above inequalities; for convenience, the proof of this lemma will be dealt with below in Section 1.3.2.

LEMMA 1.4. *The following two estimates are valid for the quantity $I_n(R)$ as defined by formula (1.3.8) for all $\lambda > 1$, $\lambda_n \geq 1$:*

(a) for all indices n such that $|\sqrt{\lambda_n} - \sqrt{\lambda}| \leq 1$,

$$\left|\overset{\lambda}{I}_n(R)\right| \leq C_9 \lambda^{-1/4} \lambda_n^{-1/4}; \tag{1.3.22}$$

(b) for all indices n such that $|\sqrt{\lambda_n} - \sqrt{\lambda}| \geq 1$,

$$\left|\overset{\lambda}{I}_n(R)\right| \leq C_{10} \lambda^{-1/4} \lambda_n^{-1/4} |\sqrt{\lambda_n} - \sqrt{\lambda}|^{-1} R^{-1}; \tag{1.3.23}$$

the constants C_9 and C_{10} in these estimates are independent of $\lambda \geq 1$, $\lambda_n \geq 1$, and $R \leq 1$.

For the same quantity $\overset{\lambda}{I}_n(R)$ for all $N \geq 2$, $\lambda_n \leq 2$, $\lambda \geq 1$, $0 \leq R \leq 1$, the estimate

$$\left|\overset{\lambda}{I}_n(R)\right| \leq C_{11} \lambda^{-1/2} \lambda_n^{(N-2)/4} \tag{1.3.24}$$

is valid, with the constant C_{11} independent of $\lambda_n \leq 2$, $\lambda \geq 1$, $R \leq 1$.

Relying on this lemma, we turn to verifying the above estimates (1.3.16)–(1.3.21).

Making use of estimate (1.3.24), we find that the quantity on the left-hand side of (1.3.16) is not greater in value than

$$C_{11}^2 \lambda^{(N-2)/2} \sum_{\sqrt{\lambda_n} \leq 1} u_n^2(x). \tag{1.3.25}$$

By virtue of estimate (1.1.4) for $\mu = 0$, we observe that

$$\sum_{\sqrt{\lambda_n} \leq 1} u_n^2(x) \leq C \tag{1.3.26}$$

uniformly with respect to x on any compact set K of domain Ω.

With reference to (1.3.25) and to the inequalities (1.3.26) and $\lambda^{(N-2)/2} \leq \lambda^{(N-1)/2}$, one can readily see that the estimate (1.3.16) is valid.

In establishing estimate (1.3.17) we note, for one thing, that for the indices n involved in this estimate, the estimate (1.3.23) can be rewritten in the following manner:

$$\left| \overset{\lambda}{I}_n(R) \right| \leq C_{12} \lambda^{-1/4} \lambda_n^{-3/4} R^{-1}. \tag{1.3.27}$$

This inequality enables us to majorize the quantity on the left-hand side of (1.3.17) by the following sum:

$$\begin{aligned} \lambda^{N/2} &\sum_{1 < \sqrt{\lambda_n} < \sqrt{\lambda}/2} \lambda_n^{(2-N)/2} u_n^2(x) \left[\overset{\lambda}{I}_n(R) \right]^2 \\ &\leq C_{12}^2 \lambda^{(N-1)/2} R^{-2} \sum_{\sqrt{\lambda_n} \geq 1} u_n^2(x) \lambda_n^{-(N+1)/2}. \end{aligned} \tag{1.3.28}$$

Now, to establish estimate (1.3.17) we invoke Corollary 2 of Theorem 1.1 (*vide supra* Section 1.1.2), by virtue of which the sum of the series that is found on the right-hand side of (1.3.28) and is coincident with the series (1.1.24) at $\delta = 1/2 > 0$ is bounded on an arbitrary compact set K.

In verifying the two estimates (1.3.18) and (1.3.20), we observe that these estimates follow from a more general estimate,

$$\lambda^{N/2} \sum_{1 < |\sqrt{\lambda_n} - \sqrt{\lambda}| < \sqrt{\lambda}/2} \lambda_n^{(2-N)/2} u_n^2(x) \left[\overset{\lambda}{I}_n(R) \right]^2 \leq C_{13} R^{-2} \lambda^{(N-1)/2}, \tag{1.3.29}$$

uniform with respect to x on an arbitrary compact set K.

To verify estimate (1.3.29), we denote by p the least number such that $2^p \geq \sqrt{\lambda}/2$. Then the segment $[1, \sqrt{\lambda}/2]$ will accommodate a system of segments $[2^{\ell-1}, 2^\ell]$, $\ell = 1, 2, \ldots, p$.

Making use of the estimate (1.3.23) and taking into account that for all numbers n in the summation of the left-hand side of (1.3.29) the inequality $\sqrt{\lambda_n}/\sqrt{\lambda} \le 2$ holds, we can majorize the quantity on the left-hand side of (1.3.29) in the following manner:

$$\lambda^{N/2} \sum_{1<|\sqrt{\lambda_n}-\sqrt{\lambda}|<\sqrt{\lambda}/2} \lambda_n^{(2-N)/2} u_n^2(x) \left[\overset{\lambda}{I}_n(R)\right]^2$$

$$\le C_{10}^2 2^{N-1} R^{-2} \sum_{1\le|\sqrt{\lambda_n}-\sqrt{\lambda}|\le\sqrt{\lambda}/2} u_n^2(x)(\sqrt{\lambda_n}-\sqrt{\lambda})^{-2}$$

$$\le C_{14} R^{-2} \sum_{\ell=1}^{p} \Big[\sum_{2^{\ell-1}\le|\sqrt{\lambda_n}-\sqrt{\lambda}|\le 2^{\ell}} u_n^2(x)(\sqrt{\lambda_n}-\sqrt{\lambda})^{-2} \Big] \tag{1.3.30}$$

$$\le C_{14} R^{-2} \sum_{\ell=1}^{p} 4^{1-\ell} \Big\{ \sum_{2^{\ell-1}\le|\sqrt{\lambda_n}-\sqrt{\lambda}|\le 2^{\ell}} u_n^2(x) \Big\}.$$

To complete the proof of estimate (1.3.29), it remains only to make use of the estimate (1.1.21) (see Corollary 1 of Theorem 1.1, Section 1.1.2) for the sum enclosed within braces on the right-hand side of (1.3.30). By virtue of the above estimate (1.1.21) at $\rho = 2^{\ell}$, $\mu = \sqrt{\lambda}$, one has

$$\sum_{2^{\ell-1}\le|\sqrt{\lambda_n}-\sqrt{\lambda}|\le 2^{\ell}} u_n^2(x) \le 2^{\ell} O(\lambda^{(N-1)/2}) \tag{1.3.31}$$

uniformly on an arbitrary compact set K.

Substituting (1.3.31) into the right-hand side of (1.3.30) and taking into account the convergence of the series $\sum_{\ell=1}^{\infty} 2^{-\ell} = 1$, we complete the proof of estimate (1.3.29) and, consequently, of estimates (1.3.18) and (1.3.20).

Estimate (1.3.19) is a trivial consequence of inequality (1.3.22), the inequality $\sqrt{\lambda}/\sqrt{\lambda_n} = O(1)$ valid for all numbers n figuring in estimate (1.3.19), and relation (1.1.20), established in Section 1.1.2 for $\mu = \sqrt{\lambda}$.

Now we have to establish estimate (1.3.21). To this effect, we note that for all numbers n figuring in this estimate, inequality (1.3.23) can be rewritten in the form (1.3.27). Invoking (1.3.27), one will observe that the left-hand side of (1.3.21) is majorized by the sum

$$\lambda^{N/2} \sum_{\sqrt{\lambda_n}>3\sqrt{\lambda}/2} \lambda_n^{(2-N)/2} u_n^2(x) \left[\overset{\lambda}{I}_n(R)\right]^2$$

$$\le C_{12}^2 \lambda^{(N-1)/2} R^{-2} \sum_{\sqrt{\lambda_n}>3\sqrt{\lambda}/2} u_n^2(x) \lambda_n^{-(N+1)/2}.$$

Now, in order to establish estimate (1.3.21) we make use of the boundedness of the sum of series (1.1.24) on an arbitrary compact set at $\delta = 1/2 > 0$ (see Section 1.1.2, Corollary 2 of Theorem 1.1). On the condition that Lemma 1.4 is valid, this completes the proof of Theorem 1.8.

1.3.2. Proof of Lemma 1.4

To begin, we establish the estimate (1.3.22) assuming $|\sqrt{\lambda_n} - \sqrt{\lambda}| \le 1$, $\sqrt{\lambda_n} \ge 1$, $\sqrt{\lambda} > 1$.

Let initially the dimension of domain N be odd. Integrating the integral

$$\overset{\lambda}{I}_n(R) = \int_R^{\infty} J_{N/2}(r\sqrt{\lambda}) J_{(N-2)/2}(r\sqrt{\lambda_n}) dr \tag{1.3.32}$$

$(N-1)/2$ times by parts using the recurrence relation[27)]

$$\int r^{-\nu} J_{\nu+1}(r\sqrt{\lambda_n}) dr = -r^{-\nu}(\sqrt{\lambda})^{-1} J_\nu(r\sqrt{\lambda}),$$

$$\frac{d}{dr}\left[r^\nu J_\nu(r\sqrt{\lambda_n})\right] = r^\nu \sqrt{\lambda_n} J_{\nu-1}(r\sqrt{\lambda_n}),$$

we arrive at the equality

$$\begin{aligned} \overset{\lambda}{I}_n(R) &= \sum_{\ell=1}^{(N-1)/2} (\sqrt{\lambda_n})^{\ell-1} (\sqrt{\lambda})^{-\ell} J_{N/2-\ell}(R\sqrt{\lambda}) J_{N/2-\ell}(R\sqrt{\lambda_n}) \\ &\quad + \left(\frac{\sqrt{\lambda_n}}{\sqrt{\lambda}}\right)^{(N-1)/2} \int_R^{\infty} J_{1/2}(r\sqrt{\lambda}) J_{-1/2}(r\sqrt{\lambda_n}) dr. \end{aligned} \tag{1.3.33}$$

Since for $|\sqrt{\lambda_n} - \sqrt{\lambda}| \le 1$, $\sqrt{\lambda_n} \ge 1$, $\sqrt{\lambda} > 1$ the inequality $\sqrt{\lambda_n}/\sqrt{\lambda} < 2$ holds, and estimate (1.3.22) is equivalent to the estimate

$$\left|\overset{\lambda}{I}_n(R)\right| = O(\lambda^{-1/2}), \tag{1.3.34}$$

for each summand under the summation sign on the right-hand side of (1.3.33), estimate (1.3.22) is valid (it suffices to have each of the Bessel functions $J_{N/2-\ell}(R\sqrt{\lambda})$ and $J_{N/2-\ell}(R\sqrt{\lambda_n})$ majorized by a constant).

27) See, for example, Bateman and Erdélyi [9, p. 20, formulas (52) and (53) for $m = 1$].

It remains to prove that the last summand on the right-hand side of (1.3.33) is of order $O(\lambda^{-1/4}\lambda_n^{-1/4})$; this follows from the fact that $(\sqrt{\lambda_n}/\sqrt{\lambda})^{(N-1)/2} \leq 2^{(N-1)/2}$ and the integral in this summand is

$$\int_R^\infty J_{1/2}\left(r\sqrt{\lambda}\right) J_{-1/2}\left(r\sqrt{\lambda_n}\right) dr$$

$$= 2\pi^{-1}(\lambda\lambda_n)^{-1/4}\int_R^\infty r^{-1}\sin(r\sqrt{\lambda})\cos(r\sqrt{\lambda_n})dr$$

$$= \pi^{-1}(\lambda\lambda_n)^{-1/4}\left\{\int_R^\infty r^{-1}\sin\left[r(\sqrt{\lambda}+\sqrt{\lambda_n})\right] dr\right. \tag{1.3.35}$$

$$\left. + \int_R^\infty r^{-1}\sin\left[(\sqrt{\lambda}-\sqrt{\lambda_n})\right] dr\right\},$$

where the term within braces is the algebraic sum of two values of the integral sine for positive values of the argument, and the integral sine on the positive semi-axis is uniformly bounded. Thus, for the case in which the dimension of domain N is odd, the estimate (1.3.22) for $|\sqrt{\lambda_n}-\sqrt{\lambda}| \leq 1$, $\sqrt{\lambda_n} \geq 1$, $\sqrt{\lambda} > 1$ is established.

We now establish the estimate (1.3.22) for $|\sqrt{\lambda_n}-\sqrt{\lambda}| \leq 1$, $\sqrt{\lambda_n} \geq 1$, $\sqrt{\lambda} > 1$ for the case in which the dimension of domain N is even. Subjecting integral (1.3.32) to integration $(N-2)/2$ times by parts, we obtain, instead of (1.3.33), the following relationship:

$$\overset{\lambda}{I}_n(R) = \sum_{\ell=1}^{(N-2)/2} (\sqrt{\lambda_n})^{\ell-1}(\sqrt{\lambda})^{-\ell} J_{N/2-\ell}(R\sqrt{\lambda})J_{N/2-\ell}(R\sqrt{\lambda_n})$$

$$+\left(\frac{\sqrt{\lambda_n}}{\sqrt{\lambda}}\right)^{(N-2)/2}\int_R^\infty J_1(r\sqrt{\lambda})J_0(r\sqrt{\lambda_n})dr. \tag{1.3.33'}$$

In much the same manner as for odd N, one proves that each summand under the summation sign on the right-hand side of (1.3.33′) is of the same order as that of the summand on the right-hand side of (1.3.34). Since $\left(\frac{\sqrt{\lambda_n}}{\sqrt{\lambda}}\right)^{(N-2)/2} \leq 2^{(N-2)/2}$, it remains to prove that the integral

$$\int_R^\infty J_1(r\sqrt{\lambda})J_0(r\sqrt{\lambda_n})dr \tag{1.3.36}$$

is also of the same order. Making use of relation (1.3.7) for $N = 2$ and of the first notation in (1.3.8), we obtain

$$\int_R^\infty J_1(r\sqrt{\lambda})J_0(r\sqrt{\lambda_n})dr = \delta_n(\lambda)\lambda^{-1/2} - \int_0^R J_1(r\sqrt{\lambda})J_0(r\sqrt{\lambda_n})dr. \tag{1.3.37}$$

Thus, in order to prove that the integral (1.3.36) has an order of $O(\lambda^{-1/2})$, it suffices to show that the integral on the right-hand side of (1.3.37) is of the same order.

Since $J_0(x) = -J_1(x)$, there is, according to Lagrange's theorem, a number θ in the interval $0 < \theta < 1$ such that

$$J_0(r\sqrt{\lambda_n}) - J_0(r\sqrt{\lambda}) = -r(\sqrt{\lambda_n} - \sqrt{\lambda})J_1[r\sqrt{\lambda} + r\theta(\sqrt{\lambda_n} - \sqrt{\lambda})].$$

From this relation, with allowance made for $|\sqrt{\lambda_n} - \sqrt{\lambda}| \leq 1$, and taking into account that $|J_1(x)| \leq C_{15}x^{-1/2}$, we obtain

$$J_0(r\sqrt{\lambda_n}) - J_0(r\sqrt{\lambda})| \leq C_{15}\sqrt{r}[\sqrt{\lambda} + \theta(\sqrt{\lambda_n} - \sqrt{\lambda})]^{-1/2}. \tag{1.3.38}$$

Since

$$\sqrt{\lambda} + \theta(\sqrt{\lambda_n} - \sqrt{\lambda}) \geq \begin{cases} \sqrt{\lambda} & \text{for } \sqrt{\lambda_n} \geq \sqrt{\lambda}, \\ \sqrt{\lambda}(1-\theta) + \theta\sqrt{\lambda_n} \geq \sqrt{\lambda_n} \geq \sqrt{\lambda}/2 & \text{for } \sqrt{\lambda_n} < \sqrt{\lambda}, \end{cases}$$

the following estimate stems from (1.3.38):

$$\left|J_0(r\sqrt{\lambda}) - J_0(r\sqrt{\lambda_n})\right| \leq C_{15}\sqrt{2r}\lambda^{-1/4}. \tag{1.3.39}$$

We now represent the integral on the right-hand side of (1.3.37) as the sum of two integrals:

$$\begin{aligned} &\int_0^R J_1(r\sqrt{\lambda})J_0(r\sqrt{\lambda_n})dr \\ &= \int_0^R J_1(r\sqrt{\lambda})\left[J_0(r\sqrt{\lambda_n}) - J_0(r\sqrt{\lambda})\right] dr + \int_0^R J_1(r\sqrt{\lambda})J_0(r\sqrt{\lambda})dr. \end{aligned} \tag{1.3.40}$$

The former integral on the right-hand side of (1.3.40) has the required order of $O(\lambda^{-1/2})$ for all $R \leq 1$ by virtue of inequality (1.3.39) and the above-mentioned trivial estimate $|J_1(r\sqrt{\lambda})| \le C_{15}(r\sqrt{\lambda})^{-1/2}$. The latter integral on the right-hand side of (1.3.40) is

$$\int_0^R J_1(r\sqrt{\lambda})J_0(r\sqrt{\lambda})dr = -(\sqrt{\lambda})^{-1}\int_0^R J_0(r\sqrt{\lambda})d[J_0(r\sqrt{\lambda})]$$

$$= (2\sqrt{\lambda})^{-1}[1 - J_0^2(R\sqrt{\lambda})] \le (2\sqrt{\lambda})^{-1}.$$

The derivation of estimate (1.3.22) is thus complete.

It now remains to establish estimate (1.3.23) for all $|\sqrt{\lambda_n} - \sqrt{\lambda}| > 1$, $\sqrt{\lambda_n} \ge 1$, $\sqrt{\lambda} > 1$.

Assuming, first, that $\sqrt{\lambda} < \sqrt{\lambda_n}$, we subject the integral (1.3.32) to double integration by parts using recursion formulas[28)]

$$\int r^\nu J_{\nu-1}(r\sqrt{\lambda_n})dr = (\sqrt{\lambda_n})^{-1} r^\nu J_\nu(r\sqrt{\lambda_n}),$$

$$\frac{d}{dr}\left[r^{-\nu}J_\nu(r\sqrt{\lambda})\right] = -\sqrt{\lambda}\, r^\nu J_{\nu+1}(r\sqrt{\lambda}). \tag{1.3.41}$$

We obtain

$$\overset{\lambda}{I}_n(R) = -(\sqrt{\lambda_n})^{-1}J_{N/2}(R\sqrt{\lambda})J_{N/2}(R\sqrt{\lambda_n})$$

$$-\frac{\sqrt{\lambda}}{\lambda_n}J_{N/2+1}(R\sqrt{\lambda})J_{N/2+1}(R\sqrt{\lambda_n}) + \frac{\lambda}{\lambda_n}\int_R^\infty J_{N/2+2}(r\sqrt{\lambda})J_{N/2+1}(r\sqrt{\lambda_n})dr. \tag{1.3.42}$$

We transform the integrand on the right-hand side of (1.3.42) as

$$J_{N/2+2}(r\sqrt{\lambda})J_{N/2+1}(r\sqrt{\lambda_n}) = J_{N/2}(r\sqrt{\lambda})J_{N/2-1}(r\sqrt{\lambda_n})$$

$$+(N+2)(r\sqrt{\lambda})^{-1}J_{N/2+1}(r\sqrt{\lambda})J_{N/2+1}(r\sqrt{\lambda_n})$$

$$-N(r\sqrt{\lambda_n})^{-1}J_{N/2}(r\sqrt{\lambda})J_{N/2}(r\sqrt{\lambda_n}),$$

invoking the recursion formulas[29)]

$$J_{N/2+2}(r\sqrt{\lambda}) = -J_{N/2}(r\sqrt{\lambda}) + (N+2)(r\sqrt{\lambda})^{-1}J_{N/2+1}(r\sqrt{\lambda}),$$

[28)] See, for example, Bateman and Erdélyi [9, p. 20, formulas (52) and (53)].

[29)] See, for example, Bateman and Erdélyi [9, p. 20, formula (56)].

$$J_{N/2+1}(r\sqrt{\lambda_n}) = -J_{N/2-1}(r\sqrt{\lambda_n}) + N(r\sqrt{\lambda_n})^{-1}J_{N/2}(r\sqrt{\lambda_n}).$$

We thus obtain the resultant expression

$$\overset{\lambda}{I}_n(R)\left(1-\frac{\lambda}{\lambda_n}\right) = -(\sqrt{\lambda_n})^{-1}J_{N/2}(R\sqrt{\lambda})J_{N/2}(R\sqrt{\lambda_n})$$
$$-\sqrt{\lambda}(\sqrt{\lambda_n})^{-1}J_{N/2+1}(R\sqrt{\lambda})J_{N/2+1}(R\sqrt{\lambda_n})$$
$$+(N+2)\sqrt{\lambda}(\sqrt{\lambda_n})^{-1}\int_R^\infty r^{-1}J_{N/2+1}(R\sqrt{\lambda})J_{N/2+1}(R\sqrt{\lambda_n})dr$$
$$-N\lambda(\lambda_n)^{-3/2}\int_R^\infty J_{N/2}(r\sqrt{\lambda})J_{N/2}(r\sqrt{\lambda_n})r^{-1}dr. \tag{1.3.43}$$

Majorizing the moduli of the Bessel functions on the right-hand side of (1.3.43) with the aid of the estimate

$$|J_\nu(x)| \le C(\nu)x^{-1/2}, \tag{1.3.44}$$

which is valid at $\nu \ge -1/2$, $x > 0$, we obtain from (1.3.43) the relation

$$\left|\overset{\lambda}{I}_n(R)\right|\left(1-\frac{\lambda}{\lambda_n}\right) \le C_{16}R^{-1}\lambda_n^{-3/4}\lambda^{-1/4},$$

which provides evidence for the validity of estimate (1.3.23) in the case $1 \le \sqrt{\lambda} < \sqrt{\lambda_n}$.

Since in estimating the right-hand side of (1.3.43) we have used only the inequality (1.3.44) valid for any $\nu \ge -1/2$ and, with $N/2$ replaced by $N/2-1$, the order of all Bessel functions on the right-hand side of (1.3.23) is not smaller than $-1/2$, the estimate (1.3.23) in the case $1 \le \sqrt{\lambda} < \sqrt{\lambda_n}$ will be valid for the integral (1.3.32); for this integral, with $N/2$ replaced by $N/2-1$, that is, at $1 \le \sqrt{\lambda} < \sqrt{\lambda_n}$, the estimate

$$\left|\int_R^\infty J_{N/2-1}(r\sqrt{\lambda})J_{N/2-2}(r\sqrt{\lambda_n})dr\right| = O(\lambda^{-1/4}\lambda_n^{-1/4}|\sqrt{\lambda_n}-\sqrt{\lambda}|^{-1}R^{-1}) \tag{1.3.45}$$

holds.

Now we must prove estimate (1.3.23) for the second case, that is, when $1 \le \sqrt{\lambda_n} < \sqrt{\lambda}$.

Making use of the recursion formula

$$J_{N/2}(r\sqrt{\lambda}) = -J_{N/2-2}(r\sqrt{\lambda}) + (N-2)(r\sqrt{\lambda})^{-1}J_{N/2-1}(r\sqrt{\lambda})$$

we rewrite integral (1.3.32) in the following form:

$$\overset{\lambda}{I}_n(R) = -\int_R^\infty J_{N/2-1}(r\sqrt{\lambda_n})J_{N/2-2}(r\sqrt{\lambda})dr$$
$$+(N+2)(\sqrt{\lambda})^{-1}\int_R^\infty J_{N/2-1}(r\sqrt{\lambda})J_{N/2-1}(r\sqrt{\lambda_n})r^{-1}dr. \tag{1.3.46}$$

Inasmuch as $N/2-1 \geq -1/2$, we find, having made use of an estimate of the type (1.3.44) for $J_{N/2-1}(x)$, that the latter integral on the right-hand side of (1.3.46) has an order of $O(\lambda^{-3/4}\lambda_n^{-1/4}R^{-1})$ and, the more so, an order of $O(\lambda^{-1/4}\lambda_n^{-1/4}|\sqrt{\lambda_n}-\sqrt{\lambda}|^{-1}R^{-1})$.

It remains to prove that the former term on the right-hand side of (1.3.46) has the same order. We observe that the modulus of the former term in question coincides with the modulus of the integral (1.3.45) in which the numbers λ and λ_n have exchanged their roles. Thus, for the modulus of the former integral on the right-hand side of (1.3.46), the required estimate

$$\left|\int_R^\infty J_{N/2-1}(r\sqrt{\lambda_n})J_{N/2-2}(r\sqrt{\lambda})dr\right| = O(\lambda^{-1/4}\lambda_n^{-1/4}|\sqrt{\lambda_n}-\sqrt{\lambda}|^{-1}R^{-1})$$

is valid by virtue of the relation (1.3.45) at $1 \leq \sqrt{\lambda_n} < \sqrt{\lambda}$.

To complete the proof of the lemma, we have to establish estimate (1.3.24) for all $N \geq 2$, $\lambda_n \leq 2$, $\lambda \geq 1$, $0 \leq R \leq 1$.

To begin, let $N \geq 3$ be odd. If so, the relation (1.3.33) holds. Each summand on the right-hand side of (1.3.33) under the summation sign has the required order of $O(\lambda_n^{(N-2)/4}\lambda^{-1/2})$ with an estimate of the O-terms independent of R at all $0 \leq R \leq 1$, because for any $1 \leq \ell \leq (N-1)/2$ one has

$$\left|J_{N/2-\ell}(R\sqrt{\lambda})\right| \leq C, \qquad \left|J_{N/2-\ell}(R\sqrt{\lambda_n})\right| \leq C(R\sqrt{\lambda_n})^{N/2-\ell}.$$

The latter term on the right-hand side of (1.3.33) also has the required order of magnitude by virtue of the equality (1.3.35) and of the boundness of the quantity within the braces in (1.3.35) and because $(N-1)/2 \geq 1$ for all $N \geq 3$.

Finally, let $N \geq 2$ be even. If so, the relation (1.3.33′) is valid, and, following the same reasoning as for odd N, we easily prove that each summand on the right-hand side of (1.3.33′) under the summation sign has the required order of magnitude.

The latter term on the right-hand side of (1.3.33′) also has the required order by virtue of the relation (1.3.37) and the inequality $(N-2)/2 \geq 0$ and by virtue of the boundedness of the functions $|J_0(x)|$ and $|J_1(x)|$ for all nonnegative values of the argument x

$$\int_0^R J_1(r\sqrt{\lambda})J_0(r\sqrt{\lambda_n})dr = -(\sqrt{\lambda})^{-1}\int_0^R J_0(r\sqrt{\lambda_n})\frac{d}{dr}[J_0(r\sqrt{\lambda})]dr$$

$$= (\sqrt{\lambda})^{-1}[J_0(r\sqrt{\lambda})J_0(r\sqrt{\lambda_n})]\Big|_{r=0}^{r=R}$$

$$-\sqrt{\lambda_n}(\sqrt{\lambda})^{-1}\int_0^R J_0(r\sqrt{\lambda})J_1(r\sqrt{\lambda_n})dr = O[(\sqrt{\lambda})^{-1}].$$

This concludes completely the proof of Lemma 1.4.

1.3.3. Corollaries to the Theorem on Estimation of a Spectral Function in the Metric L_2

COROLLARY 1. *(The upper-bound estimate of a spectral function in the metric L_q for $q < 2N/(N+1)$.) For $N \geq 2$, for any q in the interval $1 \leq q < 2N/(N+1)$, any sufficiently large $\lambda > 0$, and any sufficiently small $R > 0$, the following upper-bound estimate of the spectral function $\theta(x, y, \lambda)$, uniform with respect to x on any compact set K of domain Ω, holds:*

$$||\theta(x, y, \lambda)||_{L_q(G)} \leq C(R)\lambda^{(N-1)/4}, \tag{1.3.47}$$

in which the norm in $L_q(G)$ is specified in the coordinates of the point y.

Indeed, since $q < 2$, the estimate below for the remainder term,

$$||\hat{\theta}(x, y, \lambda)||_{L_q(G)} \leq C(R)\lambda^{(N-1)/4},$$

stems from estimate (1.3.5), and all we have to do is ensure that for the main term in (1.3.3) the estimate

$$||\overset{\lambda}{v}(|x - y|)||_{L_q(G)} \leq C(R)\lambda^{(N-1)/4} \tag{1.3.48}$$

holds, uniform with respect to x on any compact set K of domain Ω. We fix an arbitrary compact set K of domain Ω, an arbitrary number $R > 0$ smaller than the distance between the compact set K and the boundary of Ω, and an arbitrary point x on the compact set K; we note that by virtue of the form (1.3.3) of the function

$\overset{\lambda}{v}(|x-y|)$ [30],

$$\|\overset{\lambda}{v}(|x-y|)\|_{L_q(G)} = \left[\int_G \left|\overset{\lambda}{v}(|x-y|)\right|^q dy\right]^{1/q}$$

$$= \left[\omega_N \int_0^R \left|\overset{\lambda}{v}(r)\right|^q r^{N-1} dr\right]^{1/q}$$

$$= \omega_N^{1/q}(2\pi)^{-N/2}\lambda^{N/4}\left[\int_0^R r^{-Nq/2}\left|J_{N/2}(r\sqrt{\lambda})\right|^q r^{N-1}dr\right]^{1/q}. \tag{1.3.49}$$

Making use of the estimate $|J_{N/2}(r\sqrt{\lambda})| \le Cr^{-1/2}\lambda^{-1/4}$ and taking into account that $N-1-Nq/2-q/2 > -1$ for $q < 2N/(N+1)$, we obtain estimate (1.3.48) from (1.3.49). Thereby the derivation of estimate (1.3.47) is complete.

REMARK. It follows from estimate (1.3.47) that for any $\lambda > 0$, for any $N \ge 2$, $1 \le q < 2N/(N+1)$, and for any sufficiently small $R > 0$, the following estimate holds for the spectral function $\theta(x, y, \lambda)$:

$$\|\theta(x,y,\lambda)\|_{L_q(G)} \le C(R)\lambda^{(N-1)/4}, \tag{1.3.47}$$

uniform likewise with respect to x on an arbitrary compact set K of subdomain Ω.

COROLLARY 2. *(The lower-bound estimate of a spectral function in the metric L_1.) By the symbol Ω_R^x we denote an N-dimensional ball of radius R centered at the point x.*

LEMMA 1.5. *Let K be an arbitrary compact set of the subdomain Ω, and x_0 an arbitrary point of the compact set K. Then one can fix a positive number R_0 so small as to satisfy simultaneously two requirements:*
(1) the N-dimensional ball $\Omega_{R_0}^{x_0}$ is contained inside the subdomain Ω;
(2) for any R_1 in the half-interval $0 < R_1 \le R_0$ and for any set E contained in the N-dimensional ball $\Omega_{R_1}^{x_0}$ and possessing a measure mes E,

$$\operatorname{mes} E > R_1^{1/4} \operatorname{mes} \Omega_{R_1}^{x_0}, \tag{1.3.50}$$

there is a constant $\beta > 0$ such that for all sufficiently large $\lambda > 0$, the following lower-bound estimate of the spectral function in the metric $L_1(E)$ is valid:

$$\int_E |\theta(x,y,\lambda)|dy \ge \beta\lambda^{(N-1)/4}. \tag{1.3.51}$$

[30] We recall that $\omega_N = 2\pi^{N/2}[\Gamma(N/2)]^{-1}$.

PROOF: We fix an arbitrary compact set K of the subdomain Ω, an arbitrary point x_0 of the compact set K, and a positive number R smaller than the distance between the compact set K and the boundary of Ω. Then, having taken the main term of the spectral function (1.3.3) with the indicated number R, we obtain estimate (1.3.5) for the remainder term of the spectral function. It is to be inferred from this estimate and from the Cauchy–Buniakowski inequality that if $0 < R_1 < R$, then for any set E interior to the N-dimensional ball $\Omega_{R_1}^{x_0}$ the following inequality holds:

$$\int_E \left|\widehat{\theta}(x, y, \lambda)\right| dy \le CR^{-1}\lambda^{(N-1)/4} \left\{\operatorname{mes}\Omega_{R_1}^{x_0}\right\}^{1/2}, \tag{1.3.52}$$

where C is the constant specified by estimate (1.3.5).

Further, with reference to the explicit form (1.3.3) of the main term of the spectral function and to the fact that at $r > 1/\sqrt{\lambda}$ the representation

$$J_{N/2}(r\sqrt{\lambda}) = \sqrt{\frac{2}{\pi r\sqrt{\lambda}}} \cos\left(r\sqrt{\lambda} - \frac{\pi(N+1)}{4}\right) + O(r^{-3/2}\lambda^{-3/4}),$$

is valid for the Bessel function[31], and the inequality

$$\left|\cos\left(r\sqrt{\lambda} - \frac{\pi(N+1)}{4}\right)\right| \ge \frac{1}{2} + \frac{1}{2}\cos\left(2r\sqrt{\lambda} - \frac{\pi(N+1)}{2}\right)$$

is valid for $\left|\cos\left(r\sqrt{\lambda} - \frac{\pi(N+1)}{4}\right)\right|$, one infers that there exists a constant $\alpha > 0$ such that for all sufficiently large positive λ and for the set E interior to the N-dimensional ball $\Omega_{R_1}^{x_0}$, the inequality

$$\begin{aligned} &\int_E \left|\overset{\lambda}{v}|(x_0 - y|)\right| dy \\ &\qquad = \lambda^{N/4}(2\pi)^{-N/2} \int_E |x_0 - y|^{-N/2} \left|J_{N/2}(|x_0 - y|\sqrt{\lambda})\right| dy \\ &\qquad \ge \alpha\lambda^{(N-1)/4} R_1^{-(N+1)/2} \operatorname{mes} E \end{aligned} \tag{1.3.53}$$

holds.

Now we fix a positive number $R_0 < R$ so small as to satisfy the inequality

$$\frac{\alpha}{2} R_0^{-1/4} \ge CR^{-1}\Gamma\left(\frac{N}{2} + 1\right)\pi^{-N/2}, \tag{1.3.54}$$

---

[31] See, for example, Bateman and Erdélyi [9, p. 98, formula (3)].

where C is the constant specified by estimates (1.3.5) and (1.3.52), and α is the constant in (1.3.53).

In view of the property $|\theta(x_0, y, \lambda)| \geq |\overset{\lambda}{v}(|x_0 - y|)| - |\widehat{\theta}(x_0, y, \lambda)|$, one obtains from inequalities (1.3.53) and (1.3.52) that

$$\int_E |\theta(x_0, y, \lambda)| dy \geq \lambda^{(N-1)/4} \left\{ \alpha R_1^{-(N+1)/2} \operatorname{mes} E - C R^{-1} \left(\operatorname{mes} \Omega_{R_1}^{x_0}\right)^{1/2} \right\}.$$

A comparison of the latter inequality with estimates (1.3.50) and (1.3.54), taking into account that $\operatorname{mes} \Omega_{R_1}^{x_0} = \pi^{N/2} R_1^N \left[\Gamma\left(\frac{N}{2} + 1\right)\right]^{-1}$, yields the estimate

$$\int_E |\theta(x_0, y, \lambda)| dy \geq \lambda^{(N-1)/4} \frac{\alpha}{2} R_0^{-1/4} \left(\operatorname{mes} \Omega_{R_1}^{x_0}\right)^{1/2},$$

which is valid for any R_1 in the half-interval $0 < R_1 \leq R_0$. The proof of Lemma 1.6 is thus complete.

COROLLARY 3. *(The upper-bound estimate of the spectral decomposition of the fractional kernel $T_{(N-1)/4}(x, y)$ in the metric L_q for $q < 2N/(N+1)$.)*

LEMMA 1.6. *For any $N \geq 2$ and $q < 2N/(N+1)$ uniformly with respect to x on any compact set of the subdomain Ω, the estimate*

$$\|E_\lambda T_{(N-1)/4}(x, y)\|_{L_q(G)} \leq C = \text{const} \tag{1.3.55}$$

is valid, in which the actions of the operator E and the norm $L_q(G)$ are confined to the coordinates of the point y.

PROOF: To begin, we ensure that the spectral decomposition $E_\lambda T_\alpha(x, y)$ of the fractional kernel $T_\alpha(x, y)$ at any $\alpha > 0$ is defined by the equality

$$E_\lambda T_\alpha(x, y) = (1 + \lambda)^{-\alpha} \theta(x, y, \lambda) + \alpha \int_0^\lambda (1 + t)^{-1-\alpha} \theta(x, y, t) dt. \tag{1.3.56}$$

We denote the quantity on the right-hand side of (1.3.56) by the symbol $g(x, y, \lambda)$. It suffices to show that

$$\int_G g(x, y, \lambda) u_n(y) dy = \begin{cases} u_n(x)(1 + \lambda_n)^{-\alpha} & \text{for} \quad \lambda_n < \lambda, \\ 0 & \text{for} \quad \lambda_n \geq \lambda; \end{cases}$$

this equality immediately follows from the relation

$$\int_G \theta(x, y, t) u_n(y) dy = \begin{cases} u_n(x) & \text{for} \quad \lambda_n < t, \\ 0 & \text{for} \quad \lambda_n \geq t, \end{cases}$$

equivalent to the definition of a spectral function.

In order to establish estimate (1.3.55), it is enough to prove that this estimate holds for either summand on the right-hand side of (1.3.56) taken at $\alpha = (N-1)/4$.

The estimate

$$\|(1+\lambda)^{-(N-1)/4}\theta(x,y,\lambda)\|_{L_q(G)} = (1+\lambda)^{-(N-1)/4}\|\theta(x,y,\lambda)\|_{L_q(G)} \le C,$$

uniform with respect to x on any compact set K of subdomain Ω, is a straightforward consequence of the above estimate (1.3.47).

It remains to prove that the estimate

$$\int_G \left| \frac{N-1}{4} \int_0^\lambda (1+t)^{-1-(N-1)/4}\theta(x,y,t)dt \right|^q dy \le C \tag{1.3.57}$$

holds uniformly with respect to x on any compact set K of the subdomain Ω.

By virtue of the relationship $\theta(x,y,t) = \overset{t}{v}(|x-y|)+\widehat{\theta}(x,y,t)$ (see equality (1.3.4)), it suffices to establish the validity of the two estimates

$$\int_G \left| \frac{N-1}{4} \int_0^\lambda (1+t)^{-1-(N-1)/4}\overset{t}{v}(|x-y|)dt \right|^q dy \le C, \tag{1.3.58}$$

$$\int_G \left| \frac{N-1}{4} \int_0^\lambda (1+t)^{-1-(N-1)/4}\widehat{\theta}(x,y,t)dt \right|^q dy \le C \tag{1.3.59}$$

uniform with respect to x on any compact set of subdomain Ω.

First, we establish the validity of estimate (1.3.58). Note that, by virtue of the definition of the function $\overset{t}{v}(|x-y|)$, the integration over domain G reduces to the integration over an N-dimensional ball $|x-y| \le R$. Therefore, the integral in (1.3.58) can be represented as the sum of two integrals

$$J_1 = \int_{|x-y|\le(\sqrt{\lambda})^{-1}} \left| \frac{N-1}{4} \int_0^\lambda (1+t)^{-1-(N-1)/4}\overset{t}{v}(|x-y|)dt \right|^q dy, \tag{1.3.60}$$

$$J_2 = \int_{(\sqrt{\lambda})^{-1}\le|x-y|\le R} \left| \frac{N-1}{4} \int_0^\lambda (1+t)^{-1-(N-1)/4}\overset{t}{v}(|x-y|)dt \right|^q dy, \tag{1.3.61}$$

To verify estimate (1.3.58), it suffices to prove the boundedness of either of the integrals (1.3.60) or (1.3.61).

By virtue of the definition of the function $\overset{t}{v}(x-y)$, the integral (1.3.60) can be rewritten as

$$J_1 = (2\pi)^{-Nq/2} \int\limits_{|x-y|\le(\sqrt{\lambda})^{-1}} \left| \frac{N-1}{4} \int\limits_0^{\lambda} (1+t)^{-1-(N-1)/4} \right. \tag{1.3.60'}$$
$$\left. \times\, t^{N/2}(|x-y|\sqrt{t})^{-N/2} J_{N/2}(|x-y|\sqrt{t})dt \right|^q dy.$$

Since at $|x-y| \le (\sqrt{\lambda})^{-1}$ everywhere under the integration sign in t the inequality $|x-y|\sqrt{t} \le 1$ holds, we have

$$\left(|x-y|\sqrt{t}\right)^{-N/2} \left|J_{N/2}(|x-y|\sqrt{t})\right| \le C(N).$$

From this inequality and from the relationship

$$(1+t)^{-1-(N-1)/4} t^{N/2} \le t^{-1+(N+1)/4}$$

valid for all $t \ge 0$, the inequality below follows:

$$\left| \frac{N-1}{4} \int\limits_0^{\lambda} (1+t)^{-1-(N-1)/4} t^{N/2} (|x-y|\sqrt{t})^{-N/2} J_{N/2}(|x-y|\sqrt{t})dt \right|^q$$
$$\le C\lambda^{(N+1)q/4} \le C\lambda^{N/2}.$$

This inequality and expression (1.3.60′) provide immediate evidence for the boundedness of the integral (1.3.60).

Now we proceed to prove the boundedness of the integral (1.3.61).

To begin, we observe that at $0 < |x-y| \le R$ for any α (even if it is complex) satisfying the condition $\operatorname{Re}\alpha > (N-3)/4$, the following relationship is valid[32]:

$$\alpha \int\limits_0^{\infty} (1+t)^{-1-\alpha} \overset{t}{v}(|x-y|)dt$$
$$= (2\pi|x-y|^{-N/2}\alpha \int\limits_0^{\infty} (1+t)^{-1-\alpha} t^{N/4} J_{N/2}(|x-y|\sqrt{t})dt = \tag{1.3.62}$$
$$= 2^{1-\alpha}(2\pi)^{-N/2}[\Gamma(\alpha)]^{-1}|x-y|^{\alpha-N/2} K_{N/2-\alpha}(|x-y|) = v_\alpha(|x-y|).$$

[32] See Bateman and Erdélyi [9, p. 110, formula (59)].

Here $v_\alpha(r)$ refers to the function defined by equality (1.2.4) (*vide supra* Section 1.2.1).

This relationship taken at $\alpha = (N-1)/4$ and the boundedness of the integral

$$\int\limits_{|x-y|\le R} |v_{(N-1)/4}\ (|x-y|)|^q dy$$

$$= \{2^{-1-(N-1)/4}(2\pi)^{-N/2}[\Gamma(N-1)/4]^{-1}\}^q$$

$$\times \int\limits_{|x-y|\le R} |x-y|^{-(N+1)q/4}\left|K_{(N+1)/4}(|x-y|)\right|^q dy$$

imply that for any $q < 2N/(N+1)$ and for all x in the previously fixed arbitrary compact set K of subdomain Ω, we have

$$J_2' = \int\limits_{(\sqrt{\lambda})^{-1}\le|x-y|\le R} \left|\frac{N-1}{4}\int\limits_0^\infty (1+t)^{-1-(N-1)/4}\overset{t}{v}(|x-y|)dt\right|^q dy \le C.$$

Therefore, in order to prove the boundedness of the integral (1.3.61) it will suffice to prove the boundedness, at $q < 2N/(N+1)$, of the integral

$$J_2'' = \int\limits_{(\sqrt{\lambda})^{-1}\le|x-y|\le R} \left|\frac{N-1}{4}\int\limits_\lambda^\infty (1+t)^{-1-(N-1)/4}\overset{t}{v}(|x-y|)dt\right|^q dy$$

$$= (2\pi)^{-Nq/2} \int\limits_{(\sqrt{\lambda})^{-1}\le|x-y|\le R} |x-y|^{-(N+1)q/2}\left|\frac{N-1}{4}\sqrt{|x-y|}\right. \tag{1.3.63}$$

$$\left.\times \int\limits_\lambda^\infty (1+t)^{-1-(N-1)/4}t^{N/4}J_{N/2}(|x-y|\sqrt{t})dt\right|^q dy.$$

Since the inequality $(N+1)q/2 < N$ holds at $q < 2N/(N+1)$, the boundedness of the integral (1.3.63) will be proved if one proves that the inner integral

$$*K* = \sqrt{|x-y|}\int\limits_\lambda^\infty (1+t)^{-1-(N-1)/4}t^{N/4}J_{N/2}(|x-y|\sqrt{t})dt \tag{1.3.64}$$

is bounded for all $|x-y| \ge (\sqrt{\lambda})^{-1}$.

We represent integral (1.3.64) in the form

$$*K* = \sqrt{|x-y|}\int_{\lambda}^{\infty}\varphi(t)\psi(t)dt,$$

where

$$\varphi(t) = (1+t)^{-1-(N-1)/4}t^{(N+2)/4}, \qquad \psi(t) = t^{-1/2}J_{N/2}(|x-y|\sqrt{t}),$$

and observe that the function $\varphi(t)$ at $t \geq N+2$ is a decreasing one and satisfies the condition $\lim\limits_{t\to\infty}\varphi(t) = 0$.

With no loss of generality, we may assume that $\lambda \geq N+2$. In this case, the function $\varphi(t)$ is decreasing on the half-line $\lambda \leq t < \infty$. Given this condition, the Bonnet mean-value theorem[33)] for the integral $*K*$ holds, according to which there is, on the half-line $[\lambda, \infty)$, a point ξ such that

$$*K* = \sqrt{|x-y|}\int_{\lambda}^{\infty}\varphi(t)\psi(t)dt = \sqrt{|x-y|}\varphi(\lambda)\int_{\lambda}^{\xi}\psi(t)dt.$$

Thus, the integral (1.3.64) is represented as

$$*K* = \sqrt{|x-y|}\lambda^{(N+2)/4}(1+\lambda)^{-(N+3)/4}\int_{\lambda}^{\xi}t^{-1/2}J_{N/2}(|x-y|\sqrt{t})dt.$$

Having performed the integration by parts on this expression using the relation[34)]

$$\int(\sqrt{t})^{N/2+1}J_{N/2}(|x-y|\sqrt{t})\frac{dt}{2\sqrt{t}} = |x-y|^{-1}(\sqrt{t})^{N/2+1}J_{N/2+1}(|x-y|\sqrt{t}),$$

we find that the integral (1.3.64) in question is equal to

$$\begin{aligned}*K* &= |x-y|^{-1/2}\lambda^{(N+2)/4}(1+\lambda)^{-(N+3)/4}\\ &\quad\times\left\{\left[2J_{N/2+1}(|x-y|\sqrt{t})\right]_{t=\lambda}^{t=\xi} + \frac{N+2}{2}\int_{\lambda}^{\nu}J_{N/2+1}(|x-y|\sqrt{t})\frac{dt}{t}\right\}.\end{aligned}$$

Making use of the estimate $|J_{N/2+1}(y)| \leq Cy^{-1/2}$, on the right-hand side of the above equality, we arrive at the inequality

$$|*K*| \leq C[|x-y|\sqrt{t}]^{-1},$$

33) See, for example, Il'in and Poznyak [46, pp. 353–354].

34) See Bateman and Erdélyi [9, p. 20].

which provides evidence, by virtue of $|x-y| \geq (\sqrt{\lambda})^{-1}$, that the integral (1.3.64) is bounded.

Now to conclude the proof of Lemma 1.6, it only remains to establish estimate (1.3.59). In lieu of this estimate, we prefer to establish the stronger estimate

$$\int_G \left| \frac{N-1}{4} \int_0^\lambda (1+t)^{-1-(N-1)/4} \widehat{\theta}(x,y,t)dt \right|^2 dy \leq C, \tag{1.3.65}$$

uniform with respect to x on an arbitrary compact set K of subdomain Ω.

We assume, without loss of generality, that $\lambda > 1$. To prove estimate (1.3.65), it is sufficient (by virtue of the representation of the inner integral on the left-hand side of (1.3.65) in the form of the sum $\int_0^\lambda = \int_0^1 + \int_1^\lambda$ and by virtue of the trivial inequality $(A+B)^2 \leq 2A^2 + 2B^2$) to establish the validity of the two estimates

$$\int_G \left| \int_0^1 (1+t)^{-1-(N-1)/4} \widehat{\theta}(x,y,t)dt \right|^2 dy \leq C_1, \tag{1.3.66}$$

$$\int_G \left| \int_1^\lambda (1+t)^{-1-(N-1)/4} \widehat{\theta}(x,y,t)dt \right|^2 dy \leq C_2. \tag{1.3.67}$$

To prove estimate (1.3.66), we make use of the fact that, by virtue of relations (1.3.4) and (1.3.1),

$$\widehat{\theta}(x,y,t) = \sum_{\lambda_n < t} u_n(x)u_n(y) - \overset{t}{v}(|x-y|),$$

where $\overset{t}{v}(r)$ is a function defined by relation (1.3.3).

The latter relation, the trivial inequality $(A+B)^2 \leq 2A^2 + 2B^2$, and the Cauchy–Buniakowski inequality enable one to majorize the left-hand side of (1.3.66) in the following manner:

$$\int_G \left| \int_0^1 (1+t)^{-1-(N-1)/4} \widehat{\theta}(x,y,t)dt \right|^2 dy$$

$$\leq 2 \int_G \left| \int_0^1 (1+t)^{-1-(N-1)/4} \sum_{\lambda_n < t} u_n(x)u_n(y)dt \right|^2 dy$$

$$+ 2 \int_G \left| \int_0^1 (1+t)^{-1-(N-1)/4} \overset{t}{v}(|x-y|)dt \right|^2 dy \qquad (1.3.68)$$

$$\leq 2 \int_0^1 (1+t)^{-2-(N-1)/2} dt \int_0^1 \left\{ \int_G \left[\sum_{\lambda_n < t} u_n(x)u_n(y) \right]^2 dy \right\} dt$$

$$+ 2 \int_0^1 (1+t)^{-2-(N-1)/2} dt \int_0^1 \left\{ \int_G \left[\overset{t}{v}(|x-y|) \right]^2 dy \right\} dt.$$

The inequality (1.3.68), the estimate $\int_0^1 (1+t)^{-2-(N-1)/2} dt \leq 1$, and the two estimates[35)]

$$\int_G \left[\sum_{\lambda_n < t} u_n(x)u_n(y) \right]^2 dy = \sum_{\lambda_n < t} u_n^2(x) = O(1),$$

$$\int_G \left[\overset{t}{v}(|x-y|) \right]^2 dy = O(1),$$

holding for any t in the segment $0 \leq t \leq 1$ for all x in the compact set K, provide evidence for the validity of estimate (1.3.66) uniformly with respect to x on the compact set K.

It remains to prove the validity of estimate (1.3.67) uniformly with respect to x on the arbitrary compact set K that we have previously fixed. To prove this estimate, it suffices, by virtue of representation (1.3.12) and the trivial inequality $|A+B|^2 \leq 2A^2 + 2B^2$, to establish the validity of the two estimates

35) The former of these estimates holds in view of the relation (1.1.4) (see Section 1.1.2) taken at $\mu = 0$; the latter estimate is implied by the form (1.3.3) of the function $\overset{t}{v}(|x-y|)$.

$$\int_G \left| \int_1^\lambda (1+t)^{-1-(N-1)/4} \sum_{\lambda_n = t} u_n(x)u_n(y)dt \right|^2 dy \leq C_3, \tag{1.3.69}$$

$$\int_G \left| \int_1^\lambda (1+t)^{-1-(N-1)/4} t^{N/4} \left[\sum_{n=1}^{\infty} \lambda_n^{(2-N)/4} \overset{t}{I}_n(R) u_n(x) u_n(y) \right] dt \right|^2 dy \leq C_4, \tag{1.3.70}$$

uniform with respect to x on the compact set K.

Denoting the left-hand side of (1.3.69) by M_1, we rewrite it in the form

$$M_1 = \int_G \left[\int_1^\lambda (1+t)^{-1-(N-1)/4} \sum_{\lambda_n = t} u_n(x)u_n(y)dt \right]$$
$$\times \left[\int_1^\lambda (1+\tau)^{-1-(N-1)/4} \sum_{\lambda_k = \tau} u_k(x)u_k(y)d\tau \right] dy.$$

Reversing the order of integration with respect to y, t, and τ, we represent the quantity M_1 in the following manner:

$$M_1 = \int_1^\lambda \int_1^\lambda (1+t)^{-1-(N-1)/4}(1+\tau)^{-1-(N-1)/4}$$
$$\times \left[\sum_{\lambda_n = t} \sum_{\lambda_k = \tau} u_n(x)u_k(x) \int_G u_n(y)u_k(y)dy \right] dt\, d\tau.$$

Now, one can readily see that the value of M_1 is in fact zero. Indeed, the expression within the brackets is, in view of the orthonormality of the system $\{u_n(y)\}$, equal to zero at any fixed t in the segment $1 \leq t \leq \lambda$ and at any τ in the segment $1 \leq \tau \leq \lambda$, distinct from t.

Thus, $M_1 = 0$, and the estimate (1.3.69) is justified.

To substantiate the estimate (1.3.70) we denote the left-hand side of (1.3.70) by M_2. Proceeding in much the same manner as previously for the quantity M_1 and invoking the orthonormality of the system $\{u_n(y)\}$, we obtain

$$M_2 = \int_G \left\{ \int_1^\lambda (1+t)^{-1-(N-1)/4} t^{N/4} \sum_{n=1}^{\infty} \lambda_n^{(2-N)/4} \overset{t}{I}_n(R) u_n(x) u_n(y)\, dt \right\}$$

$$\times \left\{ \int_1^\lambda (1+\tau)^{-1-(N-1)/4} \tau^{N/4} \sum_{k=1}^{\infty} \lambda_k^{(2-N)/4} \overset{\tau}{I}_k(R) u_k(x) u_k(y)\, d\tau \right\} dy$$

$$= \int_1^\lambda \int_1^\lambda (1+t)^{-1-(N-1)/4} t^{N/4} (1+\tau)^{-1-(N-1)/4} \tau^{N/4}$$

$$\times \left[\sum_{n=1}^{\infty} \sum_{k=1}^{\infty} \lambda_n^{(2-N)/4} \lambda_k^{(2-N)/4} \overset{t}{I}_n(R) \overset{\tau}{I}_k(R) u_n(x) u_k(x) \int_G u_n(y) u_k(y)\, dy \right] dt\, d\tau$$

$$= \int_1^\lambda \int_1^\lambda (1+t)^{-1-(N-1)/4} (1+\tau)^{-1-(N-1)/4} t^{N/4} \tau^{N/4}$$

$$\times \left[\sum_{n=1}^{\infty} \lambda_n^{(2-N)/2} u_n^2(x) \overset{t}{I}_n(R) \overset{\tau}{I}_n(R) \right] dt\, d\tau$$

$$= \sum_{n=1}^{\infty} \lambda_n^{(2-N)/2} u_n^2(x) \left[\int_1^\lambda (1+t)^{-1-(N-1)/4} t^{N/4} \overset{t}{I}_n(R)\, dt \right]^2 .$$

Now, having partitioned the sum $\sum_{n=1}^{\infty}$ into the two sums $\sum_{\lambda_n \le 2}$ and $\sum_{\lambda_n > 2}$, we reduce the proof of estimate (1.3.70) to the validation of the two estimates

$$M_2' = \sum_{\lambda_n \le 2} \lambda_n^{(2-N)/2} u_n^2(x) \left[\int_1^\lambda (1+t)^{-1-(N-1)/4} t^{N/4} \overset{t}{I}_n(R)\, dt \right]^2 \le C_5, \qquad (1.3.71)$$

$$M_2'' = \sum_{\lambda_n > 2} \lambda_n^{(2-N)/2} u_n^2(x) \left[\int_1^\lambda (1+t)^{-1-(N-1)/4} t^{N/4} \overset{t}{I}_n(R)\, dt \right]^2 \le C_6, \qquad (1.3.72)$$

uniform with respect to x on the arbitrary compact set K that we have previously fixed.

The estimate (1.3.71) is a trivial sequel to the inequality (1.3.24), to the estimate (1.1.17) taken at $\mu_0 = \sqrt{\lambda}$, and to the boundedness of the integral

$$\int_1^{\lambda} (1+t)^{-1-(N-1)/4} t^{(N-2)/4} dt \le \int_1^{\lambda} (1+t)^{-5/4} dt \le \text{const.}$$

To prove the inequality (1.3.72), it suffices to establish, for $\lambda_n \ge 2$, the estimate

$$\int_1^{\lambda} (1+t)^{-1-(N-1)/4} t^{N/4} \left| \overset{t}{I}_n(R) \right| dt \le C_7 \lambda_n^{-5/8}. \qquad (1.3.73)$$

Indeed, having established the estimate (1.3.73), we obtain from this estimate and from Corollary 2 to Theorem 1.1 (to be more exact, from the convergence of series (1.1.24), uniform on any compact set) that

$$M_2'' \le C_7^2 \sum_{\lambda_n \ge 2} u_n^2(x) \lambda_n^{-N/2-1/4} \le C_8.$$

The more so, it is sufficient to prove, for $\lambda_n \ge 2$, the estimate

$$\int_1^{\infty} t^{-1+1/4} \left| \overset{t}{I}_n(R) \right| dt \le C_7 \lambda_n^{-5/8}, \qquad (1.3.74)$$

which is a consequence of the three estimates

$$\mathcal{K}_1 = \int\limits_{1 \le \sqrt{t} \le \sqrt{\lambda_n} - \lambda_n^{1/8}} t^{-3/4} \left| \overset{t}{I}_n(R) \right| dt = O(\lambda_n^{-5/8}), \qquad (1.3.75^1)$$

$$\mathcal{K}_2 = \int\limits_{|\sqrt{t} - \sqrt{\lambda_n}| \le \lambda_n^{1/8}} t^{-3/4} \left| \overset{t}{I}_n(R) \right| dt = O(\lambda_n^{-5/8}), \qquad (1.3.75^2)$$

$$\mathcal{K}_3 = \int\limits_{\sqrt{t} \ge \sqrt{\lambda_n} + \lambda_n^{1/8}} t^{-3/4} \left| \overset{t}{I}_n(R) \right| dt = O(\lambda_n^{-5/8}). \qquad (1.3.75^3)$$

The estimation of $\mathcal{K}_2$ with the aid of inequality (1.3.22) yields

$$\mathcal{K}_2 = O(\lambda_n^{-1/4}) \int\limits_{|\sqrt{t} - \sqrt{\lambda_n}| \le \lambda_n^{1/8}} t^{-1} dt = O(\lambda_n^{-1/4}) \ln \frac{\sqrt{\lambda_n} + \lambda_n^{1/8}}{\sqrt{\lambda_n} - \lambda_n^{1/8}}$$

$$= O(\lambda_n^{-1/4}) \ln \frac{1 + \lambda_n^{-3/8}}{1 - \lambda_n^{-3/8}} = O(\lambda_n^{-5/8}).$$

Estimating $\mathcal{K}_3$ with the aid of the inequality (1.3.23), the identity $(\sqrt{t})^{-1}(\sqrt{t}-\sqrt{\lambda_n})^{-1} = (\sqrt{\lambda_n})^{-1}[(\sqrt{t}-\sqrt{\lambda_n})^{-1} - (\sqrt{t})^{-1}]$, and the relation $dt = 2\sqrt{t}\,d(\sqrt{t})$, we obtain

$$\mathcal{K}_3 = O(\lambda_n^{-1/4}) \int\limits_{\sqrt{t}\geq\sqrt{\lambda_n}+\lambda_n^{1/8}} t^{-1}(\sqrt{t}-\sqrt{\lambda_n})^{-1}dt$$

$$= O(\lambda_n^{-1/4}) \int\limits_{\sqrt{t}\geq\sqrt{\lambda_n}+\lambda_n^{1/8}} (\sqrt{t})^{-1}(\sqrt{t}-\sqrt{\lambda_n})^{-1}d(\sqrt{t})$$

$$= O(\lambda_n^{-3/4}) \int\limits_{\sqrt{t}\geq\sqrt{\lambda_n}+\lambda_n^{1/8}} \left[(\sqrt{t}-\sqrt{\lambda_n})^{-1} - (\sqrt{t})^{-1}\right] d(\sqrt{t})$$

$$= O(\lambda_n^{-3/4})\ln(1+\lambda_n^{3/8}) = O(\lambda_n^{-5/8}).$$

Finally, estimating $\mathcal{K}_1$ with the aid of the inequality (1.3.23), the identity $(\sqrt{t})^{-1}(\sqrt{\lambda_n}-\sqrt{t})^{-1} = (\sqrt{\lambda_n})^{-1}[(\sqrt{\lambda_n}-\sqrt{t})^{-1}+(\sqrt{t})^{-1}]$, and the relation $dt = 2\sqrt{t}\,d(\sqrt{t})$, we obtain

$$\mathcal{K}_1 = O(\lambda_n^{-1/4}) \int\limits_{1\leq\sqrt{t}\leq\sqrt{\lambda_n}-\lambda_n^{1/8}} t^{-1}(\sqrt{\lambda_n}-\sqrt{t})^{-1}dt$$

$$= O(\lambda_n^{-1/4}) \int\limits_{1\leq\sqrt{t}\leq\sqrt{\lambda_n}-\lambda_n^{1/8}} (\sqrt{t})^{-1}(\sqrt{\lambda_n}-\sqrt{t})^{-1}d(\sqrt{t})$$

$$= O(\lambda_n^{-3/4}) \int\limits_{1\leq\sqrt{t}\leq\sqrt{\lambda_n}-\lambda_n^{1/8}} \left[(\sqrt{\lambda_n}-\sqrt{t})^{-1} + (\sqrt{t})^{-1}\right] d(\sqrt{t})$$

$$= O(\lambda_n^{-3/4})\left[\ln(\lambda_n^{3/8}-1) + \ln(\sqrt{\lambda_n}-1)\right] = O(\lambda_n^{-5/8}).$$

Thereby the derivation of estimates $(1.3.75^1)$–$(1.3.75^3)$ is complete, and Lemma 1.6 is thus fully proved.

COROLLARY 4. *The upper-bound estimate of the spectral decomposition of the kernel $T_{(N-1)/4}(x,y)$ in the L_2-metric in a domain free of a singular point.*

LEMMA 1.7. *If $N \geq 2$ and D is an arbitrary subdomain of Ω, then the estimate below holds uniformly with respect to x in each compact set of domain D,*

$$\|E_\lambda T_{(N-1)/4}(x,y)\|_{L_2(G\setminus D)} \leq C = \text{const}, \tag{1.3.76}$$

in which the actions of the operator E_λ and the norm $L_2(C\backslash D)$ are confined to the coordinates of the point y.

PROOF: The proof of Lemma 1.7 is carried out in much the same manner as that of Lemma 1.6. The only difference is that, in place of estimate (1.3.47), we refer to the estimate below, uniform with respect to x on an arbitrary compact set K of domain D,

$$\|\theta(x,y,\lambda)\|_{L_2(G\backslash D)} \leq C(R)\lambda^{(N-1)/4},$$

whose validity for all sufficiently large $\lambda > 0$ stems from the estimate (1.3.5) and the equality for the main term

$$\|\overset{\lambda}{v}(|x-y|)\|_{L_2(G\backslash D)} = 0,$$

which immediately follows from the expression (1.3.3) if the fixed value of R is less than the distance between the compact set K and the boundary of domain D. All other arguments used in proving Lemma 1.6 should be subject to the condition $q = 2$.

COROLLARY 5. *The nonboundedness in the L_1-metric of the spectral decomposition of the kernel $T_\alpha(x,y)$ of order $\alpha < (N-1)/4$.*

LEMMA 1.8. *Let $N \geq 2$, let K be an arbitrary compact set of domain Ω, let x_0 be an arbitrary point of the compact set K, let the positive number R_0 be fixed in the same manner as in Lemma 1.5, and let the set E have the same meaning as in Lemma 1.5. Then, given any $0 \leq \alpha < (N-1)/4$, the following relation holds:*

$$\overline{\lim_{\lambda\to\infty}} \int_E |E_\lambda T_\alpha(x_0,y)|dy = \infty. \tag{1.3.77}$$

PROOF: We note first of all that, by making use of the relation

$$\int_G E_\lambda T_\alpha(x,y)u_n(y)dy = \begin{cases} u_n(x)(1+\lambda_n)^{-\alpha} & \text{for } \lambda_n < \lambda, \\ 0 & \text{for } \lambda_n \geq \lambda, \end{cases}$$

the equality

$$\theta(x_0,y,\lambda) = (1+\lambda)^\alpha E_\lambda T_\alpha(x_0,y) - \alpha\int_0^\lambda (1+t)^{\alpha-1}E_t T_\alpha(x_0,y)dt \tag{1.3.78}$$

is trivially verified.

Assume that the relation (1.3.77) is not true, that is, the set $\left\{\int_E |E_\lambda T_\alpha(x_0,y)|dy\right\}$ for certain $0 \leq \alpha < (N-1)/4$ is a bounded one:

$$\int_E |E_\lambda T_\alpha(x_0, y)|dy \le C \qquad (\text{for all} \quad \lambda > 0).$$

Then the inequality

$$\int_E |\theta(x_0, y, \lambda)|dy \le 2C(1+\lambda)^\alpha$$

is obtained from (1.3.78); this inequality is at variance with inequality (1.3.51), thereby invalidating our assumption that (1.3.77) is not true.

The proof of Lemma 1.8 is complete.

REMARK TO LEMMA 1.8. It follows from Lemma 1.8 that, *given any* $p \ge 1$ *and* $0 \le \alpha < (N-1)/4$, *the set*

$$||E_\lambda T_\alpha(x_0, y)||_{L_p(G)}$$

is not bounded with respect to $\lambda > 0$.

1.4. EXACT CONDITIONS FOR THE LOCALIZATION AND UNIFORM CONVERGENCE OF EXPANSIONS WITH RESPECT TO AN ARBITRARY FSF IN THE SOBOLEV–LIOUVILLE CLASSES

Throughout this section, we are concerned with an arbitrary FSF of the Laplace operator in any N-dimensional subdomain Ω of an arbitrary N-dimensional domain G.

We presume that the reader is familiar with the well-known Sobolev classes $W_p^n(G)$ and with the Sobolev–Liouville functional classes $L_p^\alpha(G)$, the latter generalizing the former to the case of a nonintegral order of differentiability α. A concise description of these classes is found in Section 2.1.3 (*vide infra*); for greater detail the reader is referred to the monograph of S.M. Nikol'skii[36].

In Section 1.4.1 we will establish the uniform (on any compact set of subdomain Ω) convergence of the Fourier series in an arbitrary FSF of interest for an arbitrary function $f(x)$ which is compactly supported in the subdomain Ω and belongs in this domain to the Sobolev–Liouville class L_p^α; the order of differentiability α and the degree of summability p of this class are represented by two arbitrary numbers satisfying the three inequalities

$$\alpha \ge (N-1)/2, \qquad \alpha p > N, \qquad p \ge 1. \tag{1.4.1}$$

[36] See Nikol'skii [65, Chapter 9].

In Section 1.4.2 we will verify the localization property of the Fourier series in an arbitrary FSF of interest for an arbitrary function $f(x)$ which is compactly supported in the subdomain Ω and in the Sobolev–Liouville class L_2^α of this subdomain at any $\alpha \geq (N-1)/2$.

In Section 1.4.3 we will establish for an arbitrary FSF, given arbitrary fixed $\alpha < (N-1)/2$ and $p \geq 1$, the existence of a function $f(x)$ compactly supported in the subdomain Ω and in the Sobolev–Liouville class L_p^α of this subdomain for the case in which the localization principle for the Fourier series of this function fails to hold.

A comparison of the results of Sections 1.4.1–1.4.3 and an analysis thereof will enable us, in Section 1.4.4, to infer that, for each specified FSF of the Laplace operator, the conditions (1.4.1) are definitive conditions in the Sobolev–Liouville classes L_p^α for the uniform convergence of the Fourier series of the function $f(x)$ compactly supported in the subdomain Ω, and that the condition $\alpha \geq (N-1)/2$ is a definitve condition in the Sobolev–Liouville classes L_2^α for the localization of the Fourier series of the function $f(x)$ compactly supported in the subdomain Ω.

Finally, in Section 1.4.5 we will present results on the divergence of Fourier series in an arbitrary FSF of the Laplace operator, both in the L_p-metric and on a set of positive measure.

1.4.1. Conditions for the Uniform Convergence of Fourier Series in an Arbitrary FSF

THEOREM 1.9. *Let $N \geq 2$, let $\{u_n(x)\}$ be an arbitrary FSF of the Laplace operator in an N-dimensional subdomain Ω of an arbitrary N-dimensional domain G, and let $f(x)$ be an arbitrary function compactly supported in the subdomain Ω, continued with zero outside the subdomain Ω, and belonging to the Sobolev–Liouville class $L_p^\alpha(\Omega)$ with arbitrary fixed α and p which satisfy the three inequalities (1.4.1). Then the Fourier series of the function $f(x)$ in the system $\{u_n(x)\}$ converges to this function uniformly on an arbitrary compact set of the subdomain Ω.*

PROOF: To begin, we verify the convergence, on any compact set of subdomain Ω, of the Fourier series of the function $f(x)$ which is compactly supported in Ω and in the Sobolev–Liouville class L_2^{2m} of this subdomain, where m is an integer greater than $N/4$.

We apply successively the second Green's formula to the fundamental function $u_n(x)$ and to the functions $f(x)$, $\Delta f(x)$, ... , $\Delta^{m-1} f(x)$ to ensure that the Fourier coefficients of the functions $\Delta^m f(x)$ and $f(x)$ are related as

$$(\Delta^m f)_n = (-1)^m \lambda_n^m f_n. \tag{1.4.2}$$

We write the Bessel inequalities for the functions $f(x)$ and $\Delta^m f(x)$ in the form

$$\sum_{n=1}^{\infty} f_n^2 \leq ||f||^2_{L_2(\Omega)}, \qquad \sum_{n=1}^{\infty} f_n^2 \lambda_n^{2m} \leq ||\Delta^m f||^2_{L_2(\Omega)}$$

to obtain, with reference to (1.4.2), that the numerical series on the left-hand side of these inequalities are convergent. For this reason, the numerical series

$$\sum_{n=1}^{\infty} f_n^2(1+\lambda_n^{2m}) \tag{1.4.3}$$

is also convergent.

Since for any $\lambda_n > 0$ the inequality[37)]

$$(1+\lambda_n)^{2m} \leq 2^{2m}(1+\lambda_n^{2m})$$

holds, the convergence of series (1.4.3) implies that the numerical series

$$\sum_{n=1}^{\infty} f_n^2(1+\lambda_n)^{2m} \tag{1.4.4}$$

is also convergent.

Since $m > N/4$, the integral[38)]

$$\int_G T_m^2(x,y)dy \tag{1.4.5}$$

is, invoking the remark at the end of Section 1.2.2, a continuous function of the point x on an arbitrary compact set K of subdomain Ω; therefore, in view of the Parseval equality

$$\sum_{n=1}^{\infty} u_n^2(x)(1+\lambda_n)^{-2m} = \int_G T_m^2(x,y)dy, \tag{1.4.6}$$

in view of the continuity of the function (1.4.5) and of each fundamental function $u_n(x)$ on an arbitrary compact set K of subdomain Ω, and by virtue of the Dini theorem, the series on the left-hand side of (1.4.6) is uniformly convergent with respect to x on any compact set K of subdomain Ω.

We infer, from the convergence of series (1.4.4), from the uniform convergence of the series on the left-hand side of (1.4.6) on any compact set of the subdomain the Ω, and from the Cauchy–Buniakowski inequality that the Fourier series of the function in question

$$\sum_{n=1}^{\infty} f_n u_n(x) \tag{1.4.7}$$

converges uniformly on any compact set of subdomain Ω.

---

37) Since $(1+\lambda_n)^{2m} \leq 2^{2m} \leq 2^{2m}(1+\lambda_n^{2m})$ in the case where $\lambda_n \leq 1$, and $(1+\lambda_n)^{2m} \leq (2\lambda_n)^{2m} \leq 2^{2m}(1+\lambda_n)^{2m}$ in the case where $\lambda_n > 1$.

38) See integral (1.2.44) in Section 1.2.2.

Thus, for any function $f(x)$ compactly supported in the subdomain Ω and belonging to the Sobolev–Liouville class L_2^{2m} of this subdomain (where the integer m is subject to the condition $m > N/4$), the Fourier series (1.4.7) converges uniformly on any compact set of the subdomain Ω.

This is especially true of a function compactly supported in the class $C^{(2m)}(\Omega)$ as well as of a function of class $C_0^\infty(\Omega)$.

We emphasize that if the Fourier series (1.4.7) of the function $f(x)$ is uniformly convergent on any compact set of domain Ω, then this series converges precisely to the function $f(x)$, since it converges to this function in the $L_2(G)$-metric.

Now we proceed directly to the proof of Theorem 1.9.

By virtue of the well-known embedding theorem for the Sobolev–Liouville classes[39], the class $L_p^\alpha(\Omega)$, for any be α and p satisfying the three conditions (1.4.1), is contained in the class $L_p^{(N-1)/2}(\Omega)$ at any p subject to the inequality $p > 2N/(N-1)$.

Therefore, it suffices to prove the convergence, uniform on any compact set of domain Ω, of the Fourier series of a function $f(x)$ compactly supported in Ω and belonging to the class $L_p^{(N-1)/2}(\Omega)$ at $p > 2N/(N-1)$. We continue the function $f(x)$ in zero in $E^N \backslash \Omega$.

By the definition of the Sobolev–Liouville class[40], the function $f(x)$ may be brought into correspondence with a function $h(x)$ of the class $L_p(E^N)$ such that the representation

$$f(x) = \int_{E^N} \overset{0}{T}_{(N-1)/4}(x,y)h(y)dy \tag{1.4.8}$$

holds; note that

$$\|f\|_{L_p^{(N-1)/2}(E^N)} = \|f\|_{L_p^{(N-1)/2}(\Omega)} = \|h\|_{L_p(E^N)} \tag{1.4.9}$$

by definition.

We recall that the symbol $\overset{0}{T}_{(N-1)/4}(x,y)$ in the representation (1.4.8) denotes the so-called Bessel–MacDonald kernel expressed through the cylindrical MacDonald function $K_\nu(r)$ of order ν according to the formula

$$\overset{0}{T}_{(N-1)/4}(x,y) = 2^{1-(N-1)/4}\left[(2\pi)^{N/2}\Gamma\left(\frac{N-1}{4}\right)\right]^{-1} |x-y|^{-1/2}K_{1/2}(|x-y|).$$

Since, by condition, the function $f(x)$ is compactly supported in the domain Ω, there exists a compact set K_1 of domain Ω outside which the function $f(x)$ becomes zero.

Let us ensure that the function $f(x)$ as represented by (1.4.8) belongs to the class C^∞ in $E^N \backslash K_1$. Let m be an integer such that $m > N/4$, as has been specified above.

---

39) See Nikol'skii [65, Chapter 9].

40) See Nikol'skii [65, Chapter 9]; also, the reader is referred to Section 2.1.3 (*vide infra*).

Invoking the property of composition for the Bessel–MacDonald kernels, we obtain the equality

$$\int_{E^N} \overset{0}{T}_{2m-(N-1)/4}(x,y)f(y)dy = \int_{E^N} \overset{0}{T}_{2m}(x,y)h(y)dy \qquad (1.4.10)$$

from (1.4.8).

Having applied the operator $(-\Delta+1)^m$ to both sides of equality (1.4.10) at any point x in $E^N \backslash K_1$ and having taken into account the fact that, by virtue of $f(y) \equiv 0$ outside K_1, the left-hand side of (1.4.10) belongs to the class $C^\infty(E^N \backslash K_1)$, we obtain $h(x) \in C^\infty(E^N \backslash K_1)$.

We denote by K_2, K_3, K_4, and K_5 four compact sets of the subdomain Ω such that for any $\ell = 1,2,3,4$ the compact set K_ℓ is strictly interior to the compact set $K_{\ell+1}$; also, we consider two "cutting" functions $\eta(x)$ and $\eta_1(x)$ which belong to the class $C_0^\infty(E^N)$ and are such that $\eta(x) \equiv 1$ inside the compact set K_2, $\eta(x) \equiv 0$ outside the compact set K_3, $\eta_1(x) \equiv 1$ inside the compact set K_4, and $\eta_1(x) \equiv 0$ outside the compact set K_5.

Then, recalling that $f(x) \equiv 0$ outside K_1, we can rewrite the representation (1.4.8) in the following manner:

$$\begin{aligned} f(x) &= \int_{E^N} \overset{0}{T}_{(N-1)/4}(x,y)h(y)dy = \eta_1(x) \int_{E^N} \overset{0}{T}_{(N-1)/4}(x,y)h(y)dy \\ &= \eta_1(x) \int_{E^N} \overset{0}{T}_{(N-1)/4}(x,y)h(y)\eta(y)dy + \eta_1(x) \int_{E^N} \overset{0}{T}_{(N-1)/4}(x,y)h(y)[1-\eta(y)]dy. \end{aligned} \qquad (1.4.11)$$

Since, as has been proved above, $h(y) \in C^\infty(E^N \backslash K_1)$ and $1-\eta(y) \equiv 0$ in the compact set K_2, the latter summand on the right-hand side of (1.4.11) belongs to the class $C_0^\infty(\Omega)$ and, consequently, is expandable into a uniformly convergent Fourier series in the system $\{u_n(x)\}$ on any compact set of the domain Ω.

Thus, to prove Theorem 1.9, it remains to ensure that the former summand on the right-hand side of (1.4.11) is also expandable into a uniformly convergent Fourier series in the system $\{u_n(x)\}$ on any compact set of domain Ω.

We denote by $T_{(N-1)/4}(x,y)$ a fractional kernel corresponding to the FSF of interest, and we fix a positive number R which is less than the distance between the compact set K_3 and the boundary of the compact set K_4, and is less than the distance between the compact set K_5 and the boundary of the domain Ω.

Then, in view of the results of Section 1.2.1 for all x of the compact set K_5, for all y of the domain Ω, and for any even number $2m$, the kernel $T_{(N-1)/4}(x,y)$ can be represented in the form

$$T_{(N-1)/4}(x,y) = v_{(N-1)/4}(|x-y|) - w_{(N-1)/4}(|x-y|) - \Psi_{(N-1)/4}(x,y), \qquad (1.4.12)$$

where

$$v_{(N-1)/4}(r) = \begin{cases} \overset{0}{T}_{(N-1)/4}(r) & \text{for} \quad r \le R, \\ 0 & \text{for} \quad r > R; \end{cases} \tag{1.4.13}$$

$$w_{(N-1)/4}(r) = \begin{cases} \sum\limits_{k=0}^{2m} a_k r^{2k} & \text{for} \quad r \le R, \\ 0 & \text{for} \quad r > R; \end{cases} \tag{1.4.14}$$

$$\Psi_{(N-1)/4}(x, y) = \sum_{n=1}^{\infty} \overset{2m}{\gamma_n} u_n(x) u_n(y); \tag{1.4.15}$$

the coefficients $a_0, a_1, \dots, a_{2m}$ have been chosen so as to render the function $v_{(N-1)/4}(r) - w_{(N-1)/4}(r)$ and its derivatives up to an order of $2m$ for all $r \ge 0$ continuous, and the estimate

$$\left|\overset{2m}{\gamma_n}\right| \le C(R, m, N)(1 + \lambda_n)^{-N/2-m} \tag{1.4.16}$$

remains valid for any number $\overset{2m}{\gamma_n}$.

The former summand on the right-hand side of (1.4.11) can be represented as the sum of two terms[41]:

$$\eta_1(x) \int\limits_{E^N} \overset{0}{T}_{(N-1)/4}(x, y) h(y) \eta(y) dy = \eta_1(x) \int\limits_{K_3} \Big\{ \overset{0}{T}_{(N-1)/4}(x, y) - \big[v_{(N-1)/4}(|x - y|)$$

$$- w_{(N-1)/4}(|x - y|) \big] \Big\} h(y) \eta(y) dy \tag{1.4.17}$$

$$+ \eta_1(x) \int\limits_{K_3} \big[v_{(N-1)/4}(|x - y|) - w_{(N-1)/4}(|x - y|) \big] h(y) \eta(y) dy.$$

The form of the representation of functions (1.4.13) and (1.4.14) implies that the kernel within the braces on the right-hand side belongs to the class $C^{2m}(\Omega)$, with respect to x, at any y of the compact set K_3. A straightforward consequence of this is that the former summand on the right-hand side of (1.4.17), belonging to the class $C^{(2m)}(\Omega)$ and compactly supported in Ω, is expandable into a uniformly convergent Fourier series in the system $\{u_m(x)\}$ on any compact set of domain Ω.

---

41) It should also be kept in mind that $\eta(y) \equiv 0$ outside the compact set K_3.

As to the latter summand on the right-hand side of (1.4.17), it can be rewritten as

$$\int_{K_3} \left[v_{(N-1)/4}(|x-y|) - w_{(N-1)/4}(|x-y|)\right] h(y)\eta(y)dy \qquad (1.4.18)$$

in view of the fact that (i) $\eta(y)$ is distinct from zero only on the compact set K_3, (ii) $|x-y| \leq R$, (iii) R is less than the distance between K_3 and the boundary of K_4, and (iv) the function $\eta_1(x)$ everywhere in K_4 identically equals unity.

Invoking representation (1.4.12), the function (1.4.18) can, in turn, be written as the sum of two functions:

$$f_1(x) = \int_{K_3} T_{(N-1)/4}(x,y)h(y)\eta(y)dy, \qquad (1.4.19)$$

$$f_2(x) = \int_{K_3} \Psi_{(N-1)/4}(x,y)h(y)\eta(y)dy. \qquad (1.4.20)$$

It follows from relation (1.4.15) that the Fourier expansion takes the form

$$\sum_{n=1}^{\infty} \gamma_n^{2m} u_n(x) \int_{K_3} h(y)\eta(y)u_n(y)dy. \qquad (1.4.21)$$

To establish the convergence of the Fourier series (1.4.21) uniform on any compact set of domain Ω, it suffices to make use of the Cauchy–Buniakowski inequality and to take into account that the numerical series

$$\sum_{n=1}^{\infty} \left[\int_{K_3} h(y)\eta(y)u_n(y)dy\right]^2$$

is convergent by virtue of the fact that the function $h(y)\eta(y)$ belongs in any event to the class $L_2(G)$[42] and that the functional series

$$\sum_{n=1}^{\infty} \left(\gamma_n^{2m}\right)^2 u_n^2(x)$$

is uniformly convergent on any compact set of the domain Ω, in view of the estimate (1.4.16) and Lemmas 1.2 and 1.3 (*vide supra* Section 1.2).

---

[42] In view of the fact that $h(y) \in L_p(G)$ at $p > 2N/(N-1) > 2$.

Finally, we observe that in order to establish the convergence of the function (1.4.19) in the system $\{u_n(x)\}$ uniform on any compact set of domain Ω, it suffices to prove that the estimate

$$|E_\lambda f_1(x)| \leq C\|f_1\|_{L_p^{(N-1)/2}(\Omega)} \tag{1.4.22}$$

holds uniformly with respect to x on any compact set of domain Ω for the set of spectral decompositions $\{E_\lambda f_1(x)\}$.

Indeed, having proved the estimate (1.4.22), we obtain that the set $\{E_\lambda f_1(x)\}$ is uniformly convergent with respect to x on any compact set of domain Ω as λ tends to ∞. To this end, we fix an arbitrary $\varepsilon > 0$ and a function $F(x)$ from the class $C_0^\infty(\Omega)$ such that

$$\|f(x) - F(x)\|_{L_p^{(N-1)/2}(\Omega)} < \frac{\varepsilon}{2C},$$

where C is the constant we used for the estimate (1.4.22), and make use of estimate (1.4.22) for the function $f_1(x) - F(x)$, with allowance made for the uniform convergence of the function $F(x)$ on any compact set of domain Ω.

Note that the validity of the estimate (1.4.22) uniform on any compact set of domain Ω follows immediately from the Hölder inequality, Lemma 1.6 (see Section 1.3.3), and relation (1.4.9).

This completes the proof of Theorem 1.9.

1.4.2. Conditions for Fourier Series Localization with Respect to an Arbitrary FSF

THEOREM 1.10. *Let $N \geq 2$, let $\{u_n(x)\}$ be an arbitrary FSF of the Laplace operator in an N-dimensional subdomain Ω of an arbitrary N-dimensional domain G, and let $f(x)$ be an arbitrary function compactly supported in the subdomain Ω, continued with zero outside Ω; we also assume that this function reduces to zero within a domain D interior to Ω and belongs to the class $L_2^\alpha(\Omega)$ for an arbitrary $\alpha \geq (N-1)/2$. Then the Fourier series of the function $f(x)$ converges uniformly to zero on an arbitrary compact set of the domain D.*

PROOF: Let K_1 be a compact set such that $f(x) \equiv 0$ outside this compact set. Let this compact set contain no point that belongs to domain D. We fix an arbitrary compact set K in D; following the same line of reasoning that we adhered to in the proof of Theorem 1.9, we can now fix four compact sets K_2, K_3, K_4, and K_5 of the domain Ω in such a manner that these compacts sets are free of points belonging to the compact set K; given any $\ell = 1, 2, 3, 4$, the compact set K_ℓ is strictly interior to the compact set $K_{\ell+1}$.

Resorting, for the case $p = 2$, to the same arguments that we used in the proof of Theorem 1.9, we establish that the function $f(x)$ in question can be represented as the sum of three functions: (i) a function from the class $C_0^\infty(\Omega)$; (ii) a function $f_2(x)$ whose Fourier series (1.4.21) is uniformly convergent on any compact set of domain Ω; (iii) a function $f_1(x)$ as defined by the equality (1.4.19).

To complete the proof of Theorem 1.10, it is sufficient to show that for the spectral decomposition $E_\lambda f_1(x)$ of the function $f_1(x)$ uniformly with respect to x on an arbitrary compact set K that we have fixed in domain D, the following estimate holds:

$$|E_\lambda f_1(x)| \leq C\|f_1\|_{L_2^{(N-1)/2}(\Omega)};$$

in fact, the validity of this estimate is inferred at once from the Cauchy–Buniakowski inequality, from the estimate (1.3.76) (see Lemma 1.7 in Section 1.3.3), and from the relation (1.4.9) taken at $p = 2$.

1.4.3. Conditions for the Nonlocalization of Fourier Series with Respect to an Arbitrary FSF

THEOREM 1.11. *Let $N \geq 2$, let $\{u_n(x)\}$ be an arbitrary FSF of the Laplace operator in an N-dimensional subdomain of an arbitrary N-dimensional domain G, let α be an arbitrary number subject to the condition $0 < \alpha < (N-1)/2$, and let x_0 be an arbitrary fixed point of the subdomain Ω. Then there exists a function $f(x)$ satisfying the following conditions: (i) $f(x)$ is compactly supported in the domain Ω and belongs to the Hölder class $C^{(\alpha)}(\Omega)$; (ii) $f(x)$ reduces to zero in a certain neighborhood of the point x_0; (iii) $\overline{\lim\limits_{\lambda\to\infty}} E_\lambda f(x_0) = \infty$.*

PROOF: Let the set E have the same meaning as in Lemmas 1.5 and 1.8 (see Section 1.3.3), and let the number $R_0 > 0$ be fixed in the same manner as in the aforesaid lemmas. Then, invoking the relation (1.3.77) in Lemma 1.8 and the known theorem of resonance type[43], we infer the existence of a function $h(y)$ which is bounded on the set E devoid of point x_0 and for which the following relation holds:

$$\overline{\lim_{\lambda\to\infty}} \left| \int_E E_\lambda T_\alpha(x_0, y) h(y)\, dy \right| = \infty. \tag{1.4.23}$$

We now extend the function $h(y)$ onto the entire domain G and make it equal to zero in $G \backslash E$. Then, by virtue of (1.4.23),

$$\overline{\lim_{\lambda\to\infty}} \left| \int_G E_\lambda T_\alpha(x_0, y) h(y)\, dy \right| = \infty. \tag{1.4.24}$$

[43] See, for example, Kaczmarz and Steinhaus [51, p. 31, Assertion 3].

The relation (1.4.24) establishes the nonboundedness at the point x_0 of the spectral decomposition of the function

$$F(x) = \int_G T_\alpha(x,y)h(y)dy. \tag{1.4.25}$$

We arbitrarily fix a sufficiently large number m and a sufficiently small positive number R; in view of the results of Section 1.2.1, we can represent the kernel $T_\alpha(x,y)$ for any interior point x of domain Ω in the form

$$T_\alpha(x,y) = [v_\alpha(|x-y|) - w_\alpha(|x-y|)] - \Psi_\alpha(x,y), \tag{1.4.26}$$

where

$$v_\alpha(r) = \begin{cases} \overset{0}{T}_\alpha(r) & \text{for } r \le R, \\ 0 & \text{for } r > R, \end{cases}$$

$$w_\alpha(r) = \begin{cases} \sum\limits_{k=0}^{m} a_k r^{2k} & \text{for } r \le R; \\ 0 & \text{for } r > R; \end{cases} \tag{1.4.27}$$

$$\Psi_\alpha(x,y) = \sum_{n=1}^{\infty} \overset{m}{\gamma}_n u_n(x) u_n(y).$$

The coefficients $a_0, a_1, \ldots, a_m$ have been chosen so as to render the function $v_\alpha(r) - w_\alpha(r)$ and its derivatives up to an order of m for all $r \ge 0$ continuous, and the estimate

$$\left|\overset{m}{\gamma}_n\right| \le C(R,m,N,\alpha)(1+\lambda_n)^{-N/4-m/2} \tag{1.4.28}$$

remains valid uniformly for the number $\overset{m}{\gamma}_n$ on any compact set of domain Ω.

In view of the relation (1.4.26), the function (1.4.25) can be represented as the sum of two functions

$$F(x) = f(x) + g(x),$$

where

$$f(x) = \int_G [v_\alpha(|x-y|) - w_\alpha(|x-y|)]\, h(y)dy, \tag{1.4.29}$$

$$g(x) = \int_G \Psi_\alpha(x,y)h(y)dy. \tag{1.4.30}$$

With reference to representation (1.4.27) and estimate (1.4.28), we can assert that if the number m is fixed sufficiently large, then the spectral decomposition of the function (1.4.30) converges at any interior point of the domain Ω and, consequently, converges at the point x_0. If so, we infer, invoking the representation (1.4.26) and relation (1.4.24), that the spectral decomposition of the function (1.4.29) is divergent at the point x_0.

One will observe that the boundedness of the function $h(y)$ and the well-known properties of potential-type integrals imply that the function (1.4.29) belongs to the Hölder class $C^{(\alpha-\delta)}$ for any $\delta > 0$, whereas the choice of a sufficiently small $R > 0$ in representation (1.4.26) ensures the reduction of function (1.4.29) to zero in the neighborhood of the point x_0 and everywhere outside a compact set K interior to the domain Ω and containing the set E.

This completes the proof of Theorem 1.11.

REMARK TO THEOREM 1.11. Theorem 1.11 remains all the more valid as the requirement (i) to the function $f(x)$ is replaced by the following requirement: *f(x) is compactly supported in domain Ω and belongs to the Sobolev–Liouville class $L_p^\alpha(\Omega)$ with arbitrarily fixed α and p subject to the conditions $0 < \alpha < (N-1)/2$, $p \geq 1$.*

1.4.4. On the Exactness of the Established Conditions for Uniform Convergence and Localization

Now we wish to show that, for an arbitrary FSF of the Laplace operator in an N-dimensional subdomain Ω of an arbitrary N-dimensional domain G, the three inequalities (1.4.1) are definitive conditions in the Sobolev–Liouville classes $L_p^\alpha(\Omega)$ for the uniform (on any compact set of domain Ω) convergence of the Fourier series of an arbitrary function $f(x)$ which is compactly supported in Ω.

Indeed, a comparison of Theorems 1.9. and 1.11 reveals that the first inequality (1.4.1), namely $\alpha \geq (N-1)/2$, is a definitive condition for the order of differentiability α, since for any arbitrarily fixed positive $\alpha < (N-1)/2$ and for any arbitrarily fixed $p \geq 1$ there exists (in view of the remark to Theorem 1.11), for each specified FSF, a function $f(x)$ belonging to the Sobolev–Liouville class $L_p^\alpha(\Omega)$ and compactly supported in Ω such that even the localization principle fails to hold for its Fourier series.

The second inequality (1.4.1), $\alpha p > N$, is also a definitive condition, since, given $\alpha p = N$, there exists in the Sobolev–Liouville class $L_p^\alpha(\Omega)$ a function which is compactly supported in the domain Ω and which is not continuous on a certain compact set of this domain (the Fourier series of such a function fails to converge uniformly on this compact set).

Finally, the third inequality (1.4.1) that is, $p \geq 1$, is a condition in the definition of the Sobolev–Liouville class.

In exactly the same manner one will infer via comparing Theorems 1.10 and 1.11 that in the Sobolev–Liouville classes $L_2^\alpha(\Omega)$ the inequality $\alpha \geq (N-1)/2$ is a definitive condition under which the localization principle holds for the Fourier series of a function $f(x)$ compactly supported in Ω with respect to each specified FSF.

1.4.5. Conditions for Fourier Series Divergence with Respect to an Arbitrary FSF in the L_p-Metric and on the Set of Positive Measure

THEOREM 1.12. *Let $N \geq 2$, and let $\{u_n(x)\}$ be an arbitrary FSF of the Laplace operator in an N-dimensional subdomain of an arbitrary N-dimensional domain G. Then there exists, for any p in the half-interval $1 \leq p < 2N/(N+1)$, a function in the class $L_p(\Omega)$ such that it equals zero outside the domain Ω and its Fourier series (i) diverges in the L_1-metric (especially, in the L_p-metric); (ii) diverges on the set of points of positive measure in Ω.*

Such a function may be exemplified by the function

$$f(y) = \begin{cases} T_\alpha(x_0, y) & \text{for} \quad \alpha < (N-1)/4, \quad y \in E, \\ 0 & \text{for} \quad y \in G \backslash E, \end{cases} \tag{1.4.31}$$

where x_0 is an arbitrarily fixed interior point of the domain Ω, and the set E bears the same meaning as in Lemmas 1.5 and 1.8.

PROOF: The divergence of the Fourier series of the function (1.4.31) in the L_1-metric follows immediately from Lemma 1.8 (*vide supra* Section 1.3.3, relation (1.3.77)). To prove the divergence of the Fourier series of interest on a set of positive measure, it suffices to take into account the fact that the set E in (1.3.77) is an arbitrary measurable set subject only to the condition (1.3.50), and to apply the familiar Egorov theorem to the Fourier series.

This completes the proof of Theorem 1.12.

1.5. ON THE POTENTIAL GENERALIZATION OF THE THEORY

A potential subject for investigation may be an arbitrary system $\{u_n(x)\}$ complete in an arbitrary N-dimensional domain Ω. One may concede the knowledge of each function $u_n(x)$ belonging to the class $C^{(2)}$ in an open domain Ω and satisfying the equation $\Delta u_n + \lambda_n u_n = 0$ in Ω, where λ_n is a nonnegative number. However, whether the system of interest is orthonormal or not in the domain Ω, or in any other domain G to which Ω is interior, may be an uncertain issue.

Olevskii[44] has established necessary and sufficient conditions for a feasible continuation of functions $u_n(x)$ onto a domain G enclosing its interior subdomain Ω; under these conditions, the continued system of functions $\{u_n(x)\}$ in domain G is a complete and orthonormal one, implying thereby that it is a FSF of the Laplace operator in the subdomain Ω of domain G.

---

[44] See Olevskii [66], [67].

These necessary and sufficient conditions are expressed by two requirements imposed on the matrix A with elements

$$a_{mn} = \int_{\Omega} u_m(x)u_n(x)dx.$$

The first of these requirements is the relation $A^2 = A$, and the second one is the boundedness of the matrix A in the Hilbert sense or, stated otherwise, the validity of the inequality

$$\sum_{m=1}^{s}\sum_{n=1}^{s} a_{mn}\xi_m\xi_n \leq \sum_{m=1}^{s} \xi_m^2$$

for any ordinal number s and for any real numbers $\xi_1, \xi_2, \dots, \xi_s$.

COMMENTS ON CHAPTER 1

The concept of the fundamental system of functions of the Laplace operator, both in the entire domain G and in an arbitrary subdomain Ω of G, was first introduced in 1968 by Il'in [30].

The existence of the FSF of the Laplace operator with a specified countable subset of eigenvalues was first established in the paper [43] by Il'in and Filippov. This subject received further development in the works of Gekhtman [20] and Gorbachuk [22].

The fractional kernels were first dealt with by Il'in in [25].

The point of departure for studying the convergence problem of arbitrary spectral decompositions has been the theory of multiple trigonometric series and Fourier integrals with spherical partial sums. The works of Bochner, Titchmarsh, Chandrasekharan, Minakshisundaram, Levitan, and Stein have played an important role in the development of this theory.

Bochner [12, p. 276] has shown that it suffices, for the convergence of a finite function expanded into an N-fold Fourier integral, that the mean $f_r(x)$ of this function over a sphere of radius r centered at a point x have a summable derivative of order $[(N+1)/2]$ for $r \geq 0$[45].

Levitan [59] has shown that if a function $f(x)$ and all its derivatives of order up to $[N/2]$ are continuous in a closed N-dimensional cube Q, then the spherical partial sums of the N-fold trigonometric Fourier series of the function $f(x)$ converge to this function provided that the function and its derivatives of order up to $[N/2]-1$ admit of a periodic continuation over the entire N-dimensional space with the conservation of continuity.

[45] The square brackets denote that only the integral part of the number enclosed therein is taken into account.

It has also been noted that if the requirement of continuity of the derivatives of order $[N/2]$ is replaced by the requirement that all the derivatives of order $[N/2]-1$ be absolutely continuous with respect to each of the variables for almost all values of the other variables and the derivatives of order $[N/2]$ be square-integrable over the cube Q and be bounded in the vicinity of the point x_0 in question, then one can assert that the spherical partial sums of the N-fold trigonometric Fourier series at this point x_0 are convergent.

The Fourier series with respect to the eigenfunctions of the first boundary-value problem for the Laplace operator in an N-dimensional domain G have been studied in the works of Titchmarsh [76] and Chandrasekharan and Minakshisundaram [15]. It has been shown that if, for a function $f(x)$, the series $\sum_{n=1}^{\infty} f_n^2 \lambda_n^{N/2}$ is convergent (where f_n are the Fourier coefficients of the function $f(x)$ and λ_n are the eigenvalues), then for the spectral decomposition $E_\lambda f(x)$ to be convergent to $f(x)$ at a fixed point x it is necessary and sufficient that the mean $\widehat{f}_r(x)$ of the function $f(x)$ over a ball of radius r centered at point x tend to $f(x)$ as $r \to 0$.

Conditions for the convergence of the spectral decomposition with respect to an arbitrary FSF of the Laplace operator, exact in the Sobolev classes W_p^n of integer order n, were established by Il'in as early as 1958 [26].

Conditions for the convergence of the spectral decomposition with respect to an arbitrary FSF of the Laplace operator, exact in the Sobolev–Liouville classes L_p^α, have been established in the work of Il'in [30]. A theorem of negative type was first proved in the same work [30] for each specified FSF of the Laplace operator in a subdomain Ω of an arbitrary N-dimensional domain G. This theorem provides exact bounds of requirements for the order of differentiability α $(\alpha \geq (N-1)/2)$; once out of these bounds, the localization principle for the Fourier series of a function with compact support no longer remains valid.

In Chapter 1, we have chosen not to dwell on the results concerned with the absolute convergence of decompositions with respect to the eigenfunctions. This pertinent issue has been dealt with in the works of Ladyzhenskaya [56], Krasnosel'skii and Pustyl'nik [55], Il'in [27], [28], and Kendzhaev [52], [53].

We have omitted in our exposé the result of Il'in [29] dealing with the convergence of spectral decompositions exhibiting a singular point.

In conclusion, we remark that the results of Sections 1.3 and 1.4 have been extended in the work of Il'in and Shishmarev [47] to the case of the Fourier series with respect to the FSF of a polyharmonic operator.

In the work of Goldman [21], the theorem of negative type has been extended to the case of functional classes with a more structured modulus of continuity than that of the Hölder classes.

Chapter 2

Spectral Decompositions Corresponding to an Arbitrary Self-Adjoint Nonnegative Extension of the Laplace Operator

In this chapter we establish exact conditions for the convergence of the spectral decompositions corresponding to an arbitrary self-adjoint nonnegative extension of the Laplace operator in the domain G (not necessarily a bounded one) of the space E^N.

The exact definition of the self-adjoint nonnegative extension of a general formally self-adjoint elliptic operator of second order (in particular, the Laplace operator) in domain G will be given in Section 2.1. The decompositions with respect to the FSF of the Laplace operator that have been studied in Chapter 1 for the case in which the FSF is taken over the entire domain G will correspond to a self-adjoint nonnegative extension of the Laplace operator with a purely point spectrum.

We will show that the use of the known Gårding–Browder–Mautner theorem on the ordered spectral representation of the space $L_2(G)$ with respect to an arbitrary self-adjoint extension of the elliptic operator makes it possible to transfer the method we have developed in Chapter 1 to the case of an arbitrary self-adjoint nonnegative extension of the Laplace operator, irrespective of the form of its spectrum — point, continuous, or mixed.

The next issue of interest in this chapter is to establish not only exact conditions for convergence of the spectral decompositions themselves, but also exact conditions for convergence of the so-called Riesz means of these decompositions, taking into account that the employment of the Riesz means enables one to relax the requirements of smoothness which are imposed on a function to be decomposed and which ensure uniform convergence.

Finally, a further concern of this chapter focuses on a study of the exact conditions for the uniform convergence of spectral decompositions and their Riesz means in the classes of differentiable (in the generalized sense) functions of N variables.

In Sections 2.2–2.5, we provide an exhaustive answer to the problem on the conditions for uniform convergence of the spectral decompositions and their Riesz means in the functional classes of Zygmund–Hölder, Sobolev–Liouville, Nikol'skii, and Besov.

We show that the uniform convergence conditions of the spectral decompositions and their Riesz means in each of the four classes mentioned are novel (also definitive ones) even for the expansions into an N-fold Fourier integral and into an N-fold trigonometric Fourier series (with spherical sums).

In order to improve the rationale, we modify the method of Chapter 1 so as to enable the study of spectral decompositions without preliminary asymptotic estimation of the spectral function.

To demonstrate the potential utility of our method for an arbitrary self-adjoint nonnegative Laplace operator, we estimate the remainder term of the spectral function in both the L_2-metric (Section 2.6) and the L_∞-metric (Section 2.7).

2.1. SELF-ADJOINT NONNEGATIVE EXTENSIONS OF ELLIPTIC OPERATORS. ORDERED SPECTRAL REPRESENTATIONS OF THE SPACE L_2. CLASSES OF DIFFERENTIABLE FUNCTIONS OF N VARIABLES

This section serves as an introduction. It provides a brief outlook of major concepts in the theory of the self-adjoint extensions of elliptic operators and in the theory of differentiable (in the generalized sense) functions of N variables. The topics covered here will be used in this and subsequent chapters.

2.1.1. Self-Adjoint Nonnegative Extensions of Elliptic Operators

Let G be an arbitrary N-dimensional domain, perhaps, coincident entirely with the space E^N. Suppose that a general linear formally self-adjoint elliptic operator of second order is specified in the domain G

$$Lu = -\sum_{i,j=1}^{N} \frac{\partial}{\partial x_i}\left[a_{ij}(x)\frac{\partial u}{\partial x_j}\right] + c(x)u, \tag{2.1.1}$$

where for all $x \in G$ and for all real numbers $\xi_1, \xi_2, \dots, \xi_N$ the following conditions are fulfilled:

$$a_{ij}(x) = a_{ji}(x), \qquad \sum_{i,j=1}^{N} a_{ij}(x)\xi_i\xi_j \geq \alpha \sum_{i=1}^{N} \xi_i^2 \qquad \text{for} \qquad \alpha > 0;$$

for simplicity, the coefficients $a_{ij}(x)$ and $c(x)$ of this operator are assumed to be infinitely differentiable in G.

We denote by $C_0^\infty(G)$ a space of functions which are infinitely differentiable in domain G and which are compactly supported in G. The functions which are compactly supported in domain G will be referred to as *finite* in G. Thus, a function finite in G is a function which is defined, generally speaking, in the entire space E^N and which is distinct from zero on a certain compact set K interior to G and which is removed to a positive distance from the boundary ∂G of domain G.

We denote by A an operator which acts according to the rule $Au = Lu$ in the space $L_2(G)$ and which is defined within a domain $D(A)$ coincident with the space $C_0^\infty(G)$; here $u(x) \in C_0^\infty(G)$, and Lu is the elliptic operator (2.1.1).

It is easily checked, using Green's formula, that if the operator Lu takes the form (2.1.1), then the operator A, defined with the aid of the operator L, is a symmetric one, that is, for any $u(x)$ and $v(x)$ from the class $C_0^\infty(G)$, the equality

$$(Au, v) = (u, Av)$$

holds.

The operator A is referred to as a *semi-bounded* one if there exists a real number μ (termed the *lower bound* of this operator) such that $(Au, u) \geq \mu(u, u)$ for all $u \in C_0^\infty(G)$.

In particular, the operator A is referred to as a nonnegative one, if $(Au, u) \geq 0$ for all $u \in C_0^\infty(G)$.

The remarkable theorem of K. O. Friedrichs [17] asserts that *there exists, for each symmetric semi-bounded operator A, at least one self-adjoint extension $\widehat{A}$ with the same lower bound, that is, for each symmetric semi-bounded operator A there exists at least one operator $\widehat{A}$ exhibiting the following four properties:*

(1) $\widehat{A}$ is a self-adjoint operator, that is, $(\widehat{A}u, v) = (u, \widehat{A}v)$ for any u and v from the domain of definition $D(A)$ of operator A;

(2) $(\widehat{A}u, u) \geq \mu(u, u)$ for all u from the domain of definition $D(\widehat{A})$ of operator $\widehat{A}$ (with the same μ as for operator A);

(3) the domain of definition $D(A)$ of operator A is interior to the domain of definition $D(\widehat{A})$ of operator $\widehat{A}$;

(4) the equality $\widehat{A}u = Au$ is valid for all u in the domain of definition $D(A)$ of operator $\widehat{A}$.

Let $\widehat{A}$ be an arbitrary self-adjoint semi-bounded extension of operator A defined on the functions $u(x) \in C_0^\infty(G)$ by the rule $Au = Lu$, where Lu is a formally self-adjoint elliptic operator of second order (2.1.1).

Then the known theorem of J. von Neumann [64] holds for $\widehat{A}$ as for any self-adjoint operator; according to this theorem, *this operator exhibits the so-called unit expansion, that is, possesses a family of projectors* $\{E_\lambda\}$ *so that*

$$\widehat{A} = \int_{\mu}^{\infty} \lambda \, dE_\lambda,$$

where the projectors E_λ *increase monotonically, are left-continuous, and converge strongly to a unit operator, that is, for any* u *from* $L_2(G)$ *one has*

$$\lim_{\lambda \to \infty} ||E_\lambda u - u||_{L_2(G)} = 0.$$

In our case, when the operator $\widehat{A}$ is a self-adjoint semi-bounded extension of the operator A generated by the elliptic differential operator (2.1.1), an essential contribution to the von Neumann theorem has been given by L. Gårding, who has shown that each projector E_λ is an integral operator with a kernel of Carleman type. To be more exact, Gårding [18] has proved the following statement:

For each λ *on the numerical axis* $E = (-\infty, \infty)$, *there exists a kernel* $\theta(x, y, \lambda)$ *of Carleman type exhibiting the property that for an arbitrary function* $f(x)$ *from the class* $L_2(G)$ *and for almost all points* x *of the domain* G *one has*

$$E_\lambda f(x) = \int_G \theta(x, y, \lambda) f(y) dy. \tag{2.1.2}$$

The kernel $\theta(x, y, \lambda)$ *is a Borel function on the set* $G \times G \times E$ *and exhibits a Hermitian symmetry* $\theta(x, y, \lambda) = \overline{\theta(y, x, \lambda)}$, *such that the integral*

$$\int_G |\theta(x, y, \lambda)|^2 dy$$

is a continuous function of x *in domain* G *for any* $\lambda \in E$. *Moreover, the kernel* $\theta(x, y, \lambda)$ *equals zero at* $\lambda < \mu$ *and has a bounded variation with respect to* λ *in each segment* $\mu \leq \lambda \leq \lambda_0$ *of the numerical axis* E, *this variation being uniformly bounded with respect to the ensemble* x, y *on each compact set* K *of domain* $G \times G$.

The kernel $\theta(x, y, \lambda)$ of the integral operator (2.1.2) is called the *spectral function* of the self-adjoint extension $\widehat{A}$, and the expression (2.1.2) is called the *spectral decomposition* of the function $f(x)$ corresponding to the self-adjoint extension $\widehat{A}$.

We emphasize that the properties of a spectral function, as established by the Gårding theorem, enable one to determine the spectral decomposition (2.1.2) for any function $f(x)$ from the class $L_2(G)$, as well as from any finite function of the class $L_1(G)$ in domain G.

We introduce, along with the spectral decomposition (2.1.2) of a function, the *Riesz means* of this spectral decomposition of order $s \geq 0$ and of size λ, defining them by the equality

$$E_\lambda^s f(x) = \int_{\mu}^{\lambda} \left(1 - \frac{t}{\lambda}\right)^s dE_t f. \tag{2.1.3}$$

The operator E_λ^s, like E_λ, is an integral operator

$$E_\lambda^s f(x) = \int_G \theta^s(x, y, \lambda) f(y) dy, \tag{2.1.4}$$

whose kernel $\theta^s(x, y, \lambda)$ represents, obviously, the Riesz means of order s of the spectral function $\theta(x, y, \lambda)$:

$$\theta^s(x, y, \lambda) = \int_\mu^\lambda \left(1 - \frac{t}{\lambda}\right)^s d_t\theta(x, y, t). \tag{2.1.5}$$

2.1.2. Ordered Spectral Representations of the Space $L_2(G)$ (with respect to extension $\widehat{A}$)

Let $\widehat{A}$ be an arbitrary self-adjoint semi-bounded (with lower bound μ) extension of an operator generated by the elliptic differential operator (2.1.1). If the spectrum of the closure of the aforesaid extension $\widehat{A}$ is a purely point spectrum, then there exists a complete orthonormal system of eigenfunctions $\{u_n(x)\}$, each of which is a regular solution in domain G of the equation

$$Lu_n = \lambda_n u_n \tag{2.1.6}$$

for an eigenvalue of $\lambda_n \geq \mu$.

In the case of a general self-adjoint semi-bounded extension $\widehat{A}$ of the elliptic operator (2.1.1) (with any spectrum), the role of such regular solutions of Eq. (2.1.6) is assigned to the so-called *fundamental functions* whose existence and major properties have been established by the familiar Gårding–Browder–Mautner theorem (see, for example, Dunford and Schwartz [16, pp. 875–876]). We reformulate this theorem in a more convenient form:

For each self-adjoint semi-bounded extension $\widehat{A}$ of the operator (2.1.1) in domain G there exists at least one ordered spectral representation of the space $L_2(G)$ with spectral measure $\rho(\lambda)$, sets of multiplicity e_i, fundamental functions $u_i(x, \lambda)$ ($i = 1, 2, \ldots, \widehat{m}$) and multiplicity $\widehat{m} \leq \infty$ such that the following requirements are met:

(1) the fundamental functions $u_i(x, \lambda)$ are measurable with respect to the Lebesgue measure of domain G and the spectral measure $\rho(\lambda)$, become zero on the complements of the sets e_i, and for each fixed $\lambda \geq \mu$ belong to the class $C^\infty(G)$ and satisfy inside G the elliptic differential equation

$$Lu_i(x, \lambda) = \lambda u_i(x, \lambda); \tag{2.1.7}$$

(2) for each function $f(x)$ from the class $L_2(G)$, the Fourier image

$$\widehat{f}_j(\lambda) = \int_G f(y) u_i(y, \lambda) dy, \qquad i = 1, 2, \ldots, \widehat{m}, \tag{2.1.8}$$

is defined on the set e_i as an element of the space L_2 with measure $\rho(\lambda)$ such that the spectral decomposition $E_\lambda f(x)$ of each function $f(x)$ from the class L_2 takes the form

$$E_\lambda f(x) = \sum_{i=1}^{\widehat{m}} \int_{\mu}^{\lambda} \widehat{f}_i(t) u_i(x,t) d\rho(t) \qquad (2.1.9)$$

and converges at $\lambda \to \infty$ to $f(x)$ in the $L_2(G)$ metric;

(3) for any two functions $f(x)$ and $g(x)$ from the class $L_2(G)$, the Parseval equality holds,

$$\int_G f(x)g(x)dx = \sum_{i=1}^{\widehat{m}} \int_{\mu}^{\infty} \widehat{f}_i(t)\widehat{g}_i(t) d\rho(t), \qquad (2.1.10)$$

and, in particular, for any $f(x)$ from the class $L_2(G)$ we have

$$\int_G f^2(x)dx = \sum_{i=1}^{\widehat{m}} \int_{\mu}^{\infty} \widehat{f}_i^2(t) d\rho(t). \qquad (2.1.11)$$

In what follows we will consider, for the sake of simplicity, a nonnegative self-adjoint extension $\widehat{A}$ of the elliptic operator (2.1.1) instead of a semi-bounded self-adjoint extension $\widehat{A}$ of this operator. This means that in all the assertions formulated above we must set $\mu = 0$. In particular, we must take $\mu = 0$ in relations (2.1.3), (2.1.5), and (2.1.9)–(2.1.11) and assume that in Eq. (2.1.6) all $\lambda_n \geq 0$, and in Eq. (2.1.7) all the values of λ are nonnegative.

The material of the current and the preceding sections will be used in this form in Chapter 4; in this chapter we consider, instead of the general formally self-adjoint elliptic operator (2.1.1), the simplest operator of the form $Lu = -\Delta u$, where $\Delta = \sum_{i=1}^{N} \partial^2/\partial x_i^2$ is the Laplace operator.

Thus, everywhere in this chapter the operator $\widehat{A}$ represents an arbitrary self-adjoint nonnegative extension of the Laplace operator $Lu = -\Delta u$. Accordingly, Eqs. (2.1.6) and (2.1.7) become the following:

$$\Delta u_n + \lambda_n u_n = 0, \qquad (2.1.6')$$

$$\Delta u_i(x,\lambda) + \lambda u_i(x,\lambda) = 0. \qquad (2.1.7')$$

2.1.3. Classes of Differentiable Functions of N Variables

We now give definitions and formulate the simplest properties of the main classes of differentiable functions of N variables. Everywhere in what follows the partial derivatives of functions of N variables are understood as generalized partial derivatives in the Sobolev sense.[1)]

We denote the multi-index $\bar{k} = (k_1, k_2, \dots, k_N)$ by the symbol $\bar{k}$ made up of N integral nonnegative numbers k_j and set $\bar{k} = k_1 + k_2 + \dots + k_N$.

The symbol $\partial^{\bar{k}} f(x)$ is used to denote the generalized (in the Sobolev sense) partial derivative

$$\partial^{\bar{k}} f(x) = \frac{\partial^{|\bar{k}|} f}{\partial x_1^{k_1} \partial x_2^{k_2} \dots \partial x_N^{k_N}}. \tag{2.1.12}$$

We say that a function $f(x)$, defined in domain G, belongs to the *Sobolev class* $W_p^\ell(G)$, where ℓ is an integral negative number and p satisfies the condition $1 \leq p \leq \infty$, if $f(x)$ has all generalized partial derivatives of the form (2.1.12) for $|\bar{k}| = \ell$ and if the quantity

$$\|f\|_{L_p(G)} + \sum_{|\bar{k}|=\ell} \|\partial^{\bar{k}} f(x)\|_{L_p(G)}, \tag{2.1.13}$$

called the *norm* of $f(x)$ in $W_p^\ell(G)$ and denoted $\|f\|_{W_p^\ell(G)}$, is finite.

To define the *Nikol'skii class* $H_p^\alpha(G)$ with an arbitrary real positive (not necessarily integral) differentiability index α and with an arbitrary degree of summability p on the half-line $1 \leq p \leq \infty$, we represent the differentiability index α in the form $\alpha = \ell + \kappa$, where $\ell = \alpha - 1$ if α is an integer, and $\ell = [\alpha]$ if α is a noninteger,[2)] and $\kappa = 1$ if α is an integer and $\kappa = \alpha - [\alpha]$ if α is a noninteger, so that always $0 < \kappa \leq 1$.

We adopt the convention that for any domain $G \subseteq E^N$ and for any positive number h, the subset of points G removed from the boundary ∂G of domain G to a distance greater than the number h is denoted by the symbol G_h.

We say that a function $f(x)$ belongs to the *Nikol'skii class* $H_p^\alpha(G)$ if $f(x)$ belongs to the class $L_p(G)$ and has all generalized partial derivatives of the form (2.1.12) for $\bar{k} = \ell$, and if for each partial derivative and for each vector u the relation

$$\left\| \partial^{\bar{k}} f(x+u) - 2\partial^{\bar{k}} f(x) + \partial^{\bar{k}} f(x-u) \right\|_{L_p(G_{|u|})} = o(|u|^\kappa) \tag{2.1.14}$$

holds; here the norm $L_p(G_{|u|})$ is taken with respect to the coordinates of the point x. If for the second difference within the modulus brackets in (2.1.14) we introduce

---

1) The concept of a generalized partial derivative may be treated in the following simple manner. We specify in the space E^N an open set G and denote by G_1 the orthogonal projection of this set onto a hyperplane $x_1 = 0$. Suppose that a measurable function $f(\bar{x}) = f(x_1, \bar{y})$ is given on G, where $\bar{y} = (x_2, x_3, \dots, x_N)$. If a function of a single variable x_1, derived from the above function at a fixed $\bar{y}$ on an open unidimensional set, is absolutely continuous on any closed segment belonging to the above set, we say that the function $f(x, \bar{y})$ at the specified $\bar{y}$ is locally absolutely continuous with respect to x_1. We say that a function f has a *generalized partial derivative* of the form $\partial f / \partial x_1$ on the set G if the function f is measurable on G and if there exists its equivalent function f_1 on G locally absolutely continuous with respect to x_1 for almost all $\bar{y}$ from G_1. We call any equivalent $\partial f_1 / \partial x_1$ (in the sense of an N-dimensional measure) function the generalized partial derivative of f with respect to x_1 on the set G and denote it by $\partial f / \partial x_1$.

The generalized partial derivatives of higher order are defined in a similar manner.

2) $[\alpha]$ denotes the integral part of the number α.

the notation

$$\partial^{\bar{k}} f(x+u) - 2\partial^{\bar{k}} f(x) + \partial^{\bar{k}} f(x-u) = \Delta_u^2 \partial^{\bar{k}} f(x),$$

then the norm in the Nikol'skii class, denoted $\|f\|_{H_p^\alpha(G)}$, can be defined by the equality

$$\|f\|_{H_p^\alpha(G)} = \|f\|_{L_p(G)} + \sum_{|\bar{k}|=\ell} \sup_u \left\{ |u|^{-\kappa} \|\Delta_u^2 \partial^{\bar{k}} f(x)\|_{L_p(G_{|u|})} \right\}. \tag{2.1.15}$$

We note that the Nikol'skii class $H_p^\alpha(G)$ can be obtained by closing in the norm (2.1.15) the set $C^\infty(G)$ of infinitely differentiable functions in domain G.

In the case $p = \infty$, the Nikol'skii class $H_p(G)$ is conventionally referred to as the *Zygmund–Hölder class*, denoted by the symbol $C^\alpha(G)$. If α is a noninteger, this class is coincident with the conventional Hölder class in which the norm is introduced with the aid of the second, rather than the first, differences; however, if α is an integer, the Zygmund–Hölder class is wider than the common Hölder class (since with an integral α, a function from the Zygmund–Hölder class may have neither classical nor generalized derivative of order α).

To define the *Besov class* $B_{p,\theta}^\alpha(G)$ which, apart from the real order of differentiability $\alpha > 0$ and the degree of summability p which varies within the range $1 \le p \le \infty$, is also dependent on another parameter θ which varies within the range $1 \le \theta \le \infty$, we represent the order of differentiability α (as in the case of the Nikol'skii class) as the sum $\alpha = \ell + \kappa$, where

$$\ell = \begin{cases} \alpha - 1, & \text{if } \alpha \text{ is an integer,} \\ [\alpha] & \text{if } \alpha \text{ is a noninteger} \end{cases}$$

$$\kappa = \begin{cases} 1, & \text{if } \alpha \text{ is an integer,} \\ \alpha - [\alpha] & \text{if } \alpha \text{ is a noninteger.} \end{cases}$$

We say that a function $f(x)$ belongs to the *Besov class* $B_{p,\theta}^\alpha(G)$ if $f(x)$ belongs to the class $L_p(G)$, has all the generalized partial derivatives of the form (2.1.12) for $\bar{k} = \ell$, and if the quantity

$$\|f\|_{L_p(G)} + \sum_{|\bar{k}|=\ell} \left[\int_0^\infty \left(|u|^{-\kappa} \left\| \Delta_u^2 \partial^{\bar{k}} f(x) \right\|_{L_p(G_{|u|})} \right)^\theta \frac{du}{|u|^N} \right]^{1/\theta}$$

is a finite one; this is called the *norm* in the space $B_{p,\theta}^\alpha(G)$ and is denoted by the symbol $\|f\|_{B_{p,\theta}^\alpha(G)}$.

As is readily seen, at $\theta = \infty$ the Besov class $B_{p,\theta}^\alpha(G)$ becomes in fact the Nikol'skii class $H_p^\alpha(G)$.

One can also ensure that, given an integer α and $\theta = p$, the Besov class $B_{p,\theta}^\alpha(G)$ becomes the Sobolev class $W_p^\alpha(G)$ as defined above. For this reason, the Besov classes $B_{p,\theta}^\alpha(G)$ are one of the feasible extensions of Sobolev classes with integral orders of differentiability to arbitrary real positive orders of differentiability.

However, a more natural extension of Sobolev classes to arbitrary positive real orders of differentiability are the so-called Sobolev–Liouville classes, which now become the focus of our concern.

Since, in what follows, we shall be concerned with functions which are finite in domain G (that is, are defined in the entire space E^N but are distinct from zero only on a certain compact set K contained in G and removed from the boundary ∂G of domain G to a positive distance), we can restrict ourselves to the definition of Sobolev–Liouville classes in the entire space E^N.

To define the Sobolev–Liouville class $L_p^\alpha(E^N)$, we make use of the main part of the fractional kernel $T_{\alpha/2}(x, y)$, as it has been introduced in Section 1.2:

$$\overset{0}{T}_{\alpha/2}(x,y) = 2^{1-\alpha/2}\left[(2\pi)^{N/2}\Gamma(\alpha/2)\right]^{-1}|x-y|^{\alpha-N/2}K_{N/2-\alpha}(|x-y|). \tag{2.1.16}$$

We will refer to it as to the *Bessel–MacDonald kernel.*

We say that a function $f(x)$, defined in the entire space E^N, belongs to the Sobolev–Liouville class $L_p^\alpha(E^N)$ for $\alpha > 0$, $p \geq 1$, if there exists a function $h(x)$ from the class $L_p(G)$ such that the following equality holds:

$$f(x) = \int_{E^N} \overset{0}{T}_{\alpha/2}(x,y)h(y)dy.$$

For $\alpha = 0$, $L_p^\alpha(E^N)$ is presumed to be coincident with $L_p(E^N)$. By definition, the norm in the class $L_p^\alpha(E^N)$ is taken equal to

$$||f||_{L_p^\alpha(E^N)} = ||h||_{L_p(E^N)}. \tag{2.1.17}$$

For integral α, the $L_p^\alpha(E^N)$ class coincides with the Sobolev class $W_p^\alpha(E^N)$ introduced above.

Thus, the Sobolev–Liouville class is a natural extension of the class W_p^α, initially suggested by Sobolev for all integrals spanning the real positive orders of differentiability α. We prefer to retain the notation W_p^α, originally introduced by S. L. Sobolev for the Sobolev–Liouville class with an integral order of differentiability.

For a more detailed characterization of all of the above classes, the reader is referred to the monograph by S. M. Nikol'skii [65].

Here we discuss only some of the properties of these classes which will be of interest in what follows.

We adopt the convention to designate by $A \to B$ the fact that a class A is embedded into a class B, keeping in mind that the operator responsible for this embedding is continuous. We can assert that $C^\alpha \to H_p^\alpha$ for any $p \geq 1$.[3)]

Further, given any $\alpha > 0$ and $p \geq 1$, and any $\theta \geq 1$, we have

$$L_p^\alpha \to H_p^\alpha, \qquad B_{p,\theta}^\alpha \to H_p^\alpha. \tag{2.1.18}$$

[3)] Everywhere in the sequel it is assumed that all the classes are taken over the entire E^N, and all the functions are finite in domain $G \subseteq E^N$.

One can assert, therefore, that of the four classes H_p^α, L_p^α, $B_{p,\theta}^\alpha$, and C^α for any fixed $\alpha > 0$ and $p \geq 1$ and for any $\theta \geq 1$ class H_p^α is wider and for any $\varepsilon > 0$ the embeddings

$$H_p^{\alpha+\varepsilon} \to L_p^\alpha, \qquad H_p^{\alpha+\varepsilon} \to B_{p,\theta}^\alpha \tag{2.1.19}$$

hold.

Note that, in defining the Nikol'skii class H_p and its norm, we can take the fourth differences $\Delta_u^4 f(x)$

$$\Delta_u^4 f(x) = f(x+2u) - 4f(x+u) + 6f(x) - 4f(x-u) + f(x-2u)$$

in place of the second differences $\Delta_u^2 f(x)$.

The norm

$$\|f\|_{H_p^\alpha(G)} = \|f\|_{L_p(G)} + \sum_{|\bar{k}|=\ell} \sup_u \left\{ |u|^{-\kappa} \left\| \Delta_u^4 \partial^{\bar{k}} f(x) \right\|_{L_p(G_{2|u|})} \right\}, \tag{2.1.15'}$$

defined in terms of the fourth difference, is equivalent to the norm (2.1.15).

We note further that the functions $f(x)$ from the Nikol'skii class $H_p^\alpha(G)$ can be represented as the serial sums of integral functions of exponential type[4)]

$$f(x) = \sum_{m=0}^{\infty} Q_m(x), \tag{2.1.20}$$

in which each term $Q_m(x)$ is an integral function of exponential type 2^m with a norm in $L_p(G)$ satisfying the inequality

$$\|Q_m(x)\|_{L_p(G)} \leq C\,2^{-m\alpha} \|f\|_{H_p^\alpha(G)} \tag{2.1.21}$$

(in this inequality, the constant C is independent of either the serial term number m or the function $f(x)$).

From the representation (2.1.20), inequality (2.1.21), and the familiar Bernstein inequality for $Q_m(x)$

$$\left\| \partial^{\bar{k}} Q_m(x) \right\|_{L_p(G)} \leq 2^{m|\bar{k}|} \|Q_m(x)\|_{L_p(G)} \tag{2.1.22}$$

the following assertion stems, which we shall make use of later. *If D is an arbitrary domain in E^N, and the symbol S_r^x is used to denote an N-dimensional sphere of radius r centered at point $x \in D_{3R}$ for any fixed $R > 0$, then for any p and α subject to $p \geq 1$, $p\alpha > 1$ uniformly with respect to r on a segment $R \leq r \leq 2R$, the inequality*

$$\|f\|_{L_p(S_r^x)} \leq C_R \|f\|_{H_p^\alpha(D)} \tag{2.1.23}$$

4) Concerning the integral functions of exponential type see, for example, the monograph of S. M. Nikol'skii [65, Chapter 3].

holds, where the constant C_R is independent of either the point $x \in D_{3R}$ or the function $f(x) \in H_p^\alpha(D)$.

For functions from the class H_p^α, a closed system of embedding theorems has been established by S. M. Nikol'skii.

We give here an embedding theorem which we shall need later and which is a generalization of the embedding theorem of S. L. Sobolev for the class W_p^α with integers α.

If p and p' are any two numbers satisfying the inequalities $1 \leq p \leq p' \leq \infty$, then, given $\alpha > N[p^{-1} - (p')^{-1}]$, the embedding

$$H_p^\alpha \to H_{p'}^{\alpha - N[p^{-1} - (p')^{-1}]} \tag{2.1.24}$$

holds such that for $\alpha' = \alpha - N[p^{-1} - (p')^{-1}]$ for any function $f \in H_p^\alpha$ we have

$$||f||_{H_{p'}^{\alpha'}} \leq C||f||_{H_p^\alpha}. \tag{2.1.25}$$

In particular, for $p' = \infty$ we obtain that with $\alpha p > N$

$$H_p^\alpha \to C^{\alpha - N/p} \qquad \text{and, the more so,} \qquad L_p^\alpha \to C^{\alpha - N/p}. \tag{2.1.26}$$

All of the above embedding theorems hold when all of the classes encountered therein are defined in the entire space E^N. If the classes on the left-hand side of theorems (2.1.24), (2.1.25), and (2.1.26) are defined in an arbitrary domain G, the formulated embedding theorems remain valid if the classes on the right-hand sides of (2.1.24), (2.1.25), and (2.1.26) are taken in a subset G_R of domain G for any fixed $R > 0$.

2.2. FORMULATION AND ANALYSIS OF MAIN RESULTS

2.2.1. Formulation of the Main Theorems and the Resulting Corollaries

Let G be an arbitrary domain in the space E^N, and $\widehat{A}$ an arbitrary self-adjoint nonnegative extension of the Laplace operator $Lu = -\Delta u$ in the domain G; let $\{E_\lambda\}$ be a family of the projectors of this extension, and $\theta(x, y, \lambda)$ be the kernel E_λ, that is, the spectral function for extension $\widehat{A}$.

For an arbitrary function $f(x)$ from the class $L_2(G)$, we intend to explore its spectral decomposition

$$E_\lambda f(x) = S_\lambda(x, f) = \int_G \theta(x, y, \lambda) f(y) dy \tag{2.2.1}$$

and the Riesz means of this spectral decomposition of order $s \geq 0$

$$E^s_\lambda f(x) = \sigma^s_\lambda(x, f) = \int_0^\lambda \left(1 - \frac{t}{\lambda}\right)^s dS_t(x, f). \tag{2.2.2}$$

We point out immediately that at $s = 0$, the Riesz means reduce to the spectral decomposition (2.2.1) whose study is thereby included in the study of the Riesz means of order $s \geq 0$.

In this chapter, we restrict ourselves to a study of the Riesz means (2.2.2) of order s satisfying the inequalities $0 \leq s < (N-1)/2$.

To begin, we turn to the conditions that fail to satisfy not only the uniform convergence, but also the localization of the Riesz means of spectral decompositions of the specified order s.

THEOREM 2.1. *(On conditions that fail to provide for the localization of the Riesz means in the Zygmund–Hölder classes). Assume that $N \geq 2$, $0 \leq s < (N-1)/2$, and G is an arbitrary domain in the space E^N, $\widehat{A}$ is an arbitrary self-adjoint nonnegative extension of the Laplace operator $Lu = -\Delta u$ in the domain G, x_0 is any fixed interior point of the domain G, and α is any fixed real number satisfying the inequalities $0 < \alpha < (N-1)/2 - s$. Then there exists a function $f(x)$ satisfying the following requirements:*

(1) $f(x)$ is finite in domain G and reduces to zero in a certain vicinity D of point x_0;

(2) $f(x)$ belongs to the class Zygmund–Hölder C^α with differentiability order α (both in domain G and the entire space E^N);

(3) the Riesz means (2.2.2) of order s of the spectral decomposition of function $f(x)$ have no limit at the point x_0 as $\lambda \to \infty$.

THEOREM 2.2. *(On conditions that fail to provide for the localization of the Riesz means in the Sobolev–Liouville, Nikol'skii, and Besov classes). Assume that $N \geq 2$, $0 \leq s < (N-1)/2$, G is an arbitrary domain in the space E^N, $\widehat{A}$ is an arbitrary self-adjoint nonnegative extension of the Laplace operator $Lu = -\Delta u$ in the domain G, x_0 is a fixed interior point of the domain G, and α is a fixed real number subject to $0 < \alpha < (N-1)/2 - s$. Then there exists a function $f(x)$ satisfying the following requirements:*

(1) $f(x)$ is finite in domain G and becomes zero in a certain vicinity D of point x_0;

(2) $f(x)$ belongs (both in domain G and the entire space E^N) to each of the Sobolev–Liouville L^α_p, Nikol'skii H^α_p, and Besov $B^\alpha_{p,\theta}$ classes with order of differentiability α, with any degree of summability $p \geq 1$, and (in the case of Besov class), with any $\theta \geq 1$;

(3) the Riesz means (2.2.2) of order s of the spectral decomposition of function $f(x)$ have no limit at point x_0 as $\lambda \to \infty$.

Sections 2.3 and 2.4 will be concerned with a proof of Theorem 2.1. Here we ensure that Theorem 2.2 is a direct consequence of Theorem 2.1.

Indeed, for any s in a half-segment $0 \leq s < (N-1)/2$ we fix an arbitrary α satisfying the condition $0 < \alpha < (N-1)/2 - s$. Since $\alpha < (N-1)/2 - s$, the difference $((N-1)/2 - s) - \alpha$ is a positive number which we denote by 2ε, so that $\alpha + 2\varepsilon = (N-1)/2 - s$. If so, the number $\alpha' = \alpha + \varepsilon$ satisfies the inequalities $0 < \alpha < \alpha + \varepsilon < (N-1)/2 - s$ and, by virtue of Theorem 2.1, there is a function $f(x)$ from the Zygmund-Hölder class $C^{\alpha+\varepsilon}$ (both in domain G and the entire E^N), finite in domain G and becoming zero in a certain vicinity D of the point x_0 such that the Riesz means (2.2.2) of order s of its spectral decomposition have no limit at point x_0 as $\lambda \to \infty$.

It remains to observe that the Zygmund–Hölder class $C^{\alpha+\varepsilon}$ is contained at any $p \geq 1$ in the Nikol'skii class $H_p^{\alpha+\varepsilon}$ and, therefore, is contained both in the class H_p^{α} and, by virtue of the embedding theorems (2.1.18′), in each of the classes L_p^{α} and $B_{p,\theta}^{\alpha}$ at any $p \geq 1$ and at any $\theta \geq 1$.

Theorems 2.1 and 2.2 provide evidence that if the order of differentiability α in each of the classes C^{α}, L_p^{α}, H_p^{α}, and $B_{p,\theta}^{\alpha}$ is less than $(N-1)/2-s$, then, whatever may be the degree of summability $p \geq 1$ and the number $\theta \geq 1$, one will not expect (for any self-adjoint nonnegative extension of the Laplace operator) either uniform convergence or even localization of the Riesz means of order s of the spectral decomposition to take place for a finite-in-the-domain function that may belong to any of the above four classes.

A natural question arises if there is a route to the study of uniform convergence and localization of the Riesz means of order s of the spectral decomposition for finite-in-domain G functions belonging to one of the four classes C^{α}, L_p^{α}, H_p^{α}, or $B_{p,\theta}^{\alpha}$ with the order of differentiability α subject to the condition $(N-1)/2 - s$.

In a definite sense, an exaustive answer to this question is given by the following three theorems.

THEOREM 2.3. *(On conditions that provide for the localization and on conditions that provide for the uniform convergence of the Riesz means of the Nikol'skii classes). Assume that $N \geq 2$, $0 \leq s < (N-1)/2$, and G is an arbitrary domain in the space E^N, $\widehat{A}$ is an arbitrary self-adjoint nonnegative extension of the Laplace operator $Lu = -\Delta u$ in the domain G, $f(x)$ is an arbitrary function satisfying the following requirements:*

(1) $f(x)$ becomes zero outside the set G_{h_0} at a certain $h_0 > 0$[5]*;*

(2) in the entire domain G the function $f(x)$ belongs to the Nikol'skii class H_2^{α} at $\alpha \geq (N-1)/2 - s$;

(3) in a certain domain D interior to G (which, accidentally, may coincide with G), the function $f(x)$ belongs to the Nikol'skii class H_p^{α} at certain α and p satisfying

[5] We recall that for any $h > 0$ the symbol D_h denotes a subset of points of an arbitrary domain D removed from the boundary ∂D of domain D to a distance greater than h.

the conditions

$$\alpha \geq (N-1)/2 - s, \qquad p\alpha > N, \quad p \geq 1. \tag{2.2.3}$$

Then for each $h > 0$, the Riesz means (2.2.2) of order s of the spectral decomposition of a function $f(x)$ converge to $f(x)$ as $\lambda \to \infty$ uniformly on the set D_h.

A comparison of Theorem 2.2 with Theorem 2.3 leads to the following conclusions:

I. A definitive condition for localization of the Riesz means of order s ($0 \leq s < (N-1)/2$) of the spectral decomposition of a finite-in-G function of the Nikol'skii class $H_2^\alpha(G)$ is the requirement $\alpha \geq (N-1)/2 - s$ (at $\alpha \geq (N-1)/2 - s$, according to Theorem 2.3, localization of the said Riesz means takes place; at $\alpha < (N-1)/2 - s$, according to Theorem 2.2, no localization of the Riesz means occurs).

II. A definitive condition for uniform convergence of the Riesz means of order s ($0 \leq s < (N-1)/2$) of the spectral decomposition of a finite-in-G function $f(x)$ of the Nikol'skii class $H_p^\alpha(G)$ is the fulfillment of the three inequalities (2.2.3).

Indeed, if the first inequality in (2.2.3) fails to hold, that is, if $\alpha < (N-1)/2 - s$, and at any $p \geq 1$, then neither uniform convergence nor localization of the Riesz means takes place for any finite-in-G function of the Nikol'skii class $H_p^\alpha(G)$ by virtue of Theorem 2.2.

The second inequality in (2.2.3), that is, the inequality $p\alpha > N$, is also definitive since there exists, at $p\alpha = N$ and at $p \geq 1$, in the Nikol'skii class an unbounded function (or, to be more exact, a function with singularity at an interior point) for which the Riesz means fail *a fortiori* to uniformly converge in a domain containing this point.

The third inequality in (2.2.3), that is, $p \geq 1$, is self-evident and need not be discussed.

THEOREM 2.4. *(On the conditions that provide for localization and on conditions that provide for uniform convergence of the Riesz means in the Sobolev–Liouville class and in the Besov classes). Let $N \geq 2$, $0 \leq s < (N-1)/2$, let G be an arbitrary domain in the space E^N, let $\hat{A}$ be an arbitrary self-adjoint extension of the Laplace operator $Lu = -\Delta u$ in the domain G, and let $f(x)$ be an arbitrary function satisfying the following requirements:*

(1) $f(x)$ becomes zero outside the set G_{h_0} for certain $h_0 > 0$;

(2) in the entire domain G, the function belongs to the Sobolev–Liouville class L_2^α [accordingly, to the Besov class $B_{2,\theta}^\alpha$] at $\alpha \geq (N-1)/2 - s$ [and at any $\theta \geq 1$];

(3) in a certain domain D interior to G (in particular, accidentally coincident with G), the function $f(x)$ belongs to the Sobolev–Liouville class L_p [accordingly, to the Besov class $B_{p,\theta}^\alpha$] at certain α and p satisfying the three inequalities from (2.2.3) [and in the case of the Besov class at any $\theta \geq 1$]. Then for each $h > 0$, the Riesz means (2.2.2) of order s of the spectral decomposition of the function $f(x)$ converge to $f(x)$ as $\lambda \to \infty$ uniformly on the set D_h.

A comparison of Theorem 2.4 with Theorem 2.2 yields the same inferences as those for the Nikol'skii class. The condition $\alpha \geq (N-1)/2 - s$ is a definitive condition for localization of the Riesz means of order s ($0 \leq s < (N-1)/2$) of the spectral decomposition of a finite-in-G function from the Sobolev–Liouville class $L_2^\alpha(G)$ and the Besov class $B_{2,\theta}^\alpha(G)$ (for any $\theta \geq 1$). The three inequalities in (2.2.3) are a definitive condition for uniform convergence of the Riesz means of order s ($0 \leq s < (N-1)/2$) of the spectral decomposition of a finite-in-G function from the Sobolev–Liouville class $L_p^\alpha(G)$ and the Besov class $B_{p,\theta}^\alpha(G)$ (for any $\theta \geq 1$).

THEOREM 2.5. *(On the conditions that provide for localization and uniform convergence of the Riesz means in the Zygmund–Hölder class). Let $N \geq 2$, $0 \leq s < (N-1)/2$, let G be an arbitrary domain in the space E^N, $\hat{A}$ be an arbitrary self-adjoint nonnegative extension of the operator $Lu = -\Delta u$ in the domain G, and $f(x)$ be an arbitrary function satisfying the following conditions:*

(1) $f(x)$ goes to zero outside the set G_{h_0} at certain $h_0 > 0$;

(2) $f(x)$ belongs to the Zygmund–Hölder class $C^\alpha(G)$ at certain $\alpha \geq (N-1)/2-s$.

Then, for each $h > 0$, the Riesz means (2.2.2) of order s of the spectral decomposition of the function $f(x)$ converge to $f(x)$ as $\lambda \to \infty$ uniformly on the set G_h.

A comparison of Theorem 2.5 with Theorem 2.1 enables one to conclude that the requirement $\alpha \geq (N-1)/2 - s$ is a definitive condition for both the localization and uniform convergence of the Riesz means of order s ($0 \leq s < (N - 1/2)$ of the spactral decomposition of a finite-in-domain G function $f(x)$ from the Zygmund–Hölder class $C^\alpha(G)$ (at $\alpha < (N-1)/2-s$, according to Theorem 2.1, the localization and, the more so, the uniform convergence of the Riesz means will never hold; however, in accordance with Theorem 2.5, the uniform convergence and, the more so, the localization of the Riesz means in question will take place at $\alpha \geq (N-1)/2 - s$).

Thus, the Zygmund–Hölder classes are incapable of resolving the "gap" between the definitive conditions of localization and the definitive conditions for uniform convergence: at $\alpha < (N-1)/2-s$, there is neither localization nor uniform convergence; by contrast, at $\alpha \geq (N-1)/2 - s$, there is uniform convergence and, the more so, localization of the Riesz means of order s.

Section 2.5 will be dedicated to a proof of Theorem 2.3. Here we wish to point out that Theorems 2.4 and 2.5 are merely corollaries to Theorem 2.3, since at any fixed $\alpha > 0$ and $p \geq 1$ and at any $\theta \geq 1$ by virtue of the embeddings (2.1.18) (see Section 2.1.3) each of the classes $L_p^\alpha(G)$ and $B_{p,\theta}^\alpha(G)$ is contained in $H_p^\alpha(G)$ and, apart from this, the Zygmund–Hölder class $C^\alpha(G)$ is contained in the Nikol'skii class $H_p^\alpha(G)$ at any $p \geq 1$.

2.2.2. Brief Analysis of Results

We denote by the symbol $A_p^\alpha(G)$ any of the three Nikol'skii $H_p^\alpha(G)$, Sobolev–Liouville $L_p^\alpha(G)$, or Besov $B_{p,\theta}^\alpha(G)$ classes with an arbitrary $\theta \geq 1$.

The main results of the present chapter are as follows:

(1) A definitive condition for the localization of the Riesz means of order s ($0 \leq s < (N-1)/2$) of the spectral decomposition of a finite-in-G function from the class

$A_2(G)$ has been established as expressed by the inequality $\alpha \geq (N-1)/2 - s$;

(2) A definitive condition for the uniform convergence of the Riesz means of order s $(0 \leq s < (N-1)/2)$ of the spectral decomposition of a finite-in-G function from the class $A_p^\alpha(G)$ has been established as expressed by the three inequalities

$$\alpha \geq \frac{N-1}{2} - s, \qquad p\alpha > N, \qquad p \geq 1; \tag{2.2.3}$$

(3) A definitive condition for both the localization and uniform convergence of the Riesz means of order s $(0 \leq s < (N-1)/2)$ of the spectral decomposition of a finite-in-G function from the Zygmund–Hölder class $C^\alpha(G)$ has been established as expressed by the inequality $\alpha \geq (N-1)/2 - s$.

All these results are definitive not only in the class of all self-adjoint nonnegative extensions of the Laplace operator, but also for each individual self-adjoint nonnegative extension of this operator[6] (and, in particular, for the expansion into an N-fold Fourier integral corresponding to the self-adjoint nonnegative extension of the Laplace operator in the entire space E^N).

We will now proceed as follows. In Section 2.3 we establish certain useful supplementary properties of the fundamental functions of an arbitrary ordered representation of the space L_2; in Section 2.4, we give a proof of Theorem 2.1, and in Section 2.5, a proof of Theorem 2.3.

2.3. CERTAIN PROPERTIES OF THE FUNDAMENTAL FUNCTIONS OF AN ARBITRARY ORDERED SPECTRAL REPRESENTATION IN THE SPACE L_2

We consider an arbitrary self-adjoint nonnegative extension $\widehat{A}$ of the Laplace operator $Lu = -\Delta u$ in an arbitrary N-dimensional domain G and an arbitrary ordered spectral representation of the space $L_2(G)$ with respect to the extension $\widehat{A}$ with spectral measure $\rho(\lambda)$, sets of multiplicity e_i, fundamental functions $u_i(x, \lambda)$ $(i = 1, 2, \ldots, \widehat{m})$ and multiplicity $\widehat{m} \leq \infty$.

In this section we establish certain useful properties of the fundamental functions $u_i(x, \lambda)$ which resemble the properties of elements $u_n(x)$ of an arbitrary FSF of the Laplace operator that we derived in Sections 1.1 and 1.2.

2.3.1. The Mean-Value Formula. Expression for the Fourier Image of a Function from the Class of Radial Functions

Let x be any point of an open domain G, r be any positive number such that an N-dimensional ball of radius r centered at point x is interior to domain G. Then

[6] This follows from Theorem 2.1.

for each fundamental function $u_i(x,\lambda)$ corresponding to the value of λ, the following mean-value formula holds:[7)]

$$\int_{\omega}\cdots\int u_i(x+r\omega,\lambda)d\omega = (2\pi)^{N/2}u_i(x,\lambda)\left(r\sqrt{\lambda}\right)^{(2-N)/2} J_{(N-2)/2}\left(r\sqrt{\lambda}\right). \quad (2.3.1)$$

The derivation of formula (2.3.1), based merely on the fact that the fundamental function $u_i(x,\lambda)$ is a regular solution of the equation $\Delta u_i + \lambda u_i = 0$ in domain G, repeats word for word the derivation of the mean-value formula as expounded in Section 1.1.1.

Assume that f is an arbitrary function which belongs in domain G to a class of radial functions, that is, the function is dependent on the distance $r = |x-y|$ between a variable point y of domain G and a fixed point x of this domain, and is different from zero only inside an N-dimensional ball of radius R interior to domain G. Then for the Fourier image $\widehat{f}_i(\lambda)$ of this function as defined by relation (2.1.8), the following equality holds:

$$\widehat{f}_i(\lambda) = (2\pi)^{N/2}u_i(x,\lambda)\lambda^{(2-N)/4}\int_0^R f(r)r^{N/2}J_{(N-2)/2}\left(r\sqrt{\lambda}\right)dr \quad (2.3.2)$$

$$(i = 1, 2, \dots, \widehat{m}).$$

To verify (2.3.2), it suffices to observe that, by virtue of the above-mentioned relation (2.1.8), the Fourier image $\widehat{f}_i(\lambda)$ is

$$\widehat{f}_i(\lambda) = \int_G f(|x-y|)u_i(y,\lambda)dy = \int_0^R f(r)r^{N-1}\left(\int_{\omega}\cdots\int u_i(x+r\omega,\lambda)d\omega\right)dr$$

$$(i = 1, 2, \dots, \widehat{m})$$

and to make use of the mean-value formula (2.3.1).

2.3.2. Integral Estimate for the Square of the Fundamental Function

LEMMA 2.1. *For any $\mu \geq 0$ uniformly with respect to x in each subdomain G' strictly interior to domain G, the following estimate holds:*

$$\sum_{i=1}^{\widehat{m}} \int_{\mu\leq\sqrt{\lambda}\leq\mu+1} u_i^2(x,\lambda)d\rho(\lambda) = O[(\mu+1)^{N-1}]. \quad (2.3.3)$$

7) The symbol $\int_{\omega}\cdots\int u_i(x+r\omega)d\omega$, similar to that in Chapter 1, signifies an integral of the function u_i over all the angles on the surface of an N-dimensional sphere of radius r centered at point x.

The derivation of estimate (2.3.3) is completely identical to the derivation of estimate (1.1.4); therefore we restrict ourselves merely to a brief comment.

It will amply suffice to prove, by analogy with the derivation of estimate (1.1.4), the following two assertions:

(1) the validity of estimate (2.3.3) at $\mu \geq \mu_0$, where μ_0 is a fixed, sufficiently large number;

(2) the validity, at $\mu_0 \geq 1$, of a uniform-with-respect-to-x estimate in G',

$$\sum_{i=1}^{\widehat{m}} \int\limits_{\sqrt{\lambda}\leq\mu_0} u_i^2(x,\lambda)d\rho(\lambda) = O(\mu_0^N). \tag{2.3.4}$$

(1) To begin, let $\mu > \mu_0$, where μ_0 is a sufficiently large fixed number. We further fix an arbitrary subdomain G' strictly interior to domain G and denote by R a positive number smaller than the distance between G' and the boundary of domain G. We take that x is an arbitrary fixed point in subdomain G' and consider the same function $v(r)$ as defined by equality (1.1.5) (see Section 1.1.2), assuming that in this function $r = |x - y|$ is the distance between a variable point y and the fixed point x.

With the aid of relation (2.3.2), we obtain that the Fourier image $\widehat{v}_i(x,\lambda)$ of this function is

$$\widehat{v}_i(x,\lambda) = \mu^{N/2}\lambda^{(2-N)/4}u_i(x,\lambda)\int\limits_{R/2}^{R} J_{(N-2)/2}(r\mu)J_{(N-2)/2}(r\sqrt{\lambda})r\,dr \tag{2.3.5}$$

$$(i = 1, 2, \ldots, \widehat{m}).$$

Making use of relation (2.3.5) in much the same manner as in Section 2.1.2, we will easily prove that if the number μ_0 is sufficiently large, then for $\mu \geq \mu_0$ for all λ such that $\mu \leq \sqrt{\lambda} \leq (\mu + 1)$ there is a constant α such that

$$|\widehat{v}_i(x,\lambda)| \geq \alpha|u_i(x,\lambda)| \qquad (i = 1, 2, \ldots, \widehat{m}).$$

Now, having written the Parseval equality (2.1.11) (see Section 2.1.2) for the function $v(|x - y|) = f(y)$,

$$\int\limits_G v^2 dy = \sum_{i=1}^{\widehat{m}} \int\limits_0^\infty \widehat{v}_i^2(x,\lambda)d\rho(\lambda),$$

we obtain, with the aid of the latter inequality and taking into account that $\mu \geq \mu_0 \geq 1$, the estimate

$$\sum_{i=1}^{\widehat{m}} \int\limits_{\mu\leq\sqrt{\lambda}\leq\mu+1} u_i^2(x,\lambda)d\rho(\lambda) \leq \alpha^{-2}\int\limits_G v^2(|x - y|)dy$$

$$= \mu^N \omega_N (2\pi)^{-N} \alpha^{-2} \int_{R/2}^{R} r J^2_{(N-2)/2}(r\mu) dr = O[(\mu+1)^{N-1}].$$

Thereby the estimate (2.3.3) for $\mu \geq \mu_0$ is established.

(2) We ensure now that, for $\mu_0 \geq 1$, the estimate (2.3.4) holds uniformly with respect to x in subdomain G'. To that end, we fix an arbitrary subdomain G' strictly interior to domain G, an arbitrary positive number R less than the distance between subdomain G' and the boundary of domain G, and we also fix an arbitrary point x in G'; we further consider the same function $w(r)$ as defined by Eq. (1.1.18) (see Section 1.1.2), assuming that $r = |x - y|$ in this function is the distance between a variable point y and the fixed point x. With the aid of relation (2.3.2), we obtain that the Fourier image $\widehat{w}_i(x, \lambda)$ of this function is

$$\widehat{w}_i(x,\lambda) = 2^{N/2}\Gamma\left(\frac{N}{2}+1\right)\frac{J_{N/2}(R\sqrt{\lambda})}{(R\sqrt{\lambda})^{N/2}} u_i(x,\lambda) \qquad (i = 1, 2, \ldots, \widehat{m}).$$

Having written the Parseval equality (2.1.11) (see Section 2.1.2) for the function $w(|x-y|) = f(y)$, we obtain

$$\sum_{i=1}^{\widehat{m}} \int_0^\infty \left\{ 2^{N/2}\Gamma\left(\frac{N}{2}+1\right)\frac{J_{N/2}(R\sqrt{\lambda})}{(R\sqrt{\lambda})^{N/2}} \right\}^2 u_i^2(x,\lambda) d\rho(\lambda)$$
$$= \omega_N \Gamma^2\left(\frac{N}{2}+1\right) \pi^{-N} R^{-2N} \int_0^R r^{N-1} dr = \Gamma\left(\frac{N}{2}+1\right)\pi^{-N} R^{-N}. \qquad (2.3.6)$$

It has been proved in Section 1.1.2 that for $R_0 = (2\mu_0)^{-1}$ and for all $\sqrt{\lambda} \leq \mu_0$ the quantity enclosed in braces in (2.3.6) exceeds a certain positive number α_0. In this case, the equality

$$\alpha_0^2 \sum_{i=1}^{\widehat{m}} \int_{\sqrt{\lambda} \leq \mu_0} u_i^2(x,\lambda) d\rho(\lambda) = O(\mu_0^N)$$

follows from (2.3.6), equivalent to estimate (2.3.4).

This completes the derivation of estimate (2.3.3) for any $\mu \geq 0$.

Two simple corollaries follow from Lemma 2.1.

COROLLARY 1. *For any $\mu \geq 1$ and for any ρ_0 within a segment $1 \leq \rho_0 \leq \mu$, the estimate*

$$\sum_{i=1}^{\widehat{m}} \int_{|\sqrt{t}-\mu| \leq \rho_0} u_i^2(x,t) d\rho(t) = \rho_0 O(\mu^{N-1}) \qquad (2.3.7)$$

holds uniformly with respect to x in each subdomain G' strictly interior to domain G.

To prove Corollary 1, it suffices (invoking the additivity of an integral), to divide the segment $[\mu - \rho_0, \mu + \rho_0]$, within which $\sqrt{\lambda}$ is allowed to vary, into subsegments of length not exceeding unity and not sharing common interior points; the number of such segments is not larger than $[2\rho_0] + 1$, where $[2\rho_0]$ is the integer part of the number $2\rho_0$. Applying estimate (2.3.3) to each of the subsegments, we arrive at estimate (2.3.7).

COROLLARY 2. *For any $\delta > 0$ and for all $\lambda \geq 1$, the estimates*

$$\sum_{i=1}^{\widehat{m}} \int_1^{\lambda} u_i^2(x,t) t^{\delta - N/2} d\rho(t) = O(\lambda^{\delta}), \tag{2.3.8}$$

$$\sum_{i=1}^{\widehat{m}} \int_{\lambda}^{\infty} u_i^2(x,t) t^{-\delta - N/2} d\rho(t) = O(\lambda^{-\delta}) \tag{2.3.9}$$

hold uniformly with respect to x in each subdomain G' strictly interior to domain G.

In particular, one may assert that for any $\delta > 0$ the quantity

$$\sum_{i=1}^{\widehat{m}} \int_1^{\infty} u_i^2(x,t) t^{-\delta - N/2} d\rho(t), \tag{2.3.10}$$

uniform with respect to x in subdomain G', is bounded.

To prove estimate (2.3.8), we observe that the quantity on the left-hand side of (2.3.8) does not exceed

$$\sum_{i=1}^{\widehat{m}} \int_1^{([\sqrt{\lambda}]+1)^2} u_i^2(x,t) t^{\delta - N/2} d\rho(t) = \sum_{k=1}^{[\sqrt{\lambda}]} \left\{ \sum_{i=1}^{\widehat{m}} \int_{k \leq \sqrt{t} \leq k+1} u_i^2(x,t) t^{\delta - N/2} d\rho(t) \right\}, \tag{2.3.11}$$

where the symbol $[\sqrt{\lambda}]$ denotes the integral part of $\sqrt{\lambda}$.

Since, for all t satisfying the condition $k \leq \sqrt{t} \leq k+1$, the inequality $t^{\delta - N/2} \leq C_0 k^{2\delta - N}$ holds, where $C_0 = \max\{1, 2^{2\delta - N}\}$ (see Section 1.1.2), the right-hand side of (2.3.11) is majorized by the quantity

$$C_0 \sum_{k=1}^{[\sqrt{\lambda}]} k^{2\delta - N} \left[\sum_{i=1}^{\widehat{m}} \int_{k \leq \sqrt{t} \leq k+1} u_i^2(x,t) d\rho(t) \right].$$

Invoking estimate (2.3.3) for the quantity enclosed in square brackets, we obtain that the right-hand side of (2.3.11) is majorized by the sum

$$CC_0 \sum_{k=1}^{[\sqrt{\lambda}]} k^{2\delta - N} (k+1)^{N-1} = O\left(\left[\sqrt{\lambda}\right]^{2\delta} \right) = O(\lambda^{\delta}),$$

which completes the derivation of estimate (2.3.8).

To obtain estimate (2.3.9), we represent the left-hand side of (2.3.9) in the form

$$\sum_{k=0}^{\infty}\left\{\sum_{i=1}^{\widehat{m}}\int\limits_{\sqrt{\lambda}+k\le\sqrt{t}\le\sqrt{\lambda}+k+1} u_i^2(x,t)t^{-\delta-N/2}d\rho(t)\right\}. \tag{2.3.12}$$

Since at $\sqrt{\lambda}+k \le \sqrt{t} \le \sqrt{\lambda}+k+1$ the inequality $t^{-\delta-N/2} \le (\lambda+k)^{-2\delta-N}$ is valid, quantity (2.3.12) does not exceed the sum

$$\sum_{k=0}^{\infty}(\sqrt{\lambda}+k)^{-2\delta-N}\left[\sum_{i=1}^{\widehat{m}}\int\limits_{\sqrt{\lambda}+k\le\sqrt{t}\le\sqrt{\lambda}+k+1} u_i^2(x,t)d\rho(t)\right],$$

which, by virtue of estimate (2.3.3), is not larger than

$$C\sum_{k=0}^{\infty}(\sqrt{\lambda}+k)^{-2\delta-N}(\sqrt{\lambda}+k+1)^{N-1}. \tag{2.3.13}$$

In Section 1.1.2 we have shown, in proving Corollary 2 to Theorem 1.1, that the sum (2.3.13) is $O(\lambda^{-\delta})$. This completes the proof of Corollary 2.

2.3.3. Fractional Kernels

We consider once again an arbitrary self-adjoint nonnegative extension $\widehat{A}$ of the Laplace operator $Lu = -\Delta u$ in an arbitrary N-dimensional domain G; with respect to this extension, we also consider an arbitrary ordered spectral representation of the space $L_2(G)$ with spectral measure $\rho(\lambda)$, sets of multiplicity e_i, fundamental functions $u_i(y,\lambda)$, and multiplicity $\widehat{m} \le \infty$.

Further, we fix an arbitrary $h > 0$ and an arbitrary point x of the set G_h.[8] By a *kernel* of real positive order α we mean a function $T_\alpha(x,y)$ such that its Fourier image with respect to the fundamental function $u_i(y,\lambda)$ takes the form

$$u_i(x,\lambda)(1+\lambda)^{-\alpha}.$$

LEMMA 2.2. *For any real $\alpha > 0$, there exists a kernel $T_\alpha(x,y)$ representable in the form*

$$T_\alpha(x,y) = A_N^\alpha |x-y|^{\alpha-N/2}K_{N/2-\alpha}(|x-y|) + \chi_\alpha(x,y), \tag{2.3.14}$$

[8] We recall that the symbol G_h denotes a subset G all of whose points are removed from the boundary of G to a distance greater than $h > 0$.

where A_N is a constant defined by the equality

$$A_N^\alpha = 2^{1-\alpha}[(2\pi)^{N/2}\Gamma(\alpha)]^{-1},$$

and the symbol $K_\nu(r)$ denotes a MacDonald function of order ν of argument r, and $\chi_\alpha(x,y)$ is a function possessing, in an open domain G, continuous partial derivatives of arbitrary high order with respect to the coordinates of points x and y.

Lemma 2.2 is proved in much the same manner as Theorem 1.6 in Section 1.2.1. Here we restrict ourselves to a brief comment.

Having fixed an arbitrary $h > 0$, an arbitrary point x of the set G_h, and an arbitrary number m, we consider the same functions $v_\alpha(r)$ and $w_\alpha(r)$ of the distance $r = |x-y|$ between a variable point y and a fixed point x, taken at $R = h$ and defined by relations (1.2.4) and (1.2.5) in Section 1.2.1:

$$v_\alpha(r) = \begin{cases} A_N^\alpha r^{\alpha-N/2}K_{N/2-\alpha}(r) & \text{for } r \le h, \\ 0 & \text{for } r > h; \end{cases}$$

$$w_\alpha(r) = \begin{cases} \sum\limits_{k=0}^{m} a_k r^{2k} & \text{for } r \le h, \\ 0 & \text{for } r > h. \end{cases} \tag{2.3.15}$$

Setting $f(r) = v_\alpha(r) - w_\alpha(r)$ and making use of relation (2.3.2), we obtain the following expression for the Fourier image $\widehat{f}_i(x,\lambda)$ of the function $f(|x-y|)$:

$$\widehat{f}_i(x,\lambda) = (2\pi)^{N/2}\lambda^{(2-N)/4}u_i(x,\lambda)\int_0^h [v_\alpha(r) - w_\alpha(r)]r^{N/2}J_{(N-2)/2}(r\sqrt{\lambda})dr$$

$$(i = 1,2,\dots,\widehat{m}).$$

We integrate the right-hand side of the last equality $(m+1)$ times by parts and, performing the same manipulations as in the proof of Theorem 1.6 (see Section 1.2.1), we obtain, for the Fourier image $\widehat{f}_i(x,\lambda)$, the following expression:

$$\widehat{f}_i(x,\lambda) = \lambda^{-N/4-m/2}2^{1-\alpha}[\Gamma(\alpha)]^{-1}u_i(x,\lambda)$$
$$\times \int_0^h r^\alpha J_{N/2+m}(r\sqrt{\lambda})K_{N/2-\alpha+m+1}(r)dr \qquad (i = 1,2,\dots,\widehat{m}). \tag{2.3.16}$$

Making use of the known value of the integral

$$\lambda^{-N/4-m/2}2^{1-\alpha}[\Gamma(\alpha)]^{-1}\int_0^\infty r^\alpha J_{N/2+m}(r\sqrt{\lambda})K_{N/2-\alpha+m+1}(r)dr = (1+\lambda)^{-\alpha},$$

we arrive at another expression for the Fourier image $\widehat{f}_i(x,\lambda)$:

$$\widehat{f}_i(x,\lambda) = u_i(x,\lambda)(1+\lambda)^{-\alpha} - u_i(x,\lambda)\lambda^{-N/4-m/2}2^{1-\alpha}[\Gamma(\alpha)]^{-1}$$

$$\times \int_h^\infty r^\alpha J_{N/2+m}(r\sqrt{\lambda})K_{N/2-\alpha+m+1}(r)dr, \quad (i = 1,2,\dots,\widehat{m}). \tag{2.3.17}$$

Expressions (2.3.16) and (2.3.17) can be written in the form

$$\widehat{f}_i(x,\lambda) = u_i(x,\lambda)(1+\lambda)^{-\alpha} + \overset{m}{\gamma}(\lambda)u_i(x,\lambda) \qquad (i = 1,2,\dots,\widehat{m}), \tag{2.3.18}$$

where $\overset{m}{\gamma}(\lambda)$ is independent of either x, y, or i and is defined by the following two formulas:

$$\overset{m}{\gamma}(\lambda) = \lambda^{-N/4-m/2}2^{1-\alpha}[\Gamma(\alpha)]^{-1}\int_h^\infty r^\alpha J_{N/2+m}(r\sqrt{\lambda})K_{N/2-\alpha+m+1}(r)dr,$$

$$\overset{m}{\gamma}(\lambda) = -(1+\lambda)^{-\alpha} + \lambda^{-N/4-m/2}2^{1-\alpha}[\Gamma(\alpha)]^{-1}\int_0^h r^\alpha J_{N/2+m}(r\sqrt{\lambda})K_{N/2-\alpha+m+1}(r)dr.$$

From these two formulas in much the same manner as it has been done in Section 1.2.1, the following two estimates are derived for $\overset{m}{\gamma}(\lambda)$:

(a) for all $\lambda \geq 1$

$$\left|\overset{m}{\gamma}(\lambda)\right| \leq C(h,\alpha,m,N)(1+\lambda)^{-N/4-m/2}; \tag{2.3.19}$$

(b) for all $\lambda \geq 0$

$$\left|\overset{m}{\gamma}(\lambda)\right| \leq C_1(h,\alpha,m,N). \tag{2.3.20}$$

If, in the same manner as in Section 1.2.1, we introduce a function $\varphi_\alpha(|x-y|)$ defined as

$$\varphi_\alpha(|x-y|) = \begin{cases} -w_\alpha(|x-y|) & \text{for} \quad |x-y| \leq h, \\ -A_N^\alpha|x-y|^{\alpha-N/2}K_{N/2-\alpha}(|x-y|) & \text{for} \quad |x-y| > h, \end{cases} \tag{2.3.21}$$

then the function $f = v_\alpha - w_\alpha$ everywhere in domain G can be represented in the form

$$f = A_N^\alpha|x-y|^{\alpha-N/2}K_{N/2-\alpha}(|x-y|) + \varphi_\alpha(|x-y|), \tag{2.3.22}$$

where the function $\varphi_\alpha(|x-y|)$, by its construction, possesses at $x \in G_h$, $y \in G$ continuous partial derivatives both with respect to the coordinates of point x and with respect to the coordinates of point y up to order m inclusively.

It follows from relation (2.3.18) and from equality (2.3.22) that the function $T_\alpha(x,y)$, which has the quantity $u_i(x,\lambda)(1+\lambda)^{-\alpha}$ for its Fourier image, is representable in the form

$$T_\alpha(x,y) = A_N^\alpha |x-y|^{\alpha-N/2} K_{N/2-\alpha}(|x-y|) + \varphi_\alpha(|x-y|) - \Psi_\alpha(x,y), \tag{2.3.23}$$

where $\varphi_\alpha(x-y)$ is a function of the form (2.3.21) possessing at $x \in G_h$, $y \in G$ continuous partial derivatives both with respect to the coordinates x and with respect to the coordinates y up to order m inclusively; m is an arbitrary number we have fixed and $\Psi_\alpha(x,y)$ is a function whose spectral decomposition is

$$\Psi_\alpha(x,y) = \sum_{i=1}^{\widehat{m}} \int_0^\infty \overset{m}{\gamma}(\lambda) u_i(x,\lambda) u_i(y,\lambda) d\rho(\lambda). \tag{2.3.24}$$

We have placed an equality sign in (2.3.24) since it is our intention to prove that not only the integral itself on the right-hand side of (2.3.24), but also the integrals obtained by its formal differentiation (under the sign of the integral) with respect to the coordinates of points x and y up to an order growing indefinitely with increasing m converge uniformly at $x \in G_h$, $y \in G$.

This completes the proof of Lemma 2.2, since the function $\chi_\alpha(x,y)$ in (2.3.14) can be taken as the difference of two functions $\varphi_\alpha(|x-y|) - \Psi_\alpha(x,y)$, each of these having at $x \in G_h$, $y \in G_h$ continuous partial derivatives with respect to the coordinates x and y up to an order growing indefinitely with increasing m.

Thus, it remains to prove that, given arbitrarily fixed numbers p and q and a sufficiently large m, the integral

$$\sum_{i=1}^{\widehat{m}} \int_0^\infty \overset{m}{\gamma}(\lambda) u_i^{(p)}(x,\lambda) u_i^{(q)}(y,\lambda) d\rho(\lambda) \tag{2.3.25}$$

converges uniformly with respect to the ensemble (x,y) on the set $(G_h \times G_h)$; here $u_i^{(p)}(x,\lambda)$ denotes any partial derivative of the fundamental function $u_i(x,\lambda)$ with respect to the coordinates of the point x of order p, and $u_i^{(q)}(y,\lambda)$ denotes any partial derivative of the fundamental function $u_i(y,\lambda)$ with respect to the coordinates of the point y of order q.

By virtue of the Cauchy–Buniakowski inequality, to this effect it will suffice to prove two assertions:

(1) for an arbitrary fixed number p and for all sufficiently large m, the integral

$$\sum_{i=1}^{\widehat{m}} \int_0^\infty \left|\overset{m}{\gamma}(\lambda)\right| |u_i^{(p)}(x,\lambda)|^2 \, d\rho(\lambda) \tag{2.3.26}$$

converges uniformly with respect to x on the set G_h;

(2) for an arbitrary fixed number q and for all sufficiently large m, the integral $\sum_{i=1}^{\widehat{m}}\int_0^\infty \left|\overset{m}{\gamma}(\lambda)\right| |u_i^{(q)}(y,\lambda)|^2\, d\rho(\lambda)$ converges uniformly with respect to y on the set G_h.

We restrict ourselves to the proof of Assertion (1) only, since Assertion (2) is proved analogously.

Let μ_0 be a fixed sufficiently large number. By analogy with Section 1.2.1 we set

$$\overset{m}{\gamma}{}'(\lambda) = \begin{cases} \overset{m}{\gamma}(\lambda) & \text{for} \quad \lambda \geq \mu_0, \\ 0 & \text{for} \quad \lambda < \mu_0; \end{cases}$$

$$\overset{m}{\gamma}{}''(\lambda) = \begin{cases} 0 & \text{for} \quad \lambda \geq \mu_0, \\ \overset{m}{\gamma}(\lambda) & \text{for} \quad \lambda < \mu_0. \end{cases}$$

Then, to prove Assertion (1), it suffices to show that either of the two integrals

$$\sum_{i=1}^{\widehat{m}}\int_0^\infty \left|\overset{m}{\gamma}{}'(\lambda)\right| |u_i^{(p)}(x,\lambda)|^2\, d\rho(\lambda), \qquad \sum_{i=1}^{\widehat{m}}\int_0^\infty \left|\overset{m}{\gamma}{}''(\lambda)\right| |u_i^{(p)}(x,\lambda)|^2\, d\rho(\lambda) \tag{2.3.27}$$

converges at a fixed number p and at all sufficiently large m uniformly with respect to x on G_h.

This is proved in complete analogy with Section 2.2.1 with the aid of ancillary assertions:

I. for any $\delta > 0$ at sufficiently large $\mu_0 > 0$, the integral

$$\sum_{i=1}^{\widehat{m}}\int_0^\infty |u_i^{(p)}(x,\lambda)|^2 \beta^2(\lambda)\, d\rho(\lambda), \tag{2.3.28}$$

in which

$$\beta(\lambda) = \begin{cases} \frac{1}{2}(1+\lambda)^{-N/4-p/2-\delta/4} & \text{for} \quad \lambda \geq \mu_0, \\ 0 & \text{for} \quad \lambda < \mu_0, \end{cases} \tag{2.3.29}$$

converges uniformly with respect to x in G_h;

II. for any fixed number p, the integral

$$\sum_{i=1}^{\widehat{m}}\int_0^\infty |u_i^{(p)}(x,\lambda)|^2 \delta^2(\lambda)\, d\rho(\lambda), \tag{2.3.30}$$

in which, given a fixed $\mu_0 > 0$ as in Assertion I, the quantity $\delta(\lambda)$ takes the form

$$\delta(\lambda) = \begin{cases} 3/4 & \text{for} \quad \lambda \geq \mu_0, \\ 0 & \text{for} \quad \lambda < \mu_0, \end{cases}$$

converges uniformly with respect to x in G_h.

The proof of Assertions I and II is identical to the proof of Lemmas 1.2 and 1.3 in Section 1.2.2; one must merely take, in place of the Fourier coefficients of the functions $f(r)$, the Fourier images of these functions and take, in place of the Parseval equations in the form of (1.2.36) and (1.2.43), the Parseval equation in the form

$$\sum_{i=1}^{\widehat{m}} \int_0^\infty \left|\widehat{f^{(p)}}(\lambda)\right|^2 d\rho(\lambda) = \int_G \left[f^{(p)}\right]^2 dy,$$

(where the symbol $\widehat{f^{(p)}}(\lambda)$ denotes the Fourier image of the function $f^{(p)}$) and, finally, make use, in place of the Dini rule for the uniform convergence of a functional series, of the Dini rule for the uniform convergence in the improper integral parameters.

Restricting our attention to the above assertions, we conclude that, in order to prove the uniform convergence in $x \in G_h$ of the former integral (2.3.27), it suffices to observe that, by virtue of estimate (2.3.19), this integral at $m > N/2 + 2p$ to within a constant factor[9] $1/4C(h, \alpha, m, N)$ will be majorized uniformly on G_h by the integral (2.3.28), for which the quantity $\beta(\lambda)$ is defined by the integral (2.3.29) at $\delta = m - N/2 - 2p > 0$.

In much the same manner, in order to prove convergence of the latter integral (2.3.27), uniform with respect to $x \in G_h$, it suffices to observe that, by virtue of estimate (2.3.20), this integral to within a constant factor will be majorized uniformly on G_h by integral (2.3.30).

This completes the proof of Lemma 2.2.

REMARK. It follows from the equality $f(r) = v_\alpha(r) - w_\alpha(r)$ and from the Fourier image representation (2.3.18) that the kernel $T_\alpha(x, y)$ for any $h > 0$ at $x \in G_h$, $y \in G$ can be represented in the form

$$T_\alpha(x, y) = [v_\alpha(|x - y|) - w_\alpha(|x - y|)] - \Psi_\alpha(x, y), \tag{2.3.31}$$

where $v_\alpha(r)$ and $w_\alpha(r)$ are functions defined by relations (2.3.15), and $\Psi_\alpha(x, y)$ is a function whose spectral decomposition takes the form (2.3.24).

In this representation, the difference $[v_\alpha(|x - y|) - w_\alpha(|x - y|)]$ is a "smoothed" Bessel–MacDonald kernel. This kernel is the difference of a conventional Bessel–MacDonald kernel and a polynomial within the ball $|x - y| \leq h$; outside the ball $|x - y| \leq h$, it is zero and possesses continuous partial derivatives up to a fixed order m everywhere at $(x, y) \in G \times G$ except for the points $x = y$.

The function $\Psi_\alpha(x, y)$ is such that for any function $g(y)$ from the class $L_2(G)$, the function $F(x)$, as defined by the equality

$$F(x) = \int_G \Psi_\alpha(x, y) g(y) dy, \tag{2.3.32}$$

---

[9] Here $C(h, \alpha, m, N)$ is the constant in estimate (2.3.19).

has in G_h continuous partial derivatives up to an order indefinitely growing with m. (This follows from representation (2.3.24), from the Parseval equality (2.1.10), from the evidence in Section 2.3.2, and from the convergence of integral (2.3.26) uniform with respect to x in G_h.) Moreover, it follows from the uniform-in-G_h convergence of integral (2.3.26) and from the Cauchy–Buniakowski inequality at sufficiently large m that the Riesz means of any nonnegative orders s of the spectral decomposition of function (2.3.32),

$$E_\lambda^s F(x) = \sum_{i=1}^{\widehat{m}} \int_0^\lambda u_i(x,\lambda)\overset{m}{\gamma}(t)\widehat{g}_i(t)\left(1-\frac{t}{\lambda}\right)^s d\rho(t), \tag{2.3.33}$$

are uniformly convergent with respect to x in G_h, $\widehat{g}_i(\lambda)$ being the Fourier image of function $g(y)$.

2.4. PROOF OF NEGATIVE THEOREM 2.1

This section is entirely concerned with the proof of the main theorem of negative type, Theorem 2.1, that has been outlined in Section 2.2.1. We also establish a number of results of independent interest.

2.4.1. Lower-Bound Estimate for the Lebesgue Functions of the Riesz Means

LEMMA 2.3. *Assume that G is an arbitrary domain in the space E^N, $\widehat{A}$ is an arbitrary self-adjoint nonnegative extension of the Laplace operator $Lu = -\Delta u$ in the domain G; specified with respect to this extension are an arbitrary ordered spectral representation of $L_2(G)$ with spectral measure $\rho(\lambda)$, sets of multiplicity e_i, fundamental functions $u_i(x,\lambda)$ $(i = 1, 2, \dots, \widehat{m})$, and multiplicity $\widehat{m}$. Further, let x_0 be any fixed interior point of domain G, $r = |x_0 - y|$ the distance between point x_0 and a variable point y, E an annular layer of type $E = \{R^4/4 \le |x_0 - y| \le R^4\}$ entirely contained in domain G. Then, for any $s \ge 0$, a positive number R can be fixed so small that for a certain $\alpha_0 > 0$ and for all sufficiently large λ the following inequality holds:*

$$\int_E \left| \sum_{i=1}^{\widehat{m}} \int_0^\lambda u_i(x_0,t)u_i(y,t)\left(1-\frac{t}{\lambda}\right)^s d\rho(t)\right| dy \ge \alpha_0 \lambda^{(N-1)/4 - s/2}. \tag{2.4.1}$$

Prior to focusing on a proof of Lemma 2.3, we wish to clarify the meaning of inequality (2.4.1). Recalling Sections 2.1.1 and 2.1.2, we observe that with the Fourier image (2.1.8) substituted into the spectral decomposition (2.1.9) at $\mu = 0$, we obtain

$$E_\lambda f(x) = \int_G \left[\sum_{i=1}^{\widehat{m}} \int_0^\lambda u_i(x,t) u_i(y,t) d\rho(t) \right] f(y) dy, \tag{2.4.2}$$

whence, by virtue of the Gårding theorem (or, to be more exact, by virtue of relation (2.1.2)), the quantity within square brackets in (2.4.2), that is,

$$\theta(x,y,\lambda) = \sum_{i=1}^{\widehat{m}} \int_0^\lambda u_i(x,t) u_i(y,t) d\rho(t), \tag{2.4.3}$$

is a spectral function of extension $\widehat{A}$.

If so, by virtue of relation (2.1.5) taken at $\mu = 0$, the quantity

$$\theta^s(x,y,\lambda) = \sum_{i=1}^{\widehat{m}} \int_0^\lambda u_i(x,t) u_i(y,t) \left(1 - \frac{t}{\lambda}\right)^s d\rho(t) \tag{2.4.4}$$

is the Riesz means of the spectral function of extension $\widehat{A}$ of order $s \geq 0$.

Thus, the inequality (2.4.1) can be rewritten in the form

$$\int_E |\theta^s(x_0, y, \lambda)| dy \geq \alpha_0 \lambda^{(N-1)/4 - s/2}. \tag{2.4.1'}$$

Recall that the theory of functions[10)] usually has to do with functions of the type

$$L_\lambda(x) = \int_G |\theta(x,y,\lambda)| dy,$$

which are generally referred to as Lebesgue functions of the decomposition with which we are concerned. Accordingly, it seems appropriate to call functions of the type

$$L_\lambda^s(x) = \int_G |\theta^s(x,y,\lambda)| dy$$

Lebesgue functions of the Riesz means of order s of the spectral decomposition of interest.

Since

$$\int_G |\theta^s(x_0, y, \lambda)| dy \geq \int_E |\theta^s(x_0, y, \lambda)| dy,$$

[10)] See, for example, Sections 5 and 9 in the book by Kaczmarz and Steinhaus [51] and Chapter 3 in the book by Alexits [2].

inequality (2.4.1′) and, consequently, inequality (2.4.1), set a lower-bound estimate for the Lebesgue functions of the Riesz means of order s of the spectral decomposition under study.

PROOF OF LEMMA 2.3. We fix an arbitrary point x_0 inside G and assume that the positive number R is in any case smaller than the distance between x_0 and the boundary of G. Let us consider now, for any $\lambda > 0$, the following function of the distance $r = |x_0 - y|$ between a variable point y and the point x_0 we have previously fixed:

$$\overset{\lambda}{v}(r) = \begin{cases} \Gamma(s+1)2^s(2\pi)^{-N/2}\lambda^{N/4-s/2}r^{-(\frac{N}{2}+s)}J_{\frac{N}{2}+s}(r\sqrt{\lambda}) & \text{for} \quad r \leq R, \\ 0 & \text{for} \quad r < R. \end{cases} \tag{2.4.5}$$

By virtue of relation (2.3.1) in Section 2.3.1, the Fourier image $\widehat{v}_i^\lambda(x_0, t)$ of function (2.4.5) with respect to the fundamental function $u_i(y, t)$ takes the form

$$\widehat{v}_i^\lambda(x_0, t) = (2\pi)^{N/2}t^{(2-N)/4}u_i(x_0, t)\int_0^R \overset{\lambda}{v}(r)r^{N/2}J_{(N-2)/2}(r\sqrt{t})dr \tag{2.4.6}$$

$$(i = 1, 2, \dots, \widehat{m}).$$

We substitute the concrete value (2.4.5) of this function into (2.4.6) in place of $\overset{\lambda}{v}(r)$ and then transform the integral $\int_0^R$ with the aid of the relation $\int_0^R = \int_0^\infty - \int_R^\infty$.

We make use of the known value of the integral[11]

$$\Gamma(s+1)2^s\lambda^{N/4-s/2}t^{(2-N)/4}\int_0^\infty J_{N/2+s}(r\sqrt{\lambda})J_{N/2-1}(r\sqrt{t})r^{-s}dr = \delta_t^\lambda\left(1 - \frac{t}{\lambda}\right)^s,$$

where

$$\delta_t^\lambda = \begin{cases} 1 & \text{for} \quad t < \lambda, \\ 0 & \text{for} \quad t \geq \lambda; \end{cases} \tag{2.4.7}$$

therefrom, given $s = 0$ and $t = \lambda$, one must take $\delta_t^\lambda = 1/2$ and $(1 - t/\lambda)^s = 1$.

Ultimately, we obtain the following expression for the Fourier image $\widehat{v}_i^\lambda(x_0, t)$:

$$\widehat{v}_i^\lambda(x_0, t) = \delta_t^\lambda u_i(x_0, t)\left(1 - \frac{t}{\lambda}\right)^s - \Gamma(s+1)2^s\lambda^{N/4-s/2}t^{(2-N)/4}u_i(x_0, t)I_t^\lambda(R)$$

$$(i = 1, 2, \dots, \widehat{m}), \tag{2.4.8}$$

[11] See, for example, Bateman and Erdélyi [9, p. 107, formula (34)].

in which $I_t^\lambda(R)$ denotes the integral

$$I_t^\lambda(R) = \int_R^\infty J_{N/2+s}(r\sqrt{\lambda})J_{N/2-1}(r\sqrt{t})r^{-s}dr. \tag{2.4.9}$$

Note that, given $s = 0$, $t = \lambda$, we must subtract $1/2\sqrt{\lambda}$ from the right-hand side of (2.4.9).

In our subsequent reasoning, the following two estimates of the quantity (2.4.9) will play a decisive role:

(a) for any $s \geq 0$, $t \geq 1$, $\lambda \geq 1$, and $|\sqrt{\lambda} - \sqrt{t}| \leq 1$

$$|I_t^\lambda(R)| \leq C_1 R^{-s}\lambda^{-1/4}t^{-1/4}; \tag{2.4.10}$$

(b) for any $s \geq 0$, $t \geq 1$, $\lambda \geq 1$, and $|\sqrt{\lambda} - \sqrt{t}| \geq 1$,

$$|I_t^\lambda(R)| \leq C_2 R^{-1-s}\lambda^{-1/4}t^{-1/4}|\sqrt{\lambda} - \sqrt{t}|^{-1} \tag{2.4.11}$$

(the constants C_1 and C_2 are independent of λ, t, and R).

With a view to proceeding with the proof of Lemma 2.3, the proof of estimates (2.4.10) and (2.4.11) will be deferred until the proof of Lemma 2.3 has been completed.

Assuming that estimates (2.4.10) and (2.4.11) are valid, we multiply both sides of (2.4.8) by the fundamental function $u_i(y,t)$ and take the summation of the equality obtained with respect to all numbers i from 1 to $\widehat{m}$; then we integrate with respect to the spectral measure $\rho(t)$ for t going from zero to Λ, where $\Lambda(\lambda)$ is a sufficiently large number (the way it is to be chosen will be specified below). The resultant expression is[12)]

$$\begin{aligned} \sum_{i=1}^{\widehat{m}} \int_0^\Lambda \widehat{v}_i^\lambda(x_0,t)u_i(y,t)d\rho(t) &= \sum_{i=1}^{\widehat{m}} \int_0^\lambda u_i(x_0,t)u_i(y,t)\left(1 - \frac{t}{\lambda}\right)^s d\rho(t) \\ &- \Gamma(s+1)2^s\lambda^{N/4-s/2}\sum_{i=1}^{\widehat{m}} \int_0^\Lambda u_i(x_0,t)u_i(y,t)t^{(2-N)/4}I_t^\lambda(R)d\rho(t). \end{aligned} \tag{2.4.12}$$

The following inequality stems from (2.4.12) and from the inequality $|A - B| \geq |A| - |B|$:

$$\left|\sum_{i=1}^{\widehat{m}} \int_0^\lambda u_i(x_0,t)u_i(y,t)\left(1 - \frac{t}{\lambda}\right)^s d\rho(t)\right| \geq \sum_{i=1}^{\widehat{m}} \int_0^\Lambda \widehat{v}_i^\lambda(x_0,t)u_i(y,t)d\rho(t)$$

12) In doing so, we take into account equality (2.4.7) and make use of the fact that at any fixed λ and Λ the three integrals in (2.4.12) converge absolutely. To make certain of this, it will suffice to apply the Cauchy–Buniakowski inequality to each of these integrals and use the Parseval equality (2.1.11) for the function $\overset{\lambda}{v}(|x_0 - y|)$ of class $L_2(G)$, the estimate (2.3.7) taken at $\mu = \sqrt{\lambda}$, and estimates (2.4.10) and (2.4.11).

$$-\Gamma(s+1)2^s\lambda^{N/4-s/2}\left|\sum_{i=1}^{\widehat{m}}\int_0^{\Lambda}u_i(x_0,t)u_i(y,t)t^{(2-N)/4}I_t^{\lambda}(R)d\rho(t)\right|.$$

Its integration with respect to the coordinates of point y over the annular layer $E=\{R^4/4\le|x_0-y|\le R^4\}$ yields

$$\int_E\left|\sum_{i=1}^{\widehat{m}}\int_0^{\lambda}u_i(x_0,t)u_i(y,t)\left(1-\frac{t}{\lambda}\right)^s d\rho(t)\right|dy\ge\int_E\left|\sum_{i=1}^{\widehat{m}}\int_0^{\Lambda}\widehat{v}_i^{\lambda}(x_0,t)u_i(y,t)d\rho(t)\right|dy$$

$$-\Gamma(s+1)2^s\lambda^{N/4-s/2}\int_E\left|\sum_{i=1}^{\widehat{m}}\int_0^{\Lambda}u_i(x_0,t)u_i(y,t)t^{(2-N)/4}I_t^{\lambda}(R)d\rho(t)\right|dy. \tag{2.4.13}$$

Since at any fixed $\lambda>0$ the function $\overset{\lambda}{v}(|x_0-y|)$ belongs to class $L_2(G)$, by virtue of the Gårding–Browder–Mautner theorem (see Section 2.1.2) at any fixed λ one has

$$\lim_{\Lambda\to\infty}\int_E\left|\sum_{i=1}^{\widehat{m}}\int_0^{\Lambda}\widehat{v}_i^{\lambda}(x_0,t)u_i(y,t)d\rho(t)\right|dy=\int_E\left|\overset{\lambda}{v}(|x_0-y|)\right|dy. \tag{2.4.14}$$

Now we prove that there exists a positive constant β dependent only on s and N such that for any $R>0$ and for all λ larger than a positive $\lambda_0(R)$, the following inequality holds:

$$\int_E\left|\overset{\lambda}{v}(|x_0-y|)\right|dy\ge 3\beta R^{2N-2-4s}\lambda^{(N-1)/4-s/2}. \tag{2.4.15}$$

Indeed, from the definition of the function (2.4.5) and the annular layer E we obtain[13)]

$$\int_E\left|\overset{\lambda}{v}(|x_0-y|)\right|dy$$
$$=\Gamma(s+1)2^s(2\pi)^{-N/2}\omega_N\lambda^{(N-1)/4-s/2}\int_{R^4/4}^{R^4}\left|J_{N/2+s}(r\sqrt{\lambda})\right|r^{N/2-s-1}dr. \tag{2.4.16}$$

We apply the asymptotic formula for a Bessel function

$$J_{N/2+s}(r\sqrt{\lambda})=\sqrt{\frac{2}{\pi r\sqrt{\lambda}}}\cos\left[r\sqrt{\lambda}-\frac{\pi}{2}\left(\frac{N}{2}+s\right)-\frac{\pi}{4}\right]+O\left(r^{-3/2}\lambda^{-3/4}\right)$$

13) We denote by ω_N the surface area of an N-dimensional sphere of unit radius equal to $2\pi^{N/2}[\Gamma(N/2)]^{-1}$.

and the trivial inequality

$$\left|\cos\left[r\sqrt{\lambda} - \frac{\pi}{2}\left(\frac{N}{2}+s\right) - \frac{\pi}{4}\right]\right| \geq \frac{1}{2} + \frac{1}{2}\cos\left[2r\sqrt{\lambda} - \pi\left(\frac{N}{2}+s\right) - \frac{\pi}{2}\right]$$

to the right-hand side of (2.4.16).

We obtain that

$$\begin{aligned}
&\lambda^{N/4-s/2}\int_{R^4/4}^{R^4}\left|J_{N/2+s}(r\sqrt{\lambda})\right| r^{N/2-s-1}dr \\
&\geq \frac{1}{\sqrt{2\pi}}\left\{\lambda^{(N-1)/4-s/2}\int_{R^4/4}^{R^4} r^{(N-1)/2-s-1}dr\right. \\
&\left.+\lambda^{(N-1)/4-s/2}\int_{R^4/4}^{R^4} r^{(N-1)/2-s-1}\cos\left[2r\sqrt{\lambda} - \pi\left(\frac{N}{2}+s\right) - \frac{\pi}{2}\right]dr\right\} \\
&-\lambda^{(N-1)/4-s/2-1/2}\int_{R^4/4}^{R^4}\left|O(r^{(N-1)/2-s-2})\right| dr.
\end{aligned} \tag{2.4.17}$$

Now we observe that

$$\frac{1}{\sqrt{2\pi}}\lambda^{(N-1)/4-s/2}\int_{R^4/4}^{R^4} r^{(N-1)/2-s-1}dr$$

$$= \begin{cases} \dfrac{1}{\sqrt{2\pi}}\dfrac{(1-2^{2s-N+1})}{((N-1)/2-s)}R^{2N-2-4s}\lambda^{(N-1)/4-s/2} & \text{for} \quad s \neq (N-1)/2, \\[2ex] \dfrac{1}{\sqrt{2\pi}}\ln 4 & \text{for} \quad s = (N-1)/2, \end{cases} \tag{2.4.18}$$

$$\frac{1}{\sqrt{2\pi}}\lambda^{(N-1)/4-s/2}\int_{R^4/4}^{R^4} r^{(N-1)/2-s-1}\cos\left[2r\sqrt{\lambda} - \pi\left(\frac{N}{2}+s\right) - \frac{\pi}{2}\right]dr$$

$$= R^{2N-2-4s}\lambda^{(N-1)/4-s/2}O\left(\frac{R^4+1}{\sqrt{\lambda}}\right), \tag{2.4.19}$$

$$\lambda^{(N-1)/4-s/2-1/2}\int_{R^4/4}^{R^4} O\left|\left(r^{(N-1)/2-s-2}\right)\right| dr = R^{2N-2-4s}\lambda^{(N-1)/4-s/2}O\left(\frac{R^4}{\sqrt{\lambda}}\right). \tag{2.4.20}$$

From relations (2.4.17)–(2.4.20) and from equality (2.4.16) we obtain that inequality (2.4.15) is valid (for all $R > 0$ and all $\lambda \geq \lambda_0(R)$) with constant β of the form

$$\beta = \begin{cases} \Gamma(s+1)2^s(2\pi)^{-(N+1)/2}\frac{\omega_N}{6}(1-2^{2s-N+1})\left(\frac{N-1}{2}-s\right) & \text{for } s \neq \frac{N-1}{2}, \\ \Gamma(s+1)2^s(2\pi)^{-(N+1)/2}\frac{\omega_N}{6}\ln 4 & \text{for } s = \frac{N-1}{2}. \end{cases}$$

Relation (2.4.14) and inequality (2.4.15) enable us, for each sufficiently large $\lambda > 0$, to fix the number $\Lambda = \Lambda(\lambda)$ so large as to make the following inequality hold:

$$\int_E \left|\sum_{i=1}^{\widehat{m}} \int_0^{\Lambda} \widehat{v}_i^{\lambda}(x_0,t)u_i(y,t)d\rho(t)\right| dy \geq 2\beta R^{2N-2-4s}\lambda^{(N-1)/4-s/2} \tag{2.4.21}$$

(β is the same constant as in (2.4.15)).

It follows from inequalities (2.4.13) and (2.4.21) that in order to establish the desired inequality (2.4.1), it suffices to prove that a positive number $R > 0$ can be fixed so small that for all sufficiently large $\lambda > 0$ and for any Λ the inequality

$$\Gamma(s+1)2^s\lambda^{1/4}\int_E \left|\sum_{i=1}^{\widehat{m}} \int_0^{\Lambda} u_i(x_0,t)u_i(y,t)t^{(2-N)/4}I_t^{\lambda}(R)d\rho(t)\right| dy \leq \beta R^{2N-2-4s} \tag{2.4.22}$$

holds with the same constant β as that in (2.4.15).

To prove inequality (2.4.22), we divide the integral in (2.4.22) into four integrals. In this manner, we reduce the proof of inequality (2.4.22) to the proof (for sufficiently small $R > 0$, for sufficiently large $\lambda > 0$ and for any Λ) of the four inequalities below:

$$S_1 = \Gamma(s+1)2^s\lambda^{1/4}\int_E \left|\sum_{i=1}^{\widehat{m}} \int_{0\leq\sqrt{t}\leq 1} \left[u_i(x_0,t)u_i(y,t)t^{(2-N)/4}I_t^{\lambda}(R)\right] d\rho(t)\right| dy \leq \frac{\beta}{4}R^{2N-2-4s}, \tag{2.4.23}$$

$$S_2 = \Gamma(s+1)2^s\lambda^{1/4}\int_E \left|\sum_{i=1}^{\widehat{m}} \int_{\substack{1\leq\sqrt{t}\leq\sqrt{\lambda}/2 \\ 3\sqrt{\lambda}/2\leq\sqrt{t}\leq\sqrt{\Lambda}}} [\ldots]\, d\rho(t)\right| dy \leq \frac{\beta}{4}R^{2N-2-4s}, \tag{2.4.24}$$

$$S_3 = \Gamma(s+1)2^s\lambda^{1/4}\int\limits_E \left|\sum_{i=1}^{\widehat{m}} \int\limits_{|\sqrt{t}-\sqrt{\lambda}|\leq 1} [\ldots]\, d\rho(t)\right| dy \leq \frac{\beta}{4}R^{2N-2-4s}, \tag{2.4.25}$$

$$S_4 = \Gamma(s+1)2^s\lambda^{1/4}\int\limits_E \left|\sum_{i=1}^{\widehat{m}} \int\limits_{1\leq|\sqrt{t}-\sqrt{\lambda}|\leq \sqrt{\lambda}/2} [\ldots]\, d\rho(t)\right| dy \leq \frac{\beta}{4}R^{2N-2-4s} \tag{2.4.26}$$

(here β is the same constant as that in (2.4.15); the square brackets denote in (2.4.24)–(2.4.26) the same quantity as in (2.4.23)).

Now we proceed with the proof of inequalities (2.4.23)–(2.4.26).

Applying the Cauchy–Buniakowski inequality to the integrals over the annular layer E on the left-hand sides of (2.4.23)–(2.4.26) and recalling that $\int\limits_E dy = O(R^{4N})$, we obtain (invoking also the Parseval equality (2.1.11)) the following estimates for the left-hand sides of (2.4.23)–(2.4.26):

$$S_1 = O(R^{2N})\sqrt{\sum_{i=1}^{\widehat{m}} \int\limits_{0\leq\sqrt{t}\leq 1} \left\{u_i^2(x_0,t)\lambda^{1/2}t^{(2-N)/2}\left[I_t^\lambda(R)\right]^2\right\} d\rho(t)}, \tag{2.4.27}$$

$$S_2 = O(R^{2N})\sqrt{\sum_{i=1}^{\widehat{m}} \int\limits_{\substack{1\leq\sqrt{t}\leq\sqrt{\lambda}/2\\ 3\sqrt{\lambda}/2\leq\sqrt{t}\leq\sqrt{\Lambda}}} \{\ldots\}\, d\rho(t)}, \tag{2.4.28}$$

$$S_3 = O(R^{2N})\sqrt{\sum_{i=1}^{\widehat{m}} \int\limits_{|\sqrt{t}-\sqrt{\lambda}|\leq 1} \{\ldots\}\, d\rho(t)}, \tag{2.4.29}$$

$$S_4 = O(R^{2N})\sqrt{\sum_{i=1}^{\widehat{m}} \int\limits_{1\leq|\sqrt{t}-\sqrt{\lambda}|\leq\sqrt{\lambda}/2} \{\ldots\}\, d\rho(t)} \tag{2.4.30}$$

(the braces in (2.4.28)–(2.4.30) are used to denote the same quantity as in (2.4.27)).

The inequalities (2.4.23)–(2.4.26) will be proved if we show that each of the quantities S_1, S_2, S_3, and S_4 is $o(R^{2N-2-4s})$ for all $R > 0$, all Λ, and all sufficiently large $\lambda > 0$.[14)]

[14)] Then, given any $\beta > 0$, the inequalities (2.4.23)–(2.4.26) are with certainty valid for sufficiently small $R > 0$.

To begin, we shall show that $S_1 = o(R^{2N-2-4s})$. For this purpose, we consider two cases:
(1) $s \le (N-1)/2$;
(2) $s > (N-1)/2$.
In the former case, we prove that $S_1 = O(R^{2N})$, and in the latter case we prove that $S_1 = O(R^{2N+(N-1)/2-s})$. It will be ascertained thereby that in both instances $S_1 = o(R^{2N-2-4s})$.

In the case of $s \le (N-1)/2$ we infer from relation (2.4.8) that[15)]

$$\sqrt{\lambda}u_i^2(x_0,t)t^{(2-N)/2}\left\{I_t^\lambda(R)\right\}^2$$
$$= [\Gamma(s+1)2^s]^{-2}\lambda^{s-(N-1)/2}\left[\delta_t^\lambda u_i(x_0,t)\left(1-\frac{t}{\lambda}\right)^s - \widehat{v}_t^\lambda(x_0,t)\right]^2, \quad (2.4.31)$$

and from relation (2.4.6) we obtain, with reference to estimate $|J_\nu(x)| \le C(\nu)|x|^\nu$, that[16)]

$$\left|\widehat{v}_i^\lambda(x_0,t)\right| \le C_1|u_i(x_0,t)|\int_0^R \left|\overset{\lambda}{v}(r)\right| r^{N-1}dr. \quad (2.4.32)$$

A comparison of (2.4.32) with the estimate

$$\left|\overset{\lambda}{v}(r)\right| \le C_2\lambda^{(N-1)/4-s/2}r^{-((N+1)/2+s)}$$

following from (2.4.5) and from the inequality $|J_\nu(x)| \le C_1(\nu)|x|^{-1/2}$ shows that

$$\left|\widehat{v}_i^\lambda(x_0,t)\right| \le C_3|u_i(x_0,t)|\lambda^{(N-1)/4-s/2}R^{(N-1)/2-s}. \quad (2.4.33)$$

With reference to (2.4.31) and (2.4.33) and to the fact that the number R can be chosen smaller than unity and the number λ larger than unity, we obtain the inequality

$$\sqrt{\lambda}\,u_i^2(x_0,t)t^{(2-N)/2}\left[I_t^\lambda(R)\right]^2 \le C_4u_i^2(x_0,t). \quad (2.4.34)$$

Finally, from (2.4.34) and (2.4.27) we infer that for the particular case $s \le (N-1)/2$, by virtue of estimate (2.3.3) taken at $\mu = 0$,

$$S_1 = O(R^{2N})\sqrt{\sum_{i=1}^{\widehat{m}}\int_{0\le\sqrt{t}\le 1} u_i^2(x_0,t)d\rho(t)} = O(R^{2N}).$$

---

15) Recall that at $s = 0$, $t = \lambda$ we must take $\delta_t^\lambda = 1/2$ and $(1-t/\lambda)^s = 1$.

16) In what follows we denote by the symbol C_j $(j = 1,2,\ldots)$ constants dependent solely on N and s.

Let us consider now the second case $s > (N-1)/2$. In this instance, invoking (2.4.9), we obtain

$$\lambda^{1/2}t^{(2-N)/2}\left[I_t^\lambda(R)\right]^2 = \lambda^{1/2}\left\{\int_0^R J_{N/2+s}(r\sqrt{\lambda})J_{N/2-1}(r\sqrt{t})(r\sqrt{t})^{1-N/2}r^{N/2-s-1}dr\right\}^2.$$

From this relation and from the estimates for Bessel functions

$$\left|J_{N/2-1}(r\sqrt{t})\right| \le C_5(r\sqrt{t})^{N/2-1}, \qquad \left|J_{N/2+s}(r\sqrt{\lambda})\right| \le C_6 r^{-1/2}\lambda^{-1/4}$$

we obtain the inequality

$$\lambda^{1/2}t^{(2-N)/2}\left[I_t^\lambda(R)\right]^2 \le C_7 R^{N-1-2s}, \tag{2.4.35}$$

which, when substituted into (2.4.27) and using estimate (2.3.3) taken at $\mu = 0$, yields finally

$$S_1 = O\left(R^{2N+(N-1)/2-s}\right)\sqrt{\sum_{i=1}^{\widehat{m}}\int_{0\le\sqrt{t}\le 1} u_i^2(x_0,t)d\rho(t)} = O(R^{2N+(N-1)/2-s}).$$

Thus, we have proved that in both cases $S_1 = o(R^{2N-2-2s})$. Now we prove the estimate $S_2 = o(R^{2N-2-4s})$.

Since in the integral (2.4.28) there is either $1 \le \sqrt{t} \le \sqrt{\lambda}/2$ or $3/2\sqrt{\lambda} \le \sqrt{t} \le \sqrt{\Lambda}$, estimate (2.4.11) can be applied to $I_t^\lambda(R)$; in this particular case this estimate may be rewritten in the form

$$\left|I_t^\lambda(R)\right| \le C_8 t^{-3/4}\lambda^{-1/4}R^{-1-s}. \tag{2.4.36}$$

It is inferred from (2.4.36) and (2.4.28) that

$$S_2 = O\left(R^{2N-1-s}\right)\sqrt{\sum_{i=1}^{\widehat{m}}\int_{\sqrt{t}\ge 1} u_i^2(x_0,t)t^{-(N+1)/2}d\rho(t)}. \tag{2.4.37}$$

Finally, we obtain from (2.4.37) and from the boundedness of integral (2.3.10) at any $\delta > 0$ (see Section 2.3.2) that $S_2 = o(R^{2N-2-4s})$.

Let us ensure now that $S_3 = o(R^{2N-2-4s})$. Substituting into (2.4.29) the estimate (2.4.10) for the quantity $I_t^\lambda(R)$ and making use of the estimate (2.3.7) taken at $\rho_0 = 1$, we obtain

$$S_3 = O\left(R^{2N-s}\right)\sqrt{\sum_{i=1}^{\widehat{m}} \int\limits_{|\sqrt{t}-\sqrt{\lambda}|\le 1} u_i^2(x_0,t)t^{(1-N)/2}d\rho(t)}$$

$$= O\left(R^{2N-s}\right)\sqrt{\lambda^{(1-N)/2}\sum_{i=1}^{\widehat{m}} \int\limits_{|\sqrt{t}-\sqrt{\lambda}|\le 1} u_i^2(x_0,t)d\rho(t)}$$

$$= O(R^{2N-s}) = o(R^{2N-2-4s}).$$

It remains to prove that $S_4 = o(R^{2N-2-4s})$.

We denote by p the least of the indices for which $2^p \ge \sqrt{\lambda}/2$. Then the segment $[1, \sqrt{\lambda}/2]$ is covered by a set of segments $[2^{\ell-1}, 2^\ell]$, $\ell = 1, 2, \dots, p$.

Making use of estimate (2.4.11), we can majorize the quantity in (2.4.30) within the square root as

$$\sum_{i=1}^{\widehat{m}} \int\limits_{1\le|\sqrt{t}-\sqrt{\lambda}|\le\sqrt{\lambda}/2} u_i^2(x_0,t)\lambda^{1/2}t^{(2-N)/2}\left[I_t^\lambda(R)\right]^2 d\rho(t)$$

$$\le C_9\lambda^{(1-N)/2}R^{-2-2s}\sum_{i=1}^{\widehat{m}} \int\limits_{1\le|\sqrt{t}-\sqrt{\lambda}|\le\sqrt{\lambda}/2} u_i^2(x_0,t)(\sqrt{t}-\sqrt{\lambda})^{-2}d\rho(t)$$

$$\le C_9\lambda^{(1-N)/2}R^{-2-2s}\sum_{\ell=1}^{p}\left[\sum_{i=1}^{\widehat{m}} \int\limits_{2^{\ell-1}\le|\sqrt{t}-\sqrt{\lambda}|\le 2^\ell} u_i^2(x_0,t)(\sqrt{t}-\sqrt{\lambda})^{-2}d\rho(t)\right]$$

$$\le C_9\lambda^{(1-N)/2}R^{-2-2s}\sum_{\ell=1}^{p}\frac{1}{4^{\ell-1}}\left\{\sum_{i=1}^{\widehat{m}} \int\limits_{2^{\ell-1}\le|\sqrt{t}-\sqrt{\lambda}|\le 2^\ell} u_i^2(x_0,t)d\rho(t)\right\}. \tag{2.4.38}$$

Applying inequality (2.3.8) taken at $\rho_0 = 2^\ell$, $\mu = \sqrt{\lambda}$ to the quantity within braces in (2.4.38), we obtain

$$\sum_{i=1}^{\widehat{m}} \int_{1\le|\sqrt{t}-\sqrt{\lambda}|\le\sqrt{\lambda}/2} u_i^2(x_0,t)\lambda^{1/2}t^{(2-N)/2}\left[I_t^\lambda(R)\right]^2 d\rho(t) \le C_{10}R^{-2-2s}\sum_{\ell=1}^{\infty}2^{-\ell} \le C_{11}R^{-2-2s}. \tag{2.4.39}$$

Substituting (2.4.39) into (2.4.30), we finally obtain

$$S_4 = O(R^{2N-1-s}) = o(R^{2N-2-4s}).$$

This completes the proof of Lemma 2.3.

All we have left to do is to verify estimates (2.4.10) and (2.4.11) for the quantity (2.4.9).

2.4.2. Proof of Ancillary Estimates

We wish to establish estimates of the type (2.4.10) and (2.4.11) not only for the integral (2.4.9), but also for the more general integral

$$\overset{\nu}{I}{}_t^\lambda(R) = \int_R^\infty J_{\nu+s}(r\sqrt{t})J_{\nu-1}(r\sqrt{\lambda})r^{-s}dr, \tag{2.4.9'}$$

for any $\nu \ge 1/2$, $s \ge 0$, $\lambda \ge 1$, $t \ge 1$. (Integral (2.4.9) is derived from (2.4.9′) with $\nu = N/2$.)

We establish the estimate

$$\left|\overset{\nu}{I}{}_t^\lambda(R)\right| \le C_1R^{-s}\lambda^{-1/4}t^{-1/4} \tag{2.4.10'}$$

for integral (2.4.9′) at any $\nu \ge 1/2$, $s \ge 0$, $\lambda \ge 1$, $t \ge 1$, $|\sqrt{t}-\sqrt{\lambda}| \le 1$, and the estimate

$$\left|\overset{\nu}{I}{}_t^\lambda(R)\right| \le C_2R^{-1-s}\lambda^{-1/4}t^{-1/4}|\sqrt{t}-\sqrt{\lambda}|^{-1} \tag{2.4.11'}$$

for any real ν and for any $s > -1$, $\lambda \ge 1$, $t \ge 1$, $|\sqrt{t}-\sqrt{\lambda}| \le 1$ (with constants C_1 and C_2 independent of $\lambda \ge 1$, $t \ge 1$, $0 < R \le 1$).

We begin by verifying estimate (2.4.10′). For $s > 0$, this estimate stems in a trivial manner from the relations

$$|J_{\nu+s}(r\sqrt{\lambda})| \le C_3(r\sqrt{\lambda})^{-1/2}, \quad |J_{\nu-1}(r\sqrt{t})| \le C_4(r\sqrt{t})^{-1/2}, \quad \int_R^\infty r^{-s-1}dr = s^{-1}R^{-s}.$$

To establish estimate (2.4.10′) for $s = 0$, we make use of the known value of the integral[17)]

$$\int_0^\infty J_\nu(r\sqrt{\lambda})J_{\nu-1}(r\sqrt{t})dr = \begin{cases} \dfrac{(\sqrt{t})^{\nu-1}}{(\sqrt{\lambda})^\nu} \le t^{-1/4}\lambda^{-1/4} & \text{for } t < \lambda, \\ (2\sqrt{\lambda})^{-1} = 2^{-1}t^{-1/4}\lambda^{-1/4} & \text{for } t = \lambda, \\ 0 & \text{for } t > \lambda, \end{cases}$$

by virtue of which it will suffice, given $t \ge 1$, $\lambda \ge 1$, $|\sqrt{r} - \sqrt{\lambda}| \le 1$, $\nu \ge 1/2$, to establish the estimate

$$\left|\int_0^R J_\nu(r\sqrt{\lambda})J_{\nu-1}(r\sqrt{t})dr\right| \le C_5\lambda^{-1/4}t^{-1/4}. \tag{2.4.40}$$

To establish estimate (2.4.40), by virtue of the identity

$$J_{\nu-1}(r\sqrt{t}) = \left[J_{\nu-1}(r\sqrt{t}) - J_{\nu-1}(r\sqrt{\lambda})\right] + J_{\nu-1}(r\sqrt{\lambda}),$$

it will be sufficient to verify the estimates

$$\int_0^R \left|J_\nu(r\sqrt{\lambda})\right| \left|J_{\nu-1}(r\sqrt{t}) - J_{\nu-1}(r\sqrt{\lambda})\right| dr \le C_6\lambda^{-1/4}t^{-1/4}, \tag{2.4.41}$$

$$\left|\int_0^R J_\nu(r\sqrt{\lambda})J_{\nu-1}(r\sqrt{\lambda})dr\right| \le C_7\lambda^{-1/4}t^{-1/4}. \tag{2.4.42}$$

To verify estimate (2.4.41), we observe that by virtue of Lagrange's theorem of finite increments there is a number θ within the interval $0 < \theta < 1$ such that

$$\left|J_{\nu-1}(r\sqrt{t}) - J_{\nu-1}(r\sqrt{\lambda})\right| = r|\sqrt{t} - \sqrt{\lambda}|\,|J'_{\nu-1}(\xi)|,$$

where $\xi = r[\sqrt{t} + \theta(\sqrt{t} - \sqrt{\lambda})]$. Since[18)] $\sqrt{t} \ge \sqrt{\lambda}/2$, we have $\xi = r[(1-\theta)\sqrt{t} + \theta\sqrt{\lambda}] \ge r\sqrt{\lambda}/2$, and, with reference to the recurrence relation

$$J'_{\nu-1}(\xi) = \frac{\nu - 1}{\xi} J_{\nu-1}(\xi) - J_\nu(\xi)$$

17) See, for example, Bateman and Erdélyi [9, p. 107, formula (34)].

18) Indeed, it follows from the inequalities $\sqrt{t} \ge 1$, $|\sqrt{t} - \sqrt{\lambda}| \le 1$ that $\sqrt{\lambda}/\sqrt{t} = (\sqrt{\lambda} - \sqrt{t})/\sqrt{t} + 1 \le 2$, which is equivalent to the inequality $\sqrt{t} \ge \sqrt{\lambda}/2$.

and estimates $|J_{\nu-1}(\xi)| \le C_8\xi^{-1/2}$, $|J_\nu(\xi)| \le C_9\xi^{-1/2}$, it follows that

$$|J'_{\nu-1}(\xi)| \le \frac{C_8(\nu-1)}{\xi^{3/2}} + \frac{C_9}{\xi^{1/2}} \le C_{10}\left[\frac{(\nu-1)}{(r\sqrt{\lambda})^{3/2}} + \frac{1}{(r\sqrt{\lambda})^{1/2}}\right].$$

This implies, invoking $|\sqrt{\lambda} - \sqrt{t}| \le 1$, that

$$|J_{\nu-1}(r\sqrt{t}) - J_{\nu-1}(r\sqrt{\lambda})| \le C_{10}\left[\frac{(\nu-1)}{\sqrt{r}\lambda^{3/4}} + \frac{\sqrt{r}}{\lambda^{1/4}}\right].$$

Thus, the left-hand side of (2.4.41) is majorized by the sum of the integrals

$$C_{10}(\nu-1)\frac{1}{\lambda^{3/4}}\int_0^R \frac{1}{\sqrt{r}}|J_\nu(r\sqrt{\lambda})|dr + C_{10}\frac{1}{\lambda^{1/4}}\int_0^R \sqrt{r}\,|J_\nu(r\sqrt{\lambda})|dr,$$

and to establish inequality (2.4.41), it suffices, in the former integral, to make use of the estimate $|J_\nu(x)| \le C_{11}$, and in the latter integral, of the estimate $|J_\nu(x)| \le C_9x^{-1/2}$ and to take into account that $1/\sqrt{\lambda} \le C_{12}(\lambda t)^{-1/4}$.

To establish inequality (2.4.42) we make use of the recurrence relation[19)]

$$J_{\nu-1}(r\sqrt{\lambda}) = \frac{1}{\sqrt{\lambda}}\frac{d}{dr}\left[J_\nu(r\sqrt{\lambda})\right] + \frac{\nu}{r\sqrt{\lambda}}J_\nu(r\sqrt{\lambda})$$

to represent the left-hand side of (2.4.42) as the sum

$$\left|\frac{1}{\sqrt{\lambda}}\int_0^R J_\nu(r\sqrt{\lambda})d\left[J_\nu(r\sqrt{\lambda})\right] + \frac{\nu}{\sqrt{\lambda}}\int_0^R J_\nu^2(r\sqrt{\lambda})\frac{dr}{r}\right|$$

$$\le \frac{1}{2\sqrt{\lambda}}J_\nu^2(R\sqrt{\lambda}) + \frac{\nu}{\sqrt{\lambda}}\int_0^\infty J_\nu^2(r\sqrt{\lambda})\frac{dr}{r}.$$

Now, to establish the validity of estimate (2.4.42), we recall the relations[20)]

$$\left|J_\nu(R\sqrt{\lambda})\right| \le C_{11}, \qquad \int_0^\infty J_\nu^2(r\sqrt{\lambda})\frac{dr}{r} = \frac{1}{2\nu}.$$

This completes the proof of estimate (2.4.10′).

---

19) See Bateman and Erdélyi [9, p. 20, formula (54)].

20) See, for example, Bateman and Erdélyi [9, p. 107, formula (32)]

Let us to estimate (2.4.11′). We wish to establish this estimate for an arbitrary real ν and for any $s > -1$, $\lambda \geq 1$, $t \geq 1$, $|\sqrt{t} - \sqrt{\lambda}| \geq 1$. We consider separately two cases: (1) $\sqrt{\lambda} + 1 < \sqrt{t}$; (2) $1 \leq \sqrt{t} \leq \sqrt{\lambda} - 1$.

(1) In the former case, we subject (2.4.9) to two-fold integration by parts using the recurrence relations

$$\int r^{\nu} J_{\nu-1}(r\sqrt{t})dr = (\sqrt{t})^{-1} r^{\nu} J_{\nu}(r\sqrt{t}), \qquad \frac{d}{dr}\left[r^{-\nu} J_{\nu}(r\sqrt{\lambda})\right] = -\sqrt{\lambda} r^{\nu} J_{\nu+1}(r\sqrt{\lambda}).$$

The result is

$$\overset{\nu}{I}\,{}^{\lambda}_{t}(R) = (\sqrt{t})^{-1}\left[J_{\nu+s}(r\sqrt{\lambda})J_{\nu}(r\sqrt{t})r^{-s}\right]\Big|_{r=R}^{r=\infty}$$
$$+\frac{\sqrt{\lambda}}{t}\left[J_{\nu+s+1}(r\sqrt{\lambda})J_{\nu+1}(r\sqrt{t})r^{-s}\right]\Big|_{r=R}^{r=\infty}$$
$$+\frac{\lambda}{t}\int_{R}^{\infty} J_{\nu+s+2}(r\sqrt{\lambda})J_{\nu+1}(r\sqrt{t})r^{-s}\,dr.$$

Under the integration sign we make use of the relation

$$J_{\nu+s+2}(r\sqrt{\lambda})\ J_{\nu+1}(r\sqrt{t}) = J_{\nu+s}(r\sqrt{\lambda})J_{\nu-1}(r\sqrt{t})$$
$$-(2\nu+2+2s)(r\sqrt{\lambda})^{-1}J_{\nu+s+1}(r\sqrt{\lambda})J_{\nu-1}(r\sqrt{t})$$
$$+2\nu(r\sqrt{t})^{-1}J_{\nu+2+s}(r\sqrt{\lambda})J_{\nu}(r\sqrt{t}),$$

which trivially follows from the recurrence formulas[21)]

$$J_{\nu+2+s}(r\sqrt{\lambda}) = -J_{\nu+s}(r\sqrt{\lambda}) + (2\nu+2+2s)(r\sqrt{\lambda})^{-1}J_{\nu+s+1}(r\sqrt{\lambda}),$$

$$J_{\nu+1}(r\sqrt{t}) = -J_{\nu-1}(r\sqrt{t}) + 2\nu(r\sqrt{t})^{-1}J_{\nu}(r\sqrt{t}).$$

---

21) See Bateman and Erdélyi [9, p. 20, formula (56)].

The result is

$$\overset{\nu}{I}{}_t^{\lambda}(R)\left(1-\frac{\lambda}{t}\right)$$

$$= (\sqrt{t})^{-1} J_{\nu+s}(R\sqrt{\lambda})J_\nu(R\sqrt{t})R^{-s} - \frac{\sqrt{\lambda}}{t}J_{\nu+s+1}(R\sqrt{\lambda})J_{\nu+1}(R\sqrt{t})R^{-s}$$

$$-(2\nu+2+s)\frac{\sqrt{\lambda}}{t}\int_R^\infty r^{-1-s}J_{\nu+s+1}(r\sqrt{\lambda})J_{\nu-1}(r\sqrt{t})dr \tag{2.4.43}$$

$$+2\nu\frac{\lambda}{t^{3/2}}\int_R^\infty r^{-1-s}J_{\nu+2+s}(r\sqrt{\lambda})J_\nu(r\sqrt{t})dt.$$

Majorizing the moduli of all Bessel functions on the right-hand side of (2.4.43) with the aid of estimate $|J_\nu(x)| \le C(\nu)x^{-1/2}$, valid for all real ν at $x \ge R > 0$,[22] we obtain, with reference to the inequality $s > -1$ (2.4.43), the relation

$$\left|\overset{\nu}{I}{}_t^{\lambda}(R)\right|\left(1-\frac{\lambda}{t}\right) \le C_{12}R^{-1-s}\lambda^{-1/4}t^{-3/4},$$

lending support to the validity of estimate (2.4.11′) at $\sqrt{\lambda}+1<\sqrt{t}$.

(2) We now consider the second case $1 \le \sqrt{t} \le \sqrt{\lambda}-1$. We apply two-fold integration of (2.4.9′) by parts to the other side, that is, we use the recurrence relations

$$\int r^{1-\nu}J_\nu(r\sqrt{\lambda})dr = -(\sqrt{\lambda})^{-1}r^{1-\nu}J_{\nu-1}(r\sqrt{\lambda}),\quad \frac{d}{dr}\left[r^\nu J_\nu(r\sqrt{t})\right] = \sqrt{t}\,r^\nu J_{\nu-1}(r\sqrt{t}).$$

The result is

$$\overset{\nu}{I}{}_t^{\lambda}(R) = -(\sqrt{\lambda})^{-1}\left[J_{\nu+s-1}(r\sqrt{\lambda})J_{\nu-1}(r\sqrt{t})r^{-s}\right]\Bigg|_{r=R}^{r=\infty}$$

$$-\frac{\sqrt{t}}{\lambda}\left[J_{\nu+s-2}(r\sqrt{\lambda})J_{\nu-2}(r\sqrt{t})r^{-s}\right]\Bigg|_{r=R}^{r=\infty}$$

$$+\frac{\lambda}{t}\int_R^\infty J_{\nu+s-2}(r\sqrt{\lambda})J_{\nu-3}(r\sqrt{t})r^{-s}dr.$$

Under the integration sign we make use of the relations

[22] ibidem, p. 98, formula (3).

$$J_{\nu+s-2}(r\sqrt{\lambda})\ J_{\nu-3}(r\sqrt{t}) = J_{\nu+s}(r\sqrt{\lambda})J_{\nu-1}(r\sqrt{t})$$

$$+ (2\nu + 2s - 2)(r\sqrt{\lambda})^{-1}J_{\nu+s-1}(r\sqrt{\lambda})J_{\nu-3}(r\sqrt{t})$$

$$- (2\nu - 4)(r\sqrt{t})^{-1}J_{\nu+s}(r\sqrt{\lambda})J_{\nu-2}(r\sqrt{t}),$$

which trivially follows from the recurrence formulas

$$J_{\nu+s-2}(r\sqrt{\lambda}) = -J_{\nu+s}(r\sqrt{\lambda}) + (2\nu + 2s - 2)(r\sqrt{\lambda})^{-1}J_{\nu+s-1}(r\sqrt{\lambda}),$$

$$J_{\nu-3}(r\sqrt{t}) = -J_{\nu-1}(r\sqrt{t}) + (2\nu - 4)(r\sqrt{t})^{-1}J_{\nu-2}(r\sqrt{t}).$$

The result is

$$\overset{\nu}{I}{}_t^{\lambda}(R)\left(1 - \frac{\lambda}{t}\right)$$

$$= (\sqrt{\lambda})^{-1}J_{\nu+s-1}(R\sqrt{\lambda})J_{\nu-1}(R\sqrt{t})R^{-s}$$

$$+ \frac{\sqrt{t}}{\lambda}J_{\nu+s-2}(R\sqrt{\lambda})J_{\nu-2}(R\sqrt{t})R^{-s}$$

$$+ (2\nu + 2s - 2)\frac{\sqrt{t}}{\lambda^{3/2}}\int_R^{\infty} J_{\nu+s-1}(r\sqrt{\lambda})J_{\nu-3}(r\sqrt{t})r^{-1-s}dr \tag{2.4.44}$$

$$- (2\nu - 4)\frac{\sqrt{t}}{\lambda}\int_R^{\infty} J_{\nu+s}(r\sqrt{\lambda})J_{\nu-2}(r\sqrt{t})r^{-1-s}dr.$$

On the right-hand side of (2.4.44) we make use, for all Bessel functions, of the estimate $|J_\nu(x)| \leq C(R, \nu)x^{-1/2}$, which holds for $x \geq R > 0$ with a constant $C(R, \nu)$ for all real (not necessarily positive) values of ν.[23] We obtain from (2.4.44), with reference to $s > -1$, the relation

$$\left|\overset{\nu}{I}{}_t^{\lambda}(R)\right|\left(1 - \frac{t}{\lambda}\right) \leq C_{13}R^{-1-s}\lambda^{-3/4}t^{-1/4},$$

lending support to the validity of estimate (2.4.11′) at $1 \leq \sqrt{t} \leq \sqrt{\lambda} - 1$.

This completes the proof of ancillary estimates (2.4.10′) and (2.4.11′).

---

[23] See, for example, Bateman and Erdélyi [9, p. 98, formula (3)].

2.4.3. Lemma on the Relation Between the Riesz Means of Integrals $\int_0^\lambda (t+1)^\beta U(t)d\rho(t)$ and $\int_0^\lambda U(t)d\rho(t)$

LEMMA 2.4. *Assume that $s > 0$, $\beta > 0$, $s = r + \kappa$, where r is an integer, κ satisfies the inequalities $0 < \kappa \le 1$, $\rho(t)$ is a certain measure, $U(r)$ is an arbitrary function such that $\int_0^\lambda |U(t)|d\rho(t) < \infty$ for any $\lambda > 0$, the symbols S_λ, σ_λ^s, $\bar{S}_\lambda$ and $\bar{\sigma}_\lambda^s$ signify the following:*

$$\begin{cases} S_\lambda = \int_0^\lambda U(t)d\rho(t), & \sigma_\lambda^s = \int_0^\lambda \left(1 - \frac{t}{\lambda}\right)^s U(t)d\rho(t), \\ \bar{S}_\lambda = \int_0^\lambda (1+t)^\beta U(t)d\rho(t), & \bar{\sigma}_\lambda^s = \int_0^\lambda \left(1 - \frac{t}{\lambda}\right)^s (1+t)^\beta U(t)d\rho(t). \end{cases} \tag{2.4.45}$$

Then the inequality

$$\bar{\sigma}_\lambda^s = (1+\lambda)^\beta \sigma_\lambda^s$$
$$+(-1)^{r+1}\lambda^{-s} \int_0^\lambda \frac{d^{r+2}}{dt^{r+2}} \left\{(\lambda - t)^s \left[(1+t)^\beta - (1+\lambda)^\beta\right]\right\} \frac{t^{r+1}}{(r+1)!}\sigma_t^{r+1} dt \tag{2.4.46}$$

holds.

PROOF: It is obvious that

$$\bar{S}_\lambda = -\beta \int_0^\lambda (1+t)^{\beta-1} S_t\, dt + (1+\lambda)^\beta S_\lambda. \tag{2.4.47}$$

Moreover, by definition,

$$\bar{\sigma}_\lambda^s = s\lambda^{-s} \int_0^\lambda (\lambda - t)^{s-1} \bar{S}_t\, dt. \tag{2.4.48}$$

Substituting (2.4.47) into (2.4.48), we obtain

$$\bar{\sigma}_\lambda^s = -\beta s \lambda^{-s} \int_0^\lambda (\lambda - t)^{s-1} \left[\int_0^t (1+\tau)^{\beta-1} S_\tau\, d\tau\right] dt + s\lambda^{-s} \int_0^\lambda (\lambda - t)^{s-1} (1+t)^\beta S_t\, dt.$$

Changing in the first of the integrals the order of integration with respect to t and τ, we obtain

$$\bar{\sigma}^s_\lambda = -\beta s\lambda^{-s}\int\limits_0^\lambda S_\tau(1+\tau)^{\beta-1}\left[\int\limits_\tau^\lambda(\lambda-t)^{s-1}\,dt\right]d\tau + s\lambda^{-s}\int\limits_0^\lambda(\lambda-t)^{s-1}(1+t)^\beta S_t\,dt.$$

Or finally

$$\bar{\sigma}^s_\lambda = -\beta\lambda^{-s}\int\limits_0^\lambda(\lambda-t)^s(1+t)^{\beta-1}S_t\,dt + s\lambda^{-s}\int\limits_0^\lambda(\lambda-t)^{s-1}(1+t)^\beta S_t\,dt. \qquad (2.4.49)$$

Dividing the latter integral in (2.4.49) into the sum of two integrals

$$s\lambda^{-s}\int\limits_0^\lambda(\lambda-t)^{s-1}(1+t)^\beta S_t\,dt$$

$$= s\lambda^{-s}(1+\lambda)^\beta\int\limits_0^\lambda(\lambda-t)^{s-1}S_t\,dt + s\lambda^{-s}\int\limits_0^\lambda(\lambda-t)^{s-1}\left[(1+t)^\beta-(1+\lambda)^\beta\right]S_t\,dt$$

and, keeping in mind that

$$s\lambda^{-s}\int\limits_0^\lambda(\lambda-t)^{s-1}S_t\,dt = \sigma^s_\lambda, \qquad (2.4.50)$$

we rewrite formula (2.4.49) in the following manner:

$$\bar{\sigma}^s_\lambda = \lambda^{-s}\int\limits_0^\lambda\frac{d}{dt}\left\{(\lambda-t)^s[(1+t)^\beta-(1+\lambda)^\beta]\right\}S_t\,dt + (1+\lambda)^\beta\sigma^s_\lambda. \qquad (2.4.51)$$

In what follows we will use the symbol IS_t to denote the operation of integration $IS_t = \int\limits_0^t S_\tau d\tau$, and the symbol I^k to denote the result of k-fold successive application of the operation I.

Through the $(k-1)$-fold integration by parts of equality

$$\sigma^k_t = kt^{-k}\int\limits_0^t(t-\tau)^{k-1}S_\tau\,d\tau$$

we make certain that the formula

$$I^kS_t = \frac{t^k}{k!}\sigma^k_t \qquad (2.4.52)$$

applies to any positive integer k.

Recall that, according to the condition of the lemma, $s = r + \kappa$, where r is an integer and $0 < \kappa \leq 1$. With respect to equality (2.4.52), we integrate (2.4.51) $(r+1)$ times by parts.

The result is

$$\bar{\sigma}_\lambda^s = (1+\lambda)^\beta \sigma_\lambda^s$$

$$+\lambda^{-s} \sum_{k=1}^{r+1} (-1)^{k+1} \left\{ \frac{d^k}{dt^k} \left[(\lambda - t)^s [(1+t)^\beta - (1+\lambda)^\beta] \right] \frac{t^k}{k!} \sigma_t^k \right\} \Bigg|_{t=0+0}^{t=\lambda-0} \quad (2.4.53)$$

$$+(-1)^{r+1} \lambda^{-s} \int_0^\lambda \frac{d^{r+2}}{dt^{r+2}} \{ (\lambda - t)^s [(1+t)^\beta - (1+\lambda)^\beta] \} \frac{t^{r+1}}{(r+1)!} \sigma_t^{r+1} \, dt.$$

To complete the proof of the lemma, it remains to ensure that both the upper and lower substitutions in the right-hand side of (2.4.53) go to zero.

To begin, we ensure that all the upper substitutions become zero. To this end, it suffices to show that for any $k = 1, 2, \ldots, r+1$

$$\lim_{t \to \lambda - 0} \frac{d^k}{dt^k} \{ (\lambda - t)^s [(1+t)^\beta - (1+\lambda)^\beta] \} = 0. \quad (2.4.54)$$

We represent the derivative of order k for the product of two functions within braces in (2.4.54) as the sum of $(k+1)$ summands according to Leibnitz's rule.

Since $s = r + \kappa$ (where $0 < \kappa \leq 1$), we have $s > r$, and the derivative $\frac{d^\ell}{dt^\ell}[(\lambda - t)^s]$ of any order $\ell \leq r$ becomes zero at $t = \lambda$. In view of the fact that the function $[(1+t)^\beta - (1+\lambda)^\beta]$ itself and its derivative with respect to t of any order are bounded in the vicinity of the point $t = \lambda$, for $k \leq r$ according to Leibnitz's rule, in which each summand contains a derivative $\frac{d^\ell}{dt^\ell}[(\lambda - t)^s]$ of order $\ell \leq r$, the relation (2.4.54) is justified.

It remains to verify relation (2.4.54) for $k = r + 1$.

By virtue of the same Leibnitz's rule, it suffices to show that

$$\lim_{t \to \lambda - 0} \left\{ [(1+t)^\beta - (1+\lambda)^\beta] \frac{d^{r+1}}{dt^{r+1}} [(\lambda - t)^s] \right\} = 0. \quad (2.4.55)$$

For $s = r + 1$, when $\kappa = 1$, the relation (2.4.55) is obvious.

For $s = r + \kappa$, when $0 < \kappa < 1$, the left-hand side of (2.4.55) becomes

$$(-1)^{r+1} s(s-1)(s-2)\ldots(s-r) \lim_{t \to \lambda - 0} \frac{(1+t)^\beta - (1+\lambda)^\beta}{(\lambda - t)^{1-\kappa}}, \quad (2.4.56)$$

and, as is readily seen, (2.4.56) becomes zero in accordance with the l'Hospital's rule.[24)]

[24)] See, for example, Il'in and Poznyak [46, pp. 261–264].

The fact that all the lower substitutions in the right-hand side of (2.4.53) become zero is ensured by the cofactor t^k (for $k \geq 1$) and by the boundedness of the derivative $\frac{d^k}{dt^k}\left\{(\lambda-t)^s[(1+t)^\beta-(1+\lambda)^\beta]\right\}$ of any order k in the vicinity of $t=0$.

Thus, all the substitutions in the right-hand side of (2.4.53) become zero, which completes the proof of Lemma 2.4.

COROLLARY TO LEMMA 2.4. Provided that the conditions for Lemma 2.4 are fulfilled, the inequality

$$|\bar{\sigma}_\lambda^s| \leq (1+\lambda)^\beta |\sigma_\lambda^s| + \lambda^{-s}\int_0^\lambda \left|\frac{d^{r+2}}{dt^{r+2}}\left\{(\lambda-t)^s[(1+t)^\beta-(1+\lambda)^\beta]\right\}\frac{t^{r+1}}{(r+1)!}\right|\sigma_t^{r+1}\,dt \tag{2.4.57}$$

holds.

2.4.4. Lemma on the Unboundedness in L_1 of the Riesz Means of Order $s \geq 0$ of the Spectral Decompositions of Kernels of Order $\alpha < (N-1)/4 - s/2$

LEMMA 2.5. *Assume that $N \geq 2$, G is an arbitrary domain in the space E^N, $\widehat{A}$ is an arbitrary self-adjoint nonnegative extension of the Laplace operator $Lu = -\Delta u$ in domain G; taken with respect to this extension is an arbitrary ordered spectral representation of the space $L_2(G)$ with spectral measure $\rho(\lambda)$, sets of multiplicity e_i, fundamental functions $u_i(x,\lambda)$ $(i = 1, 2, \dots, \widehat{m})$ and multiplicity $\widehat{m} \leq \infty$. Assume, further, that x_0 is a fixed interior point of domain G, $r = |x_0 - y|$ is the distance between the point x_0 and a variable point y, E is an annular layer of the form $E = \{R^4/4 \leq |x_0 - y| \leq R^4\}$ strictly interior to domain G. Then for any s in a half-segment $0 \leq s < (N-1)/2$, a positive R can be fixed so small that for any δ within the interval $0 < \delta < (N-1)/4 - s/2$ the quantity*

$$\tau_\lambda^s = \int_E \left|\sum_{i=1}^{\widehat{m}}\int_0^\lambda u_i(x_0,t)u_i(y,t)(1+t)^{-((N-1)/4-s/2-\delta)}\left(1-\frac{t}{\lambda}\right)^s d\rho(t)\right| dy \tag{2.4.58}$$

is unbounded on the half-line $\lambda \geq 0$.

Prior to proceeding with the proof of Lemma 2.5, let us take a closer look at expression (2.4.58). By virtue of the definition of fractional kernel $T_{(N-1)/4-s/2-\delta}(x_0,y)$ (see Section 2.3.3), the quantity

$$\sum_{i=1}^{\widehat{m}}\int_0^\lambda u_i(x_0,t)(1+t)^{-((N-1)/4-s/2-\delta)}u_i(y,t)\,d\rho(t)$$

is a spectral decomposition $E_\lambda T_{(N-1)/4-s/2-\delta}(x_0, y)$ of this fractional kernel, and for this reason the quantity

$$\sum_{i=1}^{\widehat{m}} \int_0^\lambda u_i(x_0, t) u_i(y, t)(1+t)^{-((N-1)/4-s/2-\delta)} \left(1 - \frac{t}{\lambda}\right)^s d\rho(t)$$

is the Riesz mean $E_\lambda^s T_{(N-1)/4-s/2-\delta}(x_0, y)$ of order s of the spectral decomposition of this fractional kernel.

In this case, (2.4.58) can be rewritten in the form

$$\tau_\lambda^s = \int_E \left|E_\lambda^s T_{(N-1)/4-s/2-\delta}(x_0, y)\right| dy. \tag{2.4.58'}$$

Denoting the kernel order $(N-1)/4 - s/2 - \delta$ by α and keeping in mind that δ is an arbitrary number in the interval $0 < \delta < (N-1)/4 - s/2$, α is an arbitrary number in the interval $0 < \alpha < (N-1)/2 - s/2$; we will observe that the unboundedness of (2.4.58′) on the half-line $\lambda \geq 0$ signifies the boundedness in the $L_1(E)$ metric (and the more so, in the $L_1(G)$ metric) of the Riesz means of order s $(0 \leq s \leq (N-1)/2)$ of the spectral decomposition of the kernel $T_\alpha(x_0, y)$ of an arbitrary positive order α smaller than $(N-1)/4 - s/2$.

PROOF OF LEMMA 2.5. First, we ensure that it will suffice to prove Lemma 2.5 for any strictly positive s smaller than $(N-1)/2$. This will imply the validity of Lemma 2.5 for $s = 0$ also.

To that effect, we wish to prove the following ancillary assertion: *If, for any s in the interval $0 < s < (N-1)/2$ and for any δ in the interval $0 < \delta < (N-1)/4 - s/2$ the quantity (2.4.58) is unbounded on the half-line $\lambda \geq 0$, then for any δ' in the interval $0 < \delta' < (N-1)/4$ the quantity*

$$T_\lambda = \int_E \left| \sum_{i=1}^{\widehat{m}} \int_0^\lambda u_i(x_0, t) u_i(y, t)(1+t)^{-((N-1)/4-\delta')} d\rho(t) \right| dy \tag{2.4.59}$$

is also unbounded on the half-line $\lambda \geq 0$.

We prove the formulated assertion by contradiction.

Suppose that quantity (2.4.58) is an unbounded one of the half-line $\lambda \geq 0$ for any s in the interval $0 < s < (N-1)/2$ and for any δ in the interval $0 < \delta < (N-1)/4 - s/2$, and quantity (2.4.59) is a bounded one on the half-line $\lambda \geq 0$ for a certain δ' in the interval $0 < \delta' < (N-1)/4$.

Let us consider quantity (2.4.58) at $s = \delta'$, $\delta = \delta'/2$ (provided that the conditions that s belongs to the interval $(0, (N-1)/2)$ and δ belongs to the interval $(0, (N-1)/4 - s/2)$ are fulfilled, and that the factor $(1+t)^{-((N-1)/4-s/2-\delta)}$ reduces to $(1+$

$t)^{-((N-1)/4-\delta'))}$. We obtain, with reference to relation (2.4.50), that (2.4.58) at $s = \delta'$, $\delta = \delta'/2$ is related to (2.4.59) through the inequality

$$\tau_\lambda^s \le s\lambda^{-s} \int_0^\lambda (\lambda - t)^{s-1} T_t \, dt. \tag{2.4.60}$$

It follows immediately from (2.4.60) and from the assumed boundedness of (2.4.59) taken for the specified δ' on the half-line $\lambda \ge 0$ that (2.4.58), taken at $s = \delta'$, $\delta = \delta'/2$ on the half-line $\lambda \ge 0$, is bounded; however, this is at variance with our assumption and completes thus the proof of the ancillary assertion.

Now we focus in a straightforward manner on the proof of Lemma 2.5 for any strictly positive s smaller than $(N-1)/2$. We also proceed with this proof by contradiction.

Assume that for a certain s in the interval $0 < s < (N-1)/2$ and for a certain δ in the interval $0 < \delta < (N-1)/4 - s/2$, quantity (2.4.58) is bounded on the half-line $\lambda \ge 0$. Making use of inequality (2.4.57), we set the number β, in terms of Lemma 2.4, as

$$\beta = \frac{N-1}{4} - \frac{s}{2} - \delta > 0,$$

and the function $U(t)$ as

$$U(t) = \sum_{i=1}^{\widehat{m}} u_i(x_0, t) u_i(y, t)(1+t)^{-\beta}.$$

In this case, the σ_λ^s and $\bar{\sigma}_\lambda^s$ in inequality (2.4.57) take the form

$$\sigma_\lambda^s = \sum_{i=1}^{\widehat{m}} \int_0^\lambda u_i(x_0, t) u_i(y, t)(1+t)^{-\beta} \left(1 - \frac{t}{\lambda}\right)^s d\rho(t),$$

$$\bar{\sigma}_\lambda^s = \sum_{i=1}^{\widehat{m}} \int_0^\lambda u_i(x_0, t) u_i(y, t) \left(1 - \frac{t}{\lambda}\right)^s d\rho(t).$$

We integrate both sides of inequality (2.4.57) with respect to the coordinates of the point y over the annular layer E chosen as indicated in the formulation of Lemmas 2.5 and 2.3, keeping in mind (by virtue of the boundedness of (2.4.58)) that there is a constant $M > 0$ such that for all $\lambda \ge 0$ the inequality

$$\tau_\lambda^s = \int_E |\sigma_\lambda^s| dy \le M \tag{2.4.61}$$

holds,

By analogy with Lemma 2.4, let $s = r + \kappa$, where r is an integer, $0 < \kappa \leq 1$. Let us prove that for all $\lambda \geq 0$ the inequality

$$\tau_\lambda^{r+1} = \int_E |\sigma_\lambda^{r+1}| dy \leq M$$

also holds.

Indeed, if s is an integer, then $\kappa = 1$ and $s = r + 1$. Therefore, in this case inequality (2.4.62) is simply identical to inequality (2.4.61).

If s is not an integer, then $0 < \kappa < 1$, and the number $s' = r + 1$ satisfies the condition $s' > s$.

But for any s and s' such that $s' > s > 0$, the following expression is valid, relating the Riesz means of order s' to the Riesz means of order s (the validity of this relation is easily proved):

$$\sigma_\lambda^{s'} = \lambda^{-s'}\Gamma(s'+1)\,[\Gamma(s'-s)\Gamma(s+1)]^{-1} \int_0^\lambda (\lambda - t)^{s'-s-1} t^s \sigma_t^s \, dt. \tag{2.4.63}$$

Relation (2.4.63) and inequality (2.4.61) lend support to the validity of inequality (2.4.62) (at $s' = r + 1 > s$).

Now let us turn once again to inequality (2.4.57) with the specified values of σ_λ^s and $\bar{\sigma}_\lambda^s$ and integrated over the annular layer E. This inequality can, with reference to (2.4.61) and (2.4.62), be written in the form

$$\int_E \left| \sum_{i=1}^{\widehat{m}} \int_0^\lambda u_i(x_0,t) u_i(y,t) \left(1 - \frac{t}{\lambda}\right)^s d\rho(t) \right| dy$$
$$\leq M \left\{ (1+\lambda)^\beta + \frac{\lambda^{-s}}{(r+1)!} \int_0^\lambda t^{r+1} \left| \frac{d^{r+2}}{dt^{r+2}} \left[(\lambda - t)^s[(1+t)^\beta - (1+\lambda)^\beta]\right] \right| dt \right\}. \tag{2.4.64}$$

One readily sees that the quantity on the right-hand side of (2.4.64) enclosed in braces has an order of $O[(\lambda+1)^\beta]$, that is, has an order of $O(\lambda^\beta)$ at sufficiently large λ. It will suffice to ensure that, given sufficiently large λ,

$$\int_0^\lambda t^{r+1} \left| \frac{d^{r+2}}{dt^{r+2}} \left[(\lambda - t)^s[(1+t)^\beta - (1+\lambda)^\beta]\right] \right| dt = O(\lambda^{\beta+s}). \tag{2.4.65}$$

To prove relation (2.4.65), it will be enough to apply Leibnitz's rule to the derivative within the modulus brackets (2.4.65) and to make sure that the modulus of each summand in Leibnitz's formula, after having been multiplied by t^{r+1} and integrated with respect to t between the limits from 0 to λ, yields $O(\lambda^{\beta+s})$.

Comparing inequality (2.4.64) with relation (2.4.65), we shall see that for all sufficiently large λ, we have

$$\int_E \left| \sum_{i=1}^{\widehat{m}} \int_0^\lambda u_i(x_0,t)u_i(y,t)\left(1-\frac{t}{\lambda}\right)^s d\rho(t) \right| dy = O(\lambda^\beta), \tag{2.4.66}$$

where $\beta = (N-1)/4 - s/2 - \delta$ at certain s in the interval $0 < s < (N-1)/2$ and at a certain δ in the interval $0 < \delta < (N-1)/4 - s/2$.

Since, given a large λ, the quantity on the right-hand side of (2.4.66) is $\lambda^{(N-1)/4-s/2}$ $O(\lambda^{-\delta}) = o(\lambda^{(N-1)/4-s/2})$, relation (2.4.66) contradicts inequality (2.4.1), proved earlier (see Lemma 2.3, Section 2.4.1).

This contradiction completes the proof of Lemma 2.5.

2.4.5. Direct Proof of Theorem 2.1

The proof of Theorem 2.1 is based on the use of Lemma 2.5, the fractional kernel representation (established by us in Section 2.3.3), and on the familiar resonance-type theorem.[25]

By virtue of Lemma 2.5, quantity (2.4.58′) is unbounded on the half-line $\lambda \geq 0$ (for any fixed s in the half-segment $0 \leq s < (N-1)/2$ and for any fixed δ in the interval $0 < \delta < (N-1)/4 - s/2$). In other words, the quantity $E_\lambda^s T_{(N-1)/4-s/2-\delta}(x_0, y)$ for fixed s and δ is unbounded on the half-line $\lambda \geq 0$ in the $L_1(E)$ metric with respect to y.

It follows, from this fact and from the resonance-type theorem,[26] that there exists a function $g(y)$, continuous in the annular layer E, such that the quantity

$$\sigma_\lambda^s(x_0) = \int_E E_\lambda^s T_{(N-1)/4-s/2-\delta}(x_0, y) g(y) dy \tag{2.4.67}$$

is also unbounded on the half-line $\lambda \geq 0$.

Using for the kernel $T_{(N-1)/2-s/2-\delta}(x_0, y)$ the representation (2.3.31), we obtain that, for sufficiently small $h > 0$ subject to $h < R^4/4$ and for an arbitrary fixed number m, quantity (2.4.67) can be represented as the difference of two quantities:

$$\begin{aligned} \sigma_\lambda^s(x_0) = & \int_E E_\lambda^s \left[v_{\beta/2}(|x_0 - y|) - w_{\beta/2}(|x_0 - y|) \right] g(y) dy \\ & - \int_E E_\lambda^s \Psi_{\beta/2}(x_0, y) g(y) dy = E_\lambda^s f(x_0) - E_\lambda^s F(x_0), \end{aligned} \tag{2.4.68}$$

[25] See, for example, Kaczmarz and Steinhaus [51, Section 1.5].

[26] See ibidem, p. 31, Assertion 3.

where $\beta = (N-1)/2 - s - 2\delta$, and the functions $f(x)$ and $F(x)$ take the form

$$f(x) = \int_E [v_{\beta/2}(|x-y|) - w_{\beta/2}(|x-y|)]\, g(y)dy, \tag{2.4.69}$$

$$F(x) = \int_E \Psi_{\beta/2}(x,y)g(y)dy. \tag{2.4.70}$$

If we continue the function $g(y)$ with zero beyond the annular layer E, that is, if we introduce the function

$$\bar{g}(y) = \begin{cases} g(y) & \text{for} \quad y \in E, \\ 0 & \text{for} \quad y \in E^N \backslash N, \end{cases}$$

then functions (2.4.69) and (2.4.70) can be rewritten in the form

$$f(x) = \int_{E^N} [v_{\beta/2}(|x-y|) - w_{\beta/2}(|x-y|)]\, \bar{g}(y)dy, \tag{2.4.69$'$}$$

$$F(x) = \int_G \Psi_{\beta/2}(x,y)\bar{g}(y)dy. \tag{2.4.70$'$}$$

Since the function $\bar{g}(y)$ in all cases belongs to the class $L_2(G)$, for a sufficiently large number m in accordance with the remark in the end of Section 2.3.3, the Riesz means of any nonnegative order s of function (2.4.70′) converge uniformly with respect to x in G_h and, in particular, converge (and therefore are bounded with respect to λ) at the point x_0 that we have fixed.

If so, it follows from the unboundedness of (2.4.67) on the half-line $\lambda \geq 0$ and from equality (2.4.68) that the Riesz means $E_\lambda^s f(x)$ of order s for the function $f(x)$ defined by relation (2.4.69′) are unbounded at the point x_0 for a sufficiently large number m.

It remains to prove that, given sufficiently small $h > 0$, the constructed function $f(x)$ also satisfies other requirements of Theorem 2.1, namely: (1) it is finite in domain G; (2) it becomes zero in a certain vicinity D of the point x_0; (3) it belongs in domain G to the Zygmund–Hölder class C^α with an arbitrary positive differentiability index α subject to the condition $\alpha < \beta = (N-1)/2 - s - 2\delta$.[27]

The finiteness in G of the function $f(x)$, defined by relation (2.4.69), stems at sufficiently small $h > 0$ from the fact that the annular layer E is completely interior to G, and the function $[v_{\beta/2}(|x-y|) - w_{\beta/2}(|x-y|)]$ is distinct from zero only at $|x-y| < h$.

---

[27] Since for δ we have fixed an arbitrary number in the interval $0 < \delta < (N-1)/4 - s/2$, $\beta = (N-1)/2 - s - 2\delta$ is an arbitrary number in the interval $0 < \beta < (N-1)/2 - s$, and therefore $\alpha < \beta$ is an arbitrary number in the interval $0 < \alpha < (N-1)/2 - s$.

The function $f(x)$ becoming zero in a certain vicinity of the point x_0 results from the facts that $h < R^4/4$, the layer E is outside the ball $|x_0 - y| \le R^4/4$, and the function $[v_{\beta/2}(|x-y|) - w_{\beta/2}(|x-y|)]$ is other than zero only at $|x-y| < h$.

It remains to be proved that $f(x) \in C^\alpha(E^N)$ for any α smaller than $\beta = (N-1)/2 - s - 2\delta$.

By virtue of relation (2.3.22), integral (2.4.69′) can be written in the form $f(x) = f_1(x) - f_2(x)$, where

$$f_1(x) = \int\limits_{E^N} \overset{0}{T}_{\beta/2}(x,y)\bar{g}(y)dy, \qquad f_2(x) = \int\limits_{E^N} \varphi_{\beta/2}(x,y)\bar{g}(y)dy,$$

$\overset{0}{T}_{\beta/2}(x,y)$ is the common Bessel–MacDonald kernel defined by relation (2.1.16), and $\varphi_{\beta/2}(x,y)$ is a function of the form (2.3.21). By virtue of the properties inherent in $\varphi_{\beta/2}(x,y)$, the function $f_2(x)$ belongs to the class C^α with the index α growing indefinitely with m.

The function $f_1(x)$, by virtue of the fact that $\bar{g}(y) \in L_p(E^N)$ with an arbitrarily large p, belongs by definition to the class L_p^β with arbitrarily large p. On the basis of the embedding theorem (2.1.26) (see Section 2.1.3), the function $f_1(x)$ belongs to the class $C^{\beta - N/p}$ and, by virtue of the fact that p can be chosen arbitrarily large, it belongs to the class C^α for any $\alpha < \beta$.

This completes the proof of Theorem 2.1.

2.5. PROOF OF POSITIVE THEOREM 2.3

In proving the main theorem of positive type, Theorem 2.3, we shall in fact establish the uniform equiconvergence (on any subset G_h) of the Riesz means of spectral decompositions corresponding to various self-adjoint nonnegative extensions of the Laplace operator.

Our line of reasoning in proving Theorem 2.3 is as follows: in Section 2.5.1 we prove two lemmas of self-sustained interest on the Fourier images of a finite function in the Nikol'skii class H_2; in Section 2.5.2, we prove three lemmas on the properties of the Nikol'skii class functions themselves as well as on their averages over the sphere and over the ball; finally, Section 2.5.3 will be dedicated to a direct proof of Theorem 2.3.

2.5.1. Lemmas on the Fourier Images of a Finite Function in the Nikol'skii Class

For an arbitrary domain D of the space E^N, we adopt the convention to denote by the symbol $\overset{0}{H}{}_p^\alpha(D)$ the set of functions defined in the entire space E^N which belong to the class $H_p^\alpha(E^N)$ and which are zero outside domain D.

LEMMA 2.6. *Assume that G is an arbitrary domain in the space E^N, $\widehat{A}$ is an arbitrary self-adjoint nonnegative extension of the Laplace operator $Lu = -\Delta u$ in the domain G; taken with respect to this extension is an arbitrary ordered spectral representation of the space $L_2(G)$ with spectral measure $\rho(\lambda)$, sets of multiplicity e_i, fundamental functions $u_i(x,\lambda)$ $(i = 1, 2, \dots, \widehat{m})$, and multiplicity $\widehat{m} \leq \infty$. Then, for any positive numbers R and α, for any function $f(x)$ in the class*[28] $\overset{0}{H}{}_p^\alpha(G_R)$ *and for any positive λ, the Fourier image $\widehat{f}_i(\lambda)$ of the function $f(x)$ satisfies the inequality*

$$\sum_{i=1}^{\widehat{m}} \int_{\lambda}^{4\lambda} \left|\widehat{f}_i(t)\right|^2 t^\alpha d\rho(t) \leq C_R \|f\|^2_{H_2^\alpha}. \tag{2.5.1}$$

PROOF: Let us consider separately two possible cases: (1) $0 < \alpha \leq 2$; (2) $\alpha > 2$.

(1) To begin, let $0 < \alpha \leq 2$. For an arbitrary r within the interval $0 < r < R/2$, we consider the average of the function $f(x)$ with respect to a ball of radius r centered at point x and denote it by the symbol[29]

$$\varepsilon(x,r) = \varepsilon(x,r,f) = N\omega_N^{-1} r^{-N} \int_{|z|\leq r} f(x+z)dz. \tag{2.5.2}$$

Note that since the function $f(x)$ is zero outside G_R, *the function $\varepsilon(x,r,f)$ at $r < R/2$ is zero for all x exterior to G_r.*

Let us specify the Fourier image $\widehat{\varepsilon}_i(\lambda, r, f)$ of the function $\varepsilon(x,r,f)$.

Making use of the definition of the Fourier image and relation (2.5.2), we obtain

$$\begin{aligned}\widehat{\varepsilon}_i(\lambda,r,f) &= \int_{G_r} \varepsilon(y,r,f)u_i(y,\lambda)dy \\ &= N\omega_N^{-1} r^{-N} \int_{G_r} \left[\int_{|z|\leq r} f(y+z)u_i(y,\lambda)dz \right] dy.\end{aligned} \tag{2.5.3}$$

We introduce a characteristic function $\chi(z)$ which is unity at $|z| \leq r$ and is zero at $|z| > r$; we can rewrite the expression for the Fourier image (2.5.3) in the following manner:

$$\widehat{\varepsilon}_i(x,r,f) = N\omega_N^{-1} r^{-N} \int_{G_r} \left[\int_{E^N} \chi(z) f(y+z)u_i(y,\lambda)dz \right] dy. \tag{2.5.4}$$

---

[28] Recall that the symbol G_R denotes the set of points of an arbitrary domain G which are removed from the boundary of G to a distance greater than $R > 0$.

[29] The symbol ω_N denotes the surface area of an N-dimensional sphere of unit radius equal to $2(\pi)^{N/2}\,[\Gamma(N/2)]^{-1}$.

Further, having set $y+z=x$, we obtain, taking into account that $f(x)=0$ outside G,

$$\begin{aligned}\widehat{\varepsilon}_i(\lambda,r,f) &= N\omega_N^{-1}r^{-N}\int\limits_G\left[\int\limits_{E^N}\chi(z)f(x)u_i(x-z,\lambda)dz\right]dx\\ &= N\omega_N^{-1}r^{-N}\int\limits_G f(x)\left\{\int\limits_{|z|\le r}u_i(x-z,\lambda)dz\right\}dx.\end{aligned} \tag{2.5.5}$$

The mean value with respect to the ball of radius r centered at point x of the fundamental function $u_i(y,\lambda)$ on the right-hand side of (2.5.5), enclosed within braces, is easily determined using the mean-value formula (2.3.1). We obtain

$$\begin{aligned}\int\limits_{|z|\le r}u_i(x-z,\lambda)dz &= \int\limits_0^r\rho^{N-1}\left[\int\limits_\omega\!\!\ldots\!\int u_i(x+\omega\rho,\lambda)d\omega\right]d\rho\\ &= (2\pi)^{N/2}u_i(x,\lambda)\lambda^{(2-N)/2}\int\limits_0^r\rho^{N/2}J_{N/2-1}(\rho\sqrt{\lambda})d\rho\\ &= (2\pi)^{N/2}u_i(x,\lambda)\lambda^{-N/4}r^{N/2}J_{N/2}(r\sqrt{\lambda}).\end{aligned} \tag{2.5.6}$$

Substituting (2.5.6) into (2.5.5), we obtain the following representation for the Fourier image of the average (2.5.2)

$$\widehat{\varepsilon}_i(\lambda,r,f) = N\omega_N^{-1}(2\pi)^{N/2}\frac{J_{N/2}(r\sqrt{\lambda})}{(r\sqrt{\lambda})^{N/2}}\int\limits_G f(x)u_i(x,\lambda)dx. \tag{2.5.7}$$

Taking into account that $N\omega_N^{-1}(2\pi)^{N/2} = 2^{N/2}\Gamma\left(N/2+1\right)$, and making use of the function

$$\varphi(t) = 2^{N/2}\Gamma\left(\frac{N}{2}+1\right)\frac{J_{N/2}(t)}{t^{N/2}}, \tag{2.5.8}$$

we obtain ultimately the following expression for the Fourier image $\widehat{\varepsilon}_i(\lambda,r,f)$ of the average (2.5.2) of the function $f(x)$:

$$\widehat{\varepsilon}_i(\lambda,r,f) = \widehat{f}_i(\lambda)\varphi(r\sqrt{\lambda}) \tag{2.5.9}$$

(here $\widehat{f}_i(\lambda)$ denotes the Fourier image of the function $f(x)$).

Now if, in place of the average (2.5.2), we take a function which is the algebraic sum of three summands

$$\varepsilon(x,2r,f) - 4\varepsilon(x,r,f) + 3f(x), \tag{2.5.10}$$

then, as is easily seen, the Fourier image of this function for any r in the interval $0 < r < R/4$ takes the form

$$[\varphi(2r\sqrt{\lambda}) - 4\varphi(r\sqrt{\lambda}) + 3]\widehat{f_i}(\lambda). \tag{2.5.11}$$

Note that the double function (2.5.10) represents the average with respect to the ball of radius r centered at point x of the fourth difference $\Delta_z^4 f(x)$:

$$\Delta_z^4 f(x) = f(x+2z) - 4f(x+z) + 6f(x) - 4f(x-z) + f(x-2z).$$

This is readily checked, taking into account that

$$\int\limits_{|z|\le r} f(x+z)dz = \int\limits_{|z|\le r} f(x-z)dz \quad\text{and}\quad \int\limits_{|z|\le r} f(x+2z)dz = \int\limits_{|z|\le r} f(x-2z)dz.$$

Indeed,

$$N\omega_N^{-1}r^{-N} \int\limits_{|z|\le r} \Delta_z^4 f(x)dz = N\omega_N^{-1}r^{-N} \int\limits_{|z|\le r} [f(x+2z) + f(x-2z)$$

$$-4f(x+z) - 4f(x-z) + 6f(x)]dz = 2[\varepsilon(x,2r,f) - 4\varepsilon(x,r,f) + 3f(x)].$$

In this case, the Fourier image of the average with respect to the ball of radius r centered at point x of the fourth difference $\Delta_z^4 f(x)$ is equal to the double of (2.5.11) and, applying the Parseval equality (2.1.11), we obtain the relation

$$4\sum_{i=1}^{\widehat{m}} \int\limits_0^\infty \left|\varphi(2r\sqrt{\lambda}) - 4\varphi(r\sqrt{\lambda}) + 3\right|^2 [\widehat{f_i}(\lambda)]^2 d\rho(\lambda)$$

$$= \int\limits_G \left| N\omega_N^{-1}r^{-N} \int\limits_{|z|\le r} \Delta_z^4 f(x)dz \right|^2 dx. \tag{2.5.12}$$

It follows from the definition of the Nikol'skii class and its norm in terms of the fourth differences (see Section 2.1.3) that the following inequality holds:

$$\|\Delta_z^4 f(x)\|_{L_2} \le |z|^\alpha \|f\|_{H_2^\alpha}. \tag{2.5.13}$$

Applying the Cauchy–Buniakowski inequality to the integral on the right-hand side of (2.5.12) and making use of inequality (2.5.13), we obtain

$$\int\limits_G \left| N\omega_N^{-1}r^{-N} \int\limits_{|z|\le r} \Delta_z^4 f(x)dz \right|^2 dx \le r^{2\alpha}\|f\|_{H_2^\alpha}^2. \tag{2.5.14}$$

A comparison of inequalities (2.5.12) and (2.5.14) leads to the following inequality:

$$\sum_{i=1}^{\widehat{m}} \int_0^\infty \left|\varphi(2r\sqrt{t}) - 4\varphi(r\sqrt{t}) + 3\right|^2 [\widehat{f}_i(t)]^2 d\rho(t) \le \frac{1}{4} r^{2\alpha} \|f\|^2_{H_2^\alpha}. \tag{2.5.15}$$

Making use of the series representation of a Bessel function, we obtain that for small positive values of t, the representation

$$|\varphi(2t) - 4\varphi(t) + 3|^2 = \frac{9}{(N+4)^2(N+2)^2} t^8 + O(t^{10}) \tag{2.5.16}$$

holds for the function $\varphi(t)$ defined by relation (2.5.8).

It follows from (2.5.16) that there are fixed positive numbers a and δ such that for all t in the segment $a \le t \le 2a$ the inequality

$$|\varphi(2t) - 4\varphi(t) + 3|^2 \ge \delta \tag{2.5.17}$$

holds.

One infers from a comparison of (2.5.15) and (2.5.17) that, given any r in the interval $0 < r < R/4$ and any α in the half segment $0 < \alpha \le 2$, the inequality

$$\sum_{i=1}^{\widehat{m}} \int_{a/r \le \sqrt{t} \le 2a/r} |\widehat{f}_i(t)|^2 d\rho(t) \le C r^{2\alpha} \|f\|^2_{H_2^\alpha} \tag{2.5.18}$$

holds with a constant C independent of r.

Let us proceed now with a direct proof of inequality (2.5.1).

Assume that a is a fixed number such that the validity of inequality (2.5.17) is ensured in the segment $a \le t \le 2a$.

It will suffice to prove the validity of inequality (2.5.1) in two separate cases: (1) at $\lambda \le (4a/R)^2$; (2) at $\lambda > (4a/R)^2$.

In the case $\lambda \le (4a/R)^2$, inequality (2.5.1) is a trivial sequence to the Parseval equality (2.1.11) written for the function $f(x)$.

Indeed, making use of this equality, we obtain that

$$\sum_{i=1}^{\widehat{m}} \int_\lambda^{4\lambda} \left|\widehat{f}_i(t)\right|^2 t^\alpha d\rho(t) \le (4\lambda)^\alpha \sum_{i=1}^{\widehat{m}} \int_0^\infty \left|\widehat{f}_i(t)\right|^2 d\rho(t)$$

$$= (4\lambda)^\alpha \|f\|^2_{L_2} \le (4\lambda)^\alpha \|f\|^2_{H_2^\alpha} \le \left(\frac{64a^2}{R^2}\right)^\alpha \|f\|^2_{H_2^\alpha}.$$

Thus, for the values of $\lambda \le (4a/R)^2$, inequality (2.5.1) is valid with a constant $C_R = (64a^2/R^2)^\alpha$.

Now, with reference to relation (2.5.18), we wish to prove inequality (2.5.1) in the latter case of $\lambda > (4a/R)^2$. We write inequality (2.5.18) for $r = a/\sqrt{\lambda}$, where λ is any number satisfying the condition $\lambda > (4a/R)^2$. Note that such r always satisfies the condition $0 < r < R/4$ that has been used in deriving inequality (2.5.18).

We obtain from (2.5.18) that

$$\sum_{i=1}^{\widehat{m}} \int_{\lambda}^{4\lambda} \left|\widehat{f}_i(t)\right|^2 d\rho(t) \leq C a^{2\alpha} \lambda^{-\alpha} \|f\|^2_{H_2^\alpha},$$

which, in turn, implies that

$$\sum_{i=1}^{\widehat{m}} \int_{\lambda}^{4\lambda} \left|\widehat{f}_i(t)\right|^2 t^\alpha d\rho(t) \leq (4\lambda)^\alpha \sum_{i=1}^{\widehat{m}} \int_{\lambda}^{4\lambda} \left|\widehat{f}_i(t)\right|^2 d\rho(t) \leq C a^{2\alpha} 4^\alpha \|f\|^2_{H_2^\alpha}.$$

We have therefore shown that inequality (2.5.1) also holds in the case $\lambda > (4a/R)^2$, with a constant $C_R = C(4a^2)^\alpha$, where C is the constant defined by inequality (2.5.18).

This completes the proof of inequality (2.5.1) subject to $0 < \alpha \leq 2$.

(2) Let us focus on the case $\alpha > 2$. We define a positive integer m subject to $2m < \alpha \leq 2m+2$. Successive application of Green's formula to the fundamental function $u_i(x, \lambda)$ and to one of the functions $f, \Delta f, \Delta^2 f, \dots, \Delta^{m-1} f$ yields the following relation between the Fourier images of the functions $f(x)$ and $\Delta^m f(x)$:

$$(\widehat{\Delta^m f})_i(\lambda) = (-1)^m \lambda^m \widehat{f}_i(\lambda). \tag{2.5.19}$$

Here we point out that the function $\Delta^m f(x)$ belongs to the class $H_2^{\alpha-2m}(G_R)$ with differentiability index $(\alpha - 2m)$ satisfying the condition $0 < \alpha - 2m \leq 2$. Therefore, by virtue of case (1) dealt with above, the following inequality holds:

$$\sum_{i=1}^{\widehat{m}} \int_{\lambda}^{4\lambda} \left|\left(\widehat{\Delta^m f}\right)_i\right|^2 t^{\alpha-2m} d\rho(t) \leq C_R \|\Delta^m f\|^2_{H_2^{\alpha-2m}} \leq C_R \|f\|^2_{H_2^\alpha}. \tag{2.5.20}$$

A comparison of inequalities (2.5.20) and (2.5.19) lends support to inequality (2.5.1) for an arbitrary $\alpha > 2$.

The proof of Lemma 2.6 is thus complete.

In this section, we shall prove another lemma on Fourier images which is in fact conducive to a theorem on localization of the Riesz means of order s $(0 \leq s < (N-1)/2)$ for a function in the class $H_2^{(N-1)/2-s}(G_R)$.

Prior to formulating this lemma, let us consider the integral

$$\overset{\nu_n}{I_t^\lambda}(R) = \int_R^\infty r^{-s} J_{\nu_n+s}(r\sqrt{\lambda}) J_{\nu_n-1}(r\sqrt{t})\,dr, \tag{2.5.21}$$

where $\nu_n = N/2 - n$, and the number n can take any of the values $n = 0, 1, \dots, [(N-1)/2]$, where $[(N-1)/2]$ is the integer part of a number $(N-1)/2$.

Since for any n the quantity ν_n satisfies the inequality $\nu_n = N/2 - n \geq 1/2$, the integral (2.5.21) belongs to a class of integrals of the type (2.4.9′) that have been dealt with in Section 2.4.2 and therefore for integral (2.5.21) for any $s \geq 0$, $\lambda \geq 1$, and $t \geq 1$ the following two estimates hold:[30)]

(a) for $|\sqrt{t} - \sqrt{\lambda}| \leq 1$,

$$\left| \overset{\nu_n}{I_t^\lambda}(R) \right| \leq C_1 R^{-s} \lambda^{-1/4} t^{-1/4}; \tag{2.5.22}$$

(b) for $|\sqrt{t} - \sqrt{\lambda}| \geq 1$,

$$\left| \overset{\nu_n}{I_t^\lambda}(R) \right| \leq C_2 R^{-1-s} \lambda^{-1/4} t^{-1/4} |\sqrt{t} - \sqrt{\lambda}|^{-1} \tag{2.5.23}$$

(with constants C_1 and C_2 independent of $\lambda \geq 1$, $t \geq 1$, and $0 < R \leq 1$).

LEMMA 2.7. *Assume that G is an arbitrary domain in the space E^N, $\widehat{A}$ is an arbitrary self-adjoint nonnegative extension of the Laplace operator $Lu = -\Delta u$ in the domain G; taken with respect to $\widehat{A}$ is an arbitrary ordered spectral representation of the space $L_2(G)$ with spectral measure $\rho(\lambda)$, sets of multiplicity e_i, fundamental functions $u_i(x, \lambda)$ $(i = 1, 2, \dots, \widehat{m})$ and of multiplicity $\widehat{m} \leq \infty$. Let $\alpha \leq N/2$ and R be two positive numbers such that $\alpha = \ell + \kappa$, where ℓ is a nonnegative integer, and κ is subject to the condition $0 < \kappa \leq 1$. Then for any sufficiently large $\lambda > 0$ and for any function $f(x)$ in the class $\overset{0}{H}{}_2^\alpha(G_R)$ the estimate below holds*

$$\sum_{i=1}^{\widehat{m}} \int_0^\infty \widehat{f}_i(t) u_i(x, t) t^{\ell/2 - (N-2)/4} \lambda^{1/4} \overset{\nu_\ell}{I_t^\lambda}(R) d\rho(t) = O(\lambda^{-\kappa/2}) \|f\|_{H_2^\alpha}, \tag{2.5.24}$$

uniform with respect to x in each subdomain G' strictly interior to domain G.

PROOF: For convenience, we introduce the following notation:

$$\widehat{f}_i(t) t^{\alpha/2} = a_i(t).$$

By virtue of Lemma 2.6, for any $\lambda > 0$ the inequality

$$\sum_{i=1}^{\widehat{m}} \int_\lambda^{4\lambda} |a_i(t)|^2 \, d\rho(t) \leq A^2 \tag{2.5.25}$$

---

30) These estimates are consequences of estimates (2.4.10′) and (2.4.11′), established in Section 2.4.2.

holds, where the constant A equals the norm

$$A = \|f\|_{H_2^\alpha}.$$

Estimate (2.5.24) is established if we prove that the inequality

$$\sum_{i=1}^{\widehat{m}} \int_0^\infty \left| a_i(t) u_i(x,t) I_t^{\lambda}(R) \right|^{\nu_\ell} \lambda^{1/4} t^{(2-N)/4-\kappa/2} d\rho(t) \le C\lambda^{-\kappa/2} A \tag{2.5.26}$$

holds uniformly with respect to x in any subdomain G' strictly interior to domain G (recall that $\ell/2 - \alpha/2 = -\kappa/2$).

Prior to proceeding with the proof of inequality (2.5.26), we shall want two ancillary estimates.

First prove that, given any $\varepsilon > 0$, the estimate

$$\sum_{i=1}^{\widehat{m}} \int_{t\ge 1} |a_i(t)|^2 t^{-\varepsilon} d\rho(t) \le C(\varepsilon) A^2 \tag{2.5.27}$$

holds for quantities $a_i(t)$ satisfying inequality (2.5.25). We represent the term on the left-hand side of (2.5.27) as the sum

$$\sum_{i=1}^{\widehat{m}} \int_{t\ge 1} |a_i(t)|^2 t^{-\varepsilon} d\rho(t) = \sum_{p=0}^{\infty} \left[\sum_{i=1}^{\widehat{m}} \int_{4^p}^{4^{p+1}} |a_i(t)|^2 t^{-\varepsilon} d\rho(t) \right].$$

Further, we factor the quantity $t^{-\varepsilon}$ (whose maximal value is $4^{-\varepsilon p}$) outside the bracketed integral sign and make use of inequality (2.5.25). We obtain

$$\sum_{i=1}^{\widehat{m}} \int_{t\ge 1} |a_i(t)|^2 t^{-\varepsilon} d\rho(t)$$
$$\le \sum_{p=0}^{\infty} 4^{-\varepsilon p} \left\{ \sum_{i=1}^{\widehat{m}} \int_{4^p}^{4^{p+1}} |a_i(t)|^2 d\rho(t) \right\} \le A^2 \sum_{p=0}^{\infty} 4^{-\varepsilon p} = A^2 (1 - 4^{-\varepsilon})^{-1}.$$

Thus, inequality (2.5.27) with the constant $C(\varepsilon) = (1 - 4^{-\varepsilon})^{-1}$ is proved.

Let us ensure now that if the quantities $a_i(t)$ satisfy inequality (2.5.25), then for any $\varepsilon > 0$ the estimate

$$\sum_{i=1}^{\widehat{m}} \int_{t\ge 1} |a_i(t) u_i(x,t)| t^{-\varepsilon - N/4} d\rho(t) \le CA, \tag{2.5.28}$$

uniform with respect to x in any subdomain G' strictly interior to domain G, holds.

Indeed, we obtain, applying the Cauchy–Buniakowski inequality to the left-hand side of (2.5.28) first for the integral and then for the sum, and making use of estimate (2.5.27) and estimate (2.3.9) taken at $\lambda = 1$, $\delta = \varepsilon$ and uniform in any subdomain G' strictly interior to domain G, that

$$\sum_{i=1}^{\widehat{m}} \int_{t\geq 1} |a_i(t)u_i(x,t)|\, t^{-\varepsilon-N/4}d\rho(t)$$

$$\leq \sum_{i=1}^{\widehat{m}} \left(\int_{t\geq 1} |a_i(t)|^2 t^{-\varepsilon} d\rho(t)\right)^{1/2} \left(\int_{t\geq 1} |u_i(x,t)|^2 t^{-\varepsilon-N/2} d\rho(t)\right)^{1/2}$$

$$\leq \left[\sum_{i=1}^{\widehat{m}} \int_{t\geq 1} |a_i(t)|^2 t^{-\varepsilon} d\rho(t)\right]^{1/2} \left[\int_{t\geq 1} |u_i(x,t)|^2 t^{-\varepsilon-N/2} d\rho(t)\right]^{1/2} \leq \text{const}\, A.$$

The estimate (2.5.28) is thus established.

We now proceed directly to a proof of inequality(2.5.26) uniform in each domain G' strictly interior to domain G.

It suffices to prove that there exist constants C_3, C_4, C_5, and C_6 such that the following four inequalities hold uniformly with respect to x in each domain G' strictly interior to domain G:

$$S_1 = \sum_{i=1}^{\widehat{m}} \int_{\sqrt{t}\leq 1} \left|a_i(t)u_i(x,t)\overset{\nu_\ell}{I_t^\lambda}(R)\right| \lambda^{1/4} t^{(2-N)/4-\kappa/2} d\rho(t) \leq C_3\lambda^{-\kappa/2}A, \tag{2.5.29}$$

$$S_2 = \sum_{i=1}^{\widehat{m}} \int_{1\leq\sqrt{t}\leq\sqrt{\lambda}/2} \left|a_i(t)u_i(x,t)\overset{\nu_\ell}{I_t^\lambda}(R)\right| \lambda^{1/4} t^{(2-N)/4-\kappa/2} d\rho(t) \leq C_4\lambda^{-\kappa/2}A, \tag{2.5.30}$$

$$S_3 = \sum_{i=1}^{\widehat{m}} \int_{|\sqrt{t}-\sqrt{\lambda}|\leq\sqrt{\lambda}/2} \left|a_i(t)u_i(x,t)\overset{\nu_\ell}{I_t^\lambda}(R)\right| \lambda^{1/4} t^{(2-N)/4-\kappa/2} d\rho(t) \leq C_5\lambda^{-\kappa/2}A, \tag{2.5.31}$$

$$S_4 = \sum_{i=1}^{\widehat{m}} \int_{3\sqrt{\lambda}/2\leq\sqrt{t}<\infty} \left|a_i(t)u_i(x,t)\overset{\nu_\ell}{I_t^\lambda}(R)\right| \lambda^{1/4} t^{(2-N)/4-\kappa/2} d\rho(t) \leq C_6\lambda^{-\kappa/2}A. \tag{2.5.32}$$

To begin, we wish to establish estimate (2.5.29); to this effect, making use of the relation $a_i(t) = \widehat{f}_i(t)^{\alpha/2}$, we write the left-hand side of (2.5.29) in the form

$$S_1 = \sum_{i=1}^{\widehat{m}} \int_{\sqrt{t}\leq 1} \left|\widehat{f}_i(t)u_i(x,t)\right| \left|\overset{\nu_\ell}{I_t^\lambda}(R)\right| \lambda^{1/4} t^{(2-N)/4+\ell/2} d\rho(t). \tag{2.5.33}$$

Let us ensure with the aid of relation (2.5.33) that, in order to establish (2.5.29), it will suffice for to obtain the estimate

$$\left|\overset{\nu_\ell}{I_t^\lambda}(R)\right| = O(t^{(N-2)/4-\ell/2}\lambda^{-3/4}) \tag{2.5.34}$$

for $\overset{\nu_\ell}{I_t^\lambda}(R)$ for all $t \geq 1$, $\lambda \geq 1$.

Indeed, once estimate (2.5.34) has been established, the following inequality stems from this estimate and from (2.5.33):

$$S_1 \leq C_7 \lambda^{-1/2} \sum_{i=1}^{\widehat{m}} \int_{t\leq 1} \left|\widehat{f}_i(t)u_i(x,t)\right| d\rho(t). \tag{2.5.35}$$

In order to obtain estimate (2.5.29) from inequality (2.5.35), it will suffice to take into account that $\lambda^{-1/2} \leq \lambda^{-\kappa/2}$, to apply the Cauchy–Buniakowski inequality to the right-hand side of (2.5.35) (first to the integral and then to the sum), and to make use of the estimate

$$\left[\sum_{i=1}^{\widehat{m}} \int_{t\leq 1} \left|\widehat{f}_i(t)\right|^2 d\rho(t)\right]^{1/2} \leq \|f\|_{L_2(G)} \leq \|f\|_{H_2^\alpha} = A,$$

$$\left[\sum_{i=1}^{\widehat{m}} \int_{t\leq 1} |\widehat{u}_i(x,t)|^2 d\rho(t)\right]^{1/2} \leq C_8,$$

of which the former follows from the Parseval equality (2.1.11) taken at $\mu = 0$, and the latter is a special case of estimate (2.3.4) at $\mu_0 = 1$.

Thus, to complete the proof of inequality (2.5.29), all we have to do is to establish estimate (2.5.34) for $0 \leq t \leq 1$, $\lambda \geq 1$.

We integrate the integral (2.5.21) m times by parts using the familiar recurrence formulas

$$\int \frac{J_{\nu+s}(r\sqrt{\lambda})}{r^{\nu+s-1}}\, dr = -\frac{J_{\nu+s-1}(r\sqrt{\lambda})}{r^{\nu+s-1}\sqrt{\lambda}}, \quad \frac{d}{dr}\left[r^{\nu-1}J_{\nu-1}(r\sqrt{t})\right] = r^{\nu-1}\sqrt{t}\, J_{\nu-2}(r\sqrt{t})$$

(here m is an integer satisfying the condition[31] $m > (N-1)/2 - \ell)$ to obtain

$$\overset{\nu_\ell}{I_t^\lambda}(R) = \sum_{p=1}^{m} \frac{(\sqrt{t})^{p-1}}{(\sqrt{\lambda})^p} J_{\nu_\ell+s-p}(R\sqrt{\lambda}) J_{\nu_\ell-p}(R\sqrt{t}) R^{-s}$$

$$+ \left(\sqrt{\frac{t}{\lambda}}\right)^m \int_R^\infty J_{\nu_\ell+s-m}(r\sqrt{\lambda}) J_{\nu_\ell-m-1}(r\sqrt{t}) r^{-s} dr. \tag{2.5.36}$$

Making use of the estimates

$$\left|J_{\nu_\ell+s-p}(R\sqrt{\lambda})\right| = O(R^{-1/2}\lambda^{-1/4}), \qquad \left|J_{\nu_\ell-p}(R\sqrt{t})\right| = O[(R\sqrt{t})^{\nu_\ell-p}],$$

$$\left|J_{\nu_\ell+s-m}(r\sqrt{\lambda})\right| = O(r^{-1/2}\lambda^{-1/4}), \qquad \left|J_{\nu_\ell-m-\ell}(r\sqrt{t})\right| = O[(r\sqrt{t})^{\nu_\ell-m-\ell}],$$

for the Bessel functions on the right-hand side of (2.5.36) and taking into account that $\nu_\ell = N/2 - \ell$ and that for $m > (N-1)/2 - \ell$ and for any $s \geq 0$

$$\int_R^\infty r^{-1/2+\nu_\ell-m-1-s} dr = \int_R^\infty r^{((N-1)/2-\ell)-m-1-s} dr < \infty,$$

we find that estimate (2.5.34) holds for all $0 < t < 1$, $\lambda \geq 1$.

This completes the proof of inequality (2.5.29).

To establish inequality (2.5.30), we make use of estimate (2.5.23), which, for values $1 \leq \sqrt{t} \leq \sqrt{\lambda}/2$, can be rewritten in the form

$$\left|\overset{\nu_\ell}{I_t^\lambda}(R)\right| = O(\lambda^{-3/4} t^{-1/4}). \tag{2.5.37}$$

Making use of estimate (2.5.37) we can majorize the left-hand side of (2.5.30) by the following quantity:

$$S_2 \leq C_9 \sum_{i=1}^{\widehat{m}} \int_{1 \leq \sqrt{t} \leq \sqrt{\lambda}/2} |a_i(t) u_i(x,t)| \, \lambda^{-1/2} t^{(1-N)/4-\kappa/2} d\rho(t). \tag{2.5.38}$$

It is to be noted now that since for the values of t the inequality $\lambda^{(\kappa-1)/2} \leq 2^{\kappa-1} t^{(\kappa-1)/2}$ holds, one has

$$\lambda^{-1/2} t^{(1-N)/4-\kappa/2} = \lambda^{-\kappa/2} \lambda^{(\kappa-1)/2} t^{(1-N)/4-\kappa/2} \leq 2^{\kappa-1} \lambda^{-\kappa/2} t^{-1/4-N/4}. \tag{2.5.39}$$

---

[31] Since $\ell < \alpha \leq N/2$, we have $\ell \leq [(N-1)/2]$ and, consequently, $m \geq 1$.

A comparison of (2.5.38) and (2.5.39) shows that

$$S_2 \le C_{10}\lambda^{-\kappa/2}\sum_{i=1}^{\widehat{m}}\int_{t\ge 1}|a_i(t)u_i(x,t)|t^{-1/4-N/4}d\rho(t),$$

and, in order to establish inequalities (2.5.30), it will suffice to make use of estimate (2.5.28) taken at $\varepsilon = 1/4$.

Inequality (2.5.32) is established in a manner analogous to that applied to inequality (2.5.30). Making use of estimate (2.5.23) which, for the values $3\sqrt{\lambda}/2 \le \sqrt{t} < \infty$, can be rewritten in the form

$$\left|\overset{\nu_\ell}{I_t^\lambda}(R)\right| = O(\lambda^{-1/4}t^{-3/4}),$$

we can majorize the left-hand side of (2.5.32) as shown below:

$$S_4 \le C_{11}\sum_{i=1}^{\widehat{m}}\int_{3\sqrt{\lambda}/2\le t<\infty}|a_i(t)u_i(x,t)|t^{-1/4-N/4-\kappa/2}d\rho(t). \tag{2.5.40}$$

It remains to note that the values of t are subject to the inequality

$$t^{-\kappa/2} \le \left(\frac{3}{2}\right)^{-\kappa}\lambda^{-\kappa/2},$$

which, in comparison with (2.5.40), yields the estimate

$$S_4 \le C_{12}\lambda^{-\kappa/2}\sum_{i=1}^{\widehat{m}}\int_{t\ge 1}|a_i(t)u_i(x,t)|t^{-1/4-N/4}d\rho(t). \tag{2.5.41}$$

By virtue of estimate (2.5.28) taken at $\varepsilon = 1/4$, the inequality (2.5.32) stems from (2.5.41).

Now all we have to do is prove inequality (2.5.31).

We denote by p the least of the numbers for which $2^p - 1 \ge \sqrt{\lambda}/2$, and rewrite the left-hand side of (2.5.31) in the following manner:

$$S_3 \le \sum_{k=1}^{p}\left\{\sum_{i=1}^{\widehat{m}}\int_{2^{k-1}-1\le|\sqrt{t}-\sqrt{\lambda}|<2^k-1}\left|a_i(t)u_i(x,t)\overset{\nu_\ell}{I_t^\lambda}(R)\right|\lambda^{1/4}t^{(2-N)/4-\kappa/2}d\rho(t)\right\}. \tag{2.5.42}$$

A comparison of (2.5.22) and (2.5.23) shows that the estimate

$$\left|\overset{\nu_\ell}{I_t^\lambda}(R)\right| = O[\lambda^{-1/4}t^{-1/4}(1+|\sqrt{t}-\sqrt{\lambda}|)^{-1}] \tag{2.5.43}$$

is valid for any fixed $R > 0$ and for all $\lambda \geq 1$ and $t \geq 1$. It follows from (2.5.42) and (2.5.43) that

$$S_3 \leq C_{13} \sum_{k=1}^{p} \left\{ \sum_{i=1}^{\widehat{m}} \int\limits_{2^{k-1}-1 \leq |\sqrt{t}-\sqrt{\lambda}| \leq 2^k - 1} |a_r(t) u_r(x,t)| \frac{t^{(1-N)/4-\kappa/2}}{1+|\sqrt{t}-\sqrt{\lambda}|} d\rho(t) \right\}. \tag{2.5.44}$$

We denote by A_k the term within the braces in (2.5.44). Since for all values of t falling into the segment $\sqrt{\lambda}/2 \leq \sqrt{t} \leq 3\sqrt{\lambda}/2$ the inequality $t^{(1-N)/4-\kappa/2} \leq 4^{(N-1)/4+\kappa/2} \lambda^{(1-N)/4-\kappa/2}$ holds, the estimate

$$A_k \leq 4^{(N-1)/4+\kappa/2} \lambda^{(1-N)/4-\kappa/2} 2^{-(k-1)} \left[\sum_{i=1}^{\widehat{m}} \int\limits_{|\sqrt{t}-\sqrt{\lambda}| \leq 2^k - 1} |a_i(t) u_i(x,t)| d\rho(t) \right] \tag{2.5.45}$$

also holds for A_k.

Having applied the Cauchy–Buniakowski inequality first to the integral and then to the sum within the square brackets in (2.5.45), we obtain

$$\left[\sum_{i=1}^{\widehat{m}} \int\limits_{|\sqrt{t}-\sqrt{\lambda}| \leq 2^k - 1} |a_i(t) u_i(x,t)| d\rho(t) \right]$$

$$\leq \left[\sum_{i=1}^{\widehat{m}} \int\limits_{\sqrt{\lambda}/2 \leq \sqrt{t} \leq 3\sqrt{\lambda}/2} |a_i(t)|^2 d\rho(t) \right]^{1/2} \left[\sum_{i=1}^{\widehat{m}} \int\limits_{|\sqrt{t}-\sqrt{\lambda}| \leq 2^k} |u_i(x,t)|^2 d\rho(t) \right]^{1/2}. \tag{2.5.46}$$

By virtue of estimate (2.5.25) and estimate (2.3.7) taken at $\mu = \sqrt{\lambda}$, $\rho_0 = 2^k$, we obtain the following inequalities:

$$\sum_{i=1}^{\widehat{m}} \int\limits_{\sqrt{\lambda}/2 \leq \sqrt{t} \leq 3\sqrt{\lambda}/2} |a_i(t)|^2 d\rho(t) = \sum_{i=1}^{\widehat{m}} \int\limits_{\lambda/4 \leq t \leq 9\lambda/4} |a_i(t)|^2 d\rho(t) \leq 3A^2, \tag{2.5.47}$$

$$\sum_{i=1}^{\widehat{m}} \int\limits_{|\sqrt{t}-\sqrt{\lambda}| \leq 2^k} |u_i(x,t)|^2 d\rho(t) \leq C_{14} 2^k \lambda^{(N-1)/2}. \tag{2.5.48}$$

Finally, we obtain from inequalities (2.5.44) – (2.5.48)

$$S_3 \leq C_{13} \sum_{k=1}^{p} A_k \leq C_{15} \lambda^{(1-N)/4-\kappa/2} \sum_{k=1}^{p} 2^{-(k-1)} A\, 2^{k/2} \lambda^{(N-1)/4}$$

$$= 2C_{15}\lambda^{-\kappa/2}A\sum_{k=1}^{p}2^{-k/2} \le C_{16}\lambda^{-\kappa/2}A.$$

This completes the derivation of inequality (2.5.31).

Lemma 2.7 is thus proved.

2.5.2. Properties of Derivatives Taken with Respect to the Sphere Radius and to the Sphere Means for Functions of the Nikol'skii Class

In this section, we consider functions that belong to an arbitrary domain D of the space E^N of the Nikol'skii class H_p^α with positive differentiability index α representable in the form $\alpha = \ell + \kappa$, where ℓ is a nonnegative integer and κ belongs to the half-segment $0 < \kappa \le 1$. Here $\ell = [\alpha]$ (where $[\alpha]$ is the integral part of the number α) in the case where α is not an integer, and $\ell = \alpha - 1$ in the case where α is an integer, and $\kappa = \alpha - [\alpha]$.

We fix arbitrary $r > 0$ and $h > 0$, and introduce, in the space E^N, spherical coordinates $(x + r\omega)$ centered at an arbitrarily fixed point x of a subset D_{r+2h}; we denote by $f^{(\ell)}(r,\omega)$ the derivative of the function $f(x)$ with respect to the radius r of order ℓ in the direction determined by angles ω.

We consider now, for the function $f(r,\omega)$ and for the derivative $f^{(\ell)}(r,\omega)$, the second differences with respect to the radius r in the direction determined by the angles ω:

$$\tilde{\Delta}_h^2 f(r,\omega) = f(r+2h,\omega) - 2f(r+h,\omega) + f(r,\omega), \tag{2.5.49}$$

$$\tilde{\Delta}_h^2 f^{(\ell)}(r,\omega) = f^{(\ell)}(r+2h,\omega) - 2f^{(\ell)}(r+h,\omega) + f^{(\ell)}(r,\omega). \tag{2.5.50}$$

Let us find an estimate of the L_p-norm of the second difference (2.5.50) for a function $f(x)$ of the Nikol'skii class $H_p^\alpha(D)$.

LEMMA 2.8. *If $\alpha > 0$ is representable in the form $\alpha = \ell + \kappa$, where ℓ is a nonnegative integer, $0 < \kappa \le 1$, then for any function $f(x)$ in the Nikol'skii class $H_p^\alpha(D)$, given any $p \ge 1$ and any $r > 0$ and $h > 0$, the following estimate holds for the second difference (2.5.50):*

$$\|\tilde{\Delta}_h^2 f^{(\ell)}(r,\omega)\|_{L_p(D_{r+2h})} \le Ch^\kappa \|f\|_{H_p^\alpha(D)}. \tag{2.5.51}$$

Making use of the notation (2.1.12) for the partial derivative and denoting the product $x_1^{k_1}x_2^{k_2}\dots x_N^{k_N}$ by the symbol $x^{\bar{k}}$ (here the multi-index $\bar{k}$ refers to the integer-valued coordinates $\bar{k} = (k_1, k_2, \dots, k_N)$), we can write the ℓth partial derivative of the function $f(x)$ with respect to the radius r in the direction determined by angles ω in the following manner:

$$f^{(\ell)}(r,\omega) = \sum_{|\bar{k}|=\ell} \partial^{\bar{k}} f(x) \frac{x^{\bar{k}}}{r^{\ell}}. \tag{2.5.52}$$

Since the direction determined by the angles ω is fixed, the quantity $x^{\bar{k}}/r^{\ell}$ at $|\bar{k}| = \ell$, which is the product of direction cosines, is independent of r. It is seen, therefore, that in order to prove inequality (2.5.51), it will suffice to consider the case $0 < \alpha \leq 1$, that is, when $\ell = 0$, $0 < \kappa < 1$.

It is our intention now to verify the inequality

$$||\tilde{\Delta}_h^2 f(r,\omega)||_{L_p(D_{r+2h})} \leq Ch^{\kappa}||f||_{H_p^{\alpha}(D)} \tag{2.5.53}$$

for a function $f(x)$ in the class $H_p^{\alpha}(D)$ at $0 < \alpha \leq 1$, $\kappa = \alpha$.

As pointed out in Section 2.1.3 (see relations (2.1.20) and (2.1.21)), an arbitrary function $f(x)$ in the Nikol'skii class $H_p^{\alpha}(D)$ can be represented in the form of the serial sum

$$f(x) = \sum_{m=0}^{\infty} Q_m(x), \tag{2.5.54}$$

where the mth term $Q_m(x)$ is an integral function of exponential type 2^m with norm in $L_p(D)$ satisfying the inequality

$$||Q_m||_{L_p(D)} \leq C2^{-m\alpha}||f||_{H_p^{\alpha}(D)} \tag{2.5.55}$$

where the constant C is independent of both m and the function $f(x)$.

Representation (2.5.54) implies that in order to verify inequality (2.5.53), it suffices to prove that

$$\sum_{m=0}^{\infty} ||\tilde{\Delta}_h^2 Q_m(x)||_{L_p(D_{r+2h})} \leq Ch^{\alpha}||f||_{H_p^{\alpha}(D)}. \tag{2.5.56}$$

It can be easily shown that, for any $x \in D_{r+2h}$,

$$\begin{aligned}\tilde{\Delta}_h^2 Q_m(x) &= Q_m(r+2h,\omega) - 2Q_m(r+h,\omega) + Q_m(r,\omega)\\ &= \int_0^h \left[\int_0^h \frac{\partial^2 Q_m}{\partial r^2}(r+t+s,\omega)dt\right] ds.\end{aligned} \tag{2.5.57}$$

It follows from the above relation that

$$||\tilde{\Delta}_h^2 Q_m(x)||^p_{L_p(D_{r+2h})} = \int_{D_{r+2h}} \left|\int_0^h \left[\int_0^h \frac{\partial^2 Q_m}{\partial r^2}(r+t+s)dt\right] ds\right|^p dx. \tag{2.5.58}$$

Applying the Hölder inequality to the double integral within the modulus brackets on the right-hand side of (2.5.58), we have

$$\left|\int_0^h \left[\int_0^h \frac{\partial^2 Q_m}{\partial r^2}(r+t+s)dt\right] ds\right|^p \leq \left[\int_0^h\int_0^h \left|\frac{\partial^2 Q_m}{\partial r^2}(r+t+s)\right|^p dtds\right] h^{2p-2}. \quad (2.5.59)$$

Comparing relations (2.5.58) and (2.5.59) and changing the order of integration with respect to x and with respect to t, s, we obtain that

$$\begin{aligned} &\|\widetilde{\Delta}_h^2\, Q_m(x)\|^p_{L_p(D_{r+2h})} \\ &\leq h^{2p-2}\int_0^h dt \int_0^h ds \int_{D_{r+2h}} \left|\frac{\partial^2 Q_m}{\partial r^2}(r+t+s)\right|^p r^{N-1} dr\, d\omega \\ &\leq h^{2p-2}\int_0^h dt\int_0^h ds \|Q_m(r,\omega)\|^p_{L_p(D_{r+2h})} \leq h^{2p}\|Q_m^{(2)}(r,\omega)\|^p_{L_p(D_{r+2h})}. \end{aligned} \quad (2.5.60)$$

Further, we make use of a relation of the form (2.5.52) for the derivative $Q_m^{(2)}(r,\omega)$:

$$Q_m^{(2)}(r,\omega) = \sum_{|\bar{k}|=2} \partial^{\bar{k}} Q_m(x)\frac{x^{\bar{k}}}{r^2} \quad (2.5.61)$$

and invoke the familiar Bernstein inequality (see inequality (2.1.22) in Section 2.1.3) for estimating the L_p-norm of $\partial^{\bar{k}} Q_m(x)$

$$\|\partial^{\bar{k}} Q_m(x)\|_{L_p(D)} \leq 2^{2m}\|Q_m(x)\|_{L_p(D)}. \quad (2.5.62)$$

Taking into account that $|x^{\bar{k}}/r^2| \leq 1$, it follows from (2.5.61) and (2.5.62) that

$$\|Q_m^{(2)}(r,\omega)\|_{L_p(D_{r+2h})} \leq \sum_{|\bar{k}|=2} \|\partial^{\bar{k}} Q_m(x)\|_{L_p(D)} \leq C 2^{2m}\|Q_m(x)\|_{L_p(D)}. \quad (2.5.63)$$

Finally, comparing inequalities (2.5.60) and (2.5.63), we arrive at the estimate

$$\|\widetilde{\Delta}_h^2 Q_m(x)\|_{L_p(D_{r+2h})} \leq C h^2 2^{2m}\|Q_m(x)\|_{L_p(D)}. \quad (2.5.64)$$

Let us now prove inequality (2.5.56). For each $h > 0$, we fix a number $n = n(h)$ subject to the condition

$$2^n < h^{-1} \leq 2^{n+1} \quad (2.5.65)$$

and split the sum on the left-hand side of (2.5.56) into two sums:

$$S_1 = \sum_{m=0}^{n} \|\widetilde{\Delta}_h^2 Q_m(x)\|_{L_p(D_{2h})}, \qquad S_2 = \sum_{m=n+1}^{\infty} \|\widetilde{\Delta}_h^2 Q_m(x)\|_{L_p(D_{2h})}.$$

It suffices to prove that either sum is equal to $O(h^\alpha)\|f\|_{H_p^\alpha(D)}$.

In estimating S_1 we make use of inequalities (2.5.64) and (2.5.55) and take into account that, by virtue of (2.5.65), $2^{n(2-\alpha)} < h^{\alpha-2}$. We obtain

$$\begin{aligned} S_1 &\le Ch^2 \sum_{m=0}^{n} 2^{2m}\|Q_m(x)\|_{L_p(D)} \le Ch^2\|f\|_{H_p^\alpha(D)} \sum_{m=0}^{n} 2^{m(2-\alpha)} \\ &\le C_1 h^2 2^{n(2-\alpha)}\|f\|_{H_p^\alpha(D)} \le C_1 h^\alpha \|f\|_{H_p^\alpha(D)}. \end{aligned} \tag{2.5.66}$$

In estimating S_2 we note, first of all, that it follows for the second difference from expression (2.5.49) that

$$S_2 = \sum_{m=n+1}^{\infty} \|\widetilde{\Delta}_h^2 Q_m(x)\|_{L_p(D_{r+2h})} \le 4 \sum_{m=n+1}^{\infty} \|Q_m(x)\|_{L_p(D)}.$$

Further, making use of inequality (2.5.55) and of the right-hand inequality (2.5.65), we obtain

$$\begin{aligned} S_2 &\le 4C\|f\|_{H_p^\alpha(D)} \sum_{m=n+1}^{\infty} 2^{-m\alpha} \le C_2\|f\|_{H_p^\alpha(D)} 2^{-(n+1)\alpha} \\ &\le C_2 h^\alpha \|f\|_{H_p^\alpha(D)}. \end{aligned} \tag{2.5.67}$$

Inequalities (2.5.66) and (2.5.67) thus lend support to the validity of estimate (2.5.56).

This completes the proof of Lemma 2.8.

Suppose, by analogy with the above, that the function $f(x)$ belongs to the Nikol'skii class $H_p^\alpha(D)$, where $\alpha = \ell + \kappa$, ℓ being a nonnegative integer and $0 < \kappa \le 1$; this time, however, we invoke the additional requirement that the condition $\alpha p > N$ be fulfilled.

Then the embedding theorem ensures in any case the continuity of the function on the set D_R for any $R > 0$, and for any point $x \in D_{2R}$, given any $0 < r \le R$, we can consider the mean value of the function $f(x)$ on the surface of a sphere of radius r centered at point x. We denote the mean value by the symbol $\psi(r) = \psi_x(r)$, that is, we set by definition

$$\psi(r) = \psi_x(r) = \omega_N^{-1} \int_\omega \cdots \int f(x + r\omega)d\omega, \tag{2.5.68}$$

where, by analogy with the foregoing, the symbol $\int\limits_{\omega}\dots\int f(x+r\omega)d\omega$ denotes an integral over all the angles on the surface of an N-dimensional sphere of radius r centered at point x, and $\omega_N = 2(\sqrt{\pi})^N[\Gamma(N/2)]^{-1}$.

We now prove the following assertion.

LEMMA 2.9. *Assume that a function $f(x)$ belongs to the class $H_p^\alpha(D)$ at a positive α equal to $\ell+\kappa$, where ℓ is a nonnegative integer and $0<\kappa\le 1$; also, let $p\ge 1$, $p\alpha > N$. Then, given any positive R and any x from the set D_{3R}, for any integer m taking values $0,1,\dots,\ell$, each of the functions*[32]

$$\varphi_m(r) = r^{m+\kappa-1}\psi^{(m)}(r) \tag{2.5.69}$$

is summable on the segment $0\le r\le R$. The more so, for each h in an interval $0<h<R/3$, for the second difference of each of the functions (2.5.69), the estimate below holds

$$\int_0^R |\varphi_m(r+2h) - 2\varphi_m(r+h) + \varphi_m(r)|\,dr \le Ch^\kappa \|f\|_{H_p^\alpha(D)}, \tag{2.5.70}$$

uniform with respect to x in D_{3R}.

PROOF: The condition $f(x)\in H_p^\alpha(D)$ for $\alpha p > N$ and the familiar embedding theorem[33] imply that there is an $\varepsilon>0$ such that $f(x)$ belongs to the Zygmund–Hölder class $C^\varepsilon(D_R)$; strictly speaking, the embedding of the Zygmund–Hölder class into the Nikol'skii class holds,

$$H_p^\alpha(D) \longrightarrow C^\varepsilon(D_R) \qquad \text{for} \qquad \varepsilon > 0. \tag{2.5.71}$$

One infers from the condition $f(x)\in H_p^\alpha(D)$ for $\alpha p > N$ and from the embedding theorems (2.1.24) and (2.1.18) (see Section 2.1.3) that there exists a number $p_1 = p_1(m)$ such that the embedding of the Sobolev–Liouville class into the Nikol'skii class is valid,

$$H_p^\alpha(D) \longrightarrow L_{p_1}^m(D_R) \qquad \text{for} \qquad p_1 m > N. \tag{2.5.72}$$

Finally, the same condition $f(x)\in H_p^\alpha(D)$ for $\alpha p > N$ and the same embedding theorem imply that, for each integer m in a segment $0\le m\le \ell$, there is a number $p_2 = p_2(m)$ such that the embedding for Nikol'skii classes holds,

$$H_p^\alpha(D) \longrightarrow H_{p_2}^{m+\kappa}(D_R) \qquad \text{for} \qquad p_2(m+\kappa) > N. \tag{2.5.73}$$

First, we intend to prove Lemma 2.9 for the case $m>0$, that is, for the case where $m = 1,2,\dots,\ell$.

---

32) Here $\psi^{(m)}(r)$ symbolizes a derivative of order m of the mean value of (2.5.68).

33) See the embedding theorem (2.1.26) in Section 2.1.3, according to which the embedding $H_p^\alpha \to C^\varepsilon$ takes place at $\varepsilon = \alpha - N/p > 0$, given $p\alpha > N$.

To begin, we shall prove for this particular case the summability on a segment $[0, R]$ of each function (2.5.69) and we shall establish an estimate of the L_1-norm with respect to r on the segment $0 \leq r \leq R$ of each of the functions $\psi^{(m)}(r)r^{m-1}$ through the use of $||f||_{H_p^\alpha(D)}$.

Invoking the embedding theorem (2.5.72) and the definition of a norm in the Sobolev–Liouville class with integer-valued order of differentiability (see relation (2.1.13)) and applying the Hölder inequality with exponents p_1 and $q_1 = p_1/(p_1 - 1)$, we obtain for any m (noting that $p_1 m > N$, that is, $p_1 > N/m$, and, consequently, $q_1 < N/(N-m)$)

$$\int_0^R \left|\psi^{(m)}(r)\right| r^{m-1} dr \leq \sum_{|\bar{k}|=m} \int_{r \leq R} \left|\partial^{\bar{k}} f(y)\right| r^{m-1-(N-1)} dy$$

$$\leq \sum_{|\bar{k}|=m} ||\partial^{\bar{k}} f(y)||_{L_{p_1}(D_R)} \left[\int_{r \leq R} r^{-(N-m)q_1} dy\right]^{1/q_1}$$

$$\leq C_1 ||f||^m_{L_{p_1}(D_R)} \leq C_2 ||f||_{H_p^\alpha(D)}.$$

Thus, for $1 \leq m \leq \ell$ uniformly with respect to $x \in D_{2R}$ we have

$$\int_0^R \left|\psi^{(m)}(r)\right| r^{m-1} dr \leq C_2 ||f||_{H_p^\alpha(D)}, \tag{2.5.74}$$

and, in particular, the function $r^{m-1}\psi^{(m)}(r)$ belongs to the class $L_1[0, R]$. Since $0 < \kappa \leq 1$, the function $\varphi_m(r)$, equivalent to $r^\kappa[r^{m-1}\psi^{(m)}(r)]$, also belongs to the class $L_1[0, R]$.

We will now establish that the inequality (2.5.70) holds true for the function $\varphi_m(r)$ at $1 \leq m \leq \ell$.

We split the integral on the left-side of (2.5.70) into two integrals:

$$\int_0^R |\varphi_m(r+2h) - 2\varphi_m(r+h) + \varphi_m(r)| dr = I_1 + I_2, \tag{2.5.75}$$

where

$$I_1 = \int_0^h |\varphi_m(r+2h) - 2\varphi_m(r+h) + \varphi_m(r)| dr, \tag{2.5.76}$$

$$I_2 = \int_h^R |\varphi_m(r+2h) - 2\varphi_m(r+h) + \varphi_m(r)| dr. \tag{2.5.77}$$

It suffices to prove that an estimate of the form (2.5.70) is valid uniformly with respect to $x \in D_{3R}$ for either of the integrals (2.5.76) or (2.5.77).

Proceeding with the proof of such an estimate for integral (2.5.76), we recall that, for $0 \le r \le h$, the estimate

$$\int_0^h |\varphi_m(r+2h) - 2\varphi_m(r+h) + \varphi_m(r)|dr \le 2\int_0^{3h} |\varphi_m(r)|dr$$

holds. Making use of the condition $h < R/3$ and of the inequality (2.5.74) we obtain for any $x \in D_{3R}$

$$\begin{aligned} |I_1| &\le 2\int_0^{3h} |\varphi_m(r)|dr = 2\int_0^{3h} |\psi^{(m)}(r)|r^{m+\kappa-1}dr \\ &\le 2(3h)^\kappa \int_0^R |\psi^{(m)}(r)|r^{m-1}dr \le C_3 h^\kappa \|f\|_{H_p^\alpha(D)}. \end{aligned} \tag{2.5.78}$$

To derive an analogous estimate for I_2, we represent the integrand in (2.5.77) in the form

$$\begin{aligned} &|\varphi_m(r+2h) - 2\varphi_m(r+h) + \varphi_m(r)| \\ &= \Big|(r+2h)^{m+\kappa-1}[\psi^{(m)}(r+2h) - 2\psi^{(m)}(r+h) + \psi^{(m)}(r)] \\ &+ 2\psi^{(m)}(r+h)[(r+2h)^{m+\kappa-1} - (r+h)^{m+\kappa-1}] \\ &- \psi^{(m)}(r)[(r+2h)^{m+\kappa-1} - r^{m+\kappa-1}]\Big|. \end{aligned} \tag{2.5.79}$$

Recalling that h and r under the sign of the integral in (2.5.77) are subject to the condition $0 \le h \le r$ and applying the notation (2.5.49) to the second difference, we obtain the following inequality from (2.5.79):

$$\begin{aligned} \left|\widetilde{\Delta}_h^2 \varphi_m(r)\right| &\le (r+2h)^{m+\kappa-1}\left|\widetilde{\Delta}_h^2 \psi^{(m)}(r)\right| \\ &+ C_4 h^\kappa \left\{|\psi^{(m)}(r)|r^{m-1} + |\psi^{(m)}(r+h)|(r+h)^{m-1}\right\}. \end{aligned} \tag{2.5.80}$$

Applying inequality (2.5.80) to (2.5.77), we obtain

$$|I_2| \leq \int_h^R (r+2h)^{m+\kappa-1} \left|\widetilde{\Delta}_h^2 \psi^{(m)}(r)\right| dr$$
$$+ C_4 h^\kappa \int_h^R \left\{ |\psi^{(m)}(r)| r^{m-1} + |\psi^{(m)}(r+h)|(r+h)^{m-1} \right\} dr. \quad (2.5.81)$$

It follows immediately from estimate (2.5.74) and from the conditions $x \in D_{2R}$ and $r+h < 3R/2$ that the latter integral on the right-hand side of (2.5.81) has the required order uniformly with respect to $x \in D_{3R}$.

With a view to providing evidence that the former integral on the right-hand side of (2.5.81) has the same order, we make use of the embedding theorem (2.5.73) and of the estimate (2.5.51) taken at $\alpha = m+\kappa$, $\ell = m$, $p = p_2$. Applying the Hölder inequality with exponents p_2 and $q_2 = p_2/(p_2-1)$[34] and making use of estimate (2.5.51), we obtain for any $h < R/3$

$$\int_h^R (r+2h)^{\ m+\kappa-1} \left|\widetilde{\Delta}_h^2 \psi^{(m)}(r)\right| dr \leq 3^{m+\kappa-1} \int_{r \leq R} \left|\widetilde{\Delta}_h^2 f^{(m)}(r,\omega)\right| r^{m+\kappa-N} dr$$

$$\leq 3^{m+\kappa-1} \|\widetilde{\Delta}_h^2 f^{(m)}(r,\omega)\|_{L_{p_2}(D_{2R})} \left[\int_{r \leq R} r^{(m+\kappa-N)q_2} dy \right]^{1/q_2}$$

$$\leq C_5 h^\kappa \|f\|_{H_{p_2}^{m+\kappa}(D_{R/3})} \leq C_6 h^\kappa \|f\|_{H_p^\alpha(D)}.$$

Thus, we have proved that the two integrals on the right-hand side of (2.5.81) have the required order $O(h^\kappa)\|f\|_{H_p^\alpha(D)}$ uniformly with respect to $x \in D_{3R}$, which implies that

$$|I_2| \leq C_7 h^\kappa \|f\|_{H_p^\alpha(D)}. \quad (2.5.82)$$

A comparison of relation (2.5.75) with estimates (2.5.78) and (2.5.82) brings to completion the proof of estimate (2.5.70) for the case $1 \leq m \leq \ell$.

All we have left to do is to prove Lemma 2.9 for the case $m = 0$. In doing so, we consider the function $\bar{\psi}(r) = \psi_x(r) - f(x)$ instead of the function $\psi(r) = \psi_x(r)$, and the function

$$\bar{\varphi}_0(r) = r^{\kappa-1}\,\bar{\psi}(r) = \varphi_0(r) - f(x) r^{\kappa-1} \quad (2.5.83)$$

instead of the function $\varphi_0(r) = r^{\kappa-1}\psi(x)$.

[34] One takes into account that since $p_2(m+\kappa) > N$, that is, $p_2 > N/(m+\kappa)$, then $q_2 < N/(N-m+\kappa)$.

We observe, first of all, that since $0 < \kappa \leq 1$ and the quantity $|f(x)|$ is uniformly-in-D_R bounded by a constant of the form $C\|f\|_{H_p^\alpha(D)}$, the function $f(x)r^{\kappa-1}$ is summable with respect to r on the segment $0 \leq r \leq R$. For this reason, by virtue of (2.5.83), the summability of $\varphi_0(r)$ on the segment $0 \leq r \leq R$ immediately follows from the summability of the function $\bar{\varphi}_0(r)$ on this segment.

We now proceed to prove the summability on the segment $0 \leq r \leq R$ of not only the function $\bar{\varphi}_0(f) = r^{\kappa-1}\bar{\psi}(r)$, but also the function $r^{-1}\bar{\psi}(r)$ to obtain for the L_1-norm of the latter function with respect to r on the segment $0 \leq r \leq R$ an estimate which is an analog of estimate (2.5.74).

We note, by virtue of the relation $\bar{\psi}(r) = \psi(r) - f(x)$, that

$$\int_0^R |\bar{\psi}(r)|\, r^{-1} dr = \int_0^R \left| \frac{1}{\omega_N} \int \underset{\omega}{\cdots} \int f(r,\omega) d\omega - f(x) \frac{1}{\omega_N} \int \underset{\omega}{\cdots} \int d\omega \right| r^{-1} dr$$
$$\leq \frac{1}{\omega_N} \int_{r \leq R} |f(y) - f(x)| r^{-N} dy. \tag{2.5.84}$$

It follows from the definition of the Zygmund–Hölder class and from the embedding theorem (2.5.71) that the integral on the right-hand side of (2.5.84) for any $x \in D_{2R}$ does not exceed the quantity

$$C_8 \|f\|_{C^\varepsilon(D_R)} \int_{r \leq R} r^{\varepsilon - N} dy \leq C_9 \|f\|_{H_p^\alpha(D)}. \tag{2.5.85}$$

For any $x \in D_{2R}$, an estimate is derived from inequalities (2.5.84) and (2.5.85)

$$\int_0^R |\bar{\psi}(r)| r^{-1} dr \leq C_9 \|f\|_{H_p^\alpha(D)}, \tag{2.5.86}$$

which establishes, in particular, the summability on the segment $0 \leq r \leq R$ of the function $\bar{\psi}(r)r^{-1}$ and, consequently, of the function $\bar{\psi}_0(r) = \bar{\psi}(r)r^{\kappa-1}$.

It remains to prove estimate (2.5.70) for the function $\varphi_0(r)$. We ensure that to this effect it suffices to prove an estimate of the form (2.5.70) for the function $\bar{\varphi}_0(r)$.

Indeed, since $\tilde{\Delta}_h^2 \varphi_0(r) = \tilde{\Delta}_h^2 \bar{\varphi}_0(r) + f(x)\tilde{\Delta}_h^2(r^{\kappa-1})$ and $|f(x)| \leq C_{10}\|f\|_{H_p^\alpha(D)}$ for $x \in D_R$, it suffices to ensure that

$$\int_0^R \left| \tilde{\Delta}_h^2 (r^{\kappa-1}) \right| dr \leq C_{11} h^\kappa,$$

which poses no difficulty.

Indeed, one infers from $\widetilde{\Delta}_h^2(r^{\kappa-1}) = (r+2h)^{\kappa-1} - 2(r+h)^{\kappa-1} + r^{\kappa-1}$ that $|\widetilde{\Delta}_h^2(r^{\kappa-1})| = O(h^{\kappa-1})$ for $0 < r < 4h$ and $|\widetilde{\Delta}_h^2(r^{\kappa-1})| = O(h^2 r^{\kappa-3})$ for $4h \le r \le R$ and, consequently

$$\int_0^R \left|\widetilde{\Delta}_h^2(r^{\kappa-1})\right| dr = O(h^{\kappa-1}) \int_0^{4h} dr + O(h^2) \int_{4h}^R r^{\kappa-3} dr = O(h^\kappa).$$

Thus, it suffices to obtain an estimate of the form (2.5.70) at $m = 0$, however, for the function $\bar{\varphi}_0(r)$, rather than for the function $\varphi_0(r)$. The estimate in question is established by following the same procedure used in establishing estimate (2.5.70) at $m \ge 1$, the only distinction being that estimate (2.5.86) should be used in place of estimate (2.5.74).

The proof of Lemma 2.9 is thus complete.

To formulate and prove the next lemma, we shall need two functions. One of them is

$$V_\nu(t) = \sqrt{t}\, J_\nu(t), \tag{2.5.87}$$

which is the product of $\sqrt{t}$ and a Bessel function of the first kind of order $\nu \ge -1/2$. The other function is

$$F(r) = r^{N-1}\psi(r) = r^{N-1}\omega_N^{-1} \int \dots \int_\omega f(x + r\omega) d\omega, \tag{2.5.88}$$

which is the product of r^{N-1} and the mean value (2.5.68) of the function $f(x)$ on the surface of an N-dimensional sphere of radius r centered at point x.

We use the symbol $\mathcal{D}F(r)$ to denote the operation

$$\mathcal{D}F(r) = \frac{d}{dr}\left[\frac{1}{r}F(r)\right],$$

and the symbol $\mathcal{D}^k F(r)$ to denote the result of operation $\mathcal{D}$ successively applied k times, so that $\mathcal{D}^k F(r) = \mathcal{D}[\mathcal{D}^{k-1}F(r)]$.

LEMMA 2.10. *Assume that $N \ge 2$, D is an arbitrary domain in E^N, and the function $f(x)$ belongs to the class $H_p^\alpha(D)$; α is a positive number equal to $\ell + \kappa$, where ℓ is a nonnegative integer, $0 < \kappa < 1$; assume also that $p \ge 1$ and $\alpha p > N$. Then, for all sufficiently large $\lambda > 0$ and for any positive number R uniformly with respect to x on the set D_{3R}*[35] *the inequality*

$$\left|\lambda^{\kappa/2} \int_0^R V_\nu(r\sqrt{\lambda}) r^{2\ell-N+\kappa} \mathcal{D}^\ell F(r) dr\right| \le C\|f\|_{H_p^\alpha(D)} \tag{2.5.89}$$

[35] Recall that D_{3R} denotes a set of points of domain D which are removed from the boundary of D to a distance greater than the number $3R$.

holds.

PROOF: Having applied in succession the $\mathcal{D}$ operation to a function of the form (2.5.88), we easily convince ourselves by induction that there exist, for any number ℓ and for any $N \geq 1$, constants $A_{0\ell}, A_{1\ell}, A_{2\ell}, \ldots, A_{\ell\ell}$ such that the representation

$$\mathcal{D}^{\ell} F(r) = r^{N-1-2\ell} \sum_{m=0}^{\ell} A_{m\ell} r^{m} \psi^{(m)}(r) \tag{2.5.90}$$

holds for $\mathcal{D}^{\ell} F(r)$.

It follows from the representation (2.5.90) and from the definition (2.5.69) of the functions $\varphi_m(r)$ that

$$r^{2\ell-N+\kappa} \mathcal{D}^{\ell} F(r) = \sum_{m=0}^{\ell} A_{m\ell} \varphi_m(r). \tag{2.5.91}$$

Since each of the functions $\varphi_m(r)$, as has been shown in Lemma 2.9, is summable on the segment $0 \leq r \leq R$, the function $r^{2\ell-N+\kappa} \mathcal{D} F(r)$ is also summable on this segment by virtue of representation (2.5.91).

Equality (2.5.91) allows us to assert that, in order to prove inequality (2.5.89), it will suffice to show that for each m in the segment $0 \leq m \leq \ell$, the estimate

$$\left| \int_{0}^{R} V_{\nu}(r\sqrt{\lambda}) \varphi_m(r) dr \right| \leq C \lambda^{-\kappa/2} \|f\|_{H_p^{\alpha}(D)} \tag{2.5.92}$$

holds uniformly with respect to x on the set D_{2R}.

We note that for function (2.5.87) for all $t > 0$ the representation

$$V_{\nu}(t) = \sqrt{\frac{2}{\pi}} \cos\left(t - \frac{\pi\nu}{2} - \frac{\pi}{4}\right) + O(t^{-1}) \tag{2.5.93}$$

holds, which stems from the Bessel function representation for argument values greater than unity, and from the fact that for $\nu \geq -1/2$ for all $t > 0$ the inequality $|J_{\nu}(t)| \leq C t^{-1/2}$ holds (one will recall that this inequality has been used previously on a number of occasions).

Immediate from (2.5.93) is the relationship

$$V_{\nu}(t) + V_{\nu}(t+\pi) = O(t^{-\kappa}), \tag{2.5.94}$$

valid for any κ in the half-segment $0 < \kappa \leq 1$; in turn, (2.5.94) ensures the equality

$$V_{\nu}(t) = \frac{1}{4} \left[V_{\nu}(t+2\pi) - 2V_{\nu}(t+\pi) + V_{\nu}(t) \right] + \gamma_{\nu}(t), \tag{2.5.95}$$

in which $\gamma_{\nu}(t)$ denotes a quantity subordinate to the estimate $\gamma_{\nu}(t) = O(t^{-\kappa})$ for all $t > 0$ and for any κ in the half-segment $0 < \kappa \leq 1$.

Equality (2.5.95) allows one to rewrite the integral on the left-hand side of (2.5.92) in the following form:

$$\int_0^R V_\nu(r\sqrt{\lambda})\varphi_m(r)dr = \frac{1}{4}\int_0^R \left[V_\nu(r\sqrt{\lambda}+2\pi) - 2V_\nu(r\sqrt{\lambda}+\pi) + V_\nu(r\sqrt{\lambda})\right]\varphi_m(r)dr + \int_0^R \gamma_\nu(r\sqrt{\lambda})\varphi_m(r)dr. \tag{2.5.96}$$

We now show that each of the integrals on the right-hand side of (2.5.96) has an order $O(\lambda^{-\kappa/2})\|f\|_{H_p^\alpha(D)}$; once done, this completes the proof of Lemma 2.10.

We start by estimating the latter integral on the right-hand side of (2.5.96). For $m > 0$, that is, for $1 \le m \le \ell$, we refer to inequality (2.5.74) to obtain uniformly with respect to $x \in D_{2R}$

$$\left|\int_0^R \gamma_\nu(r\sqrt{\lambda})\varphi_m(r)dr\right| \le C_1\int_0^R (r\sqrt{\lambda})^{-\kappa}|\psi_m(r)|r^{m+\kappa-1}dr \le C_1\lambda^{-\kappa/2}\int_0^R |\psi_m(r)|r^{m-1}dr \le C_2\lambda^{-\kappa/2}\|f\|_{H_p^\alpha(D)}. \tag{2.5.97}$$

Given $m = 0$, we represent the latter integral on the right-hand side of (2.5.96), making use of the function (2.5.83), in the form

$$\int_0^R \gamma_\mu(r\sqrt{\lambda})\varphi_0(r)dr = \int_0^R \gamma_\nu(r\sqrt{\lambda})\bar{\varphi}_0(r)dr + f(x)\int_0^R \gamma_\nu(r\sqrt{\lambda})r^{\kappa-1}dr. \tag{2.5.98}$$

For the former integral on the right-hand side of (2.5.98) we obtain, with the aid of estimate (2.5.86), the inequality

$$\left|\int_0^R \gamma_\nu(r\sqrt{\lambda})\bar{\varphi}_0(r)dr\right| \le \lambda^{-\kappa/2}\int_0^R r^{-1}|\bar{\psi}(r)|dr = O(\lambda^{-\kappa/2})\|f\|_{H_p^\alpha(D)} \tag{2.5.99}$$

uniformly with respect to $x \in D_{2R}$.

To estimate the latter integral on the right-hand side of (2.5.98), we take into account that $|f(x)| \le C_3\|f\|_{H_p^\alpha(D)}$ (uniformly with respect to $x \in D_R$) by virtue of the embedding theorem; we remark that, by virtue of the estimate $|J_\nu(t)| \leq C_\nu t^{-1/2}$, the function $V_\nu(t)$ and, by virtue of relation (2.5.95), the function $\gamma_\nu(t)$ are uniformly bounded on the half-line $t > 0$. On this basis, having made the change of variable $t = r\sqrt{\lambda}$, we obtain

$$\left|\int_0^R \gamma_\nu(r\sqrt{\lambda}) r^{\kappa-1} dr\right| = \lambda^{-\kappa/2} \left|\int_0^{R\sqrt{\lambda}} \gamma_\nu(t) t^{\kappa-1} dt\right| \leq C_4 \lambda^{-\kappa/2}.$$

Thus, we have

$$\left|f(x) \int_0^R \gamma_\nu(r\sqrt{\lambda}) r^{\kappa-1} dr\right| \leq C_5 \lambda^{-\kappa/2} ||f||_{H_p^\alpha(D)} \tag{2.5.100}$$

uniformly with respect to $x \in D_R$.

We obtain from relation (2.5.98) and from estimates (2.5.99) and (2.5.100) that at $m = 0$ uniformly with respect to $x \in D_{2R}$

$$\left|\int_0^R \gamma_\nu(r\sqrt{\lambda}) \varphi_0(r) dr\right| \leq C_6 \lambda^{-\kappa/2} ||f||_{H_p^\alpha(D)}. \tag{2.5.101}$$

Combining estimates (2.5.97) and (2.5.101), we obtain that, given any $m = 0, 1, \ldots, \ell$, the inequality

$$\left|\int_0^R \gamma_\nu(r\sqrt{\lambda}) \varphi_m(r) dr\right| \leq C_7 \lambda^{-\kappa/2} ||f||_{H_p^\alpha(D)} \tag{2.5.102}$$

holds uniformly with respect to x in D_{2R}.

The former integral on the right-hand side of (2.5.98) remains to be estimated. To this end, we represent it in a more convenient form.

As is readily shown,

$$\int_0^R [V_\nu(r\sqrt{\lambda}+2\pi) - 2V_\nu(r\sqrt{\lambda}+\pi) + V_\nu(r\sqrt{\lambda})]\varphi_m(r)dr$$

$$= \int_0^R V_\nu(r\sqrt{\lambda}+2\pi)\left[\varphi_m\left(r+\tfrac{2\pi}{\sqrt{\lambda}}\right) - 2\varphi_m\left(r+\tfrac{\pi}{\sqrt{\lambda}}\right) + \varphi_m(r)\right]dr$$

$$+\int_0^R V_\nu(r\sqrt{\lambda})\varphi_m(r)dr - 2\int_0^R V_\nu(r\sqrt{\lambda}+\pi)\varphi_m(r)dr \tag{2.5.103}$$

$$-\int_0^R V_\nu(r\sqrt{\lambda}+2\pi)\varphi_m\left(r+\frac{2\pi}{\sqrt{\lambda}}\right)dr$$

$$+2\int_0^R V_\nu(r\sqrt{\lambda}+2\pi)\varphi_m\left(r+\frac{\pi}{\sqrt{\lambda}}\right)dr.$$

Making the change of variable $\rho = r + 2\pi/\sqrt{\lambda}$ in the next to last integral and $\rho = r + \pi/\sqrt{\lambda}$ on the last integral on the right-hand side of (2.5.103), and comparing the first four integrals on the right-hand side of (2.5.103), we arrive at the following equation:

$$\int_0^R \left[V_\nu(r\sqrt{\lambda}+2\pi) - 2V_\nu(r\sqrt{\lambda}+\pi) + V_\nu(r\sqrt{\lambda})\right]\varphi_m(r)dr$$

$$= \int_0^R V_\nu(r\sqrt{\lambda}+2\pi)\left[\varphi_m\left(r+\frac{2\pi}{\sqrt{\lambda}}\right) - 2\varphi_m\left(r+\frac{\pi}{\sqrt{\lambda}}\right) + \varphi_m(r)\right]dr$$

$$+\int_0^{2\pi/\sqrt{\lambda}} V_\nu(r\sqrt{\lambda})\varphi_m(r)dr - 2\int_0^{\pi/\sqrt{\lambda}} V_\nu(r\sqrt{\lambda}+\pi)\varphi_m(r)dr \tag{2.5.104}$$

$$-\int_R^{R+2\pi/\sqrt{\lambda}} V_\nu(r\sqrt{\lambda})\varphi_m(r)dr + 2\int_R^{R+\pi/\sqrt{\lambda}} V_\nu(r\sqrt{\lambda}+\pi)\varphi_m(r)dr.$$

We will make use of this equality for estimating the first integral on the right-hand side of (2.5.96).

In order to estimate the first integral on the right-hand side of (2.5.104), we make use of estimate (2.5.70) at $h = \pi/\sqrt{\lambda}$ (see Lemma 2.9), keeping in mind that the function $V_\nu(t)$ is bounded for all positive values of its argument. We find that uniformly with respect to $x \in D_{3R}$ for all $\lambda > (3\pi/R)^2$ the following inequality holds:

$$\left|\int_0^R V_\nu(r\sqrt{\lambda}+2\pi)\left[\varphi_m\left(r+\frac{2\pi}{\sqrt{\lambda}}\right)-2\varphi_m\left(r+\frac{\pi}{\sqrt{\lambda}}\right)+\varphi_m(r\sqrt{\lambda})\right]dr\right| \tag{2.5.105}$$

$$\leq C_8\lambda^{-\kappa/2}\|f\|_{H_p^\alpha(D)}.$$

By virtue of the boundedness of the function $V_\nu(t)$ on the half-line $t > 0$, the estimation of the second and third integrals on the right-hand side of the (2.5.104) reduces to the estimation of the integral

$$\int_0^{2\pi/\sqrt{\lambda}} |\varphi_m(r)|dr. \tag{2.5.106}$$

In the case $1 \leq m \leq \ell$, the integral (2.5.108) is easily estimated with the aid of inequality (2.5.74):

$$\int_0^{2\pi/\sqrt{\lambda}} |\varphi_m(r)|dr = \int_0^{2\pi/\sqrt{\lambda}} r^\kappa|\psi^{(m)}(r)|r^{m-1}dr$$

$$\leq (2\pi)^\kappa\lambda^{-\kappa/2}\int_0^{2\pi/\sqrt{\lambda}} |\psi^{(m)}(r)|r^{m-1}dr \leq C_9\lambda^{-\kappa/2}\|f\|_{H_p^\alpha(D)} \tag{2.5.107}$$

(uniformly with respect to $x \in D_{2R}$).

In the case $m = 0$, we obtain with reference to equality (2.5.83) that

$$\int_0^{2\pi/\sqrt{\lambda}} |\varphi_0(r)|dr \leq \int_0^{2\pi/\sqrt{\lambda}} |\bar{\varphi}_0(r)|dr + |f(x)|\int_0^{2\pi/\sqrt{\lambda}} r^{\kappa-1}dr. \tag{2.5.108}$$

Since, by virtue of estimate (2.5.86) and in accordance with the embedding theorem uniformly with respect to $x \in D_{2R}$,

$$\int_0^{2\pi/\sqrt{\lambda}} |\bar{\varphi}_0(r)|dr = \int_0^{2\pi/\sqrt{\lambda}} r^\kappa r^{-1}|\bar{\psi}_m(r)|dr \leq C_{10}\lambda^{-\kappa/2}\|f\|_{H_p^\alpha(D)}, \quad |f(x)| \leq C_{11}\|f\|_{H_p^\alpha(D)}$$

and since $\int_0^{2\pi/\sqrt{\lambda}} r^{\kappa-1}dr \leq C_{12}\lambda^{-\kappa/2}$ (for any $0 < \kappa \leq 1$), the inequality

$$\int_0^{2\pi/\sqrt{\lambda}} |\varphi_0(r)|dr \leq C_{13}\lambda^{-\kappa/2}\|f\|_{H_p^\alpha(D)} \tag{2.5.109}$$

follows from (2.5.108).

One infers from inequalities (2.5.107) and (2.5.109) that for any $m = 0, 1, \ldots, \ell$ uniformly with respect to $x \in D_{2R}$ the second and third integrals on the right-hand side of (2.5.104) have an order of $O(\lambda^{-\kappa/2})\|f\|_{H_p^\alpha(D)}$.

Now, all we have to do is prove that the last two integrals on the right-hand side of (2.5.104) have the same order of magnitude. By virtue of the boundedness of $V_\nu(t)$ on the half-line $t > 0$ it will suffice to prove that the estimate

$$\int\limits_R^{R+2\pi/\sqrt{\lambda}} |\varphi_m(r)|dr \le C_{14}\lambda^{-\kappa/2}\|f\|_{H_p^\alpha(D)} \tag{2.5.110}$$

is valid uniformly with respect to $x \in D_{3R}$.

The estimate (2.5.110) is equivalent to the estimate

$$\int\limits_0^{R+h} |\varphi_m(r)|dr \le C_{15}h^\kappa\|f\|_{H_p^\alpha(D)}; \tag{2.5.111}$$

thus, to complete the proof of Lemma 2.10, we have to convince ourselves that estimate (2.5.111) holds for all x in D_{3R}.

We consider first the case $0 < \kappa < 1$. In this instance, the inequality (2.5.70) that has been established in Lemma 2.9 is equivalent to the inequality[36)]

$$\int\limits_0^{R} |\varphi_m(r+h) - \varphi_m(r)|dr \le Ch^\kappa\|f\|_{H_p^\alpha(D)}, \tag{2.5.112}$$

which also holds uniformly with respect to $x \in D_{3R}$.

It follows from (2.5.112) that

$$\int\limits_0^{R} |\varphi_m(r+h)|dr - \int\limits_0^{R} |\varphi_m(r)|dr = O(h^\kappa)\|f\|_{H_p^\alpha(D)} \tag{2.5.113}$$

uniformly with respect to $x \in D_{3R}$.

Carrying out the change of variable $\rho = r+h$ in the former integral on the left-hand side of (2.5.113), we transform (2.5.113) as

$$\int\limits_h^{R+h} |\varphi_m(\rho)|d\rho - \int\limits_0^{R} |\varphi_m(r)|dr = O(h^\kappa)\|f\|_{H_p^\alpha(D)}.$$

---

36) See, for example, Zygmund [79, Vol. 1, Chapter 2, p. 77].

One sees that

$$\int_R^{R+h} |\varphi_m(r)|dr = \int_0^h |\varphi_m(r)|dr + O(h^\kappa)\|f\|_{H_p^\alpha(D)} \tag{2.5.114}$$

uniformly with respect to $x \in D_{3R}$.

Since for any $m = 0, 1, \ldots, \ell$ the inequality

$$\int_0^h |\varphi_m(r)|dr = O(h^\kappa)\|f\|_{H_p^\alpha(D)} \tag{2.5.115}$$

holds uniformly with respect to $x \in D_{3R}$, and this inequality is equivalent to inequalities (2.5.107) and (2.5.109), it follows for the case $0 < \kappa < 1$ from (2.5.114) and (2.5.115) that inequality (2.5.111) holds uniformly with respect to $x \in D_{3R}$.

It remains to prove that in the case $\kappa = 1$ the estimate (2.5.111) is valid with respect to x on the set D_{3R}.

To that end, it will suffice to show that in the case $\kappa = 1$ the function $|\varphi_m(r)|$ uniformly on the segment $R \le r \le 2R$ is bounded by the quantity

$$O\left(\|f\|_{H_p^\alpha(D)}\right). \tag{2.5.116}$$

This assertion stems from the fact that, by virtue of the embedding theorem (2.5.73) for any $m = 0, 1, \ldots, \ell$, the function $f^{(m)}(r, \omega)$ belongs to the class $H^1_{p_2(m)}(D_R)$ for $p_2(m)(m+1) > N$, and for this reason, in accordance with the embedding theorem (2.1.23) (see Section 2.1.3), it also belongs to the class $L_{p_2(m)}$ on an N-dimensional sphere S_r^x of radius r centered at the point $x \in D_{3R}$; one observes that the L_{p_2}-norm of the function $f^{(m)}(r, \omega)$ on the aforesaid sphere uniformly with respect to r on the segment $R \le r \le 2R$ and uniformly with respect to x in D_{3R} is bounded by the quantity (2.5.116).

Finally, one notes that since

$$\varphi_m(r) = r^{m-1+\kappa}\psi^{(m)}(r) = \omega_N^{-1}\, r^{m-N+\kappa} \int_{S_r^x} f^{(m)}(r, \omega)dS,$$

the boundedness of the L_{p_2}-norm of the function $f^{(m)}(r, \omega)$ on the sphere S_r^x, uniform at $R \le r \le 2R$ and $x \in D_{3R}$, entails the boundedness of the function $|\varphi_m(r)|$ by a quantity of the same order (2.5.116) uniformly with respect to r on the segment $R \le r \le 2R$ and with respect to $x \in D_{3R}$.

Lemma 2.10 is thus completely proved.

2.5.3. Main Estimate for the Riesz Means of a Spectral Decomposition

In this section, the following ancillary assertion, which is also of independent interest, will be proved.

LEMMA 2.11. *Assume that $N \geq 2$, $0 \leq s < (N-1)/2$, G is an arbitrary domain in the space E^N, $\widehat{A}$ is an arbitrary self-adjoint nonnegative extension of the Laplace operator $Lu = -\Delta u$ in the domain G, $f(x)$ is an arbitrary function satisfying two requirements: (1) given a certain $R > 0$, the function $f(x)$ belongs to the class $\overset{0}{H}{}_2^\alpha(G_R)$[37] at $\alpha \geq (N-1)/2 - s$; (2) $f(x)$ belongs to the class $H_p^\alpha(D)$ in a certain domain D interior to G at $\alpha \geq (N-1)/2 - s$, $p\alpha > N$, $p \geq 1$.*[38]

Then for any point x of the set D_{3R} for the Riesz means of order s of the spectral decomposition of the function $f(x)$ the estimate

$$|\sigma_\lambda^s(x,f)| \leq C\left\{\|f\|_{H_2^\alpha(G)} + \|f\|_{H_p^\alpha(D)}\right\} \tag{2.5.117}$$

holds, uniformly with respect to x in D_{3R}.

PROOF: To begin, we shall prove the validity of estimate (2.5.117) for the function $f(x)$ in the space $C_0^\infty(G)$ of functions which are infinitely differentiable and finite in domain G.

We note, first of all, that it will suffice to have estimate (2.5.117) proved merely for $\alpha = (N-1)/2 - s$, $p\alpha > N$, since the validity of this estimate at $\alpha \geq (N-1)/2 - s$, $p\alpha > N$, $p \geq 1$ follows from the embedding theorem (2.1.24).

Thus, we assume in what follows that

$$\alpha = (N-1)/2 - s, \qquad p\alpha > N, \qquad p \geq 1.$$

We fix an arbitrary point x in the set G_{3R} and designate by r the distance $r = |x-y|$ between a fixed point x and a variable point y; further, we consider the function $\overset{\lambda}{v}(r)$ as defined by formula (2.4.5) (see Section 2.4.1).

Since the Fourier image of this function is defined by relation (2.4.8) with $x_0 = x$, we obtain, having written the Parseval equality (2.1.10) for this function and for the

---

37) We recall that $\overset{0}{H}{}_2^\alpha(G_R)$ is a set of functions that are well defined in the entire space E^N; they belong to the class $H_2^\alpha(E^N)$ and become zero outside G_R.

38) Domain D, in particular, may be coincident with domain G.

function $f(x)$ in the space $C_0^\infty(G)$, that

$$\sigma_\lambda^s(x,f) = 2^s\Gamma(s+1)(2\pi)^{-N/2}\lambda^{N/2-s/2}\int\limits_{r\le R} f(y)J_{N/2+s}(r\sqrt{\lambda})r^{-N/2-s}dy$$

$$+2^s\Gamma(s+1)\lambda^{N/4-s/2}\sum_{i=0}^{\widehat{m}}\int\limits_0^\infty \widehat{f}_i(t)u_i(x,t)t^{(2-N)/4}\overset{\nu_0}{I_t^\lambda}(R)d\rho(t), \tag{2.5.118}$$

where the symbol $\overset{\nu_n}{I_t^\lambda}(R)$ is used to denote an integral of the form (2.5.21) (see Section 2.5.1).

Making use of the function $F(r)$ as defined by relation (2.5.88) in the foregoing section, we can rewrite equality (2.5.118) in the following manner:

$$\sigma_\lambda^s(x,f) = 2^s\Gamma(s+1)(2\pi)^{-N/2}\omega_N\lambda^{N/4-s/2}\int\limits_0^R J_{N/2+s}(r\sqrt{\lambda})r^{-N/2-s}F(r)dr$$

$$+2^s\Gamma(s+1)\lambda^{N/4-s/2}\sum_{i=0}^{\widehat{m}}\int\limits_0^\infty \widehat{f}_i(t)u_i(x,t)t^{(2-N)/4}\,\overset{\nu_0}{I_t^\lambda}(R)d\rho(t). \tag{2.5.119}$$

We integrate the former integral on the right-hand side of (2.5.119) by parts ℓ times, making use of the operation $\mathcal{D}F(r) = \frac{d}{dr}\left[\frac{1}{r}F(r)\right]$ and its repetitions[39] as well as of the recurrence relation

$$\int r^{-\nu+1}J_\nu(r\sqrt{\lambda})dr = -\frac{1}{\sqrt{\lambda}}r^{-\nu+1}J_{\nu-1}(r\sqrt{\lambda}).$$

---

[39] Since $F(r) = r^{N-1}\psi(r)$ and the mean value of $\psi(r)$ and all of its derivatives are continuous on the segment $0 \le r \le R$, and since $2\ell < N-1$, for all $n \le \ell$ the functions $\frac{1}{r}\mathcal{D}^{n-1}F(r)$ are continuous on the segment $0 \le r \le R$ and become zero at $r = 0$. We recall that $\alpha = 1+\kappa$, where $0 < \kappa \le 1$.

Assuming that $\mathcal{D}^0 F(r) \equiv F(r)$, we obtain

$$2^s\Gamma(s+1)(2\pi)^{-N/2}\omega_N \lambda^{N/4-s/2} \int_0^R J_{N/2+s}(r\sqrt{\lambda}) r^{-N/2-s} F(r)dr$$

$$= -2^s\Gamma(s+1)(2\pi)^{-N/2}\omega_N \lambda^{N/4-s/2-n/2}$$

$$\times \sum_{n=1}^{\ell} \left\{ \left[\frac{1}{r} \mathcal{D}^{n-1} F(r) \right] r^{-N/2-s+n} J_{N/2+s-n}(r\sqrt{\lambda}) \right\} \Bigg|_{r=0}^{r=R} \tag{2.5.120}$$

$$+ 2^s\Gamma(s+1)(2\pi)^{-N/2}\omega_N \lambda^{N/4-s/2-\ell/2}$$

$$\times \int_0^R J_{N/2+s-\ell}(r\sqrt{\lambda}) r^{\ell-N/2-s} \mathcal{D}^\ell F(r) dr.$$

Note that all the substitutions at $r = 0$ on the right-hand side of (2.5.120) become zero, since by virtue of the fact that $N/2 + s - n > 0$ for any $n = 1, 2, \ldots, \ell$, the estimate $\rho^{-N/2+s+n}|J_{N/2+s-n}(\rho)| \leq C$ holds, and the functions $\frac{1}{r}\mathcal{D}^{n-1}F(r)$ reduce to zero at $r = 0$.

Thus, the equality (2.5.120) can be rewritten in the form

$$2^s\Gamma(s+1)(2\pi)^{-N/2}\omega_N \lambda^{N/4-s/2} \int_0^R J_{N/2+s}(r\sqrt{\lambda}) r^{-N/2-s} F(r)dr$$

$$= -2^s\Gamma(s+1)(2\pi)^{-N/2}\omega_N \lambda^{N/4-s/2-n/2} \sum_{n=1}^{\ell} R^{-N/2-s+n-1}$$

$$\times J_{N/2+s-n}(R\sqrt{\lambda}) \mathcal{D}^{n-1} F(r) \tag{2.5.121}$$

$$+ 2^s\Gamma(s+1)(2\pi)^{-N/2}\omega_N \lambda^{N/4-s/2-\ell/2}$$

$$\times \int_0^R J_{N/2+s-\ell}(r\sqrt{\lambda}) r^{\ell-N/2-s} \mathcal{D}^\ell F(r) dr.$$

We integrate the integral

$$\overset{\nu_0}{I_0^\lambda}(R) = \int_R^\infty J_{N/2+s}(r\sqrt{\lambda}) J_{N/2-1}(r\sqrt{t}) r^{-s} dr$$

in the last term of the right-hand side of (2.5.119) ℓ times by parts, making use of the recurrence relations

$$\int r^{-s-\nu+1} J_{\nu+s}(r\sqrt{\lambda})dr = -\frac{r^{-s-\nu+1}}{\sqrt{\lambda}} J_{\nu+s-1}(r\sqrt{\lambda}),$$

$$\frac{d}{dr}\left[r^{\nu-1}J_{\nu-1}(r\sqrt{t})\right] = \sqrt{t}\, r^{\nu-1} J_{\nu-2}(r\sqrt{t}),$$

to obtain (taking into account that the substitutions at $r = \infty$ reduce to zero)

$$\overset{\nu_0}{I_t^{\lambda}}(R) = \sum_{n=1}^{\ell} t^{(n-1)/2}\lambda^{-n/2} J_{N/2+s-n}(R\sqrt{\lambda}) J_{N/2-n}(R\sqrt{t}) R^{-s} + \left(\frac{t}{\lambda}\right)^{\ell/2} \overset{\nu_\ell}{I_t^{\lambda}}(R), \tag{2.5.122}$$

where the symbol $\overset{\nu_\ell}{I_t^{\lambda}}(R)$ signifies the integral (2.5.21) taken at $n = \ell$, that is, at $\nu_\ell = N/2 - \ell$.

Substituting (2.5.121) and (2.5.122) into the right-hand side of (2.5.119) and making use, at $\nu = \nu_\ell + s$, of the previously adopted notation[40)]

$$V_\nu(r\sqrt{\lambda}) = r^{1/2}\lambda^{1/4} J_\nu(r\sqrt{\lambda}),$$

and recalling that at $\alpha = (N-1)/2 - s = \ell + \kappa$ the equalities $(N-1)/4 - s/2 - \ell/2 = \kappa/2$, $\ell - (N+1)/2 - s = 2\ell - N + \kappa$, and $N/4 - s/2 - \ell/2 = 1/4 + \kappa/2$ hold, we obtain for $\sigma_\lambda^s(x, f)$ the following expression:

$$\begin{aligned}
\sigma_\lambda^s(x,f) &= 2^s\Gamma(s+1)(2\pi)^{-N/2}\omega_N \lambda^{\kappa/2} \int_0^R V_{\nu_\ell+s}(r\sqrt{\lambda}) r^{2\ell-N+\kappa} \mathcal{D}^\ell F(r)dr \\
&+ 2^s\Gamma(s+1)(2\pi)^{-N/2}\lambda^{1/4+\kappa/2} \sum_{i=1}^{\widehat{m}} \int_0^\infty \widehat{f}_i(t) u_i(x,t) t^{\ell/2+(2-N)/4} \overset{\nu_\ell}{I_t^{\lambda}}(R) d\rho(t) \\
&+ 2^s\Gamma(s+1)(2\pi)^{-N/2}\omega_N \lambda^{N/4-s/2} \sum_{n=1}^{\ell} \lambda^{-n/2} R^{-N/2-s+n-\ell} J_{N/2+s-n}(R\sqrt{\lambda}) \\
&\times \left\{ -\mathcal{D}^{n-1}F(r) + \frac{(2\pi)^{N/2}}{\omega_N} R^{N/2-n+1} \sum_{i=1}^{\widehat{m}} \int_0^\infty \widehat{f}_i(t) u_i(x,t) t^{n/2-N/4} \right. \\
&\left. \times J_{N/2-n}(R\sqrt{t}) d\rho(t) \right\}.
\end{aligned} \tag{2.5.123}$$

40) See the notation (2.5.87) in Section 2.5.2.

We now prove that, for any $n = 1, 2, \dots, \ell$, the expression in braces in (2.5.123) equals zero. Having fixed an arbitrary $R > 0$ and an arbitrary point x of the set G_R, and denoted by $\rho = |x - y|$ the distance separating an arbitrary point y of domain G from the point x, we consider the function

$$v(x-y) = \begin{cases} \omega_N^{-1} & \text{for} \quad |x-y| \le r, \\ 0 & \text{for} \quad |x-y| > r \end{cases}$$

for any r in the half-segment $0 < r \le R$.

By virtue of relation (2.3.2) the Fourier image $\widehat{v}_i(x, \lambda)$ of the function $v(|x-y|)$ takes the form

$$\widehat{v}_i(x,\lambda) = (2\pi)^{N/2}\omega_N^{-1} u_i(x,\lambda)\lambda^{(2-N)/4} \int_0^r \rho^{N/2} J_{N/2-1}(\rho\sqrt{\lambda})d\rho$$

$$= (2\pi r)^{N/2}\omega_N^{-1}\lambda^{-N/4} u_i(x,\lambda) J_{N/2}(r\sqrt{\lambda}), \qquad (i = 1, 2, \dots, \widehat{m}).$$

Considering that for any x from G_R the function $v(|x-y|)$ belongs, with respect to y, to the class $L_2(G)$, we obtain, having written the Parseval equation (2.1.10) for the two functions $v(|x-y|)$ and $f(y)$, the equality

$$\omega_N^{-1} \int_{|x-y|\le r} f(y)dy = \sum_{i=1}^{\widehat{m}} (2\pi r)^{N/2}\omega_N^{-1} \int_0^\infty t^{-N/4} J_{N/2}(r\sqrt{\lambda}) \widehat{f}_i(t) u_i(x,t) d\rho(t), \tag{2.5.124}$$

which is valid for all $x \in G_R$ and for all $r \le R$.

We have, by definition of the function (2.5.88),

$$\frac{d}{dr}\left[\omega_N^{-1} \int_{|x-y|\le r} f(y)dy\right] = \frac{d}{dr}\left[\int_0^r \left(\rho^{N-1}\omega_N^{-1} \int \dots \int_\omega f(x+\rho\omega)d\omega\right) d\rho\right]$$

$$= \frac{d}{dr}\left[\int_0^r F(\rho)d\rho\right] = F(r).$$

Therefore, the formal differentiation of equality (2.5.124) with respect to r gives the relation

$$F(r) = (2\pi)^{N/2}\omega_N^{-1} \sum_{i=1}^{\widehat{m}} \int_0^\infty r^{N/2} t^{(2-N)/4} J_{N/2-1}(r\sqrt{t}) \widehat{f}_i(t) u_i(x,t) d\rho(t), \tag{2.5.125}$$

and the formal application of operation $\mathcal{D}^{n-1}$ to both parts of (2.5.125) with respect to variable r yields

$$\mathcal{D}^{n-1}F(r) = (2\pi)^{N/2}\omega_N^{-1}\sum_{i=1}^{\widehat{m}}\int_0^\infty r^{N/2-n+1}t^{n/2-N/4}J_{N/2-n}(r\sqrt{t})\widehat{f}_i(t)u_i(x,t)d\rho(t). \tag{2.5.126}$$

If we prove that the formal application of operation $\frac{d}{dr}\mathcal{D}^{n-1}$ with respect to variable r on the segment $0 \le r \le R$ to equality (2.5.124) is possible for all $n = 1, 2, \dots, \ell$, we can show, letting $r = R$ in (2.5.126), that the expressions enclosed within braces on the right-hand side of (2.5.123) are equal to zero. To verify the eventual application of the formal operation $\frac{d}{dr}\mathcal{D}^{n-1}$ to equality (2.5.124) on the segment $0 \le r \le R$, it suffices to prove that, given any $n = 1, 2, \dots, \ell$, the integral on the right-hand side of (2.5.126) converges uniformly in r on the segment $0 \le r \le R$.

Inasmuch as the inequality

$$r^{N/2-n+1}t^{n/2-N/4}\left|J_{N/2-n}(r\sqrt{t})\right| \le Cr^{N-2n+1} \le CR^{N-2n+1}$$

holds for any $n = 1, 2, \dots, \ell$ for all $t > 0$ and for all r in the segment $0 \le r \le R$, the integral on the right-hand side of (2.5.126) for all r in the segment $0 \le r \le R$ is majorized by the integral

$$C_1\sum_{i=1}^{\widehat{m}}\int_0^\infty \left|\widehat{f}_i(t)\right| |u_i(x,t)|d\rho(t), \tag{2.5.127}$$

and by virtue of Weierstrass' test for convergence it suffices to prove the convergence of the integral (2.5.127).

We represent this integral as the sum of two integrals $I_1 + I_2$, where

$$I_1 = C_1\sum_{i=1}^{\widehat{m}}\int_0^1 \left|\widehat{f}_i(t)\right| |u_i(x,t)|d\rho(t), \tag{2.5.128}$$

$$I_2 = C_1\sum_{i=1}^{\widehat{m}}\int_1^\infty \left|\widehat{f}_i(t)\right| |u_i(x,t)|d\rho(t). \tag{2.5.129}$$

In order to ensure the convergence of the integral (2.5.128), we apply the Cauchy–Buniakowski inequality first to the integral and then to the sum and make use of the Parseval equality (2.1.9) for the function $f(x)$ and of the estimate (2.3.4) taken at $\mu_0 = 1$.

In doing so, for any point x in G_R we obtain

$$
\begin{aligned}
|I_1| &\le C_1 \sum_{i=1}^{\widehat{m}} \left(\int_0^1 \left|\widehat{f}_i(t)\right|^2 d\rho(t) \right)^{1/2} \left(\int_0^1 |u_i(x,t)|^2 \, d\rho(t) \right)^{1/2} \\
&\le C_1 \left\{ \sum_{i=1}^{\widehat{m}} \int_0^1 \left|\widehat{f}_i(t)\right|^2 d\rho(t) \right\}^{1/2} \left\{ \sum_{i=1}^{\widehat{m}} \int_0^1 |\widehat{u}_i(x,t)|^2 \, d\rho(t) \right\}^{1/2} \qquad (2.5.130) \\
&\le C_2 \|f\|_{L_2(G)}.
\end{aligned}
$$

To substantiate the convergence of the integral (2.5.129), we note first of all that, inasmuch as the function $f(x)$ belongs to the class $C_0^\infty(G)$, the function $\Delta^m f(x)$ for any m belongs to the class $C_0^\infty(G)$. Therefore, the successive application of Green's formula to the function $u_i(x,\lambda)$ and to one of the functions $f(x)$, $\Delta f(x)$, $\Delta^2 f(x)$, $\ldots$ $\Delta^{[N/4]+1} f(x)$ leads to the following relation between the Fourier images of the functions $f(x)$ and $\Delta^{[N/4]+1} f(x)$:

$$\left(\widehat{\Delta^{[N/4]+1} f}\right)_i(t) = (-1)^{[N/4]+1} t^{[N/4]+1} \widehat{f}_i(t). \qquad (2.5.131)$$

The relation (2.5.131) enables us to rewrite (2.5.129) in the form

$$I_2 = C_1 \sum_{i=1}^{\widehat{m}} \int_1^\infty \left| \left(\widehat{\Delta^{[N/4]+1} f}\right)_i(t) \right| t^{-[N/4]-1} |u_i(x,t)| d\rho(t). \qquad (2.5.129')$$

Now, applying the Cauchy–Buniakowski inequality on the right-hand side of (2.5.129′) first to the integral and then to the sum, making use of the Parseval equality (2.1.9) for the function $\Delta^{[N/4]+1} f(x)$, and taking into account the boundedness of integral (2.3.10) in G_R for any $\delta > 0$ (see Section 2.3.2), we obtain

$$
\begin{aligned}
|I_2| \le C_1 &\left\{ \sum_{i=1}^{\widehat{m}} \int_1^\infty \left| \left(\widehat{\Delta^{[N/4]+1} f}\right)_i(t) \right|^2 d\rho(t) \right\}^{1/2} \\
&\times \left\{ \sum_{i=1}^{\widehat{m}} \int_1^\infty t^{-2([N/4]+1)} |u_i(x,t)|^2 d\rho(t) \right\}^{1/2} \le C_3 \left\| \Delta^{[N/4]+1} f \right\|_{L_2(G)}
\end{aligned}
\qquad (2.5.132)
$$

for any point x in G_R.

The estimates (2.5.130) and (2.5.132) complete the proof of the convergence of the integral (2.5.127) and thereby complete the convergence of each of the integrals

(2.5.126) uniformly with respect to r on the segment $0 \leq r \leq R$. The reduction to zero of each of the terms in braces on the right-hand side of (2.5.123) is thus proved, and the equality (2.5.123) for an arbitrary function $f(x)$ in the class C_0^∞ takes the form

$$\sigma_\lambda^s(x,f) = 2^s\Gamma(s+1)(2\pi)^{-N/2}\omega_N\lambda^{\kappa/2}\int_0^R V_{\nu_\ell+s}(r\sqrt{\lambda})r^{2\ell-N+\kappa}\mathcal{D}^\ell F(r)dr$$
$$+\,2^s\Gamma(s+1)(2\pi)^{-N/2}\omega_N\lambda^{1/4+\kappa/2}\sum_{i=1}^{\widehat{m}}\int_0^R \widehat{f}_i(t)u_i(x,t)t^{\ell/2+(2-N)/4}I_t^{\nu_\ell}{}^{\lambda}(R)d\rho(t). \tag{2.5.133}$$

To obtain the desired estimate (2.5.117) for an arbitrary function from the class $C_0^\infty(G)$, one observes that the first term on the right-hand side of (2.5.133) has an order $O(\|f\|_{H_p^\alpha(D)})$ by virtue of estimate (2.5.89) (see Lemma 2.10), and the second term on the right-hand side of (2.5.133) has an order $O(\|f\|_{H_2^\alpha(G)})$ by virtue of estimate (2.5.24) (see Lemma 2.7).

This completes thereby the proof of estimate (2.5.117) for an arbitrary function $f(x)$ in the class $C_0^\infty(G)$.

It remains for us to prove estimate (2.5.117) for an arbitrary function $f(x)$ in the class $\overset{0}{H}{}_2^\alpha(G)\cap H_p^\alpha(D)$ at $\alpha=(N-1)/2-s$, $p\alpha>N$, $p\geq 1$. We make use of the fact that the functional set in $C_0^\infty(G)$ is dense in the above-mentioned class.

Let us convince ourselves that for an arbitrary function $f(x)$ from the class

$$\overset{0}{H}{}_2^\alpha(C_R)\cap H_p^\alpha(D) \qquad \text{for} \qquad \alpha=\frac{N-1}{2}-s,\quad p\alpha>N,\quad p\geq 1 \tag{2.5.134}$$

for arbitrary fixed $\varepsilon>0$ and $\lambda>0$ and for each point $x\in D_{3R}$ the inequality

$$|\sigma_\lambda^s(x,f)|\leq C\left\{\|f\|_{H_2^\alpha(G)}+\|f\|_{H_p^\alpha(D)}\right\}+\varepsilon \tag{2.5.135}$$

holds.

Having fixed arbitrarily $\varepsilon>0$ and $\lambda>0$, we specify a function $g(x)$ in the space $C_0^\infty(G)$ such that

$$\|f-g\|_{H_2^\alpha(G)}+\|f-g\|_{H_p^\alpha(D)}\leq\frac{\varepsilon}{2C}, \tag{2.5.136}$$

$$\|f-g\|_{L_2(G)}\leq\frac{\varepsilon}{2C_5\lambda^{2N}}, \tag{2.5.137}$$

where C is a constant from estimate (2.5.117) for a function in $C_0^\infty(G)$, and C_5 is a constant from the estimate (2.3.4), as established in Section 2.3.2.

Making use of estimate (2.5.137) and of the expression for the Riesz means of order $s\geq 0$

$$\sigma_\lambda^s(x,f-g)=\sum_{i=1}^{\widehat{m}}\int_0^\lambda\left(1-\frac{t}{\lambda}\right)^s\left[\widehat{f}_i(t)-\widehat{g}_i(t)\right]u_i(x,t)d\rho(t)$$

and applying the Cauchy–Buniakowski inequality in turn to the integral and to the sum, and further making use of the Parseval equality (2.1.9) for the function $[f(x) - g(x)]$, of the inequality $(1 - t/\lambda)^s \le 1$, and of the estimate (2.3.4) taken at $\mu_0 = \sqrt{\lambda}$, we obtain that

$$|\sigma_\lambda^s(x, f - g)| < \frac{\varepsilon}{2} \tag{2.5.138}$$

for any point $x \in D_{3R}$.

Inasmuch as for the function $g(x)$ in the space $C_0^\infty(G)$ the estimate

$$|\sigma_\lambda^s(x, g)| \le C\left\{||g||_{H_2^\alpha(G)} + ||g||_{H_p^\alpha(D)}\right\}$$

holds for any point $x \in D_{3R}$, we easily see from this estimate, with reference to estimates (2.5.138) and (2.5.136) and the triangle inequality, that the inequality (2.5.135) is valid indeed.

Since the inequality (2.5.135) has been established for arbitrary $\varepsilon > 0$ and $\lambda > 0$, the proof of estimate (2.5.117) for an arbitrary function in the class (2.5.134) is thus complete.

This completes the proof of Lemma 2.11.

2.5.4. Direct Proof of Theorem 2.3

Assume that $f(x)$ is an arbitrary function satisfying entirely the conditions of Theorem 2.3, that is, $f(x)$ belongs, for certain $h_0 > 0$, to the class

$$\overset{0}{H}{}_2^\alpha(G_{h_0}) \cap H_p^\alpha(D) \qquad \text{for} \qquad \alpha = \frac{N-1}{2} - s, \quad p\alpha > N, \quad p \ge 1.$$

We fix an arbitrary $\varepsilon > 0$ and an arbitrary $h = 3R > 0$, and we denote by $g(x)$ a function in the space $C_0^\infty(G)$ such that[41)]

$$||f - g||_{H_2^\alpha(D)} + ||f - g||_{H_p^\alpha(G)} < \frac{\varepsilon}{3C}, \tag{2.5.139}$$

$$\max_{x \in D_{3R}} |f(x) - g(x)| < \frac{\varepsilon}{3}, \tag{2.5.140}$$

where C is a constant as encountered in estimate (2.5.117).

---

41) A function such as $g(x)$ can always be found by virtue of the denseness of the functional set from $C_0^\infty(G)$ in the class $\overset{0}{H}{}_2^\alpha(G_{h_0}) \cap H_p^\alpha(D)$ at $\alpha \ge (N-1)/2 - s$, $p\alpha > N$, $p \ge 1$ and by virtue of the embedding theorem (2.1.26) (see Section 2.1.3), which states that $\max_{x \in D_{3R}} |f(x) - g(x)| \le C_1 ||f - g||_{H_p^\alpha(D)}$.

Further, we make use of the fact that the Riesz means of any nonnegative order s of a function $g(x)$ in the class $C_0^\infty(G)$

$$\sigma_\lambda^s(x,g) = \sum_{i=1}^{\widehat{m}} \int_0^\lambda \left(1 - \frac{t}{\lambda}\right)^s [\widehat{g}_i(t)]\, u_i(x,t)d\rho(t) \tag{2.5.141}$$

converge to $f(x)$ uniformly with respect to x in G_h.

To prove this, it will suffice to majorize the right-hand side of (2.5.141) by the quantity

$$\sum_{i=1}^{\widehat{m}} \int_0^\lambda |\widehat{g}_i(t)|\, |u_i(x,t)| d\rho(t),$$

and then to adhere to the same argumentation as in the preceding section in proving the convergence of the integral (2.5.127).

It follows from the convergence of the Riesz means (2.5.141) to $g(x)$ uniformly in G_h that for fixed $\varepsilon > 0$ there is a $\Lambda(\varepsilon)$ such that for all $x \in G_h$ and for all $\lambda \geq \Lambda(\varepsilon)$ one has

$$|\sigma_\lambda^s(x,g) - g(x)| \leq \frac{\varepsilon}{3}. \tag{2.5.142}$$

We note finally that for all x in the G_h we have

$$|\sigma_\lambda^s(x, f-g)| \leq \frac{\varepsilon}{3} \tag{2.5.143}$$

which stems from inequality (2.5.139) and from the estimate (2.5.117).

It follows from the identity

$$\sigma_\lambda^s(x,f) - f(x) \equiv \sigma_\lambda^s(x, f-g) + [\sigma_\lambda^s(x,g) - g(x)] + [g(x) - f(x)]$$

and from estimates (2.5.143), (2.5.142), (2.5.140) that for all $x \in D_h = D_{3R}$ and for all $\lambda \geq \Lambda(\varepsilon)$ one has $|\sigma_\lambda^s(x,f) - f(x)| < \varepsilon$, which completes the proof of Theorem 2.3.

REMARK. In order to define, in terms of Theorem 2.3, the conditions for localization of the Riesz means of spectral decompositions of order s $(0 \leq s < (N-1)/2)$, it suffices to set the function $f(x)$ equal to zero in the subdomain D (such a function $f(x)$ satisfies condition (3) of Theorem 2.3, since it belongs to the class $H_p^\alpha(D)$ for any $\alpha > 0$, $p \geq 1$).

2.6. ESTIMATE FOR THE REMAINDER TERM OF THE RIESZ MEANS OF A SPECTRAL FUNCTION IN THE METRIC L_2

In Chapter 1, for an arbitrary FSF of the Laplace operator we developed a method enabling one to gain insight into the spectral decompositions of finite-in-domain G functions of the Sobolev–Liouville class $L_p^\alpha(G)$. This method is based on a preliminary upper-bound estimation of the remainder term of a spectral function in the L_2 metric and is readily transferred to an arbitrary self-adjoint nonnegative extension of the Laplace operator. It also enables one to establish conditions for the uniform convergence of both the spectral decompositions and their Riesz means of order s $(0 \leq s < (N-1)/2)$ for finite functions in domain G of the Sobolev–Liouville class $L_p^\alpha(G)$.

We restrict ourselves to establishing, for an arbitrary self-adjoint nonnegative extension of the Laplace operator, an upper bound of the remainder term of the Riesz means of a spectral function in the metric L_2.

THEOREM 2.6. *Assume that G is an arbitrary domain in the space E^N, $\widehat{A}$ is an arbitrary self-adjoint nonnegative extension of the Laplace operator $Lu = -\Delta u$ in the domain G. Then, given any $R > 0$ and any sufficiently large $\lambda > 0$ for the Riesz means $\theta^s(x, y, \lambda)$ of any nonnegative order s of a spectral function $\theta(x, y, \lambda)$ of the extension $\widehat{A}$ uniformly with respect to x in G_R, the estimate below holds*

$$\|\theta^s(x, y, \lambda - \overset{\lambda}{v}(|x-y|)\|_{L_2(G)} \leq C\lambda^{(N-1)/4-s/2}, \tag{2.6.1}$$

in which the $L_2(G)$ norm is taken with respect to the coordinates of a point y, the symbol $|x-y|$ denotes the distance between points x and y, and the symbol $\overset{\lambda}{v}(r)$ refers to the familiar function

$$\overset{\lambda}{v}(r) = \begin{cases} \Gamma(s+1)2^s(2\pi)^{-N/2}\lambda^{N/4-s/2}r^{-N/2-s}J_{N/2+s}(r\sqrt{\lambda}) & \text{for } r \leq R, \\ 0 & \text{for } r > R \end{cases} \tag{2.6.2}$$

(see relation (2.4.5) in Section 2.4.1).

PROOF: Let us consider an arbitrary ordered spectral representation of the space $L_2(G)$ with respect to the extension $\widehat{A}$ with spectral measure $\rho(\lambda)$, sets of multiplicity e_i, fundamental functions $u_i(x, \lambda)$ $(i = 1, 2, \ldots, \widehat{m})$ of multiplicity $\widehat{m} \leq \infty$. Having fixed an arbitrary $R > 0$ and an arbitrary point x in G_R, we determine the Fourier image $\widehat{v}_i^\lambda(x, t)$ of the function $v(|x-y|)$ with respect to the fundamental function $u_i(y, \lambda)$. Following the line of reasoning as in Section 2.4.1, we obtain that the desired Fourier image is defined by the relation[42]

$$\widehat{v}_i^\lambda(x, t) = \delta_t^\lambda u_i(x, t)\left(1 - \frac{t}{\lambda}\right)^s - \Gamma(s+1)2^s\lambda^{N/4-s/2}t^{(2-N)/4}u_i(x, t)I_t^\lambda(R) \tag{2.6.3}$$

$$(i = 1, 2, \ldots, \widehat{m}),$$

where

$$I_t^\lambda(R) = \int_R^\infty J_{N/2+s}(r\sqrt{\lambda})J_{N/2-1}(r\sqrt{t})r^{-s}dr, \tag{2.6.4}$$

[42] See relation (2.4.6)–(2.4.9) in Section 2.4.1.

$$\delta_t^\lambda = \begin{cases} 1 & \text{for} \quad t < \lambda, \\ 0 & \text{for} \quad t \geq \lambda. \end{cases} \tag{2.6.5}$$

One will recall, given $s = 0$ and $t = \lambda$, that $\delta_t^\lambda = 1/2$, $(1 - t/\lambda)^s = 1$.

We multiply (2.6.3) by a fundamental function $u_i(y,t)$ and integrate the obtained equality with respect to the spectral measure $\rho(t)$ for t going from 0 to ∞, and then sum over all the numbers i from 1 to $\widehat{m}$.

Taking into account that, for any fixed $x \in G_R$ and $\lambda > 0$, the function $\overset{\lambda}{v}(|x-y|)$ belongs in y to the class $L_2(G)$ and making use of relation (2.6.5) and expression (2.4.4) for the Riesz means of a spectral function, we obtain the following equation:

$$\overset{\lambda}{v}(|x-y|) - \theta^s(x,y,\lambda)$$
$$= \Gamma(s+1)2^s\lambda^{N/4-s/2}\sum_{i=1}^{\widehat{m}}\int_0^\infty t^{(2-N)/4}u_i(x,t)u_i(y,t)I_t^\lambda(R)d\rho(t). \tag{2.6.6}$$

This equation, given any fixed $x \in G_R$ and $\lambda > 0$, implies the equality of elements of the space $L_2(G)$ with respect to the coordinates of the point y. It follows from (2.6.6) and from the Parseval equality (2.1.9) that, in order to prove the desired estimate (2.6.1), it suffices to establish the estimate

$$\sum_{i=1}^{\widehat{m}}\int_0^\infty t^{(2-N)/2}\left[u_i(x,t)I_t^\lambda(R)\right]^2 d\rho(t) \leq C\lambda^{-1/2}, \tag{2.6.7}$$

uniform with respect to x in G_R. To this end, it suffices, in turn, to verify the validity of the following six estimates, uniform with respect to x in G_R:

$$S_1 = \sum_{i=1}^{\widehat{m}}\int_{\sqrt{t}\leq 1}\left\{t^{(2-N)/2}\left[u_i(x,t)I_t^\lambda(R)\right]^2\right\}d\rho(t) \leq C_1\lambda^{-1/2}, \tag{2.6.8}$$

$$S_2 = \sum_{i=1}^{\widehat{m}}\int_{1\leq\sqrt{t}\leq\sqrt{\lambda}/2}\{\dots\}\,d\rho(t) \leq C_2\lambda^{-1/2}, \tag{2.6.9}$$

$$S_3 = \sum_{i=1}^{\widehat{m}}\int_{\sqrt{\lambda}/2\leq\sqrt{t}\leq\sqrt{\lambda}-1}\{\dots\}\,d\rho(t) \leq C_3\lambda^{-1/2}, \tag{2.6.10}$$

$$S_4 = \sum_{i=1}^{\widehat{m}}\int_{|\sqrt{t}-\sqrt{\lambda}|\leq 1}\{\dots\}\,d\rho(t) \leq C_4\lambda^{-1/2}, \tag{2.6.11}$$

$$S_5 = \sum_{i=1}^{\widehat{m}} \int\limits_{\sqrt{\lambda}+1 \le \sqrt{t} \le 3\sqrt{\lambda}/2} \{\dots\}\, d\rho(t) \le C_5 \lambda^{-1/2}, \tag{2.6.12}$$

$$S_6 = \sum_{i=1}^{\widehat{m}} \int\limits_{\sqrt{t} \le 3\sqrt{\lambda}/2} \{\dots\}\, d\rho(t) \le C_6 \lambda^{-1/2}, \tag{2.6.13}$$

(the braces in (2.6.9)–(2.6.13) enclose the same term as in (2.6.8)).

With a view to proving estimate (2.6.8), we make use of the inequality (2.4.34), established in Section 2.4.1 for any $\sqrt{t} \le 1$, $\lambda \ge 1$, and write it in the form

$$t^{(2-N)/2} \left[u_i(x,t) I_t^\lambda(R)\right]^2 \le C_7 u_i^2(x,t) \lambda^{-1/2}.$$

The validity of estimate (2.6.8) uniform with respect to x in G_R follows immediately from the last inequality and from estimate (2.3.3) at $\mu = 0$ uniform with respect to x in any subdomain G' strictly interior to domain G (see Section 2.3.1).

Proceeding with the proof of relationships (2.6.9)–(2.6.13), we make use of the following estimates, established in Sections 2.4.1 and 2.4.2 for expressions (2.6.4)[43]:

$$\left|I_t^\lambda(R)\right| \le C_8 \lambda^{-1/4} t^{-1/4}, \tag{2.6.14}$$

$$\left|I_t^\lambda(R)\right| \le C_9 \lambda^{-1/4} t^{-1/4} |\sqrt{t} - \sqrt{\lambda}|^{-1}. \tag{2.6.15}$$

Of these two inequalities, the former holds for any $s \ge 0$, $t \ge 1$, $\lambda \ge 1$, $|\sqrt{t} - \sqrt{\lambda}| \le 1$, and the latter holds for any $s \ge 0$, $t \ge 1$, $\lambda \ge 1$, $|\sqrt{t} - \sqrt{\lambda}| \ge 1$.

In verifying estimate (2.6.9), we shall observe that for the values of t involved in this estimate, expression (2.6.15) can be rewritten in the following manner:

$$\left|I_t^\lambda(R)\right| \le C_{10} \lambda^{-1/4} t^{-3/4}. \tag{2.6.16}$$

Inequality (2.6.16) makes it possible to majorize the left-hand side of (2.6.9) in the following manner:

$$S_2 \le C_{11} \lambda^{-1/2} \sum_{i=1}^{\widehat{m}} \int\limits_{1 \le \sqrt{t} \le \sqrt{\lambda}/2} u_i^2(x,t) t^{-1/2-N/2}\, d\rho(t).$$

Now, to establish estimate (2.6.9) it suffices to make use of the boundedness of the integral (2.3.10) uniform with respect to x in G_R at any $\delta > 0$ and, in particular, at $\delta = 1/2$ (see Section 2.3.2).

---

[43] See estimates (2.4.10) and (2.4.11) in Section 2.4.1.

To verify estimates (2.6.10) and (2.6.12), we note that these two estimates stem from the more general estimate

$$\sum_{i=1}^{\widehat{m}} \int_{1\le|\sqrt{t}-\sqrt{\lambda}|\le\sqrt{\lambda}/2} \left\{t^{(2-N)/2}\left[u_i(x,t)I_t^{\lambda}(R)\right]^2\right\} d\rho(t) \le C_{12}\lambda^{-1/2}, \tag{2.6.17}$$

uniform with respect to x in G_R.

To establish estimate (2.6.17) we denote by p the least of the numbers such that $2^p > \sqrt{\lambda}/2$. Then, the segment $[1,\sqrt{\lambda}/2]$ is covered by a system of segments $[2^{\ell-1}, 2^{\ell}]$, $\ell = 1, 2, \ldots, p$.

Making use of estimate (2.6.15) and taking into account that, under the sign of the integral in (2.6.17), $1/\sqrt{t} \le 2\sqrt{\lambda}$, we can majorize the left-hand side of (2.6.17) in the following manner:

$$\begin{aligned}
&\sum_{i=1}^{\widehat{m}} \int_{1\le|\sqrt{t}-\sqrt{\lambda}|\le\sqrt{\lambda}/2} \left\{t^{(2-N)/2}\left[u_i(x,t)I_t^{\lambda}(R)\right]^2\right\} d\rho(t) \\
&\le C_9^2\, 2^{N-1}\lambda^{-N/2} \sum_{i=1}^{\widehat{m}} \int_{1\le|\sqrt{t}-\sqrt{\lambda}|\le\sqrt{\lambda}/2} u_i^2(x,t)(\sqrt{t}-\sqrt{\lambda})^2 d\rho(t) \\
&\le C_9^2\, 2^{N-1}\lambda^{-N/2} \sum_{\ell=1}^{p} \left\{ \sum_{i=1}^{\widehat{m}} \int_{2^{\ell-1}\le|\sqrt{t}-\sqrt{\lambda}|\le 2^{\ell}} u_i^2(x,t)(\sqrt{t}-\sqrt{\lambda})^2 d\rho(t) \right\} \\
&\le C_9^2\, 2^{N-1}\lambda^{-N/2} \sum_{\ell=1}^{p} 4^{1-\ell} \left[\sum_{i=1}^{\widehat{m}} \int_{|\sqrt{t}-\sqrt{\lambda}|\le 2^{\ell}} u_i^2(x,t) d\rho(t) \right].
\end{aligned} \tag{2.6.18}$$

To estimate the quantity in square brackets on the right-hand side of (2.6.18), we make use of estimate (2.3.7) (see Corollary to Lemma 2.1 in Section 2.3.2). The estimate (2.3.7) having been taken at $\mu = \sqrt{\lambda}$, $\rho_0 = 2^{\ell}$, we obtain from (2.6.18) that the left-hand side of (2.6.17) does not exceed

$$C_{13}\lambda^{-1/2} \sum_{\ell=1}^{\infty} 2^{-\ell} = C_{13}\lambda^{-1/2}.$$

This completes the proof of estimates (2.6.10) and (2.6.12).

The estimate (2.6.11) is a trivial consequence of the inequality (2.6.14), the inequality $1/\sqrt{\lambda} \leq 2\sqrt{\lambda}$ (valid for all $\lambda \geq 4$, $|\sqrt{t} - \sqrt{\lambda}| \leq 1$), and the estimate (2.3.4) taken at $\mu = \sqrt{\lambda}$.

All we have to do now is to verify estimate (2.6.13). Keeping in mind that for the values of $\sqrt{t} \geq (3/2)\sqrt{\lambda}$ the estimate (2.6.15) may be written in the form (2.6.16), we obtain that

$$S_6 \leq C_{10}^2 \lambda^{-1/2} \sum_{i=1}^{\widehat{m}} \int\limits_{\sqrt{t} \geq 3\sqrt{\lambda}/2} u_i^2(x,t) t^{-1/2-N/2} d\rho(t). \qquad (2.6.19)$$

In order to obtain estimate (2.6.13) from inequality (2.6.19), it suffices to make use of the boundedness of the integral (2.3.10) uniform with respect to x in G_R at any $\delta > 0$ and, in particular, at $\delta = 1/2$ (see Section 2.3.2).

We have dealt with the derivation of all the estimates of interest (2.6.8)–(2.6.13), and the proof of Theorem 2.6 is thus complete.

REMARK. Let us consider the function

$$\theta_0^s(x,y,\lambda) = \Gamma(s+1)2^s(2\pi)^{-N/2}\lambda^{N/4-s/2}|x-y|^{-N/2-s}J_{N/2+s}(|x-y|\sqrt{\lambda}), \quad (2.6.20)$$

which is the Riesz mean of order s and dimension λ of the spectral function

$$\theta_0(x,y,\lambda) = (2\pi)^{-N/2}\lambda^{N/4}|x-y|^{-N/2}J_{N/2}(|x-y|\sqrt{\lambda}),$$

corresponding to a unique self-adjoint nonnegative extension of the Laplace operator $Lu = -\Delta u$ over the entire space E^N and coincident with its expansion into an N-fold Fourier integral.

If the function (2.6.2) is taken at $r = |x-y|$, one will observe that, by virtue of the inequality $|J_\nu(r\sqrt{\lambda})| \leq C_\nu(r\sqrt{\lambda})^{-1/2}$ for all $s > -1/2$ uniform with respect to x in G_R, the following estimate holds:

$$\|\overset{\lambda}{v}(|x-y|) - \theta_0^s(x,y,\lambda)\|_{L_2(G)} \leq C_{14}\lambda^{(N-1)/4-s/2}, \qquad (2.6.21)$$

where the $L_2(G)$ norm is taken with respect to the coordinates of the point y.

It follows from inequalities (2.6.1), (2.6.21) and from the triangle inequality that, given any $R > 0$ and $s \geq 0$, the following estimate is valid uniformly with respect to x in G_R:

$$\|\theta^s(x,y,\lambda) - \theta_0^s(x,y,\lambda)\|_{L_2(G)} \leq C_{15}\lambda^{(N-1)/4-s/2}. \qquad (2.6.22)$$

This estimate shows that the Riesz means of order $s \geq 0$ for the spectral functions of an arbitrary self-adjoint nonnegative extension $\widehat{A}$ of the Laplace operator is close, in the metric $L_2(G)$, to the expansion into an N-fold Fourier integral.

Note that as the order s of the Riesz means grows to infinity, we obtain in the right-hand side of (2.6.22) an arbitrarily high order of smallness in powers of $1/\lambda$.

2.7. ESTIMATE FOR THE REMAINDER TERM OF THE RIESZ MEANS OF A SPECTRAL FUNCTION IN THE METRIC L_∞

We have shown in the preceding sections of this chapter and in Section 1.3.4 that, in order to define exact conditions for the uniform convergence, localization, or lack of localization of spectral decompositions and their Riesz means, we need not necessarily know the estimate of the remainder term of a spectral function and its Riesz means in the L_∞ metric.

However, historically the course of things has been such that starting from the known work of T. Carleman [13], numerous well-reputed mathematicians (V. G. Avakumovich [7], B. M. Levitan [57], K. I. Babenko [8], L. Gårding [19], L. Hörmander [24], and others) explored the estimation of the remainder term and the Riesz means of a spectral function precisely in the L_∞ metric. All those researchers, with a view to obtaining the estimate in L_∞ of the remainder term and the Riesz means of a spectral function, employed various modifications of the original Carleman method, which was essentially based on the study of the kernel of a certain function of the operator $\widehat{A}$ with subsequent application of one or another theorem of the Tauber type.

In this section, we shall set forth a new simple method of obtaining an estimate in the metric L_∞ of the remainder term of a spectral function and its Riesz means in line with the conceptual strategy of the present monograph without resorting to the Carleman method or to the technique of Tauber-type theorems.

2.7.1. Proof of the Main Theorem

THEOREM 2.7. *Assume that G is an arbitrary domain in the space E^N, $\widehat{A}$ is an arbitrary self-adjoint nonnegative extension of the Laplace operator $Lu = -\Delta u$ in the domain G. Then, given any $R_0 > 0$ and any sufficiently large $\lambda > 0$, the estimate*

$$\left|\theta^s(x,y,\lambda) - \Gamma(s+1)2^s(2\pi)^{-N/2}\lambda^{N/4-s/2}|x-y|^{-N/2-s}J_{N/2+s}(|x-y|\sqrt{\lambda})\right| \leq C\lambda^{(N-1)/2-s/2} \tag{2.7.1}$$

holds for the Riesz means $\theta^s(x,y,\lambda)$ of any nonnegative order s of a spectral function $\theta(x,y,\lambda)$ of the extension $\widehat{A}$ uniformly with respect to the ensemble (x,y) on the set $G_{R_0} \times G_{R_0}$.[44)]

PROOF: We fix an arbitrary $R_0 > 0$ and an arbitrary point x of the subset G_{R_0}, and for any R belonging to the segment $R_0/2 \leq R \leq R_0$ we consider the function

[44)] Recall that G_{R_0} denotes a subset of points of the domain G removed from the G boundary to a distance greater than $R_0 > 0$.

$\overset{\lambda}{v}(r,R) = \overset{\lambda}{v}(r)$ with its argument $r = |x-y|$ as defined by relation (2.6.2) (see Section (2.6)), that is, the function

$$\overset{\lambda}{v}(r,R) = \overset{\lambda}{v}(r)$$
$$= \begin{cases} \Gamma(s+1)2^s(2\pi)^{-N/2}\lambda^{N/4-s/2}r^{-N/2-s}J_{N/2+s}(r\sqrt{\lambda}) & \text{for} \quad r \le R, \\ 0 & \text{for} \quad r > R. \end{cases} \tag{2.7.2}$$

For any function $f(r)$ of argument r we denote by the symbol $Df(r)$ the Bessel derivative defined as $Df(r) = \frac{1}{r}f'(r)$, and by the symbol $D^n f(r)$, the Bessel derivative of order n.

From function (2.7.2), we subtract a "smoothing" polynomial $\overset{\lambda}{w}(r,R) = \overset{\lambda}{w}(r)$ of the form[45)]

$$\overset{\lambda}{w}(r,R) = \overset{\lambda}{w}(r) = \begin{cases} \sum\limits_{m=0}^{[(N-1)/2]} a_m(r^2-R^2)^m & \text{for} \quad r = |x-y| \le R, \\ 0 & \text{for} \quad r = |x-y| > R, \end{cases} \tag{2.7.3}$$

with its coefficients a_m chosen such that at $r = R$ the following relations hold:

$$\overset{\lambda}{w}(R) = \overset{\lambda}{v}(R), \quad D\overset{\lambda}{w}(R) = D\overset{\lambda}{v}(R), \quad \dots, D^{[(N-1)/2]}\overset{\lambda}{w}(R) = D^{[(N-1)/2]}\overset{\lambda}{v}(R). \tag{2.7.4}$$

Relations (2.7.4) define unambiguously the coefficients $a_0, a_1, \dots, a_{[(N-1)/2]}$ of polynomial (2.7.3), since for all r within the segment $0 \le r \le R$ and for any $k = 0, 1, \dots, [(N-1)/2]$ we have

$$D^k\left[\overset{\lambda}{v}(r)\right] = \Gamma(s+1)2^s(2\pi)^{-N/2}\lambda^{N/4-s/2}(-1)^k\lambda^{k/2}r^{-(N/2+s+k)}J_{N/2+s+k}(r\sqrt{\lambda}), \tag{2.7.5}$$

$$D^k\left[\overset{\lambda}{w}(r)\right] = \sum_{m=k}^{[(N-1)/2]} a_m m(m-1)\dots(m-k+1)(r^2-R^2)^{m-k}\,2^k. \tag{2.7.6}$$

Consequently, from the kth equality (2.7.4) we obtain that

$$a_k = (-1)^k\Gamma(s+1)2^s(2\pi)^{-N/2}\frac{1}{2^k k!}\lambda^{N/4-s/2+k/2}R^{-(N/2+s+k)}J_{N/2+s+k}(R\sqrt{\lambda}) \tag{2.7.7}$$
$$(k = 0, 1, \dots, [(N-1)/2]).$$

Let us turn to the function $f(r,R) = f(r)$, which is the difference of functions (2.7.2) and (2.7.3):

$$f(r,R) = f(r) = \overset{\lambda}{v}(r) - \overset{\lambda}{w}(r). \tag{2.7.8}$$

[45)] The symbol $[(N-1)/2]$ is henceforth used to denote the integer part of the number $(N-1)/2$.

One observes that, by virtue of conditions (2.7.4), the relations

$$f(R) = 0, \quad Df(R) = 0, \ \dots \ , D^{[(N-1)/2]} f(R) = 0 \tag{2.7.9}$$

hold for this function.

For an arbitrary function $F(R)$ of argument R, well defined on the segment $R_0/2 \le R \le R_0$, we introduce the average $S_{R_0}[F(R)]$ of this function with weight R on the above said segment defined as

$$S_{R_0}[F(R)] = \frac{3}{8R_0^2} \int\limits_{R_0/2}^{R_0} R\, F(R)\, dR. \tag{2.7.10}$$

(The coefficient $3/(8R_0^2)$ is chosen such that if $F(R) = C$ =const, then $S_{R_0}[F(R)] = C$.)

Note that the average operation we have introduced features the following property: If the estimate $|F(R)| \le C_1$ holds for the function $F(R)$ uniformly with respect to R on the segment $R_0/2 \le R \le R_0$, then the estimate $S_{R_0}[F(R)] \le C_1$ holds for the average of this function.

We consider now an arbitrary ordered spectral representation of the space $L_2(G)$ with respect to the extension $\widehat{A}$ with spectral measure $\rho(t)$, sets of multiplicity e_i, fundamental functions $u_i(x,t)$ $(i = 1, 2, \dots , \widehat{m})$ of multiplicity $\widehat{m} \le \infty$, and we denote by the symbol $\widehat{f}_i(x, y, R)$ the Fourier image of function (2.7.8) with respect to the fundamental function $u_i(y,t)$, and by the symbol $(\widehat{S_{R_0} f})_i(x,t)$, the Fourier image of the average $S_{R_0}[f(r, R)]$. One easily observes that the following relation holds:

$$(\widehat{S_{R_0} f})_i(x,t) = S_{R_0}[\widehat{f}_r(x, t, R)]. \tag{2.7.11}$$

Our aim is to determine the Fourier image $(\widehat{S_{R_0} f})_i(x,t)$. To that end, it suffices, by virtue of relation (2.7.11), to determine the Fourier image of the function (2.7.8).

By virtue of relation (2.3.2) (see Section 2.3.1)

$$\widehat{f}_i(x, t, R) = (2\pi)^{N/2} u_i(x,t) t^{(2-N)/4} \int\limits_0^R f(r) r^{N/2} J_{N/2-1}(r\sqrt{t})\, dr. \tag{2.7.12}$$

We integrate the right-hand side of (2.7.12) ($[(N-1)/2] + 1$) times by parts using the recurrence relation

$$\int r^{\mu} J_{\mu-1}(r\sqrt{t})\, dr = (\sqrt{t})^{-1} r^{\mu} J_{\mu}(r\sqrt{t})$$

and carry out the operations of Bessel differentiation D to obtain

$$\widehat{f}_i(x,t,R)$$

$$= (2\pi)^{N/2} u_i(x,t) t^{(2-N)/4} \left\{ \sum_{k=0}^{[(N-1)/2]} \left[(-1)^k D^k[f(r)] t^{-(k+1)/2} r^{N/2+k} J_{N/2+k}(r\sqrt{t})\right] \Big|_{r=0}^{r=R} \right.$$

$$+ (-1)^{[(N-1)/2]+1} (\sqrt{t})^{-[(N-1)/2]-1}$$

$$\left. \times \int_0^R D^{[(N-1)/2]+1}[f(r)] r^{N/2+[(N-1)/2]+1} J_{N/2+[(N-1)/2]}(r\sqrt{t}) dr \right\}. \quad (2.7.13)$$

We note that all the substitutions on the right-hand side of (2.7.13) reduce to zero. Indeed, the reduction to zero of all the substitutions at $r = 0$ stems from the fact that the values of $D^k f(r)$ at $k = 0, 1, \ldots, [(N-1)/2]$ are bounded in the vicinity of the point $r = 0$ by virtue of relations (2.7.5) and (2.7.6), and the reduction to zero of all the substitutions at $r = R$ follows from relations (2.7.9).

We note further that the relations (2.7.5) and (2.7.6) taken at $k = [(N-1)/2]$ yield the equalities

$$D^{[(N-1)/2]+1}\left[\overset{\lambda}{v}(r)\right] = \Gamma(s+1) 2^s (2\pi)^{-N/2} \lambda^{N/4-s/2} (-1)^{[(N-1)/2]+1} \sqrt{\lambda}^{[(N-1)/2]+1}$$

$$\times r^{-(N/2+s+[(N-1)/2]+1)} J_{N/2+s+[(N-1)/2]+1}(r\sqrt{\lambda}),$$

$$D^{[(N-1)/2]+1}\left[\overset{\lambda}{w}(r)\right] \equiv 0,$$

which, in turn, lead to

$$D^{[(N-1)/2]+1} f(r) = \Gamma(s+1) 2^s (2\pi)^{-N/2} \lambda^{N/4-s/2} (-1)^{[(N-1)/2]+1}$$

$$\times \sqrt{\lambda}^{[(N-1)/2]+1} r^{-(N/2+s+[(N-1)/2]+1)} J_{N/2+s+[(N-1)/2]+1}(r\sqrt{\lambda}). \quad (2.7.14)$$

We define the number ν as

$$\nu = \frac{N}{2} + \left[\frac{N-1}{2}\right], \quad (2.7.15)$$

and, taking into account the reduction to zero of all the substitutions in (2.7.13) and making use of relation (2.7.14), we rewrite equality (2.7.13) in the following manner:

$$\widehat{f}_i(x,t,R) = \Gamma(s+1) 2^s u_i(x,t) (\sqrt{\lambda})^{\nu+1-s} (\sqrt{t})^{-\nu}$$

$$\times \int_0^R r^{-s} J_{\nu+1+s}(r\sqrt{\lambda}) J_\nu(r\sqrt{t}) dr. \quad (2.7.16)$$

Next, we carry out a transformation of the form $\int_0^R = \int_0^\infty - \int_R^\infty$ on the right-hand side of (2.7.16), keeping in mind that the relation[46]

$$\Gamma(s+1)2^s(\sqrt{\lambda})^{\nu+1-s}(\sqrt{t})^{-\nu}\int_0^\infty r^{-s}J_{\nu+1+s}(r\sqrt{\lambda})J_\nu(r\sqrt{t})dr = \delta_t^\lambda\left(1-\frac{t}{\lambda}\right)^s$$

holds for any $\nu > -1$, $s \geq 0$; in this relation

$$\delta_t^\lambda = \begin{cases} 1 & \text{for} \quad t < \lambda, \\ 0 & \text{for} \quad t \geq \lambda. \end{cases} \tag{2.7.17}$$

By convention, given $s = 0$, $t = \lambda$, we have $\delta_t^\lambda = 1/2$ and $(1 - t/\lambda)^s = 1$.

We introduce the quantity

$$I_t^\lambda(R) = \overset{\nu+1}{I_t^\lambda} = \int_0^\infty r^{-s}J_{\nu+1+s}(r\sqrt{\lambda})J_\nu(r\sqrt{t})dr, \tag{2.7.18}$$

taking into account that, given $s = 0$, $t = \lambda$, the number 1/2 should be subtracted from the right-hand side of (2.7.8). (To avoid confusion, we recall that the same symbol has previously been used in expression (2.4.9′).)

Making use of the integral transformation for the right-hand side of (2.7.16) and referring to notations (2.7.17) and (2.7.18), we obtain an expression for the Fourier image

$$\widehat{f}_i(x,t,R) = \delta_t^\lambda\, u_i(x,t)\left(1-\frac{t}{\lambda}\right)^s - \Gamma(s+1)2^s(\sqrt{\lambda})^{\nu+1-s}(\sqrt{t})^{-\nu}\overset{\nu+1}{I_t^\lambda}(R)u_i(x,t). \tag{2.7.19}$$

It follows from relations (2.7.19) and (2.7.11) that the Fourier image of the average $S_{R_0}[f(r,R)]$ takes the form[47]

$$\begin{aligned}\left(\widehat{S_{R_0}f}\right)_i(x,t) &= \delta_t^\lambda\, u_i(x,t)\left(1-\frac{t}{\lambda}\right)^s \\ &\quad -\Gamma(s+1)2^s(\sqrt{\lambda})^{\nu+1-s}(\sqrt{t})^{-\nu}S_{R_0}\left[\overset{\nu+1}{I_t^\lambda}(R)\right]u_i(x,t).\end{aligned} \tag{2.7.20}$$

We multiply inequality (2.7.20) by the fundamental function $u_i(y,t)$ and integrate it for spectral measure $\rho(t)$ with t going from 0 to ∞; then we take the summation over all numbers i from 1 to $\widehat{m}$.

[46] See, for example, Bateman and Erdélyi [9, p. 107, formula (34)].

[47] We take into account that the average S_{R_0} of a quantity independent of R is identical with this quantity.

Keeping in mind that for any fixed $x \in G_{R_0}$ and $\lambda > 0$ the average $S_{R_0}[f(|x-y|, R)]$ belongs, with respect to y, to the class $L_2(G)$ and making use of relation (2.7.17) and expression (2.4.4) for the Riesz means of a spectral function, we obtain the following equality:

$$\theta^s(x, y, \lambda) - S_{R_0}[f(|x-y|, R)]$$

$$= \Gamma(s+1)2^s(\sqrt{\lambda})^{\nu+1-s} \sum_{i=1}^{\widehat{m}} \int_0^\infty (\sqrt{t})^{-\nu} u_i(x,t) u_i(y,t) S_{R_0}\left[\overset{\nu+1}{I_t^\lambda}(R)\right] d\rho(t). \tag{2.7.21}$$

One will observe that this equality, given any fixed $x \in G_{R_0}$ and $\lambda > 0$, is, in any case, to be regarded as the equality of elements of the space $L_2(G)$ with respect to the coordinates of the point y.

Now we see, by virtue of equality (2.7.3), that

$$S_{R_0}[f(|x-y|, R)] = S_{R_0}\left[\overset{\nu}{v}(|x-y|, R)\right] - S_{R_0}\left[\overset{\lambda}{w}(|x-y|, R)\right]. \tag{2.7.22}$$

Further, by definition (2.7.2) of the function $\overset{\lambda}{v}(r, R)$ and with reference to the estimate $|J_{N/2+s}(r\sqrt{\lambda})| \le C(r\sqrt{\lambda})^{-1/2}$, it follows that uniformly in x, y, R (where $x \in G_{R_0}$, $y \in G$, $R \in [R_0/2, R_0]$), the following estimate holds:

$$\overset{\lambda}{v}(|x-y|, R) - \Gamma(s+1)2^s(2\pi)^{-N/2}\lambda^{N/4-s/2}|x-y|^{-N/2-s} J_{N/2+s}(|x-y|\sqrt{\lambda})$$

$$= O(\lambda^{(N-1)/4-s/2}) = O(\lambda^{(N-1)/2-s/2}).$$

Therefore, uniformly in (x, y) for $x \in G_{R_0}$, $y \in G$, the estimate

$$S_{R_0}[\overset{\lambda}{v}(|x-y|, R)] - \Gamma(s+1)2^s(2\pi)^{-N/2}$$

$$\times \lambda^{N/4-s/2}|x-y|^{-N/2-s} J_{N/2+s}(|x-y|\sqrt{\lambda}) = O(\lambda^{(N-1)/2-s/2}) \tag{2.7.23}$$

is valid.

Finally, we note that, by virtue of the inequality $|J_\mu(R\sqrt{\lambda})| \le C(R\sqrt{\lambda})^{-1/2}$, the following estimate is derived from relation (2.7.3) and from the representation (2.7.7) of the coefficients $a_0, a_1, \dots, a_{[(N-1)/2]}$ uniformly in x, y, R for $x \in G_{R_0}$, $y \in G$, $R \in [R_0/2, R_0]$:

$$\overset{\lambda}{w}(|x-y|, R) = O(\lambda^{(N-1)/2-s/2}),$$

implying that

$$S_{R_0}\left[\overset{\lambda}{w}(|x-y|, R)\right] = O(\lambda^{(N-1)/2-s/2}) \tag{2.7.24}$$

uniformly with respect to x, y for $y \in G_{R_0}$, $y \in G$.

The following estimate stems from relations (2.7.22), (2.7.23), and (2.7.24):

$$S_{R_0}[f(|x-y|,R)] - \Gamma(s+1)2^s(2\pi)^{-N/2} \times \lambda^{N/4-s/2}|x-y|^{-N/2-s}J_{N/2+s}(|x-y|\sqrt{\lambda}) = O(\lambda^{(N-1)/2-s/2}), \tag{2.7.25}$$

uniformly in x, y for $x \in G_{R_0}$, $y \in G$.

In examining relations (2.7.21) and (2.7.25), one observes that in order to prove the desired estimate (2.7.1), it is sufficient to establish that the estimate

$$(\sqrt{\lambda})^{\nu+1} \sum_{i=1}^{\widehat{m}} \int_0^R (\sqrt{t})^{-\nu} \left| u_i(x,t)u_i(y,t)S_{R_0}\left[\overset{\nu+1}{I_t^\lambda}(R)\right]\right| d\rho(t) = O(\lambda^{(N-1)/2}) \tag{2.7.26}$$

is valid uniformly in x, y on the set $G_{R_0} \times G_{R_0}$.

It suffices, in turn, to prove that uniformly with respect to x, y on the set $G_{R_0} \times G_{R_0}$, the following five estimates hold:

$$(\sqrt{\lambda})^{\nu+1} \sum_{i=1}^{\widehat{m}} \int_{\sqrt{t}\le 1} (\sqrt{t})^{-\nu} \left\{\left| u_i(x,t)u_i(y,t)S_{R_0}\left[\overset{\nu+1}{I_t^\lambda}(R)\right]\right|\right\} d\rho(t) = O(\lambda^{(N-1)/2}), \tag{2.7.27}$$

$$(\sqrt{\lambda})^{\nu+1} \sum_{i=1}^{\widehat{m}} \int_{1\le\sqrt{t}\le\sqrt{\lambda}/2} (\sqrt{t})^{-\nu} \{\ldots\}\, d\rho(t) = O(\lambda^{(N-1)/2}), \tag{2.7.28}$$

$$(\sqrt{\lambda})^{\nu+1} \sum_{i=1}^{\widehat{m}} \int_{1\le|\sqrt{t}-\sqrt{\lambda}|\le\sqrt{\lambda}/2} (\sqrt{t})^{-\nu} \{\ldots\}\, d\rho(t) = O(\lambda^{(N-1)/2}), \tag{2.7.29}$$

$$(\sqrt{\lambda})^{\nu+1} \sum_{i=1}^{\widehat{m}} \int_{|\sqrt{t}-\sqrt{\lambda}|\le 1} (\sqrt{t})^{-\nu} \{\ldots\}\, d\rho(t) = O(\lambda^{(N-1)/2}), \tag{2.7.30}$$

$$(\sqrt{\lambda})^{\nu+1} \sum_{i=1}^{\widehat{m}} \int_{\sqrt{t}\ge 3\sqrt{\lambda}/2} (\sqrt{t})^{-\nu} \{\ldots\}\, d\rho(t) = O(\lambda^{(N-1)/2}), \tag{2.7.31}$$

(the braces in (2.7.28)–(2.7.31) have the same meaning as those in (2.7.27)).

We shall be relying on the following lemma.

LEMMA 2.12. *For the average* $S_{R_0}\left[\overset{\nu+1}{I_t^\lambda}(R)\right]$ *of integral (2.7.18), the following three estimates hold, for any* $\nu \ge 1/2$, $s \ge 0$:

(a) for $\lambda \geq 1$, $t \geq 1$, $|\sqrt{t} - \sqrt{\lambda}| \leq 1$, *the estimate is*

$$\left| S_{R_0} \left[\overset{\nu+1}{I_t^\lambda} (R) \right] \right| = O(\lambda^{-1/4} t^{-1/4}); \tag{2.7.32}$$

(b) for $\lambda \geq 1$, $t \geq 1$, $|\sqrt{t} - \sqrt{\lambda}| \geq 1$, *the estimate is*

$$\left| S_{R_0} \left[\overset{\nu+1}{I_t^\lambda} (R) \right] \right| = O(\lambda^{-1/4} t^{-1/4} |\sqrt{t} - \sqrt{\lambda}|^{-2}); \tag{2.7.33}$$

(c) for $\lambda > 1$, $0 \leq t \leq 1$, *the estimate is*

$$\left| S_{R_0} \left[\overset{\nu+1}{I_t^\lambda} (R) \right] \right| = O(\lambda^{-3/4} t^{\nu/4}). \tag{2.7.34}$$

We defer the proof of Lemma 2.12 to the next section, and now proceed to complete the proof of Theorem 2.7.

In order to establish relation (2.7.27), it will suffice to invoke relation (2.7.34) in estimating the left-hand side of (2.7.27), to take into account that the inequality $\nu - 1/2 \leq N - 1$ is valid by virtue of (2.7.15), and to make use of the estimate

$$\sum_{i=1}^{\widehat{m}} \int_{\sqrt{t} \leq \mu_0} |u_i(x,t) u_i(y,t)| d\rho(t) = O(\mu_0^N), \tag{2.7.35}$$

at $\mu_0 = 1$ uniformly with respect to $x \in G_{R_0}$, $y \in G_{R_0}$, which is trivially obtained from estimate (2.3.4) and from the Cauchy–Buniakowski inequality applied initially to the integral and then to the sum on the left-hand side of (2.7.35).

The estimate (2.7.30) stems in a straightforward manner from relation (2.7.32), from the inequality $\left(\sqrt{\frac{\lambda}{t}}\right)^{\nu+1/2} \leq 2^{\nu+1/2}$ valid for all $t \geq 1$, $\lambda \geq 1$, $|\sqrt{t} - \sqrt{\lambda}| \leq 1$, from the estimate

$$\sum_{i=1}^{\widehat{m}} \int_{|\sqrt{t} - \mu| \leq \rho_0} |u_i(x,t) u_i(y,t)| d\rho(t) = \rho_0 O(\mu^{N-1}), \tag{2.7.36}$$

for $\mu = \sqrt{\lambda}$, $\rho_0 = 1$ uniformly with respect to $x \in G_{R_0}$, $y \in G_{R_0}$ (see 2.3.7), and from the Cauchy–Buniakowski inequality applied to the integral and to the sum on the left-hand side of (2.7.36).

The estimate (2.7.31) is readily obtained from relation (2.7.33) which, given $\sqrt{t} \geq 3\sqrt{\lambda}/2$, may be rewritten as

$$\left| S_{R_0} \left[\overset{\nu+1}{I_t^\lambda}(R) \right] \right| = O(\lambda^{-1/4} t^{-5/4}),$$

from the inequality $(\sqrt{\lambda}/t)^{\nu+1/2} \leq (\sqrt{\lambda}/t)^{N-1/2}$, and from the estimate

$$\sum_{i=1}^{\hat{m}} \int_{\lambda}^{\infty} |u_i(x,t)u_i(y,t)| t^{-\delta-N/2} d\rho(t) = O(\lambda^{-\delta}), \tag{2.7.37}$$

for $\delta = 3/4$ uniformly with respect to $x \in G_{R_0}$, $y \in G_{R_0}$, which, in turn, is obtained from estimate (2.3.9) by applying the Cauchy–Buniakowski inequality first to the integral and then to the sum on the left-hand side (2.7.37).

The estimate (2.7.28) stems from the relation (2.7.33) which, given $1 \leq \sqrt{t} \leq \sqrt{\lambda}/2$, can be rewritten in the form

$$\left| S_{R_0} \left[\overset{\nu+1}{I_t^\lambda}(R) \right] \right| = O(\lambda^{-3/4} t^{-3/4}),$$

invoking the inequality $(\sqrt{\lambda}/t)^{\nu-1/2} \leq (\sqrt{\lambda}/t)^{N-1}$ and the (uniform with respect to $x \in G_{R_0}$, $y \in G_{R_0}$) boundedness at $\delta = 1/2$ of the quantity

$$\sum_{i=1}^{\hat{m}} \int_{0}^{\infty} |u_i(x,t)u_i(y,t)| t^{-\delta-N/2} d\rho(t). \tag{2.7.38}$$

The boundedness of (2.7.38) follows from the boundedness of (2.3.10) and from the Cauchy–Buniakowski inequality applied successively to the integral and to the sum of the left-hand side of (2.7.38).

Now we have to prove estimate (2.7.29).

We denote by p the least number such that $2^{p-1} \geq \sqrt{\lambda}$ and then construct a system of segments $[2^{\ell-1}, 2^{\ell}]$ $(\ell = 1, 2, \ldots, p)$ covering the segment $[1, \sqrt{\lambda}/2]$. Making use of the inequality $(\sqrt{\lambda}/t)^{\nu+1} \leq C$ =const[48] and of estimates (2.7.33) and (2.7.35), we

[48] This inequality holds with certainty for values of $1 \leq |\sqrt{t} - \sqrt{\lambda}| \leq \sqrt{\lambda}/2$.

obtain that the left-hand side of (2.7.29) is majorized by the quantity

$$(\sqrt{\lambda})^{\nu+1}\sum_{\ell=1}^{p}\left\{\sum_{i=1}^{\widehat{m}}\int\limits_{2^{\ell-1}\le|\sqrt{t}-\sqrt{\lambda}|\le 2^{\ell}}(\sqrt{t})^{-\nu}\left|u_i(x,t)u_i(y,t)\,S_{R_0}\left[\overset{\nu+1}{I_t^{\lambda}}(R)\right]\right|d\rho(t)\right\}$$

$$\le C\sum_{\ell=1}^{p}\left[\sum_{i=1}^{\widehat{m}}\int\limits_{2^{\ell-1}\le|\sqrt{t}-\sqrt{\lambda}|\le 2^{\ell}}|\sqrt{t}-\sqrt{\lambda}|^{-2}|u_i(x,t)u_i(y,t)|d\rho(t)\right]$$

$$\le C\sum_{\ell=1}^{p}4^{1-\ell}\left[\sum_{i=1}^{\widehat{m}}\int\limits_{|\sqrt{t}-\sqrt{\lambda}|\le 2^{\ell}}|u_i(x,t)u_i(y,t)|d\rho(t)\right]$$

$$\le C_1\lambda^{(N-1)/2}\sum_{\ell=1}^{p}2^{-p}\le C_1\lambda^{(N-1)/2}$$

uniformly with respect to $x \in G_{R_0}$ and $y \in G_{R_0}$.

This completes the derivation of estimate (2.7.29) and, provided that Lemma 2.12 holds, Theorem 2.7 is thus proved.

We now come to the proof of Lemma 2.12.

2.7.2. Proof of Ancillary Lemma 2.12

We note, first of all, that the validity of estimate (2.7.32) follows immediately at $\lambda \ge 1$, $t \ge 1$, $|\sqrt{t}-\sqrt{\lambda}| \le 1$ from the fact that the estimate below (see also (2.4.10′))

$$\left|\overset{\nu+1}{I_t^{\lambda}}(R)\right| \le C_1 R^{-s}\lambda^{-1/4}t^{-1/4}$$

holds, given any $\nu \ge 1/2$, $s \ge 0$, $\lambda \ge 1$, $t \ge 1$, $|\sqrt{t}-\sqrt{\lambda}| \le 1$, where the constant C_1 is independent not only of $\lambda \ge 1$, $t \ge 1$, but also of R for all $0 < R \le 1$.

We now prove that, for any $\nu \ge -1/2$, $s \ge 0$, $\lambda \ge 1$, $t \ge 1$, $|\sqrt{t}-\sqrt{\lambda}| \ge 1$, the estimate (2.7.33) remains valid. With this aim in mind, it will suffice to prove the estimate

$$\left|S_{R_0}\left[\overset{\nu}{I_t^{\lambda}}(R)\right]\right| = O(\lambda^{-1/4}t^{-1/4}|\sqrt{t}-\sqrt{\lambda}|^{-2}) \qquad (2.7.39)$$

for any $\nu \ge 1/2$. We shall verify this estimate successively for $\sqrt{\lambda}+1 \le \sqrt{t}$ and $1 \le \sqrt{t} \le \sqrt{\lambda}-1$.

In the case $\sqrt{\lambda}+1 \le \sqrt{t}$, the equality (2.4.43) has been established (see Section 2.4.2). We rewrite this equality in the form

$$\overset{\nu}{I_t^\lambda}(R) = -\frac{\sqrt{t}}{(\sqrt{t}+\sqrt{\lambda})(\sqrt{t}-\sqrt{\lambda})} J_{\nu+s}(R\sqrt{\lambda})J_\nu(R\sqrt{t})R^{-s}$$

$$-\frac{\sqrt{\lambda}}{(\sqrt{t}+\sqrt{\lambda})(\sqrt{t}-\sqrt{\lambda})} J_{\nu+s+1}(R\sqrt{\lambda})J_{\nu+1}(R\sqrt{t})R^{-s}$$

$$-(2\nu+2+s)\frac{\sqrt{\lambda}}{(\sqrt{t}+\sqrt{\lambda})(\sqrt{t}-\sqrt{\lambda})}\int\limits_R^\infty r^{-1-s}J_{\nu+s+1}(r\sqrt{\lambda}) \tag{2.7.40}$$

$$\times J_{\nu-1}(r\sqrt{t})dr + 2\nu\frac{\lambda}{\sqrt{t}(\sqrt{t}+\sqrt{\lambda})(\sqrt{t}-\sqrt{\lambda})}\int\limits_R^\infty r^{-1-s}$$

$$\times J_{\nu+s+2}(r\sqrt{\lambda})J_\nu(r\sqrt{t})dr.$$

Each of the two integrals on the right-hand side of (2.7.40) is preceded by a cofactor[49] of the form $O(|\sqrt{t}-\sqrt{\lambda}|^{-1})$.

One observes that both integrals are of the form $\overset{\nu}{I_t^\lambda}(R)$. We have shown in Section 2.4.2 that, given any real ν and any $s>-1$, $\lambda \ge 1$, $t \ge 1$, $|\sqrt{t}-\sqrt{\lambda}| \ge 1$, the estimate

$$\left|\overset{\nu}{I_t^\lambda}(R)\right| \le CR^{-1-s}\lambda^{-1/4}t^{-1/4}|\sqrt{t}-\sqrt{\lambda}|^{-1} \tag{2.7.41}$$

holds, with the constant C independent of R $(0<R\le 1)$ (see also estimate (2.4.11′) in Section 2.4.2).

Therefore, having applied the averaging operation S_{R_0} to each of the last two terms on the right-hand side of (2.7.40), we obtain the desired order

$$O(\lambda^{-1/4}t^{-1/4}|\sqrt{t}-\sqrt{\lambda}|^{-2}). \tag{2.7.42}$$

It remains to prove that the averaging operation as applied to each of the first two terms on the right-hand side of (2.7.40) also leads to a magnitude of order (2.7.42).

Since each of the cofactors $\frac{\sqrt{t}}{(\sqrt{t}+\sqrt{\lambda})(\sqrt{t}-\sqrt{\lambda})}$ and $\frac{\sqrt{\lambda}}{(\sqrt{t}+\sqrt{\lambda})(\sqrt{t}-\sqrt{\lambda})}$ has an order of $O(|\sqrt{t}-\sqrt{\lambda}|^{-1})$, it suffices to prove that each of the integrals

$$\int\limits_{R_0/2}^{R_0} R^{-(s-1)}J_{\nu+s}(R\sqrt{\lambda})J_\nu(R\sqrt{t})dR,$$

[49] We take into account that $\frac{\sqrt{\lambda}}{\sqrt{t}+\sqrt{\lambda}} \le 1$, $\frac{\lambda}{\sqrt{t}(\sqrt{t}+\sqrt{\lambda})} \le 1$ (for $\sqrt{\lambda} \le \sqrt{t}$).

$$\int_{R_0/2}^{R_0} R^{-(s-1)} J_{\nu+s+1}(R\sqrt{\lambda}) J_{\nu+1}(R\sqrt{t})dR \tag{2.7.43}$$

has an order

$$O(\lambda^{-1/4} t^{-1/4} |\sqrt{t} - \sqrt{\lambda}|^{-1}). \tag{2.7.44}$$

Either of the two integrals (2.7.43) can be represented as the difference $\int\limits_{R_0/2}^{R_0} = \int\limits_{R_0/2}^{\infty} - \int\limits_{R_0}^{\infty}$. If $s > 0$, we obtain that both integrals are represented as the difference of two integrals of the $\overset{\nu}{I_t^{\lambda}}(R)$ type at $s > -1$, and for this reason each of the two integrals (2.7.43) has the required order (2.7.44).

Given $s = 0$, each of the integrals (2.7.43) is expressed as[50)]

$$\int_{R_0/2}^{R_0} R J_\nu(R\sqrt{\lambda}) J_\nu(R\sqrt{t}) dR$$
$$= \frac{1}{t-\lambda}\left[R\sqrt{t} J_{\nu+1}(R\sqrt{t}) J_\nu(R\sqrt{\lambda}) - R\sqrt{\lambda} J_\nu(R\sqrt{t}) J_{\nu+1}(R\sqrt{\lambda})\right]\Bigg|_{R=R_0/2}^{R=R_0},$$

$$\int_{R_0/2}^{R_0} R J_{\nu+1}(R\sqrt{\lambda}) J_{\nu+1}(R\sqrt{t}) dR$$
$$= \frac{1}{t-\lambda}\left[R\sqrt{t} J_{\nu+2}(R\sqrt{t}) J_{\nu+1}(R\sqrt{\lambda}) - R\sqrt{\lambda} J_{\nu+1}(R\sqrt{t}) J_{\nu+2}(R\sqrt{\lambda})\right]\Bigg|_{R=R_0/2}^{R=R_0}.$$

One immediately observes that each of the integrals (2.7.43) subject to $s = 0$ has the required order (2.7.44).

This completes the proof of estimate (2.7.33) for the first case $\sqrt{\lambda} + 1 \le \sqrt{t}$.

The proof of estimate (2.7.33) for the second case $1 \le \sqrt{t} \le \sqrt{\lambda} - 1$ is carried out in much the same manner, the only distinction being that the equality (2.7.40)

---

50) See Bateman and Erdélyi [9, p. 104, formula (9)].

should be replaced by the following equality:

$$\overset{\nu}{I_t^\lambda}(R) = -\frac{\sqrt{\lambda}}{(\sqrt{\lambda}+\sqrt{t})(\sqrt{\lambda}-\sqrt{t})} J_{\nu+s-1}(R\sqrt{\lambda})J_{\nu-1}(R\sqrt{t})R^{-s}$$

$$+\frac{\sqrt{t}}{(\sqrt{\lambda}+\sqrt{t})(\sqrt{\lambda}-\sqrt{t})} J_{\nu+s-2}(R\sqrt{\lambda})J_{\nu-2}(R\sqrt{t})R^{-s}$$

$$+(2\nu+2s-2)\frac{t}{\sqrt{\lambda}(\sqrt{\lambda}+\sqrt{t})(\sqrt{\lambda}-\sqrt{t})}\int_R^\infty J_{\nu+s-1}(r\sqrt{\lambda})$$

$$\times J_{\nu-s}(r\sqrt{t})r^{-1-s}dr - (2\nu-4)\frac{\sqrt{t}}{(\sqrt{\lambda}+\sqrt{t})(\sqrt{\lambda}-\sqrt{t})}$$

$$\times \int_R^\infty J_{\nu+s}(r\sqrt{\lambda})J_{\nu-2}(r\sqrt{t})r^{-1-s}dr,$$

which is equivalent to equality (2.4.44) (see Section 2.4.2).

The estimate (2.7.33) is thereby established.

Now we wish to establish estimate (2.7.34) for any $\nu \geq 1/2$, $s \geq 0$, $\lambda \geq 1$, $0 \leq t \leq 1$. It suffices to prove that the estimate

$$\left|\overset{\nu+1}{I_t^\lambda}(R)\right| = O(\lambda^{-3/4}t^{\nu/2}) \tag{2.7.45}$$

holds uniformly with respect to R on the segment $R_0/2 \leq R \leq R_0$.

We integrate the integral $\overset{\nu+1}{I_t^\lambda}(R)$, defined by relation (2.7.18), $[\nu+1]$ times[51] by parts, using the recursive relations

$$\int r^{-\nu-s}J_{\nu+s+1}(r\sqrt{\lambda})dr = \frac{-r^{-s-\nu}}{\sqrt{\lambda}}J_{\nu+s}(r\sqrt{\lambda}),$$

$$\frac{d}{dr}\left[r^\nu J_\nu(r\sqrt{t})\right] = \sqrt{t}\,r^\nu J_{\nu-1}(r\sqrt{t}).$$

We thus obtain the relation

$$\begin{aligned}\overset{\nu+1}{I_t^\lambda}(R) = &-\sum_{n=1}^{[\nu+1]} t^{(n-1)/2}\lambda^{-n/2}J_{\nu+s+1-n}(R\sqrt{\lambda})J_{\nu+1-n}(R\sqrt{t})R^{-s}\\ &+\left(\sqrt{\frac{t}{\lambda}}\right)^{[\nu+1]}\int_R^\infty r^{-s}J_{\nu+s+1-[\nu+1]}(r\sqrt{\lambda})J_{\nu-[\nu+1]}(r\sqrt{t})dr.\end{aligned} \tag{2.7.46}$$

[51] The symbol $[\nu+1]$ denotes the integer part of a number $\nu+1$.

Each of the nonintegral summands on the right-hand side of (2.7.46) is estimated using the inequalities

$$\left|J_\mu(R\sqrt{\lambda})\right| \le C_1(\mu)(R\sqrt{\lambda})^{-1/2}, \qquad \left|J_\mu(R\sqrt{t})\right| \le C_2(\mu)(R\sqrt{t})^{-1/2}.$$

We obtain from these inequalities, with reference to $\lambda > 1$, that each of the nonintegral summands on the right-hand side of (2.7.41) has the required order

$$O(\lambda^{-3/4}t^{\nu/2}). \tag{2.7.47}$$

It remains to prove that the last (integral-containing) term on the right-hand side of (2.7.46) likewise has the same order.

Since, by virtue of relation (2.7.15),

$$[\nu+1] = \begin{cases} \nu+1, & \text{for even } N, \\ \nu+1/2, & \text{for odd } N, \end{cases}$$

the last term on the right-hand side of (2.7.46) is

$$\begin{cases} \left(\sqrt{\dfrac{t}{\lambda}}\right)^{\nu+1} \displaystyle\int_R^\infty r^{-s}J_s(r\sqrt{\lambda})J_{-1}(r\sqrt{t})dr & \text{for even } N, \\ \left(\sqrt{\dfrac{t}{\lambda}}\right)^{\nu+1/2} \displaystyle\int_R^\infty r^{-s}J_{1/2+s}(r\sqrt{\lambda})J_{-1/2}(r\sqrt{t})dr & \text{for odd } N. \end{cases} \tag{2.7.48}$$

The fact that, for odd N, the term (2.7.48) has the required order (2.7.47) stems trivially from the inequality $\nu + 1/2 \ge 1$ and from the estimate[52)]

$$\int_R^\infty r^{-s}J_{1/2+s}(r\sqrt{\lambda})J_{-1/2}(r\sqrt{t})dr = O(\lambda^{-1/4}t^{-1/4}), \tag{2.7.49}$$

uniform with respect to R on the segment $R_0/2 \le R \le R_0$.

We now have to prove that, for even N, the term (2.7.48) has the needed order (2.7.47) uniformly with respect to R on the segment $[R_0/2, R_0]$.

If $s > 0$, this follows immediately from the estimates

$$\left|J_s(r\sqrt{\lambda})\right| \le C_3(r\sqrt{\lambda})^{-1/2}, \qquad \left|J_{-1}(r\sqrt{t})\right| \le C_4(r\sqrt{t})^{-1/2}.$$

[52)] For $s > 0$, the estimate (2.7.49) follows from the inequalities $|J_{1/2+s}(r\sqrt{\lambda})| < C_1(r\sqrt{\lambda})^{-1/2}$, $|J_{-1/2}(r\sqrt{t})| \le C_2(r\sqrt{t})^{-1/2}$; for $s = 0$, this estimate has been established earlier in Section 1.3.2 (see proof of Lemma 1.4).

If $s = 0$, then, given $t \leq 1$, $\lambda > 1$, the integral below equals zero (see Bateman and Erdélyi [9, p. 107, formula (34)]

$$\int_0^\infty J_0(r\sqrt{\lambda})J_{-1}(r\sqrt{t})dr = -\int_0^\infty J_0(r\sqrt{\lambda})J_1(r\sqrt{t})dr = 0,$$

and it follows therefore that

$$\int_R^\infty J_0(r\sqrt{\lambda})J_{-1}(r\sqrt{t})dr = -\int_0^R J_0(r\sqrt{\lambda})J_1(r\sqrt{t})dr = O(1).$$

One infers from this relation and from the inequality $\nu + 1 \geq 2$ that, for even N also, the term (2.7.48) has indeed the required order.

This completes the proof of Lemma 2.12.

COMMENTS ON CHAPTER 2

The study of the Riesz means of spectral decompositions had its origin in a study of the Riesz means of multiple trigonometric series and Fourier integrals with spherical partial sums.

Among the first results was the paper by S. Bochner [11] published in 1936. In this work, an exact order of the Riesz means, $s = (N-1)/2$, ensuring the validity of the localization principle was established for the Riesz means of order s of the decompositions of an arbitrary function $f(x)$ of the class L_2 into an N-fold trigonometric Fourier series and into an N-fold Fourier integral with spherical partial functions.

It was also shown in the same work by S. Bochner that for an arbitrary function $f(x)$ from the class L_1 the localization principle of the Riesz means of order $s = (N-1)/2$ holds for the decomposition into an N-fold Fourier integral and does not hold for the decomposition into an N-fold trigonometric Fourier series.

Bochner, based on this result, has coined the order $s = (N-1)/2$ a *critical* order.

A deeper insight into the Bochner result on the N-fold trigonometric Fourier series was made in 1958 by E. Stein [75], who showed that the localization principle of the Riesz means of the critical order $s = (N-1)/2$ remains valid for an arbitrary function from the $L \log^+ L$ class and, consequently, remains valid for an arbitrary function in the class L_p for any $p > 1$.

It has been shown in the work of B. M. Levitan [58] (1955) that for the localization principle of the Riesz means of an integer order s ($s = 0, 1, \dots, [N/2]$) of an N-fold trigonometric Fourier series with spherical partial functions to hold, it suffices for all the partial derivatives of decomposed function of order up to $[N/2] - s$ to be periodic and square-integrable over an N-dimensional cube. A comparison with the main theorem of negative type that has been proved in Section 2.4 lends support to the Levitan results for an odd number N of measurements.

Also, it has been shown by Levitan [58] that the additional requirement of continuity or, at least, of boundedness of the partial derivatives of order $[N/2] - s$ guarantees a uniform convergence for the Riesz means of order s of an N-fold trigonometric Fourier series. A comparison with the main theorem of positive type (Section 2.5) reveals that for an odd number N of measurements and for an N-fold trigonometric Fourier series the Levitan results are quite close to the exact conditions of uniform convergence (it is sufficient to require the summability of the derivatives with degree $p > \frac{2N}{N-1}$ to hold instead of the boundedness of the partial derivatives of order $\left[\frac{N}{2}\right] - s$ in an N-dimensional cube).

In an earlier paper, published in 1954, Levitan [57] showed that the above-mentioned result of S. Bochner [1] on the validity of the localization principle of the Riesz means of critical order $s = (N-1)/2$ for a function of the class L_2 can be extended to the case of eigenfunction expansion of the main boundary-value problems for the Laplace operator in an arbitrary N-dimensional domain G. It was proved in that work that, subject to the additional requirement of continuity of a function $f(x)$ at a given point interior to the domain G, the Riesz means of critical order $s = (n-1)/2$ converge at this point to a value of $f(x)$.

The exact conditions for uniform convergence and localization of the Riesz means of any order s subject to the condition $0 \leq s < (N-1)/2$ for the spectral decompositions associated with an arbitrary self-adjoint nonnegative extension of the Laplace operator in an arbitrary N-dimensional domain G (see Sections 2.4 and 2.5) were first reported by V. A. Il'in and Sh. A. Alimov [41] in 1971.

In a paper by V. A. Il'in and N. Yu. Kapustin [44] (1984), these exact conditions were transferred to the case of an arbitrary non-half-bounded self-adjoint extension of the Laplace operator, and in the paper by Z. V. Shoniya [72] (1986), to the case of an arbitrary non-half-bounded self-adjoint extension of the Schroedinger operator (allowing for the occurence of Coulomb-type singularities in the potential at separate points).

The method for estimating the remainder term of the Riesz means of any nonnegative order for the spectral function makes no use of the traditional Carleman technique and the Tauber theorem formalism (Section 2.7). This method, suggested by V. A. Il'in, for $s = 0$ was first published in 1977 (see Sh. A. Alimov, V. A. Il'in, and E. M. Nikishin [6]; Section 2.4).

Another method for estimating the remainder term of a spectral function without resort to the Carleman or Tauber techniques was reported by V. A. Il'in [39] in 1987.

The modified method for estimating the remainder term of a spectral function

(Section 2.6) makes it possible to estimate the spectral function of a non-self-adjoint ordinary differential operator (see V. A. Il'in [37], [38].

In 1986, Ya. Sh. Salimov [69], [70], making use of a modification of the method that has been discussed in Section 2.6, proposed an estimation for the remainder term of the Riesz means of a spectral function associated with a non-self-adjoint extension of the Laplace operator.

Chapter 3

On the Riesz Equisummability of Spectral Decompositions in the Classical and the Generalized Sense

In proving Theorem 2.3 (Section 2.2) we have established a *uniform* (on an arbitrary compact set of domain G) *equiconvergence* of the Riesz means of order s of two arbitrary self-adjoint nonnegative extensions of the Laplace operator in domain G for a finite-in-G function $f(x)$ belonging to one of the four classes $H_2^\alpha(G)$, $L_2^\alpha(G)$, $B_{p,\theta}^\alpha(G)$, or $G^\alpha(G)$ with order of differentiability $\alpha > (N-1)/2 - s$, where $0 \leq s < (N-1)/2$ (and in the case of the Besov class for any $\theta \geq 1$). This fact ensures a uniform (on an arbitrary compact set K of domain G) tendency to zero of the difference of the Riesz means of order s of the spectral decompositions of this function which correspond to two arbitrary self-adjoint nonnegative extensions of the Laplace operator (in domain G, or in a domain to which G is interior).

Otherwise stated, the fact that a function $f(x)$, finite in domain G, belongs to one of the aforementioned classes with α and s subject to the conditions $\alpha \leq (N-1)/2-s$, $0 \leq s < (N-1)/2$ ensures a *uniform* (on an arbitrary compact set K of domain G) *equisummability* of the Riesz means of order s of the spectral decompositions of this function which correspond to two arbitrary self-adjoint nonnegative extensions of the Laplace operator (in domain G or in a domain to which G is interior).

It is natural to call the equisummability, understood in the sense of a uniform (on any compact set) or at least of a pointwise convergence to zero of the difference of the Riesz means of two spectral decompositions, *an equisummability in the classical sense.*

It is expedient to continue a function, finite in domain G, with zero beyond domain G over the entire space E^N and to consider, for the function continued, the Riesz means of order s of its spectral decompositions into an N-dimensional Fourier integral.

Next, given the N-dimensional Fourier integral expansion as defined above, we can focus on the equiconvergence of the Riesz means of order s of two decompositions, one of which is generated by an arbitrary self-adjoint nonnegative extension of the Laplace operator in domain G, and the other of which is generated by expansion into an N-dimensional Fourier integral of the function continued with zero beyond domain G.

Now the following question arises: Does equisummability in the classical sense of the Riesz means of order s of two arbitrary spectral decompositions hold for a function, finite in domain G and belonging to one of the above classes, with positive order of differentiability subject to the conditions $\alpha \geq (N-1)/2 - s$ rather than to its opposite $\alpha < (N-1)/2 - s$?

We show in Section 1 of the present chapter that the answer to this question is not in the affirmative; we provide evidence that, given a positive α such that $\alpha < (N-1)/2 - s$, there exists for a function $f(x)$, finite in domain G and belonging to the Zygmund–Hölder class $C^{\alpha}(G)$,[1)] generally speaking, no equisummability (either pointwise, or uniform on any compact set) of the Riesz means of order s of two spectral decompositions, namely, the eigenfunction expansion of the first boundary-value problem in domain G, and the expansion of the same function, continued with zero beyond domain G, into an N-dimensional Fourier integral.

This result signifies that for the function $f(x)$, finite in domain G, continued with zero beyond G and belonging to one of the classes $H_p^{\alpha}(G)$, $L_p^{\alpha}(G)$, $B_{p,\theta}^{\alpha}(G)$, or $C^{\alpha}(G)$, at $0 < \alpha < (N-1)/2 - s$ (for any $p \geq 1$ and $\theta \geq 1$), the divergence of the Riesz means of order s of the expansion into N-dimensional Fourier integral provides, generally speaking, no information about the divergence of the Riesz means of order s of the spectral decomposition corresponding to self-adjoint nonnegative extension of the Laplace operator in domain G.

It follows therefrom that Theorems 2.1 and 2.2 are, in principle, impossible to obtain via establishing, first, the divergence of the Riesz means of the expansion into an N-fold Fourier integral and, next, the Riesz equisummability.

Thus, in the classical treatment of the Riesz equisummability of spectral decompositions, the ultimate condition of equisummability of the Riesz means of order s $(0 \leq s < (N-1)/2)$ for a function $f(x)$, finite in an N-dimensional domain G and belonging to one of the classes $H_2^{\alpha}(G)$, $L_2^{\alpha}(G)$, $B_{2,\theta}^{\alpha}(G)$, or $G^{\alpha}(G)$, is the condition $\alpha \geq (N-1)/2 - s > 0$.

The Riesz equisummability of two spectral decompositions is intimately related to the establishment of an estimate (to within an order of magnitude and uniform on any compact set of the main domain G) of the difference of the Riesz means of two spectral decompositions of the same order s for an arbitrary function $f(x)$ from a certain class.

[1)] The more so for a function $f(x)$, finite in domain G, belonging to one of the classes $H_p^{\alpha}(G)$, $L_p^{\alpha}(G)$, $B_{p,\theta}^{\alpha}(G)$, given a positive $\alpha < N - 1/2 - s$, an arbitrary $p \geq 1$, and any $\theta \geq 1$ (in the case of the Besov class).

In Section 3.1.4 we give an exhaustive solution to this problem for an arbitrary function from the class $L_2(G)$.

In Section 3.2, we renounce the classical approach to the property of equisummability of the Riesz means of order s. We say that the Riesz means of order s of two arbitrary self-adjoint nonnegative extensions of the Laplace operator are *equisummable in domain G in the generalized sense* if the difference between the Riesz means specified tends to zero almost everywhere in domain G.

In Section 3.2 we provide evidence that if a function $f(x)$ belongs, in an arbitrary N-dimensional domain G, to the class $L_2(G)$ only, then the Riesz means of any positive order s of two arbitrary self-adjoint nonnegative extensions of the Laplace operator in this domain are not only equisummable in domain G in the generalized sense, but also the difference of the Riesz means specified of order s and size λ has, almost everywhere in domain G, an order of smallness $o(\lambda^{-s/2})$ infinitely growing with increasing order s.

We also show in Section 3.2 that if a function $f(x)$ belongs, in an arbitrary N-dimensional domain G, to the class $L_2(G)$ and, what is more, becomes zero almost everywhere in a domain D interior to G, then the Riesz means of any positive order s of an arbitrary self-adjoint nonnegative extension of the Laplace operator in domain G have, almost everywhere in domain D, an order of smallness $o(\lambda^{-s/2})$, where λ is the size of the Riesz means in question.

Thus, the tendency of the Riesz means to zero almost everywhere in D grows infinitely with increasing order s of the Riesz means.

3.1. ON THE RIESZ EQUISUMMABILITY OF SPECTRAL DECOMPOSITIONS IN THE CLASSICAL SENSE

It is a well-known fact that, given $s \geq (N-1)/2$, the theorem on the uniform (on any compact set of an arbitrary N-dimensional domain G) equisummability of the Riesz means of order s of two self-adjoint nonnegative extensions of the Laplace operator in domain G holds for any function $f(x)$ in the class $L_2(G)$.[2)]

Therefore, we restrict ourselves to a study of the equisummability of the Riesz means of order s subject to $0 \leq s < (N-1)/2$ which is uniform, or at least pointwise, on any compact set.

As has already been pointed out in the introductory part of this chapter, the proof of Theorem 2.3 (see Chapter 2) implies, for any s in the half-interval $0 \leq s < (N-1)/2$, the fact that a finite function $f(x)$ belonging, in an arbitrary N-dimensional domain G, to one of four classes, viz., Nikol'skii $H_2^\alpha(G)$, Sobolev–Liouville $L_2^\alpha(G)$, Besov $B_{2,\theta}^\alpha(G)$, or Zygmund–Hölder $C^\alpha(G)$ with an order of differentiability $\alpha \geq (N-1)/2-s$, ensures a uniform (on any compact set K of domain G) equisummability

[2)] This result was first reported by B. M. Levitan [57].

of the Riesz means of order s of any two self-adjoin nonnegative extensions of the Laplace operator in domain G (or in any domain to which G is in interior).

A natural question arises here: Is the requirement for the order of differentiability $\alpha \geq (N-1)/2 - s$ in each of the above four classes definitive for the equisummability, uniform on each compact, or at least pointwise, of the Riesz means of order s ($0 \leq s < (N-1)/2$) of two arbitrary self-adjoint nonnegative extensions of the Laplace operator.

It will be shown in the theorem below that, given any s in the half-interval $0 \leq s < (N-1)/2$ and any α such that $0 < \alpha < (N-1)/2 - s$, the fact that a function $f(x)$, finite in domain G, belongs to any of the classes $H_2^\alpha(G)$, $L_2^\alpha(G)$, $B_{2,\theta}^\alpha(G)$, $G^\alpha(G)$, with order of differentiability α and an arbitrary degree of summability $p \geq 1$, does not, generally speaking, ensure either a uniform (on any point x_0 interior to domain G) equisummability of the Riesz means of order s — even under the additional condition that the function $f(x)$ becomes zero in a certain vicinity of the point x_0.

THEOREM 3.1. *Let s be an arbitrary fixed number in the half interval $0 \leq s < (N-1)/2$, and let the domain G represent an N-dimensional ball of sufficiently small radius R_0 centered at the point x_0; let for an arbitrary function $f(x)$, defined within G and continued with zero beyond the boundary of domain G, the symbol $E_\lambda^s(x,f)$ denote the Riesz means of order s taken at point x and associated with the Laplace operator eigenfunction expansion in the ball G with the homogeneous first-kind boundary condition at the ball boundary, and let the symbol $\overset{0}{E}{}_\lambda^s(x,f)$ denote the Riesz means of order s taken at point x and corresponding to the N-fold Fourier integral expansion. Then there exists, for any α in the interval $0 < \alpha < (N-1)/2 - s$, a function $f(x)$ satisfying the following requirements:*

(1) $f(x)$ belongs to the Zygmund–Hölder class $C^\alpha(G)$;

(2) $f(x)$ is compactly supported inside the ball G and, moreover, becomes zero at a certain vicinity of the ball center x_0;

(3) the difference $E_\lambda^s(x_0,f) - \overset{0}{E}{}_\lambda^s(x_0,f)$ of the Riesz means of order s of the expansions of the function $f(x)$ taken at the point x_0 is unbounded as $\lambda \to \infty$.

REMARK. Theorem 3.1 remains valid if the requirement that the function $f(x)$ belongs to the Zygmund–Hölder class $C^\alpha(G)$ for any arbitrary fixed α in the interval $0 < \alpha < (N-1)/2 - s$ is replaced by the requirement of this function belonging to any of the Nikol'skii $H_p^\alpha(G)$, Sobolev–Liouville $L_p^\alpha(G)$ or Besov $B_{p,\theta}^\alpha(G)$ classes for any fixed α in the interval $0 < \alpha < (N-1)/2 - s$ and for any $p \geq 1$ and (in the case of the Besov class) for any $\theta \geq 1$.

Indeed, one observes, given any fixed α in the interval $0 < \alpha < (N-1)/2 - s$ and a fixed α' such that $0 < \alpha < \alpha' < (N-1)/2 - s$, that the assignment of $f(x)$ to the class $C^{\alpha'}$ guarantees the assignment of this function to the class $H_p^{\alpha'}$ for any $p \geq 1$, and, therefore, guarantees the assignment of $f(x)$ both to the class H_p^α and (by virtue of the embedding theorem (2.1.18), see Section 2.1.3) to either of the classes L_p^α or $B_{p,\theta}^\alpha$, for any $p \geq 1$ and $\theta \geq 1$.

The proof of Theorem 3.1 will be preceded by the proof of a lemma on the lower-bound estimate of the difference of the Riesz means of spectral functions and of a lemma on the unboundedness of the difference of the fractional-kernel Riesz means for the two decompositions under study in the L_1 metric.

3.1.1. Lemma on the Estimation in L_1 of the Difference of the Riesz Means for Spectral Functions

We denote by the symbol $\theta^s(x, y, \lambda)$ the Riesz means of order s of a spectral function corresponding to the expansion in the Laplace operator eigenfunctions in a ball G with homogeneous first-kind boundary condition at the ball G boundary. We denote by the symbol $\theta_0^s(x, y, \lambda)$ the Riesz means of order s of this spectral function corresponding to the expansion into an N-fold Fourier integral.

Further, we denote by the symbol $r = |x_0 - y|$ the distance between a variable point y and the center x_0 of the ball G; we set $R = R_0/2$, where R_0 is the radius of the ball G. We also take into consideration a ring layer, defined as

$$E_R = \{R/4 \le |x_0 - y| \le R/2\}. \tag{3.1.1}$$

We shall only be concerned with radially symmetric normed eigenfunctions of the first boundary-value problem for the ball G, since all other eigenfunctions for this problem become zero at the center x_0 of G and are, therefore, omitted from consideration.

The eigenfunctions of interest will be denoted by the symbols $\{u_n(y)\}$, and their respective eigenvalues (numbered in increasing order) will be denoted by the symbols $\{\lambda_n\}$. Note that

$$u_n(y) = \bar{u}_n(r) = \sqrt{\frac{2}{\omega_N}}\,\frac{1}{R_0 J_{(N-1)/2}(\mu_n)}\, r^{(2-N)/2} J_{(N-2)/2}\left(\frac{r}{R}\mu_n\right), \tag{3.1.2}$$

$$\lambda_n = \frac{\mu_n^2}{R_0^2}, \tag{3.1.3}$$

where $\omega_N = 2(\sqrt{\pi})^N[\Gamma(N/2)]^{-1}$, and μ_n are the zeros of a Bessel function $J_{(N-2)/2}(x)$ numbered in increasing order.

For large zeros μ_n the following asymptotic formula holds[3]:

$$\mu_n = \pi\left(n + \frac{N-3}{4}\right) + O\left(\frac{1}{n}\right). \tag{3.1.4}$$

LEMMA 3.1. *Given a sufficiently small radius $R_0 = 2R > 0$ of the ball G, for any strictly positive s there exists a constant $A > 0$ such that for all sufficiently large*

[3] See, for example, Watson [77, p. 558].

numbers n the inequality

$$\int_{E_R} |\theta^s(x_0, y, \lambda_n) - \theta_0^s(x_0, y, \lambda_n)|\, dy \geq A\lambda_n^{(N-1)/4-s/2} \tag{3.1.5}$$

holds.

PROOF: We consider the Riesz means of order s, taken at $x = x_0$, of a spectral function corresponding to the expansion into N-fold Fourier integral (see expressions (2.6.20) in Section 2.6):

$$\theta_0^s(x_0, y, \lambda) = \Gamma(s+1)2^s(2\pi)^{-N/2}\lambda^{N/4-s/2}|x_0-y|^{-N/2-s}J_{N/2+s}(|x_0-y|\sqrt{\lambda}). \tag{3.1.6}$$

Making use of these Riesz means, we consider the function $\overset{\lambda}{v}(|x_0 - y|)$ similar to that dealt with in Section 2.4.1 (see relation (2.4.5)). This function may be written as

$$\overset{\lambda}{v}(|x_0 - y|) = \begin{cases} \theta_0^s(x_0, y, \lambda) & \text{for} \quad |x_0 - y| \leq R, \\ 0 & \text{for} \quad |x_0 - y| < R. \end{cases} \tag{3.1.7}$$

In Section 2.4.1, the spectral decomposition of this function, corresponding to an arbitrary self-adjoint nonnegative extension of the Laplace operator, has been given.[4] In particular, in expanding in the eigenfunctions $u_n(y)$ of the first boundary-value problem for the Laplace operator in an N-dimensional ball G, the spectral decomposition of the function (3.1.7) takes the form

$$\sum_{\lambda_k<\lambda} u_k(x_0)u_k(y)\left(1 - \frac{\lambda_k}{\lambda}\right)^s$$
$$-\Gamma(s+1)2^s\lambda^{N/4-s/2}\sum_{k=1}^{\infty}\lambda_k^{(2-N)/4}u_k(x_0)u_k(y)I_{\lambda_k}^{\lambda}(R), \tag{3.1.8}$$

where

$$I_{\lambda_k}^{\lambda}(R) = \int_R^{\infty} J_{N/2+s}(r\sqrt{\lambda})J_{N/2-1}(r\sqrt{\lambda_k})r^{-s}\,dr. \tag{3.1.9}$$

Since, for any fixed λ, the function (3.1.7) belongs in y to the class $L_2(G)$, the Fourier series (3.1.8) converges at the least to the function (3.1.7) in the $L_2(G)$ metric. Now, noting by definition that

$$\theta^s(x_0, y, \lambda) = \sum_{\lambda_k<\lambda} u_k(x_0)u_k(y)\left(1 - \frac{\lambda_k}{\lambda}\right)^s,$$

---

[4] For the Fourier image of the function $v(|x_0 - y|)$, see formula (2.4.8).

we arrive at the conclusion that the series

$$\Gamma(s+1)2^s\lambda^{N/4-s/2}\sum_{k=1}^{\infty}\lambda_k^{(2-N)/4}u_k(x_0)u_k(y)I_{\lambda_k}^{\lambda}(R) \tag{3.1.10}$$

for any fixed λ converges in the $L_2(G)$ metric to the difference $\theta^s(x_0,y,\lambda)-\overset{\lambda}{v}(|x_0-y|)$. Since at $|x_0-y|\le R$ the function $\overset{\lambda}{v}(|x_0-y|)$ coincides with $\theta_0^s(x_0,y,\lambda)$, within the ball $|x_0-y|\le R$ the series (3.1.10) converges in y in the L_2 metric to the difference $[\theta^s(x_0,y,\lambda)-\theta_0^s(x_0,y,\lambda)]$. The more so, the given series (3.1.10) converges (to the same difference $[\theta^s(x_0,y,\lambda)-\theta_0^s(x_0,y,\lambda)]$) in the L_1 metric both within the ball $|x_0-y|\le R$ and within the ring layer (3.1.1) belonging to this ball. A moment's reflection shows that

$$\int_{E_R}|\theta^s(x_0,y,\lambda)-\theta_0^s(x_0,y,\lambda)|\,dy$$

$$=\Gamma(s+1)2^s\lambda^{N/4-s/2}\int_{E_R}\left|\sum_{k=1}^{\infty}\lambda_k^{(2-N)/4}u_k(x_0)u_k(y)I_{\lambda_k}^{\lambda}(R)\right|dy. \tag{3.1.11}$$

Relation (3.1.11) implies that in order to prove inequality (3.1.5) it will suffice to establish the existence of a constant $B>0$ such that for all sufficiently small $R_0>0$ and for all sufficiently large n the inequality

$$\int_{E_R}\left|\sum_{k=1}^{\infty}\lambda_k^{(2-N)/4}u_k(x_0)u_k(y)I_{\lambda_k}^{\lambda_n}(R)\right|dy\ge B\lambda_n^{-1/4} \tag{3.1.12}$$

holds.

We note, for one thing, that by virtue of the fact that the modulus of the difference of two magnitudes is not smaller than the difference of the moduli of these magnitudes, the following inequality holds:

$$\int_{E_R}\left|\sum_{k=1}^{\infty}\lambda_k^{(2-N)/4}u_k(x_0)u_k(y)I_{\lambda_k}^{\lambda_n}(R)\right|dy\ge$$

$$\ge\int_{E_R}\left|\sum_{|\sqrt{\lambda_k}-\sqrt{\lambda_n}|\le R^{-2}}\lambda_k^{(2-N)/4}u_k(x_0)u_k(y)I_{\lambda_k}^{\lambda_n}(R)\right|dy \tag{3.1.13}$$

$$-\int_{E_R}\left|\sum_{|\sqrt{\lambda_k}-\sqrt{\lambda_n}|>R^{-2}}\lambda_k^{(2-N)/4}u_k(x_0)u_k(y)I_{\lambda_k}^{\lambda_n}(R)\right|dy.$$

Let us estimate from below the first integral and from above the second integral on the right-hand side of (3.1.13).

To prove inequality (3.1.12), it will suffice to show that for all sufficiently small $R_0 > 0$ there is a number $n_0(R_0)$ such that for all the numbers n that are larger than $n_0(R_0)$ the inequalities

$$\int\limits_{E_R} \left| \sum_{|\sqrt{\lambda_k}-\sqrt{\lambda_n}| \le R^{-2}} \lambda_k^{(2-N)/4} u_k(x_0) u_k(y) I_{\lambda_k}^{\lambda_n}(R) \right| dy \ge C_1 R^{(N-1)/2-s} \lambda_n^{-1/4}, \quad (3.1.14)$$

$$\int\limits_{E_R} \left| \sum_{|\sqrt{\lambda_k}-\sqrt{\lambda_n}| > R^{-2}} \lambda_k^{(2-N)/4} u_k(x_0) u_k(y) I_{\lambda_k}^{\lambda_n}(R) \right| dy \le C_2 R^{N/2-s} \lambda_n^{-1/4} \quad (3.1.15)$$

hold with certain positive constants C_1 and C_2.[5)]

Indeed, to obtain inequalities (3.1.12) from relations (3.1.13), (3.1.14), and (3.1.15), one merely has to fix a sufficiently small $R = R_0/2$.

First, we start by proving the upper-bound estimate (3.1.15). Applying the Cauchy–Buniakowski inequality to the integral on the left-hand side of (3.1.15) and taking into account that the eigenfunctions $\{u_k(y)\}$ form an orthonormal system in the ball G, we obtain

$$\int\limits_{E_R} \left| \sum_{|\sqrt{\lambda_k}-\sqrt{\lambda_n}| > R^{-2}} \lambda_k^{(2-N)/4} u_k(x_0) u_k(y) I_{\lambda_k}^{\lambda_n}(R) \right| dy$$

$$\le \left\{ \int\limits_{E_R} dy \int\limits_{G} \left[\sum_{|\sqrt{\lambda_k}-\sqrt{\lambda_n}| \le R^{-2}} \lambda_k^{(2-N)/4} u_k(x_0) u_k(y) I_{\lambda_k}^{\lambda_n}(R) \right]^2 dy \right\}^{1/2} \quad (3.1.16)$$

$$= O(R^{N/2}) \left\{ \sum_{|\sqrt{\lambda_k}-\sqrt{\lambda_n}| \le R^{-2}} \lambda_k^{(2-N)/2} u_k^2(x_1) \left[I_{\lambda_k}^{\lambda_n}(R) \right]^2 \right\}^{1/2}.$$

5) In what follows we make use of the symbols C_ℓ ($\ell = 1, 2, \ldots$) to denote positive constants dependent on both N and s.

Next, we employ the estimate

$$\lambda_k^{(1-N)/4}|u_k(x_0)| \le C_3 R_0^{-1/2}, \tag{3.1.17}$$

which immediately follows from (3.1.2) and (3.1.3); we shall also need the two estimates

$$\left|I_{\lambda_k}^{\lambda_n}(R)\right| \le C_4 R^{-s} \lambda_n^{-1/4} \lambda_k^{-1/4},$$

$$\left|I_{\lambda_k}^{\lambda_n}(R)\right| \le C_5 R^{-1-s} \lambda_n^{-1/4} \lambda_k^{-1/4} |\sqrt{\lambda_k} - \sqrt{\lambda_n}|^{-1} \tag{3.1.18}$$

that have been established in Section 2.4.2.

From estimates (3.1.16)–(3.1.18), from relation (3.1.3), and from the equality $R_0 = 2R$, we obtain the following inequality[6]:

$$\int_{E_R} \left| \sum_{|\sqrt{\lambda_k} - \sqrt{\lambda_n}| > R^{-2}} \lambda_k^{(2-N)/4} u_k(x_0) u_k(y) I_{\lambda_k}^{\lambda_n}(R) \right| dy$$

$$\le C_6 R^{(N-1)/2-s} \lambda_n^{-1/4} \left[\sum_{|\mu_k - \mu_n| \ge 2/R} |\mu_k - \mu_n|^{-2} \right]^{1/2}.$$

To obtain the upper-bound estimate (3.1.15) from the last inequality, we note that the bracketed sum

$$\sum_{|\mu_k - \mu_n| \ge 2/R} |\mu_k - \mu_n|^{-2}$$

has order $O(R)$ by virtue of the asymptotic formula (3.1.4).

Now, to prove Lemma 3.1 all we have to do is to establish the lower-bound estimate (3.1.14).

We introduce a condition which will simplify drastically our subsequent reasoning: Since, given $R_0 = 2R > 0$, we can fix the number $n_0(R_0)$ arbitrarily large, and inequality (3.1.14) must be proved only for n in excess of $n_0(R_0)$, in our argumentation we may neglect the terms containing a multiplier $1/n$ or $1/\mu_n$ (irrespective of what increasing order in the powers of $1/R = 2/R_0$ these terms may have).

Since, by virtue of (3.1.2), (3.1.3), and (3.1.4) the relations

[6] We also take into account that the inequality $|\sqrt{\lambda_k} - \sqrt{\lambda_n}| \ge R^{-2}$ is equivalent to the inequality $|\mu_k - \mu_n| \ge 2/R$ by virtue of (3.1.3) and $R_0 = 2R$.

$$
\begin{cases}
\lambda^{(1-N)/4} u_n(x_0) = \dfrac{C_7}{\sqrt{R}}(-1)^n + O\left(\dfrac{1}{n}\right), \\
u_n(y) = \bar{u}_n(r) = C_8 R^{-1/2}(-1)^n r^{(1-N)/2} \sin\left(n\pi \dfrac{r}{R_0}\right) + O\left(\dfrac{1}{n}\right)
\end{cases}
\tag{3.1.19}
$$

hold (we recall that $r = |x_0 - y|$), it will suffice to establish the estimate

$$
\int\limits_{E_R} \left| \sum_{|\mu_k - \mu_n| \leq 2/R} \lambda_k^{1/4} r^{(1-N)/2} \sin\left(k\pi \frac{r}{R_0}\right) I_{\lambda_k}^{\lambda_n}(R) \right| dy
\geq C_9 R^{(N+1)/2 - s} \lambda_n^{-1/4}
\tag{3.1.20}
$$

(where C_9 is a positive constant) for any sufficiently small $R_0 = 2R > 0$ and for all numbers n in excess of a sufficiently large $n_0(R_0)$.

From the familiar asymptotic formula for a Bessel function[7)]

$$
J_\nu(x) = \sqrt{\frac{2}{\pi x}} \cos\left(x - \frac{\pi}{4} - \frac{\pi}{2}\nu\right) + O(x^{-3/2})
$$

the following representation for the integral (3.1.9) is derived, valid for any fixed R, for all sufficiently large $n \geq n_0(R_0) = n_0(2R)$, and for $|\sqrt{\lambda_k} - \sqrt{\lambda_n}| < R^{-2}$:

$$
\lambda_n^{1/4} \lambda_k^{1/4} I_{\lambda_k}^{\lambda_n}(R) = \frac{1}{\pi} \int\limits_R^\infty \rho^{-1-s} \sin\left[\rho(\sqrt{\lambda_n} - \sqrt{\lambda_k}) - \frac{\pi}{2}s\right] d\rho
$$
$$
- \frac{1}{\pi} \int\limits_R^\infty \rho^{-1-s} \sin\left[\rho(\sqrt{\lambda_n} + \sqrt{\lambda_k}) - \frac{n\pi}{2} - \frac{\pi}{2} - \frac{\pi s}{2}\right] d\rho + O(R^{-s}\mu_n^{-1}).
\tag{3.1.21}
$$

Making use of the Bonet mean-value formula for a definite integral, we arrive at the conclusion that the second integral on the right-hand side of (3.1.21) also has the order of $O(R^{-s}\mu_n^{-1})$. If so, then the relation (3.1.21) and the above remark enable us to assert that in order to prove the estimate (3.1.20), it suffices to establish the

---

7) See, for example, Bateman and Erdélyi [9, p. 98, formula (3)].

inequality

$$\int_{E_R}\left|\sum_{|\mu_k-\mu_n|\le 2/R} r^{(1-N)/2}\sin\left(k\pi\frac{r}{R_0}\right)\int_R^\infty \rho^{-1-s}\sin\left[\rho(\sqrt{\lambda_n}-\sqrt{\lambda_k})-\frac{\pi}{2}s\right]d\rho\right|dy$$
$$\ge C_{10}R^{(N+1)/2-s}. \tag{3.1.22}$$

This inequality must be established for any sufficiently small $R_0 = 2R > 0$ and for any numbers n in excess of a sufficiently large $n_0(R_0)$.

To prove inequality (3.1.22), we first divide the integral on the left-hand side of (3.1.22) into the sum of two integrals:

$$\begin{aligned}\int_R^\infty &\rho^{-1-s}\sin\left[\rho(\sqrt{\lambda_n}-\sqrt{\lambda_k})-\frac{\pi s}{2}\right]d\rho\\ &=\int_R^{R+1}\rho^{-1-s}\sin\left[\rho(\sqrt{\lambda_n}-\sqrt{\lambda_k})-\frac{\pi s}{2}\right]d\rho\\ &\quad+\int_{R+1}^\infty \rho^{-1-s}\sin\left[\rho(\sqrt{\lambda_n}-\sqrt{\lambda_k})-\frac{\pi s}{2}\right]d\rho.\end{aligned} \tag{3.1.23}$$

For the second integral on the right-hand side of (3.1.23), making use of the Bonet mean-value formula for $k \neq n$ and majorizing the sine modulus with unity at $k = n$, we obtain the estimate

$$\int_{R+1}^\infty \rho^{-1-s}\sin\left[\rho(\sqrt{\lambda_n}-\sqrt{\lambda_k})-\frac{\pi s}{2}\right]d\rho = \begin{cases} O(R|\mu_k-\mu_n|^{-1} & \text{for} \quad k\neq n,\\ O(1) & \text{for} \quad k=n.\end{cases} \tag{3.1.24}$$

From the estimate (3.1.24), from the Cauchy–Buniakowski inequality, and from the property that the system of functions $\left\{\sqrt{\frac{2}{R_0}}\sin\left(k\pi\frac{r}{R_0}\right)\right\}$ is orthonormal on the segment $0 \le r \le R_0$, we obtain

$$\int_{E_R} \Big| \sum_{|\mu_k - \mu_n| \le 2/R} r^{(1-N)/2} \sin\left(k\pi \frac{r}{R_0}\right)$$

$$\times \int_{R+1}^{\infty} \rho^{-1-s} \sin\left[\rho(\sqrt{\lambda_n} - \sqrt{\lambda_k}) - \frac{\pi s}{2}\right] d\rho \Big| dy$$

$$\le C_{11} \Big\{ \int_{E_R} dy \int_0^{R_0} \Big[\sum_{|\mu_k - \mu_n| \le 2/R} r^{(1-N)/2} \sin\left(k\pi \frac{r}{R_0}\right)$$

$$\times \int_{R+1}^{\infty} \rho^{-1-s} \sin\left[\rho(\sqrt{\lambda_n} - \sqrt{\lambda_k}) - \frac{\pi s}{2}\right] d\rho \Big]^2 r^{N-1} dr \Big\}^{1/2}$$

$$\le C_{12} R^{(N+1)/2} \left[\sum_{|\mu_k - \mu_n| \le 2/R} \left(\int_{R+1}^{\infty} \rho^{-1-s} \sin\left[\rho(\sqrt{\lambda_n} - \sqrt{\lambda_k}) - \frac{\pi s}{2}\right] d\rho \right)^2 \right]^{1/2}$$

$$\le C_{13} R^{(N+1)/2} \left[1 + R^2 \sum_{\substack{|\mu_k - \mu_n| \le 2/R \\ k \ne n}} |\mu_k - \mu_n|^{-2} \right]^{1/2} \le C_{14} R^{(N+1)/2}. \tag{3.1.25}$$

One infers from the obtained estimate (3.1.25) and from equality (3.1.23) that, in order to establish the desired estimate (3.1.22), it suffices to show that, for any sufficiently small $R_0 = 2R > 0$ and for all numbers n in excess of a sufficiently large $n_0(R_0)$, the following inequality holds:

$$\int_{E_R} \Big| \sum_{|\mu_k - \mu_n| \le 2/R} r^{(1-N)/2} \sin\left(k\pi \frac{r}{R_0}\right) \int_R^{R+1} \rho^{-1-s} \sin\left[\rho(\sqrt{\lambda_n} - \sqrt{\lambda_k}) - \frac{\pi s}{2}\right] d\rho \Big| dy$$

$$\ge C_{15} R^{(N+1)/2 - s}, \tag{3.1.26}$$

where C_{15} is a positive constant.

In turn, it follows from relations (3.1.3), (3.1.4) and from the above remark that, in order to prove the inequality (3.1.26), it suffices to establish the inequality below for any sufficiently small $R_0 = 2R > 0$ and for all numbers n in excess of a sufficiently large $n_0(R_0)$,

$$\int\limits_{R/4}^{R/2} \left| \sum_{|k-n|<2/(\pi R)} r^{(1-N)/2} \sin\left(k\pi \frac{r}{R_0}\right) \int\limits_{R}^{R+1} \rho^{-1-s} \sin\left[(n-k)\pi\frac{\rho}{R_0} - \frac{\pi s}{2}\right] d\rho \right| dr$$

$$\geq C_{16} R^{(N+1)/2-s}, \tag{3.1.27}$$

where C_{16} designates a positive constant.

To estimate the left-hand side of (3.1.27), we make the change of variable $\rho = \tau + R$ in the inner integral. Consequently, the left-hand side of (3.1.27) is transformed as

$$\int\limits_{R/4}^{R/2} \left| \sum_{|k-n|\leq 2/(\pi R)} r^{(1-N)/2} \sin\left(k\pi \frac{r}{R_0}\right) \int\limits_{0}^{1} (\tau+R)^{-1-s} \sin\left[(n-k)\pi\frac{\tau+R}{R_0} - \frac{\pi s}{2}\right] d\tau \right| dr. \tag{3.1.28}$$

With no loss of generality, we can assume that $R_0 = 1/2m_0$, where m_0 is an integer which can be fixed sufficiently large. The inner integral in (3.1.28) can be represented in the form[8)]

$$\int\limits_{0}^{1} (\tau+R)^{-1-s} \sin\left[(n-k)\pi\frac{\tau+R}{R_0} - \frac{\pi s}{2}\right] d\tau$$

$$= \sum_{p=0}^{m_0-1} \int\limits_{2R_0 p}^{2R_0(p+1)} (\tau+R)^{-1-s} \sin\left[(n-k)\pi\frac{\tau+R}{R_0} - \frac{\pi s}{2}\right] d\tau$$

$$= \sum_{p=0}^{m_0-1} \int\limits_{0}^{2R_0} (u+R+2R_0 p)^{-1-s} \sin\left[(n-k)\pi\frac{u+R}{R_0} - \frac{\pi s}{2}\right] du$$

$$= \int\limits_{0}^{2R_0} \left[\sum_{p=0}^{m_0-1} (u+R+2R_0 p)^{-1-s}\right] \sin\left[(n-k)\pi\frac{\tau+R}{R_0} - \frac{\pi s}{2}\right] du.$$

8) We take into account that $2m_0R_0 = 1$, and partition the segment $[0,1]$ into the sum of m_0 subsegments $[R_0, p, 2R_0(p+1)]$, $p = 0, 1, \ldots, m_0 - 1$, not sharing a common inner point.

Thus, the expression (3.1.28) now becomes

$$\int_{R/4}^{R/2} \Bigg| \sum_{|k-n|\le 2/(\pi R)} r^{(1-N)/2} \sin\left(k\pi \frac{r}{R_0}\right) \int_0^{2R_0} \left[\sum_{p=0}^{m_0-1} (u+R+2R_0p)^{-1-s} \right] \times \sin\left[(n-k)\pi\frac{\tau+R}{R_0} - \frac{\pi s}{2}\right] du \Bigg| dr. \tag{3.1.28'}$$

We denote the integer part of $\frac{2}{\pi R}$ by m. Making the simple change $n-k=\ell$ and taking into account that $\cos x$ is an even function and $\sin x$ is an odd function, we arrive at the following equality:

$$\begin{aligned}
&\sum_{|k-n|\le 2/(\pi R)} \sin\left(k\pi\frac{r}{R_0}\right) \sin\left[(n-k)\pi\frac{u+R}{R_0} - \frac{\pi s}{2}\right] \\
&= -\sin\left(n\pi\frac{r}{R_0}\right)\sin\left(\frac{\pi s}{2}\right)\left[D_m\left(\pi\frac{u+R-r}{R_0}\right) + D_m\left(\pi\frac{u+R+r}{R_0}\right)\right] \\
&= -\cos\left(n\pi\frac{r}{R_0}\right)\cos\left(\frac{\pi s}{2}\right)\left[D_m\left(\pi\frac{u+R-r}{R_0}\right) - D_m\left(\pi\frac{u+R+r}{R_0}\right)\right],
\end{aligned} \tag{3.1.29}$$

where $D_m(t)$ denotes the familiar Dirichlet kernel,

$$D_m(t) = \sin[(m+\tfrac{1}{2})t]\,\{2\sin\tfrac{t}{2}\}^{-1}.$$

By virtue of (3.1.28′) and (3.1.29), the left-hand side of (3.1.27) is transformed as

$$\begin{aligned}
\int_{R/4}^{R/2} \Bigg| & r^{(N-1)/2}\sin\left(n\pi\frac{r}{R_0}\right)\sin\left(\frac{\pi s}{2}\right)\left\{\int_0^{2R_0}\left[\sum_{p=0}^{m_0-1}(u+R+2R_0p)^{-1-s}\right]\right. \\
&\left.\times\left[D_m\left(\pi\frac{u+R-r}{R_0}\right) + D_m\left(\pi\frac{u+R+r}{R_0}\right)\right] du\right\} \\
&+ r^{(N-1)/2}\cos\left(n\pi\frac{r}{R_0}\right)\cos\left(\frac{\pi s}{2}\right)\left\{\int_0^{2R_0}\left[\sum_{p=0}^{m_0-1}(u+R+2R_0p)^{-1-s}\right]\right. \\
&\left.\times\left[D_m\left(\pi\frac{u+R-r}{R_0}\right) - D_m\left(\pi\frac{u+R+r}{R_0}\right)\right] du\right\}\Bigg| dr.
\end{aligned} \tag{3.1.30}$$

We shall use the symbol $S_m(x, F)$ to denote the partial sum of the trigonometric Fourier series of a function $F(u)$ taken at the point x, that is,

$$S_m(x, F) = \frac{1}{R_0} \int_0^{2R_0} D_m\left(\pi \frac{u-x}{R_0}\right) F(u)du.$$

Having defined the function $F(u)$ as

$$F(u) = \sum_{p=0}^{m_0-1} (u + R + 2R_0 p)^{-1-s}, \tag{3.1.31}$$

we can rewrite expression (3.1.30) in the following manner:

$$\begin{aligned} R_0 \int_{R/4}^{R/2} \Big| & r^{(N-1)/2} \sin\left(n\pi \frac{r}{R_0}\right) \sin\left(\frac{\pi s}{2}\right) \{S_m(2R_0 + r - R, F) \\ & + S_m(2R_0 - r - R, F)\} + r^{(N-1)/2} \cos\left(n\pi \frac{r}{R_0}\right) \cos\left(\frac{\pi s}{2}\right) \\ & \times \{S_m(2R_0 + r - R, F) - S_m(2R_0 - r - R, F)\}\Big| dr. \end{aligned} \tag{3.1.30'}$$

Inequality (3.1.27) is proved if we establish that for a sufficiently small fixed $R_0 = 2R > 0$ (or, what is the same, for a sufficiently large fixed m_0) and for all n's exceeding a sufficiently large $n_0(R_0)$, the value of (3.1.30′) exceeds $C_{17} R^{(N+1)/2-s}$, where C_{17} is a positive constant. To accomplish this, it will suffice, in turn, to show that, for a sufficiently large fixed m_0 and for all n's exceeding a sufficiently large $n_0(m_0)$, the following three estimates hold[9]:

$$\begin{aligned} \int_{R/4}^{R/2} r^{(N-1)/2} \Big| & \cos\left(n\pi \frac{r}{R_0} - \frac{\pi s}{2}\right) F(3R + r) \\ & - \cos\left(n\pi \frac{r}{R_0} + \frac{\pi s}{2}\right) F(3R - r)\Big| dr \geq C_{18} R^{(N-1)/2-s}, \end{aligned} \tag{3.1.32}$$

$$\int_{R/4}^{R/2} r^{(N-1)/2} |S_m(3R + r, F) - F(3R + r)| \, dr \leq C_{19} R^{N/2-s}, \tag{3.1.33}$$

[9] We also take into account that $2R_0 - R = 3R$.

$$\int_{R/4}^{R/2} r^{(N-1)/2}\,|S_m(3R-r,F)-F(3R-r)|\,dr \le C_{20}R^{N/2-s}, \tag{3.1.34}$$

where C_{18}, C_{19}, and C_{20} are positive constants.

First, we wish to establish inequalities (3.1.33) and (3.1.34). One will observe that in order to prove them, it suffices to show by virtue of the Cauchy–Buniakowski inequality that, given a sufficiently small $R_0 = 2R > 0$,

$$\int_{0}^{2R_0} [S_m(u,F)-F(u)]^2 du \le C_{21}R^{-2s}. \tag{3.1.35}$$

The left-hand side of (3.1.35) is equal to the sum

$$\sum_{h>2/(\pi r)} (a_k^2+b_k^2), \tag{3.1.36}$$

where a_k and b_k are the Fourier coefficients in an expansion of the function $F(u)$ in the system

$$\left\{\frac{1}{\sqrt{R_0}}\cos\left(k\pi\frac{u}{R_0}\right)\right\}, \qquad \left\{\frac{1}{\sqrt{R_0}}\sin\left(k\pi\frac{u}{R_0}\right)\right\},$$

orthonormalized on the segment $[0, 2R_0]$.

By the simple expedient of integrating by parts with the function $F(u)$ defined as (3.1.31) and keeping in mind that $s > 0$, we obtain

$$a_k = O(R^{-1/2-s}k^{-1}), \qquad b_k = O(R^{-1/2-s}k^{-1}). \tag{3.1.37}$$

The estimate (3.1.35) follows immediately from estimates (3.1.37) and from the fact that the left-hand side of (3.1.35) is equal to the sum (3.1.36). This completes the proof of estimates (3.1.33) and (3.1.34).

It only remains for us to prove the lower-bound estimate (3.1.32). We note, for one thing, that the left-hand side of (3.1.32) can be rewritten in the form

$$\begin{aligned}\int_{R/4}^{R/2} r^{(N-1)/2}\Big|&\cos\left(n\pi\frac{r}{R_0}\right)\cos\left(\frac{\pi s}{2}\right)[F(3R+r)-F(3R-r)]\\ &+\sin\left(n\pi\frac{r}{R_0}\right)\sin\left(\frac{\pi s}{2}\right)[F(3R+r)-F(3R-r)]\Big|dr.\end{aligned} \tag{3.1.38}$$

Now, we make use of $I(r,s)$ to denote

$$\begin{aligned}I(r,s) = \Big\{&\cos^2\left(\frac{\pi s}{2}\right)[F(3R-r)-F(3R+r)]^2\\ &+\sin^2\left(\frac{\pi s}{2}\right)[F(3R-r)-F(3R+r)]^2\Big\}^{1/2},\end{aligned} \tag{3.1.39}$$

and $\alpha(r, s)$ to denote the angle as defined by the equalities

$$\begin{cases} \cos \alpha(r,s) = \dfrac{\cos\left(\frac{\pi s}{2}\right)[F(3R+r) - F(3R-r)]}{I(r,s)}, \\ \sin \alpha(r,s) = \dfrac{\sin\left(\frac{\pi s}{2}\right)[F(3R+r) - F(3R-r)]}{I(r,s)}. \end{cases} \tag{3.1.40}$$

In terms of these notations, the expression (3.1.38) can be rewritten in the form

$$\int_{R/2}^{R/4} r^{(N-1)/2} I(r,s) \left| \cos\left[n\pi \frac{r}{R_0} - \alpha(r,s)\right]\right| dr. \tag{3.1.38'}$$

In estimating (3.1.38′) from below, one will observe that the function $F(u)$ as defined by (3.1.31) implies the existence of a constant $C_{22} > 0$ such that for all r in the segment $[R/4, R/2]$ the inequality

$$F(3R-r) - F(3R+r) \geq C_{22} R^{-1-s} \tag{3.1.41}$$

holds, and the more so the inequality

$$F(3R-r) + F(3R-r) \geq C_{22} R^{-1-s}. \tag{3.1.42}$$

With reference to (3.1.39), (3.1.41), and (3.1.42), we conclude that

$$I(r,s) \geq C_{22} R^{-1-s}. \tag{3.1.43}$$

Using (3.1.43), we obtain the following lower-bound estimate of (3.1.38′):

$$\begin{aligned} &\int_{R/4}^{R/2} r^{(N-1)/2} I(r,s) \left| \cos\left[n\pi \frac{r}{R_0} - \alpha(r,s)\right]\right| dr \\ &\geq C_{23} R^{(N-1)/2-1-s} \int_{R/4}^{R/2} \left| \cos\left[n\pi \frac{r}{R_0} - \alpha(r,s)\right]\right| dr. \end{aligned} \tag{3.1.44}$$

One infers from the left-hand side of (3.1.32) being equal to (3.1.38′) and from inequality (3.1.44) that in order to establish the lower-bound estimate (3.1.32), it suffices to show that, given a fixed arbitrary small $R_0 = 2R > 0$, for all sufficiently large n's the lower-bound estimate

$$\int_{R/4}^{R/2} \left| \cos\left[n\pi \frac{r}{R_0} - \alpha(r,s)\right]\right| \geq C_{24} R \tag{3.1.45}$$

holds with a positive constant C_{24}.

Since

$$\left|\cos\left[n\pi\frac{r}{R_0}-\alpha(r,s)\right]\right| \geq \cos^2\left[n\pi\frac{r}{R_0}-\alpha(r,s)\right]$$

$$\geq \frac{1}{2}+\frac{1}{2}\cos\left[2n\pi\frac{r}{R_0}-2\alpha(r,s)\right],$$

one has

$$\int\limits_{R/2}^{R/4}\left|\cos\left[n\pi\frac{r}{R_0}-\alpha(r,s)\right]\right| dr$$

$$\geq \frac{R}{8}+\frac{1}{2}\int\limits_{R/4}^{R/2}\cos\left[2n\pi\frac{r}{R_0}-2\alpha(r,s)\right] dr. \tag{3.1.46}$$

Inequality (3.1.46) can be rewritten as

$$\int\limits_{R/2}^{R/4}\left|\cos\left[n\pi\frac{r}{R_0}-\alpha(r,s)\right]\right| dr$$

$$\geq \frac{R}{8}+\frac{1}{2}\int\limits_{R/4}^{R/2}\cos\left[2\alpha(r,s)\right]\cos\left(2n\pi\frac{r}{R_0}\right) dr \tag{3.1.47}$$

$$+\frac{1}{2}\int\limits_{R/4}^{R/2}\sin\left[2\alpha(r,s)\right]\sin\left(2n\pi\frac{r}{R_0}\right) dr.$$

It is seen that the second and third terms on the right-hand side of (3.1.47) when multiplied by $1/\sqrt{R_0}$ are in fact the Fourier coefficients of an expansion on the segment $[0, 2R_0]$ in the trigonometric system

$$\left\{\frac{1}{\sqrt{R_0}}\cos\left(n\pi\frac{u}{R_0}\right)\right\}, \qquad \left\{\frac{1}{\sqrt{R_0}}\sin\left(n\pi\frac{u}{R_0}\right)\right\},$$

of the respective functions

$$\varphi_1(r) = \begin{cases} \frac{1}{2}\cos[2\alpha(r,s)] & \text{for} \quad R/4 \leq r \leq R/2. \\ 0 & \text{elsewhere in} \quad [0, 2R_0], \end{cases}$$

$$\varphi_2(r) = \begin{cases} \frac{1}{2}\sin[2\alpha(r,s)] & \text{for} \quad R/4 \le r \le R/2. \\ 0 & \text{elsewhere in} \quad [0, 2R_0], \end{cases}$$

piecewise continuous on this segment.

Since, given a fixed arbitrary small $R_0 = 2R > 0$, the Fourier coefficients tend to zero as $n \to \infty$, the moduli of the second and third terms on the right-hand side of (3.1.47) do not exceed $R/24$ for all n's exceeding a sufficiently large $n_0(R_0)$.

In this case, inequality (3.1.45) with the constant $C_{24} = 1/24$ immediately follows from relation (3.1.47).

This completes the proof of Lemma 3.1.

3.1.2. Lemma on the Unboundedness of the Difference of the Fractional-Kernel Riesz Means

Following the same line of reasoning as in Lemma 3.1, we consider two spectral decompositions one of which is associated with an expansion in the eigenfunctions of the first boundary-value problem for the Laplace operator in an N-dimensional ball G of radius R_0 centered at point x_0, and the other of which is associated with an expansion in an N-fold Fourier integral of a function originally continued with zero beyond G.

We use the symbols $T_\beta(x,y)$ and $\overset{0}{T}_\beta(x,y)$, respectively, to denote the kernels of order $\beta > 0$ corresponding to these two spectral decompositions.[10] We fix a point x interior to these kernels at the center x_0 of the ball G and construct, in the variable point y, the Riesz means of order $s \ge 0$ of the two spectral decompositions for the two kernels specified. We denote the Riesz means by $E^s_\lambda T_\beta(x_0, y)$ and $\overset{0}{E}{}^s_\lambda \overset{0}{T}_\beta(x_0, y)$, respectively.

LEMMA 3.2. *Assume that E_R is a ring layer similar to that defined in Lemma 3.1, s is a number in the half-segment $0 \le s < \frac{N-1}{2}$, and β is a number such that $0 < \beta < \frac{N-1}{4} - \frac{s}{2}$. Then the radius $R_0 = 2R > 0$ of the ball G can be fixed so small that the function of the argument λ*

$$\tau^s_\beta(\lambda) = \int\limits_{E_R} \left| E^s_\lambda T_\beta(x_0, y) - \overset{0}{E}{}^s_\lambda \overset{0}{T}_\beta(x_0, y) \right| dy \tag{3.1.48}$$

is unbounded on the half-line $\lambda \ge 1$.

PROOF: This lemma is proved in a manner very similar to that adopted in proving Lemma 2.5 in Section 2.4.4.

---

10) For the definition of the kernel of any positive order β for an arbitrary spectral decomposition, see Section 2.3.3.

First, let us ensure that it suffices to prove Lemma 3.2 for any *strictly positive* s smaller than $(N-1)/2$ (this will show that the lemma holds for $s=0$ also). To this end, we shall prove the following assertion: *If, given a sufficiently small* $R_0 = 2R > 0$ *for any* s *in the interval* $0 < s < (N-1)/2$ *and for any* δ *in the interval* $0 < \delta < (N-1)/4 - s/2$, *the function of the argument* λ

$$\tau^{s}_{(N-1)/4-s/2-\delta}(\lambda) = \int\limits_{E_R} \left| E^{s}_{\lambda} T_{(N-1)/4-s/2-\delta}(x_0, y) - \overset{0}{E}{}^{s}_{\lambda} \overset{0}{T}_{(N-1)/4-s/2-\delta}(x_0, y) \right| dy \tag{3.1.49}$$

is unbounded on the half-line $\lambda \geq 1$, *then for any* δ' *in the interval* $0 < \delta' < (N-1)/4$ *the function of the argument* λ

$$\overset{0}{\tau}_{(N-1)/4-\delta'}(\lambda) = \int\limits_{E_R} \left| E_{\lambda} T_{(N-1)/4-\delta'}(x_0, y) - \overset{0}{E}_{\lambda} \overset{0}{T}_{(N-1)/4-\delta'}(x_0, y) \right| dy \tag{3.1.50}$$

is also unbounded on the half-line $\lambda \geq 1$.

The formulated statement is easily proved by contradiction. Assume that function (3.1.49) is unbounded on the half-line $\lambda \geq 1$ for any s in the interval $0 < s < (N-1)/2$ and for any δ in the interval $0 < \delta < (N-1)/4 - s/2$, whereas the function (3.1.50) is bounded on the half-line $\lambda \geq 1$ for a certain δ' in the interval $0 < \delta' < (N-1)/4$.

We consider the function(3.1.49) at $s = \delta'$, $\delta = \delta'/2$, assuming that s belongs to the interval $(0, (N-1)/2)$ and δ belongs to the interval $(0, (N-1)/4 - s/2)$. If so, we can assert that the function

$$\tau^{\delta'}_{(N-1)/4-\delta'}(\lambda) = \int\limits_{E_R} \left| E^{\delta'}_{\lambda} T_{(N-1)/4-\delta'}(x_0, y) - \overset{0}{E}{}^{\delta'}_{\lambda} \overset{0}{T}_{(N-1)/4-\delta'}(x_0, y) \right| dy \tag{3.1.51}$$

is unbounded on the half-line $\lambda \geq 1$.

Invoking relation (2.4.48) in Section 2.4.3, we obtain the equality

$$E^{\delta'}_{\lambda} T_{(N-1)/4-\delta'}(x_0, y) - \overset{0}{E}{}^{\delta'}_{\lambda} \overset{0}{T}_{(N-1)/4-\delta'}(x_0, y)$$
$$= \delta' \lambda^{-\delta'} \int\limits_0^{\lambda} (\lambda - t)^{\delta'-1} \left[E_t T_{(N-1)/4-\delta'}(x_0, y) - \overset{0}{E}_t \overset{0}{T}_{(N-1)/4-\delta'}(x_0, y) \right] dt. \tag{3.1.52}$$

Taking both sides of (3.1.52) in absolute value and keeping in mind that the modulus of an integral does not exceed the integral of the integrand modulus, we carry out integration over the ring layer E_R to obtain the inequality

$$\tau^{\delta'}_{(N-1)/4-\delta'}(\lambda) \leq \overset{0}{\tau}_{(N-1)/4-\delta'}(\lambda).$$

This inequality is at variance with our assumption that the function in the right-hand side is bounded on the half-line $\lambda \geq 1$, whereas the function in the left-hand side, identical to (3.1.51), is unbounded on the half-line $\lambda \geq 1$.

Thus, the formulated assertion has been proved, and it remains for us to establish Lemma 3.2 only for a strictly positive s smaller than $(N-1)/2$.

A proof of this will also be sought by contradiction.

Assume that function (3.1.48) is bounded on the half-line $\lambda \geq 1$ for a certain s in the interval $0 < s < (N-1)/2$ and for a certain δ in the interval $0 < \delta < (N-1)/4 - s/2$.

Let us make use of Lemma 2.4, proved in Section 2.4.3. We consider an arbitrary self-adjoint nonnegative extension $\widehat{A}$ of the Laplace operator in an arbitrary N-dimensional domain G and, accordingly, an arbitrary ordered spectral representation of the space $L_2(G)$ with a spectral measure $\rho(\lambda)$, with sets of multiplicity e_i, with fundamental functions $u_i(x,\lambda)$ $(i = 1, 2, \dots, \widehat{m})$ of multiplicity $\widehat{m} \leq \infty$. We assume, under the condition of Lemma 2.4 (Section 2.4.3), that the measure $\rho(t)$ coincides with the spectral measure, and the function $U(t)$ is expressed as

$$U(t) = \sum_{i=1}^{\widehat{m}} u_i(x,t) u_i(y,t)(1+t)^{-\beta}.$$

We obtain that the quantities S_λ, σ_λ^s, $\bar{S}_\lambda$, and $\bar{\sigma}_\lambda^s$ involved in this lemma take the form

$$S_\lambda = E_\lambda T_\beta(x,y), \qquad \sigma_\lambda^s = E_\lambda^s T_\beta(x,y),$$

$$\bar{S}_\lambda = \theta(x,y,\lambda), \qquad \sigma_\lambda^s = \theta^s(x,y,\lambda),$$

where $\theta(x,y,\lambda)$ is the spectral function and $T_\beta(x,y)$ is the kernel of order $\beta > 0$ associated with the self-adjoint extension $\widehat{A}$.

In this case the relation (2.4.46) as established by Lemma 2.4 assumes the form[11)]

$$\theta^s(x,y,\lambda) = (1+\lambda)^\beta E_\lambda^s T_\beta(x,y) + (-1)^{r+1}\lambda^{-s}$$

$$\times \int_0^\lambda \frac{d^{r+2}}{dt^{r+2}} \{(\lambda - t)^s[(1+t)^\beta - (1+\lambda)^\beta]\} \frac{t^{r+1}}{(r+1)!} E_t^{r+1} T_\beta(x,y)\,dt. \tag{3.1.53}$$

We shall make use of (3.1.53) to demonstrate a contradiction with the assumption that the function (3.1.48) is bounded on the half-line $\lambda \geq 1$ for a certain s in the interval $0 < s < (N-1)/2$ and for a certain δ in the interval $0 < \delta < (N-1)/4 - s/2$.

First, we write relation (3.1.53) at $x = x_0$ for the spectral function $\theta(x,y,\lambda)$ and the kernel $T_\beta(x,y)$ corresponding to the expansion in the eigenfunctions of the first boundary-value problem in an N-dimensional ball G; next, we write relation (3.1.53) for the spectral function $\theta_0(x,y,\lambda)$ and the kernel $\overset{0}{T}_\beta(x,y)$ corresponding to

11) We recall that, in terms of Lemma 2.4, $s = r + \kappa$, where r is a nonnegative integer and κ is subject to $0 < \kappa \leq 1$.

the expansion into an N-fold Fourier integral. Subtracting the latter from the former, we obtain

$$[\theta^s(x_0,y,\lambda)-\theta_0^s(x_0,y,\lambda)]=(1+\lambda)^\beta\left[E_\lambda^sT_\beta(x_0,y)-\overset{0}{E_\lambda^s}\overset{0}{T}_\beta(x_0,y)\right]$$

$$+(-1)^{r+1}\lambda^{-s}\int\limits_0^\lambda\frac{d^{r+2}}{dt^{r+2}}\{(\lambda-t)^s[(1+t)^\beta-(1+\lambda)^\beta]\}\tag{3.1.54}$$

$$\times\frac{t^{r+1}}{(r+1)!}\left[E_t^{r+1}T_\beta(x_0,y)-\overset{0}{E_t^{r+1}}\overset{0}{T}_\beta(x_0,y)\right]dt.$$

Having placed the point x_0 at the center of the ball G, we observe that in relation (3.1.54) the modulus of the integral does not exceed the integral of the integrand modulus. Next, carrying out integration with respect to y over the ring layer E_R and invoking (3.1.48), we obtain the inequality

$$\int\limits_{E_R}|\theta^s(x_0,y,\lambda)-\theta_0^s(x_0,y,\lambda)|\,dy\le(1+\lambda)^\beta\tau_\beta^s(\lambda)$$

$$+\lambda^{-s}\int\limits_0^\lambda\left|\frac{d^{r+2}}{dt^{r+2}}\{(\lambda-t)^s[(1+t)^\beta-(1+\lambda)^\beta]\}\right|\frac{t^{r+1}}{(r+1)!}\tau_\beta^{r+1}(t)dt.\tag{3.1.55}$$

By the assumption made, there exists a constant $M>0$ such that

$$\tau_\beta^s\le M\tag{3.1.56}$$

holds everywhere on the half-line $\lambda\ge1$ and, consequently, everywhere on the half-line $\lambda\ge0$.

Let us ensure, first of all, that the inequality

$$\tau_\beta^{r+1}\le M,\tag{3.1.57}$$

valid on the entire half-line $\lambda\ge0$, stems from the inequality (3.1.56).

If s is an integer, then $\kappa=1$, $r=s-1$, so that $s=r+1$, and inequality (3.1.57) is in fact identical with (3.1.56).

If s is not an integer, then $0<\kappa\le1$, and $s<r+1$. But, for any s and $s'=r+1$ subject to $0<s<s'$, the formula (2.4.63), given in Section 2.4.3, remains valid for the Riesz means of order s and s'. The inequality

$$\tau_\beta^{r+1}\le\lambda^{-(r+1)}\Gamma(r+2)[\Gamma(r+1-s)\Gamma(s+1)]^{-1}\int\limits_0^\lambda(\lambda-t)^{r-s}t^s\tau_\beta^s(t)dt\tag{3.1.58}$$

follows immediately from this formula, written for the Riesz means of the two spectral decompositions in question.

Inequality (3.1.57) follows from inequalities (3.1.58) and (3.1.56).

Now, applying inequalities (3.1.56) and (3.1.57) to the right-hand side of (3.1.55) and taking into account the estimate

$$\int_0^\lambda \left| \frac{d^{r+2}}{dt^{r+2}} \{(\lambda - t)^s [(1+t)^\beta - (1+\lambda)^\beta]\} \right| \frac{t^{r+1}}{(r+1)!} dt = O[(1+\lambda)^{\beta+s}]$$

established in Section 2.4.4, we obtain the estimate

$$\int_{E_R} |\theta^s(x_0, y, \lambda) - \overset{0}{\theta}{}^s(x_0, y, \lambda)| \, dy = O[(1+\lambda)^\beta]$$

from (3.1.55). This estimate, by virtue of $\beta < (N-1)/4 - s/2$, is in contradiction with inequality (3.1.5), established by Lemma 3.1 in the preceding section.

The contradiction that we have arrived at completes the proof of Lemma 3.2.

3.1.3. Direct Proof of Theorem 3.1

The fact that the function (3.1.48) is unbounded on the half-line $\lambda \geq 1$, as established by Lemma 3.2, and the familiar resonance-type theorem[12] imply that there exists a function $h(y)$ continuous within the ring layer E_R such that the function of argument λ

$$\int_{E_R} \left[E_\lambda^s T_\beta(x_0, y) - \overset{0}{E}{}_\lambda^s \overset{0}{T}_\beta(x_0, y) \right] h(y) dy \tag{3.1.59}$$

is also unbounded on the half-line $\lambda \geq 1$.

Our further line of reasoning is the same as that pursued in Section 2.4.5. Therefore we shall not linger over details and rather focus on major argumentative points.

We continue the function $h(y)$ to the entire space E^N and put it equal to zero outside the ring layer E_R. Next, we introduce the two functions

$$F(x) = \int_{E^N} T_\beta(x, y) h(y) dy,$$
$$\overset{0}{F}(x) = \int_{E^N} \overset{0}{T}_\beta(x, y) h(y) dy. \tag{3.1.60}$$

The unboundedness of the function (3.1.59) on the half-line $\lambda \geq 1$ is equivalent to the unboundedness on this half-line of the difference

$$E_\lambda^s(x_0, F) - \overset{0}{E}{}_\lambda^s \left(x_0, \overset{0}{F} \right) \tag{3.1.61}$$

[12] See, for example, Kaczmarz and Steinhaus [51, p. 31, Assertion 3^0].

of the Riesz means of order s of the two spectral compositions of the respective functions $F(x)$ and $\overset{0}{F}(x)$ taken at the point x_0.

As shown in Section 2.3.3, the kernels $\overset{0}{T}_\beta(x,y)$ and $T_\beta(x,y)$ of the two spectral decompositions in question for an arbitrarily small $h > 0$ and for arbitrarily large m can be represented in the form (see (2.3.31), Section 2.3.3)

$$\begin{cases} T_\beta(x,y) = [v_\beta(x,y) - w_\beta(x,y)] - \Psi_\beta(x,y), \\ \overset{0}{T}_\beta(x,y) = [v_\beta(x,y) - w_\beta(x,y)] - \overset{0}{\Psi}_\beta(x,y), \end{cases} \tag{3.1.62}$$

where $[v_\beta(x,y) - w_\beta(x,y)]$ is a smoothed Bessel–MacDonald kernel distinct from zero only within the N-dimensional ball $|x-y| \le h$ and exhibiting everywhere continuous partial derivatives with respect to all variables of order up to m inclusively. The $\Psi_\beta(x,y)$ and $\overset{0}{\Psi}_\beta(x,y)$ are such that for any function $h(y)$ from $L_2(G)$, continued with zero to the entire space E^N, the Riesz means of any nonnegative order s

$$E^s_\lambda(x,g) \qquad \text{and} \qquad \overset{0}{E}{}^s_\lambda(x,\overset{0}{g}), \tag{3.1.63}$$

composed of the functions g and $\overset{0}{g}$, respectively,

$$g(x) = \int \Psi_\beta(x,y)h(y)dy, \qquad \overset{0}{g}(x) = \int \overset{0}{\Psi}_\beta(x,y)h(y)dy, \tag{3.1.64}$$

converge as $\lambda \to \infty$ uniformly with respect to x in G_h, given a sufficiently large number m; in particular, they converge at the center x_0 of the ball G.

The function (3.1.60) with reference to (3.1.62) and (3.1.64) can be represented in the form

$$F(x) = f(x) - g(x), \qquad \overset{0}{F}(x) = f(x) - \overset{0}{g}(x), \tag{3.1.65}$$

where

$$f(x) = \int [v_\beta(x,y) - w_\beta(x,y)]h(y)dy. \tag{3.1.66}$$

Since at the point x_0 the Riesz means are convergent and for this reason are bounded in λ on the half-line $\lambda \ge 1$, the unboundedness of the difference (3.1.61) on this half-line implies the unboundedness on this half-line of the difference of the Riesz means

$$E^s_\lambda(x_0,f) - \overset{0}{E}{}^s_\lambda(x_0,f), \tag{3.1.67}$$

composed for the function $f(x)$ as defined by relation (3.1.66) and taken at point x_0.

It remains to prove that, given a sufficiently small $h > 0$, the function $f(x)$ belongs, within the ball G, to the Zygmund–Hölder class C^α for an arbitrary α smaller than β, and that this function becomes zero in a certain vicinity of the point x_0 and is compactly supported in G.

These facts are easily proved following the reasoning in Section 2.4.5.

This completes thereby the proof of Theorem 3.1.

3.1.4. Uniform and Accurate, to within an Order of Magnitude, Estimate of the Difference of the Riesz Means for Two Spectral Decompositions

The methods developed above enable one not only to establish a uniform equiconvergence of the Riesz means for functions of a specified class, but also to obtain accurate to within an order of magnitude and uniform (on any compact set of the main domain) estimates of the difference of the Riesz means for two arbitrary spectral decompositions of a particular class.

In this section we establish an accurate to within an order of magnitude and uniform (on any compact set of the main domain G) estimate of the difference of the Riesz means of two arbitrary spectral decompositions for an arbitrary function in the class $L_2(G)$.

We consider two completely arbitrary self-adjoint nonnegative extensions $\widehat{A}'$ and $\widehat{A}''$ of the Laplace operator $Lu = -\Delta u$ in an arbitrary N-dimensional domain G and their two corresponding arbitrary ordered spectral representations in the space $L_2(G)$: one representation with a spectral measure $\rho'(t)$, with sets of multiplicity e'_i, with fundamental functions $u'_i(x,t)$ $(i = 1, 2, \ldots m')$ of multiplicity $m' \leq \infty$; the other one with a spectral measure $\rho''(t)$, with sets of multiplicity e''_i, with fundamental functions $u''_i(x,t)$ of multiplicity $m'' \leq \infty$.

We make use of the symbols $\widehat{f}'_i(t)$ and, respectively, $\widehat{f}''_i(t)$ to denote the Fourier images of an arbitrary function $f(x)$ in $L_2(G)$ with respect to the fundamental functions $u'_i(x,t)$ and $u''_i(x,t)$, respectively, and we construct for the $f(x)$ in $L_2(G)$ the Riesz means of size λ and order s corresponding to the two extensions of interest. These take the form (see Chapter 2)

$$\overset{s}{\sigma}{}'_\lambda(x,f) = \sum_{i=1}^{m'} \int_0^\lambda \widehat{f}'_i(t) u'_i(x,t) \left(1 - \frac{t}{\lambda}\right)^s d\rho'(t), \tag{3.1.68}$$

$$\overset{s}{\sigma}{}''_\lambda(x,f) = \sum_{i=1}^{m''} \int_0^\lambda \widehat{f}''_i(t) u''_i(x,t) \left(1 - \frac{t}{\lambda}\right)^s d\rho''(t). \tag{3.1.69}$$

THEOREM 3.2. *For two arbitrary self-adjoint nonnegative extensions $\widehat{A}'$ and $\widehat{A}''$ of the Laplace operator in an arbitrary N-dimensional domain G and for an arbitrary function $f(x)$ in the class $L_2(G)$, the estimate below of the difference of the Riesz means (3.1.68) and (3.1.69) of any nonnegative order s holds,*

$$\left|\overset{s}{\sigma}{}'_\lambda(x,f) - \overset{s}{\sigma}{}''_\lambda(x,f)\right| = \lambda^{(N-1)/4-s/2}\, o(1), \tag{3.1.70}$$

uniform with respect to x on any compact set of domain G.

REMARK. Theorem 3.2 remains valid if we continue the function $f(x)$ with zero beyond the domain G and take, for one of the two extensions above, an arbitrary self-adjoint nonnegative extension of the Laplace operator in a domain enclosing G (for example, an extension over the entire space E^N corresponding to the expansion into an N-fold Fourier integral). This fact immediately follows from the proof of Theorem 3.2 given below.

A natural question arises: Is it possible to specify infinitesimal $o(1)$ in estimate (3.1.70) as $\lambda \to \infty$. Now we shall convince ourselves that the estimate (3.1.70) is not a final one in the sense that the function $o(1)$ can tend to zero as $\lambda \to \infty$ at an *arbitrarily slow rate*. This fact stems from the following assertion.

Let now the Riesz means (3.1.68) correspond to the expansion of the function $f(x)$ in the eigenfunctions of the first boundary-value problem for the Laplace operator in an N-dimensional ball G of radius $R_0 > 0$ centered at x_0, and the Riesz means (3.1.69) correspond to the expansion of the same function $f(x)$, continued with zero beyond G, in an N-fold Fourier integral.

THEOREM 3.3. *If we fix the radius $R_0 > 0$ of an N-dimensional ball G in the manner specified in Lemma 3.1, then for any $s \geq 0$ and for any preassigned infinitesimal function $\varphi(\lambda)$ decreasing on the half-line $\lambda \geq 1$ as $\lambda \to \infty$ there exists a function $f(x)$ which (i) belongs to the class $L_\infty(G)$, (ii) is compactly suported in the ball G, (iii) is equal to zero outside the ball G and within a small vicinity of the center x_0 of the ball G, and which is such that the function of argument λ*

$$\frac{\left|\overset{s}{\sigma}{}'_\lambda(x_0,f) - \overset{s}{\sigma}{}''_\lambda(x_0,f)\right|}{\lambda^{(N-1)/4-s/2}\varphi(\lambda)} \tag{3.1.71}$$

is not a bounded one as $\lambda \to \infty$.

Theorem 3.3 shows that the estimate (3.1.70) can hardly be improved (even if the requirement that $f(x) \in L_2(G)$ is replaced by the requirement that $f(x) \in L_\infty(G)$).

PROOF OF THEOREM 3.2. We fix an arbitrary compact set K of domain G and an arbitrary positive number R smaller than the distance between compact K and the boundary of G. As has been shown in Section 2.5.3, for any point x of compact set K and for any self-adjoint nonnegative extension of the Laplace operator in domain G, the Riesz means of any order $s \geq 0$ of an arbitrary function $f(x)$ in

$L_2(G)$ are defined by equality (2.5.118). Having written equality (2.5.118) for the two arbitrary extensions $\widehat{A}'$ and $\widehat{A}''$, we obtain

$$\overset{s}{\sigma}{}'_\lambda(x,f) = 2^s\Gamma(s+1)(2\pi)^{-N/2}\lambda^{N/2-s/2}$$
$$\times \int\limits_{|x-y|\le R} f(y)J_{N/2+s}(|x-y|\sqrt{\lambda})|x-y|^{-N/2-s}dy \tag{3.1.72}$$
$$+2^s\Gamma(s+1)\lambda^{N/4-s/2}\sum_{i=1}^{m'}\int\limits_0^\infty \widehat{f}'_i(t)u'_i(x,t)t^{(2-N)/4}I_t^\lambda(R)d\rho'(t),$$

$$\overset{s}{\sigma}{}''_\lambda(x,f) = 2^s\Gamma(s+1)(2\pi)^{-N/2}\lambda^{N/4-s/2}$$
$$\times \int\limits_{|x-y|\le R} f(y)J_{N/2+s}(|x-y|\sqrt{\lambda})|x-y|^{-N/2-s}dy \tag{3.1.73}$$
$$+2^s\Gamma(s+1)\lambda^{N/4-s/2}\sum_{i=1}^{m''}\int\limits_0^\infty \widehat{f}''_i(t)u''_i(x,t)t^{(2-N)/4}I_t^\lambda(R)d\rho''(t),$$

where (see Section 2.4.1)

$$I_t^\lambda(R) = \int\limits_R^\infty J_{N/2+s}(r\sqrt{\lambda})J_{N/2-1}(r\sqrt{t})r^{-s}dr.$$

Subtracting (3.1.73) from (3.1.72) term by term, we arrive at

$$\overset{s}{\sigma}{}'_\lambda(x,f) - \overset{s}{\sigma}{}''_\lambda(x,f)$$
$$= 2^s\Gamma(s+1)\lambda^{N/4-s/2}\left[\sum_{i=1}^{m'}\int\limits_0^\infty \widehat{f}'_i(t)u'_i(x,t)t^{(2-N)/4}I_t^\lambda(R)d\rho'(t)\right. \tag{3.1.74}$$
$$\left.-\sum_{i=1}^{m''}\int\limits_0^\infty \widehat{f}''_i(t)u''_i(x,t)t^{(2-N)/4}I_t^\lambda(R)d\rho''(t)\right].$$

The first of our major aims is to establish the estimate

$$\left|\overset{s}{\sigma}{}'_\lambda(x,f) - \overset{s}{\sigma}{}''_\lambda(x,f)\right| = \lambda^{(N-1)/4-s/2}O(\|f|_{L_2(G)}) \tag{3.1.75}$$

uniform with respect to x on the compact set K.

To this effect, it will suffice to show that the bracketed expression in (3.1.74) has an order of

$$\lambda^{-1/4} O(\|f\|_{L_2(G)}) \tag{3.1.76}$$

uniformly with respect to x on the compact set K.

The more so, it will suffice to show that either of the two sums within brackets in (3.1.74) is of the same order. We shall restrict ourselves to the proof of the former sum (the proof of the latter sum will be an easy exercise for the reader).

Applying the Cauchy–Buniakowski inequality first to the integral and next to the sum, we obtain

$$\left| \sum_{i=1}^{m'} \int_0^\infty \hat{f}_i'(t) u_i'(x,t) t^{(2-N)/4} I_t^\lambda(R) d\rho'(t) \right|$$

$$\leq \left\{ \sum_{i=1}^{m'} \int_0^\infty \left|\hat{f}_i'(t)\right|^2 d\rho'(t) \right\}^{1/2} \left\{ \sum_{i=1}^{m'} \int_0^\infty \left[u_i'(x,t) t^{(2-N)/4} I_t^\lambda(R) \right]^2 d\rho'(t) \right\}^{1/2}. \tag{3.1.77}$$

From inequality (3.1.77) and from the Parseval equality written for the function $f(x)$ in $L_2(G)$ (see (2.1.11), Section 2.1.2) one infers that for the left-hand side of (3.1.77) to have the order of (3.1.76), it suffices to ensure that the estimate

$$\left\{ \lambda^{1/2} \sum_{i=1}^{m'} \int_0^\infty \left[u_i'(x,t) t^{(2-N)/4} I_t^\lambda(R) \right]^2 d\rho'(t) \right\}^{1/2} \leq C(R,s) \tag{3.1.78}$$

holds uniformly with respect to x on the compact set K. In fact, a similar estimate has already been established by us in Section 2.4.1, where we were concerned with a proof of Lemma 2.3. Indeed, in proving this lemma, we have shown that each of the four terms within the square root on the right-hand sides of (2.4.27)–(2.4.30) is bounded by a constant which is dependent only on R and on s. Thereby the proof of estimate (3.1.75) is complete.

All we have left to do is to establish, based on estimate (3.1.75) uniform on the compact set K, the validity of estimate (3.1.70) uniform on the compact set K.

For an arbitrary function $f(x)$ from the class $L_2(G)$ and for an arbitrary $\varepsilon > 0$ there is a function $f_1(x)$ in the class $C_0^\infty(G)$ such that

$$\|f(x) - f_1(x)\|_{L_2(G)} \leq \frac{\varepsilon}{3C_1}, \tag{3.1.79}$$

where C_1 is a constant setting a limit to the growth of the O-terms on compact K in (3.1.75).

Since

$$\left[\overset{s}{\sigma}{}'_\lambda(x,f) - \overset{s}{\sigma}{}''_\lambda(x,f)\right] = \left[\overset{s}{\sigma}{}'_\lambda(x,f-f_1) - \overset{s}{\sigma}{}''_\lambda(x,f-f_1)\right]$$

$$+ \left[\overset{s}{\sigma}{}'_\lambda(x,f_1) - f_1(x)\right] + \left[f_1(x) - \overset{s}{\sigma}{}''_\lambda(x,f_1)\right],$$

the following inequality holds:

$$\left|\overset{s}{\sigma}{}'_\lambda(x,f) - \overset{s}{\sigma}{}''_\lambda(x,f)\right| \le \left|\overset{s}{\sigma}{}'_\lambda(x,f-f_1) - \overset{s}{\sigma}{}''_\lambda(x,f-f_1)\right|$$

$$+ \left|\overset{s}{\sigma}{}'_\lambda(x,f_1) - f_1(x)\right| + \left|f_1(x) - \overset{s}{\sigma}{}''_\lambda(x,f_1)\right|. \tag{3.1.80}$$

From estimate (3.1.75) as applied to the function $[f(x)-f_1(x)]$ and from inequality (3.1.79) one infers that

$$\left|\overset{s}{\sigma}{}'_\lambda(x,f-f_1) - \overset{s}{\sigma}{}''_\lambda(x,f-f_1)\right| < \frac{\varepsilon}{3}\lambda^{(N-1)/4-s/2} \tag{3.1.81}$$

(uniform with respect to x on the compact set K).

We now observe that for the function $f_1(x)$ in the class $C_0^\infty(G)$, each of the differences $[\overset{s}{\sigma}{}'_\lambda(x,f_1) - f_1(x)]$ and $[\overset{s}{\sigma}{}''_\lambda(x,f_1) - f_1(x)]$ for any $s \ge 0$ uniformly on the compact set K has the order $o(\lambda^{(N-1)/4-s/2})$,[13] and one infers therefore that for an arbitrarily fixed $\varepsilon > 0$ there is a positive number $\Lambda(\varepsilon)$ such that, given $\lambda \ge \Lambda(\varepsilon)$ uniformly on a compact set K, the following inequalities hold:

$$\begin{aligned} &\left|\overset{s}{\sigma}{}'_\lambda(x,f_1) - f_1(x)\right| < \frac{\varepsilon}{3}\lambda^{(N-1)/4-s/2}, \\ &\left|\overset{s}{\sigma}{}''_\lambda(x,f_1) - f_1(x)\right| < \frac{\varepsilon}{3}\lambda^{(N-1)/4-s/2}. \end{aligned} \tag{3.1.82}$$

Inequalities (3.1.80), (3.1.81), and (3.1.82) imply that the inequality

$$\left|\overset{s}{\sigma}{}'_\lambda(x,f) - \overset{s}{\sigma}{}''_\lambda(x,f)\right| < \varepsilon\lambda^{(N-1)/4-s/2}$$

holds for $\lambda \ge \Lambda(\varepsilon)$ uniformly on the compact set K, which completes the proof of Theorem 3.2.

PROOF OF THEOREM 3.3. We fix $R_0 = 2R > 0$ in order to satisfy inequality (3.1.5) (see Lemma 3.1, Section 3.1.1). Earlier, we have proved this inequality for the case $s > 0$. The case $s = 0$ will require a somewhat different and even less sophisticated approach. A proof of inequality (3.1.5) for the case $s = 0$ has been given elsewhere (see V.A. Il'in [34]). By virtue of inequality (3.1.5), for an arbitrary preassigned function $\varphi(\lambda)$, which is decreasing on the half-line $\lambda \ge 1$ and infinitesimal as $\lambda \to \infty$, the function of argument λ

[13] This stems from the fact that for a function $f_1(x)$ in the class $C_0^\infty(G)$ for any m the Fourier images of the function $\Delta^m f_1(x)$ in $C_0^\infty(G)$ and of the function $f_1(x)$ are related via (2.5.19) (see Section 2.5.1).

$$F(\lambda) = \left[\lambda^{(N-1)/4-s/2}\varphi(\lambda)\right]^{-1} \int\limits_{E_R} |\theta^s(x_0, y, \lambda) - \theta_0^s(x_0, y, \lambda)|\, dy$$

is unbounded on the half-line $\lambda \geq 1$.

It follows from the unboundedness function and from the resonance-type theorem (mentioned in Section 3.1.3) that there exists a function $f(y)$, continuous in the ring layer E_R, such that the function of argument λ

$$\left[\lambda^{(N-1)/4-s/2}\varphi(\lambda)\right]^{-1} \int\limits_{E_R} [\theta^s(x_0, y, \lambda) - \theta_0^s(x_0, y, \lambda)]\, f(y) dy \tag{3.1.83}$$

is also unbounded on the half-line $\lambda \geq 1$. We continue the function $f(y)$ beyond the ring layer E_R, assuming that the function becomes zero in $E^N - E_R$.

Next, making use of the above notations for the Riesz means of the two Laplace operator extensions in question, we can rewrite function (3.1.83) in the following manner:

$$\left[\lambda^{(N-1)/4-s/2}\varphi(\lambda)\right]^{-1} \left[\overset{s}{\sigma}{}'_\lambda(x_0, f) - \overset{s}{\sigma}{}''_\lambda(x_0, f)\right].$$

This completes the proof of Theorem 3.3.

The technique worked out in this section enables one to estimate the difference of the Riesz means not only for a function in the class $L_2(G)$, but also for functions in other classes (for example, for a compactly supported function in the Nikol'skii class $H_2^\alpha(G)$ for any $\alpha > 0$).

3.2. ON THE RIESZ EQUISUMMABILITY OF SPECTRAL DECOMPOSITIONS IN THE GENERALIZED SENSE

In this section we shall explore the Riesz means of any positive order s for spectral decompositions associated with two arbitrary self-adjoint nonnegative extensions of the Laplace operator in an arbitrary domain G of the space E^N. We shall prove that the difference of the Riesz means of order $s > 0$ of the two spectral decompositions for an arbitrary function in the class $L_2(G)$ not only tends to zero almost everywhere in domain G, but also has an order $o(\lambda^{-s/2})$ almost at each point of G, where λ is the size of the Riesz means in question.

3.2.1. Statement of Results

We consider two arbitrary self-adjoint nonnegative extensions $\widehat{A}'$ and $\widehat{A}''$ of the Laplace operator in domain G and, respectively, two arbitrary ordered spectral representations[14)] of the space $L_2(G)$: one, with a spectral measure $\rho'(t)$, with sets of multiplicity e'_i, with fundamental functions $u'_i(x,t)$, $(i = 1, 2, \dots, m')$, of multiplicity $m' \le \infty$; the other one, with a spectral measure $\rho''(t)$, with sets of multiplicity e''_i, with fundamental functions $u''_i(x,t)$, $i = 1, 2, \dots, m''$, of multiplicity $m'' \le \infty$.

We make use of the symbol $\widehat{f}'_i(t)$ [respectively, $\widehat{f}''_i(t)$] to denote the Fourier images of an arbitrary function $f(x)$ in the class $L_2(G)$ with respect to the fundamental function $u'_i(x,t)$ [respectively, $u''_i(x,t)$] and construct for this arbitrary function $f(x)$ the Riesz means of size λ and order $s > -1/2$ corresponding to the two extensions of interest. These, by definition, take the form

$$\overset{s}{\sigma}{}'_\lambda(x,f) = \sum_{i=1}^{m'} \int_0^\lambda \widehat{f}'_i(t) u'_i(x,t) \left(1 - \frac{t}{\lambda}\right)^s d\rho'(t), \tag{3.2.1}$$

$$\overset{s}{\sigma}{}''_\lambda(x,f) = \sum_{i=1}^{m''} \int_0^\lambda \widehat{f}''_i(t) u''_i(x,t) \left(1 - \frac{t}{\lambda}\right)^s d\rho''(t). \tag{3.2.2}$$

THEOREM 3.4. *The estimate*

$$\left| \overset{s}{\sigma}{}'_\lambda(x,f) - \overset{s}{\sigma}{}''_\lambda(x,f) \right| = o(\lambda^{-s/2}) \tag{3.2.3}$$

holds for two quite arbitrary self-adjoint nonnegative extensions $\widehat{A}'$ and $\widehat{A}''$ of the Laplace operator in an arbitrary N-dimensional domain G and for an arbitrary function $f(x)$ in the class $L_2(G)$ for the difference of the Riesz means (3.2.1) and (3.2.2) of size λ and order $s > 0$ for all sufficiently large $\lambda > 0$ almost everywhere in the domain G.

With the additional requirement that the function $f(x)$ not only belong to the class $L_2(G)$, but also become zero almost everywhere in a domain D interior to G, the estimates

$$\overset{s}{\sigma}{}'_\lambda(x,f) = o(\lambda^{-s'2}), \qquad \overset{s}{\sigma}{}''_\lambda(x,f) = o(\lambda^{-s/2}) \tag{3.2.3'}$$

hold for each of the Riesz means (3.2.1) and (3.2.2) of size λ and order $s > 0$ for all sufficiently large $\lambda > 0$ almost everywhere in D.

3.2.2. Proof of Ancillary Statements

The proof of Theorem 3.4 is based on the proof of a few ancillary statements.

By analogy with the notations in Chapter 2, we make use of the symbol $C_0^\infty(G)$ to denote the set of compactly supported functions in domain G possessing continuous

---

14) The existence of such ordered spectral representations is borne out by the Gårding–Browder–Mautner theorem (see Section 2.1.2)

partial derivatives of arbitrarily high order in this domain, and of the symbol G_R to denote the set of points in domain G removed from the boundary of G to a distance greater than $R > 0$.

We fix an arbitrary $R > 0$, an arbitrary point x in G_R, and denote by r an arbitrary distance such that $o \le r \le R$. As previously (see Section 2.5), we consider, for an arbitrary function $f(x)$ in the class C_0^∞, the mean value of this function on the surface of an N-dimensional sphere of radius r centered at x. We denote this mean value by the symbol $\Psi_x(r,f) = \Psi_x(r)$, that is,

$$\Psi_x(r) = \omega_N^{-1} \int \cdots \int_{\omega} f(x + r\omega)d\omega, \tag{3.2.4}$$

where the symbol $\int \ldots \int_{\omega} f(x + r\omega)d\omega$ refers to an integral of function f taken over all the angles on the surface of an N-dimensional sphere of radius r centered at point x, and the symbol ω_N refers to the surface area of an N-dimensional sphere of unit radius equal to $2\pi^{N/2}[\Gamma(N/2)]^{-1}$.

We introduce the function

$$F(r) = F_x(r) = r^{N-1}\Psi_x(r) \tag{3.2.5}$$

and make use of the symbol $\mathcal{D}F(r)$ to denote the operation

$$\mathcal{D}F(r) = \frac{d}{dr}\left[\frac{1}{r}F(r)\right], \tag{3.2.6}$$

and the symbol $\mathcal{D}^n F(r)$ to denote the result of a repeated n-fold application of the operation $\mathcal{D}$. It is assumed, by formal convention, that $\mathcal{D}^0 F(r) \equiv F(r)$.

LEMMA 3.3. *For an arbitrary function $f(x)$ in the class $C_0^\infty(G)$, for an arbitrary sufficiently small $R > 0$, for an arbitrary point x in the set G_R, and for an arbitrary number n satisfying the condition $n \le (N+1)/2$, each of the expansions*

$$(2\pi)^{N/2}R^{N/2-n+1}\omega_N^{-1}\sum_{i=1}^{m'}\int_0^\infty \hat{f}_i'(t)u_i'(x,t)t^{n/2-N/4}J_{N/2-n}(R\sqrt{t})d\rho'(t), \tag{3.2.7}$$

$$(2\pi)^{N/2}R^{N/2-n+1}\omega_N^{-1}\sum_{i=1}^{m''}\int_0^\infty \hat{f}_i''(t)u_i''(x,t)t^{n/2-N/4}J_{N/2-n}(R\sqrt{t})d\rho''(t) \tag{3.2.8}$$

converges to the value of $\mathcal{D}^{n-1}F_x(R)$, where the symbol $F_x(R)$ denotes function (3.2.5), and the symbol $\mathcal{D}$ denotes operation (3.2.6).

We recall that the statement of Lemma 3.3 has been proved earlier by us (see the proof of Lemma 2.11, Section 2.5.3[15]).

LEMMA 3.4. *Given any $s > -1/2$, any sufficiently small $R > 0$, and an arbitrary function $f(x)$ in the class $C_0^\infty(G)$, the estimate*

$$\int\limits_{G_R}\int\limits_0^\lambda \lambda^{s-1/2}\left|\overset{s}{\sigma}{}'_\lambda(x,f)-\overset{s}{\sigma}{}''_\lambda(x,f)\right|^2 dx d\lambda \leq C\|f\|_{L_2(G)} \tag{3.2.9}$$

holds for the Riesz means (3.2.1) and (3.2.2) of any two self-adjoint nonnegative extensions $\hat{A}'$ and $\hat{A}''$ of the Laplace operator in an arbitrary N-dimensional domain G, with a constant C dependent only on $R > 0$, $s > -1/2$, and dimension N.

PROOF OF LEMMA 3.4. We note, for one thing, that it suffices to examine the case of an arbitrary domain with an *odd dimension* N. Indeed, with Lemma 3.4 proved for an arbitrary domain of odd dimension, in the case of an arbitrary domain G of even dimension N we construct a tube domain $\widetilde{G}$ of odd dimension $(N+1)$ as the product of the domain G and the segment $-\pi \leq x_{N+1} \leq \pi$ and consider, on this segment, a self-adjoint nonnegative extension A_0 of the operator $\frac{-d^2}{dx_{N+1}^2}$ corresponding to the expansion in a trigonometric Fourier series. For the two arbitrary self-adjoint nonnegative extensions $\widehat{A}'$ and $\widehat{A}''$ of the Laplace operator $-\Delta_N$ in domain G, we consider the two self-adjoint nonnegative extensions $\widetilde{A}'$ and $\widetilde{A}''$ of the Laplace operator $-\Delta_{N+1}$ in the tube domain $\widetilde{G}$ which are produced via a union of extension $\widehat{A}'$ or, respectively, $\widehat{A}''$ with extension A_0. Note that, for an arbitrary function f, dependent on $x_1, x_2, \ldots, x_N$ only and independent of x_{N+1}, the Riesz means of extensions $\widehat{A}'$ and $\widehat{A}''$ coincide with the Riesz means of extensions $\widetilde{A}'$ and $\widetilde{A}''$, respectively.

Thus, with no loss of generality, we focus on two arbitrary self-adjoint nonnegative extensions $\widehat{A}'$ and $\widehat{A}''$ of the Laplace operator in an arbitrary domain G of odd dimension N. For two arbitrary ordered spectral representations of the space $L_2(G)$ with respect to the extensions $\widehat{A}'$ and $\widehat{A}''$ we shall estimate the difference of the Riesz means (3.2.1) and (3.2.2), taking an arbitrary sufficiently small positive R and an arbitrary fixed point x in G_R.

For the sake of definiteness we shall confine our reasoning to the extension $\widehat{A}'$ only, considering that for $\widehat{A}''$ the argumentation remains virtually the same.

---

[15] Indeed, it has been established in Section 2.5.3 that the expression within the braces in (2.5.123) is equal to zero for each n satisfying the inequality $n \leq (N+1)/2$.

Computing the Fourier image $\widehat{v}_i'(x,t)$ of the function

$$\overset{\lambda}{v}(|x-y|)$$

$$= \begin{cases} 2^s\Gamma(s+1)(2\pi)^{-N/2}\lambda^{N/4-s/2}|x-y|^{-N/2-s}J_{N/2+s}(|x-y|\sqrt{\lambda}) & \text{for} \quad |x-y| \le R, \\ 0 & \text{for} \quad |x-y| > R \end{cases}$$

with respect to the fundamental function $u_i'(x,t)$ and making use of the mean-value formula (2.3.1) (see Section 2.3.1), we arrive by virtue of equality (2.3.2) at the following relation:

$$\widehat{v}_i'(x,t)$$

$$= 2^s\Gamma(s+1)u_i'(x,t)\lambda^{N/4-s/2}t^{(2-N)/4}\int_0^R J_{N/2+s}(r\sqrt{\lambda})J_{N/2-1}(r\sqrt{t})r^{-s}dr \tag{3.2.10}$$

$$(i = 1, 2, \dots, m').$$

Integrating the integral on the right-hand side of (3.2.10) $(N-1)/2$ times by parts with the use of the recurrence relations

$$\int r^{-\nu}J_{\nu+1}(r\sqrt{\lambda})dr = -(\sqrt{\lambda})^{-1}J_\nu(r\sqrt{\lambda}),$$

$$\frac{d}{dr}[r^\nu J_\nu(r\sqrt{t})] = r^\nu\sqrt{t}J_{nu-1}(r\sqrt{t}),$$

and taking into account that all the substitutions at $r=0$ become zero, we arrive at the relation

$$\widehat{v}_i'(x,t) = -2^s\Gamma(s+1)$$

$$\times u_i'(x,t)\sum_{n=1}^{(N-1)/2}(\sqrt{\lambda})^{N/2-s-n}(\sqrt{t})^{n-N/2}J_{N/2+s-n}(R\sqrt{\lambda})J_{N/2-n}(R\sqrt{t})R^{-s}$$

$$+ 2^s\Gamma(s+1)u_i'(x,t)(\sqrt{\lambda})^{1/2-s}(\sqrt{t})^{1/2}\int_0^R J_{1/2+s}(r\sqrt{\lambda})J_{-1/2}(r\sqrt{t})r^{-s}dr \tag{3.2.11}$$

$$(i = 1, 2, \dots, m').$$

Multiplying both sides of (3.2.11) by the Fourier image $\widehat{f}_i'(t)$ of an arbitrary function $f(x)$ in the class $C_0^\infty(G)$ and then integrating the obtained equality over a spectral measure $\rho'(t)$ between the limits from 0 to ∞ with respect to t and summing it over all values of i from 1 to m', we obtain the relation

$$\sum_{i=1}^{m'} \int_0^\infty \widehat{f}_i'(t)\widehat{v}_i{}'(t)d\rho'(t)$$

$$= -2^s\Gamma(s+1)(2\pi)^{-N/2}\omega_N \sum_{n=1}^{(N-1)/2} \lambda^{N/4-s/2-n/2}R^{n-N/2-1-s}J_{N/2+s-n}(R\sqrt{\lambda})$$

$$\times \left\{ (2\pi)^{N/2}R^{N/2-n+1}\omega_N^{-1} \sum_{i=1}^{m'} \int_0^\infty \widehat{f}_i'(t)u_i'(x,t)(\sqrt{t})^{n-N/2}J_{N/2-n}(R\sqrt{t})d\rho'(t) \right\}$$

$$+2^s\Gamma(s+1)\lambda^{-s/2} \sum_{i=1}^{m'} \int_0^\infty \widehat{f}_i'(t)u_i'(x,t)$$

$$\times \left[\lambda^{1/4}t^{1/4} \int_0^R J_{1/2+s}(r\sqrt{\lambda})J_{-1/2}(r\sqrt{t})r^{-s}dr \right] d\rho'(t). \tag{3.2.12}$$

We recall now that the functions $f(y)$ and $\overset{\lambda}{v}(|x-y|)$ belong in all cases to the class $L_2(G)$; therefore, by virtue of the Parseval equality (see equality (2.1.10), Section 2.1.2) we have

$$\sum_{i=1}^{m'} \int_0^\infty \widehat{f}_i'(t)\widehat{v}_i'(t)d\rho'(t) = \int_G \overset{\lambda}{v}(|x-y|)f(y)dy. \tag{3.2.13}$$

Next, the expression within the braces in (3.2.12) is equal, by virtue of Lemma 3.3, to

$$\left\{ (2\pi)^{N/2}R^{N/2-n+1}\omega_N^{-1} \sum_{i=1}^{m'} \int_0^\infty \widehat{f}_i'(t)u_i'(x,t)(\sqrt{t})^{n-N/2}J_{N/2-n}(R\sqrt{t})d\rho'(t) \right\}$$

$$= \mathcal{D}^{n-1}F_x(R). \tag{3.2.14}$$

Finally, the integral within the square brackets in (3.1.12) is subjected to a transformation of the form $\int_0^R = \int_0^\infty - \int_R^\infty$. With the quantity $I_t^\lambda(R)$ defined as

$$I_t^\lambda(R) = \begin{cases} \lambda^{1/4}t^{1/4} \int_R^\infty J_{1/2+s}(r\sqrt{\lambda})J_{-1/2}(r\sqrt{t})r^{-s}dr & \text{for } t \neq \lambda, \\ \lambda^{1/4}t^{1/4} \int_0^R J_{1/2+s}(r\sqrt{\lambda})J_{-1/2}(r\sqrt{t})r^{-s}dr & \text{for } t = \lambda, \end{cases} \tag{3.2.15}$$

and keeping in mind that the relation

$$\begin{aligned} &2^s\Gamma(s+1)\lambda^{1/4-s/2}t^{1/4} \int_0^\infty J_{1/2+s}(r\sqrt{\lambda})J_{-1/2}(r\sqrt{t})r^{-s}dr \\ &\qquad = \begin{cases} (1-\frac{t}{\lambda})^s & \text{for } t < \lambda, \\ 0 & \text{for } t > \lambda \end{cases} \end{aligned} \tag{3.2.16}$$

holds for any $s > -1$ and for $\lambda \neq t$,[16] we obtain from relations (3.2.12)–(3.2.16) with reference to (3.2.1) the following equality:

$$\begin{aligned} &\int_G v(|x-y|)f(y)dy \\ &= -2^s\Gamma(s+1)(2\pi)^{-N/2}\omega_N \sum_{n=1}^{(N-1)/2} \lambda^{N/4-s/2-n/2}R^{n-N/2-1-s} \\ &\times J_{N/2+s-n}(R\sqrt{\lambda})\mathcal{D}^{n-1}F_x(R) + \overset{s}{\sigma}{}'_\lambda(x,f) \\ &-2^s\Gamma(s+1)\lambda^{-s/2}\sum_{i=1}^{m'}\int_0^\infty \hat{f}'_i(t)u'_i(x,t)I_t^\lambda(R)d\rho'(t). \end{aligned} \tag{3.2.17}$$

In much the same manner, the following equality is established for the second

[16] See, for example, Bateman and Erdélyi [9, p. 107, formula (34)].

extension $\widehat{A}''$:

$$\int_G v(|x-y|)f(y)dy$$

$$= -2^s\Gamma(s+1)(2\pi)^{-N/2}\omega_N \sum_{n=1}^{(N-1)/2} \lambda^{N/4-s/2-n/2}R^{n-N/2-1-s}$$

$$\times J_{N/2+s-n}(R\sqrt{\lambda})\mathcal{D}^{n-1}F_x(R) + \overset{s}{\sigma}{}''_\lambda(x,f) \tag{3.2.18}$$

$$-2^s\Gamma(s+1)\lambda^{-s/2}\sum_{i=1}^{m''}\int_0^\infty \widehat{f}_i''(t)u_i''(x,t)I_t^\lambda(R)d\rho''(t).$$

Subtracting (3.2.18) from (3.2.17), we arrive at the relation

$$\overset{s}{\sigma}{}'_\lambda(x,f) - \overset{s}{\sigma}{}''_\lambda(x,f)$$

$$= 2^s\Gamma(s+1)\lambda^{-s/2}[\sum_{i=1}^{m'}\int_0^\infty \widehat{f}_i'(t)u_i'(x,t)I_t^\lambda(R)d\rho'(t) \tag{3.2.19}$$

$$-\sum_{i=1}^{m''}\int_0^\infty \widehat{f}_i''(t)u_i''(x,t)I_t^\lambda(R)d\rho''(t)],$$

which holds at any point x of the set G_R. In turn, it follows from (3.2.19) that

$$\int_{G_R}\int_0^\infty \lambda^{s-1/2}\left|\overset{s}{\sigma}{}'_\lambda(x,f) - \overset{s}{\sigma}{}''_\lambda(x,f)\right|^2 dx d\lambda$$

$$\leq 2^{2s+1}\Gamma^2(s+1)\int_0^\infty\left\{\int_G\left[\sum_{i=1}^{m'}\int_0^\infty \widehat{f}_i'(t)u_i'(x,t)I_t^\lambda(R)d\rho'(t)\right]^2 dx\right\}\frac{d\lambda}{\sqrt{\lambda}}$$

$$+2^{2s+1}\Gamma^2(s+1)\int_0^\infty\left\{\int_G\left[\sum_{i=1}^{m''}\int_0^\infty \widehat{f}_i''(t)u_i''(x,t)I_t^\lambda(R)d\rho''(t)\right]^2 dx\right\}\frac{d\lambda}{\sqrt{\lambda}}. \tag{3.2.20}$$

Applying the Parseval equality (see (2.1.11), Section 2.1.2) to the terms within the braces on the right-hand side of (3.2.20) and reversing the order of integration with respect to t and λ, we obtain

$$\int_{G_R}\int_0^\infty \lambda^{s-1/2}\left|\overset{s}{\sigma'_\lambda}(x,f)-\overset{s}{\sigma''_\lambda}(x,f)\right|^2 dx d\lambda$$

$$\le 2^{2s+1}\Gamma^2(s+1)\int_0^\infty\left\{\sum_{i=1}^{m'}\int_0^\infty\left|\widehat{f}'_i(t)I_t^\lambda(R)\right|^2 d\rho'(t)\right\}\frac{d\lambda}{\sqrt{\lambda}}$$

$$+2^{2s+1}\Gamma^2(s+1)\int_0^\infty\left\{\sum_{i=1}^{m''}\int_0^\infty\left|\widehat{f}''_i(t)I_t^\lambda(R)\right|^2 d\rho''(t)\right\}\frac{d\lambda}{\sqrt{\lambda}} \tag{3.2.21}$$

$$\le 2^{2s+1}\Gamma^2(s+1)\sum_{i=1}^{m'}\int_0^\infty\left|\widehat{f}'_i(t)\right|^2\left\{\int_0^\infty[I_t^\lambda(R)]^2 d(\sqrt{\lambda})\right\}d\rho'(t)$$

$$\le 2^{2s+1}\Gamma^2(s+1)\sum_{i=1}^{m''}\int_0^\infty\left|\widehat{f}''_i(t)\right|^2\left\{\int_0^\infty[I_t^\lambda(R)]^2 d(\sqrt{\lambda})\right\}d\rho''(t).$$

Having established the estimate

$$\int_0^\infty[I_t^\lambda(R)]^2 d(\sqrt{\lambda})\le C_1, \tag{3.2.22}$$

uniform with respect to t on the half-line $0\le t<\infty$, we obtain, in terms of (3.2.22) and the Parseval equality, that the right-hand side of (3.2.21) does not exceed the quantity

$$8\,2^{2s}\Gamma^2(s+1)C_1\|f\|^2_{L_2(G)},$$

which will complete the proof of Lemma 3.4.

We have seen, therefore, that to prove Lemma 3.4 it suffices to prove the following ancillary assertion.

LEMMA 3.5. *For $I_t^\lambda(R)$ defined by (3.2.15), given any $s>-1/2$, uniform with respect to t on the half-line $0\le t<\infty$, the estimate (3.2.22) holds.*

The proof of Lemma 3.5 will be carried out in two steps.

1^0. We show first that for (3.2.15), given fixed $R>0$ and $s>-1$ and any nonnegative λ and t, the following two estimates hold:

(a) in the case $|\sqrt{t}-\sqrt{\lambda}|\ge 1$ and any $s>-1$

$$|I_t^\lambda(R)|=O(|\sqrt{t}-\sqrt{\lambda}|^{-1}); \tag{3.2.23}$$

(b) in the case of $0 < |\sqrt{t} - \sqrt{\lambda}| < 1$

$$|I_t^\lambda(R)| = \begin{cases} O(1) & \text{for} \quad s \geq 0, \\ O(|\sqrt{t} - \sqrt{\lambda}|^s) & \text{for} \quad -1 < s < 0. \end{cases} \tag{3.2.24}$$

To provide a complete proof of the formulated statement, it will suffice, given a fixed $C > 0$, to establish: (a) the estimate (3.2.23) on the condition that at least one of the two $\sqrt{t}$ and $\sqrt{\lambda}$ does not exceed C; (b) the estimate (3.2.24) on the condition that both $\sqrt{t}$ and $\sqrt{\lambda}$ are distinct from each other and are not in excess of C; (c) the estimate (3.2.24) on the condition that both $\sqrt{t}$ and $\sqrt{\lambda}$ are in excess of C (note that this stems from the fact that the estimates (3.2.23) and (3.2.24) are equivalent subject to $1/C \leq |\sqrt{t} - \sqrt{\lambda}| \leq C$).

(a) First, we ensure that the estimate (3.2.23) holds on the condition that of the two $\sqrt{t}$ and $\sqrt{\lambda}$ (subject to $|\sqrt{t} - \sqrt{\lambda}| \geq 1$) at least one is larger than a fixed $C > 0$. If $C \geq 2$, both $\sqrt{t}$ and $\sqrt{\lambda}$ are in excess of a $C - 1 \geq 1$; then the estimate (2.4.10′) holds (see Section 2.4.2), which becomes equivalent to estimate (3.2.23) at $\nu = 1/2$.

(b) Next we verify the validity of estimate (3.2.24) on the condition that $\sqrt{t}$ and $\sqrt{\lambda}$ are different from each other and either is not in excess of C.

We note, first of all, that for $s > 0$ the estimate (3.2.24) follows immediately from the familiar inequality $|J_\nu(x)| \leq C(\nu)x^{-1/2}$, valid for all $\nu \geq -1/2$ and $x \geq 0$. Indeed, this inequality and expression (3.2.15) imply that

$$|I_t^\lambda(R)| \leq C_1 \int_R^\infty r^{-1-s} dr = C\frac{R^{-s}}{s} = O(1). \tag{3.2.25}$$

It now remains to establish the relation (3.2.24) for $\sqrt{t}$ and $\sqrt{\lambda}$ different from each other and either not in excess of C, provided that $-1 < s \leq 0$.

In this case, given $\sqrt{t} \neq \sqrt{\lambda}$, we infer from relations (3.2.15) and (3.2.16) that

$$I_t^\lambda(R) + \lambda^{1/4} t^{1/4} \int_0^R J_{1/2+s}(r\sqrt{\lambda}) J_{-1/2}(r\sqrt{t}) r^{-s} dr$$
$$= \begin{cases} [2^s\Gamma(s+1)]^{-1} \lambda^{s/2} \left(1 - \dfrac{t}{\lambda}\right)^s & \text{for} \quad \sqrt{t} < \sqrt{\lambda}, \\ 0 & \text{for} \quad \sqrt{t} > \sqrt{\lambda}, \end{cases} \tag{3.2.26}$$

The inequality

$$\lambda^{s/2} \left(1 - \frac{t}{\lambda}\right)^s \leq (\sqrt{\lambda} - \sqrt{t})^s$$

holds for $-1 < s \leq 0$ and $0 \leq \sqrt{t} < \sqrt{\lambda}$; therefore, by virtue of (3.2.26), it suffices to prove estimate (3.2.24) for $-1 < s \le 0$ in order to show that the estimate

$$\lambda^{1/4}t^{1/4}\int_0^R J_{1/2+s}(r\sqrt{\lambda})J_{-1/2}(r\sqrt{t})r^{-s}\,dr = O(1) \qquad (3.2.27)$$

holds for $\sqrt{t} \le C$, $\sqrt{\lambda} \le C$.

In view of the last estimate, we note that, inasmuch as the nonnegative arguments of both Bessel functions under the integration sign in (3.2.27) do not exceed the value of CR, the estimate $|J_\nu(x)| \le C(\nu)x^\nu$ holds for the moduli of these Bessel functions. Making use of this estimate, we obtain

$$\left|\lambda^{1/4}t^{1/4}\int_0^R J_{1/2+s}(r\sqrt{\lambda})J_{-1/2}(r\sqrt{t})r^{-s}\,dr\right|$$
$$\le C\left(\frac{1}{2}+s\right)C\left(-\frac{1}{2}\right)(\sqrt{\lambda})^{1+s}R = O(1).$$

This completes the derivation of estimate (3.2.24) for any $\sqrt{t}$ and $\sqrt{\lambda}$ different from each other and not in excess of C.

(c) It now remains to establish estimate (3.2.24) for the case where both $\sqrt{t}$ and $\sqrt{\lambda}$ are in excess of C and satisfy the condition $0 < |\sqrt{t} - \sqrt{\lambda}| < 1$.

The case $s > 0$ requires little attention, since the validity of estimate (3.2.24) stems immediately from the relation (3.2.25) established above.

We turn now to the case $-1 < s \le 0$. Making use of the exact expression for $J_{-1/2}(x)$ and of the asymptotic formula at $x \ge 1$ for $J_{1/2+s}(x)$, we obtain that integral (3.2.15) is equal to [17)]

$$I_t^\lambda(R) = \frac{2}{\pi}\int_R^\infty r^{-1-s}\sin\left(r\sqrt{\lambda}-\frac{\pi s}{2}\right)\cos(r\sqrt{t})\,dr + O\left(\frac{1}{\sqrt{\lambda}}\right)\int_R^\infty r^{-1-s}\,dr. \qquad (3.2.28)$$

Considering that $\sqrt{\lambda} \ge C$, $s > -1$, the addend on the right-hand side of (3.2.28) is $O(1)$ and, the more so, $O(|\sqrt{t}-\sqrt{\lambda}|^s)$. The augend on the right-hand side of (3.2.28)

---

17) The exact expression and the asymptotic formula referred to are, respectively, $J_{-1/2}(x) = \sqrt{2/(\pi x)}\cos x$ and $J_{1/2+s}(x) = \sqrt{2/(\pi x)}\sin(x - \pi s/2) + O(x^{-3/2})$ (see, for example, Bateman and Erdélyi [9, p. 98, formula (3)].

can be rewritten in the form

$$
\begin{aligned}
&\frac{2}{\pi}\int_R^\infty r^{-1-s}\sin\left(r\sqrt{\lambda}-\frac{\pi s}{2}\right)\cos(r\sqrt{t})dr \\
&= \frac{2}{\pi}\cos\left(\frac{\pi s}{2}\right)\int_R^\infty r^{-1-s}\sin\left(r\sqrt{\lambda}\right)\cos(r\sqrt{t})dr \\
&\quad -\frac{2}{\pi}\sin\left(\frac{\pi s}{2}\right)\int_R^\infty r^{-1-s}\cos\left(r\sqrt{\lambda}\right)\cos(r\sqrt{t})dr \\
&= \frac{1}{\pi}\cos\left(\frac{\pi s}{2}\right)\left\{\int_R^\infty r^{-1-s}\sin[(\sqrt{\lambda}-\sqrt{t})r]dr\right. \\
&\quad \left. +\int_R^\infty r^{-1-s}\sin[(\sqrt{\lambda}+\sqrt{t})r]dr\right\} \\
&\quad -\frac{1}{\pi}\sin\left(\frac{\pi s}{2}\right)\left\{\int_R^\infty r^{-1-s}\cos[(\sqrt{\lambda}-\sqrt{t})r]dr\right. \\
&\quad \left. +\int_R^\infty r^{-1-s}\cos[(\sqrt{\lambda}+\sqrt{t})r]dr\right\} \\
&= \frac{1}{\pi}\cos\left(\frac{\pi s}{2}\right)\left\{|(\sqrt{\lambda}-\sqrt{t}|^s\operatorname{sgn}(\sqrt{\lambda}-\sqrt{t})\int_{R|\sqrt{t}-\sqrt{\lambda}|}^\infty \rho^{-1-s}\sin\rho\, d\rho\right. \\
&\quad \left. +(\sqrt{\lambda}+\sqrt{t})^s\int_{R|\sqrt{\lambda}+\sqrt{t}|}^\infty \rho^{-1-s}\sin\rho\, d\rho\right\} \\
&\quad -\frac{1}{\pi}\sin\left(\frac{\pi s}{2}\right)\left\{|(\sqrt{\lambda}-\sqrt{t}|^s\int_{R(\sqrt{\lambda}-\sqrt{t})}^\infty \rho^{-1-s}\cos\rho\, d\rho\right. \\
&\quad \left. +(\sqrt{\lambda}+\sqrt{t})^s\int_{R(\sqrt{\lambda}+\sqrt{t})}^\infty \rho^{-1-s}\cos\rho\, d\rho\right\}.
\end{aligned}
\tag{3.2.29}
$$

Note that at $s = 0$ the terms on the right-hand side of (3.2.9) containing $\sin(\pi s/2)$ for a cofactor are equal to zero.

Since $-1 < s \leq 0$, the relation (3.2.29) immediately implies that

$$
\frac{2}{\pi}\int_R^\infty r^{-1-s}\sin\left(r\sqrt{\lambda}-\frac{\pi s}{2}\right)\cos(r\sqrt{t})dr = O(|\sqrt{t}-\sqrt{\lambda}|^s).
$$

Whence it follows that the estimate (3.2.24) for this particular case has been proved.

Thus, the proof of estimates (3.2.23) and (3.2.24) is complete.

2^0. We now prove the main statement of Lemma 3.5 on the validity of estimate (3.2.22) uniform in t on the half-line $0 \le t < \infty$ for any fixed $s > -1/2$.

We represent the integral on the left-hand side of (3.2.22) as the sum of two integrals:

$$\int_0^\infty |I_t^\lambda(R)|^2 d(\sqrt{\lambda}) = \int_{|\sqrt{t}-\sqrt{\lambda}|\ge 1} |I_t^\lambda(R)|^2 d(\sqrt{\lambda}) + \int_{|\sqrt{t}-\sqrt{\lambda}|\le 1} |I_t^\lambda(R)|^2 d(\sqrt{\lambda}). \tag{3.2.30}$$

Making use of relation (3.2.23) to estimate the former integral and of relation (3.2.24) to estimate the latter integral on the right-hand side of (3.2.30), we obtain[18)]

$$\int_{|\sqrt{t}-\sqrt{\lambda}|\ge 1} |I_t^\lambda(R)|^2 d(\sqrt{\lambda}) \le C_1 \int_{|\sqrt{t}-\sqrt{\lambda}|\ge 1} |\sqrt{t}-\sqrt{\lambda}|^{-2} d(\sqrt{\lambda}) \le 2C_1 \int_1^\infty \tau^{-2} d\tau = C_2, \tag{3.2.31}$$

$$\int_{|\sqrt{t}-\sqrt{\lambda}|\le 1} |I_t^\lambda(R)|^2 d(\sqrt{\lambda}) \le \begin{cases} C_3 \displaystyle\int_{|\sqrt{t}-\sqrt{\lambda}|\le 1} d(\sqrt{\lambda}) \le 2C_3 & \text{for } s \ge 0, \\ C_4 \displaystyle\int_{|\sqrt{t}-\sqrt{\lambda}|\le 1} |\sqrt{t}-\sqrt{\lambda}|^{2s} d(\sqrt{\lambda}) \le 2C_4 \int_0^1 \tau^{2s} d\tau = C_5 & \text{for } -1/2 < s < 0. \end{cases} \tag{3.2.32}$$

A comparison of relations (3.2.30)–(3.2.22) lends support to the validity of estimate (3.2.22) uniform in t on the half-line $0 < t < \infty$.

This completes the proof of Lemma 3.5.

LEMMA 3.6. *For any $s > -1/2$, for any sufficiently small $R > 0$ and for any function $f(x)$ that belongs to the class $L_2(G)$ only, for the Riesz means (3.2.1) and*

18) In what follows the symbols $C_1, C_2, \ldots$ will be used to denote constants dependent on s and R only.

(3.2.2) of any two self-adjoint nonnegative extensions $\widehat{A}'$ and $\widehat{A}''$ of the Laplace operator in an arbitrary N-dimensional domain G, the estimate (3.2.9) holds with a constant C dependent on $R > 0$, $s > -1/2$, and dimension N only.

PROOF: It suffices to show that for an arbitrary $\Lambda \geq 1$ and an arbitrary function $f(x)$ in $L_2(G)$ the inequality

$$\int_{G_R}\int_0^\Lambda \lambda^{s-1/2}\left|\overset{s}{\sigma}{}'_\lambda(x,f) - \overset{s}{\sigma}{}''_\lambda(x,f)\right|^2 dx d\lambda \leq C_2\|f\|^2_{L_2(G)} \tag{3.2.33}$$

holds with a constant C_2 independent of Λ and f.

Note that for an arbitrary $s > -1/2$, an arbitrary function $f(x)$ in $L_2(G)$, and an arbitrary $\Lambda \geq 1$ there is a function $f_1(x)$ in $C_0^\infty(G)$ such that

$$\|f - f_1\|_{L_2(G)} \leq (\sqrt{\Lambda})^{-s-1/2}\|f\|_{L_2(G)}. \tag{3.2.34}$$

One infers from the equality $f_1 = (f_1 - f) + f$, from the triangle inequality, from relation (3.2.34), and from the fact that $\Lambda \geq 1$, $s + 1/2 > 0$ that

$$\|f_1\|_{L_2(G)} \leq \|f_1 - f\|_{L_2(G)} + \|f\|_{L_2(G)} \leq 2\|f\|_{L_2(G)}. \tag{3.2.35}$$

Let us estimate the integral

$$\begin{aligned}
&\int_{G_R}\int_0^\Lambda \lambda^{s-1/2}\left|\overset{s}{\sigma}{}'_\lambda(x,f) - \overset{s}{\sigma}{}''_\lambda(x,f)\right|^2 dx d\lambda \\
&\leq 2\int_{G_R}\int_0^\Lambda \lambda^{s-1/2}\left|\overset{s}{\sigma}{}'_\lambda(x,f_1) - \overset{s}{\sigma}{}''_\lambda(x,f_1)\right|^2 dx d\lambda \\
&+2\int_{G_R}\int_0^\Lambda \lambda^{s-1/2}\left|\overset{s}{\sigma}{}'_\lambda(x,f-f_1) - \overset{s}{\sigma}{}''_\lambda(x,f-f_1)\right| dx d\lambda,
\end{aligned} \tag{3.2.36}$$

making use of the inequality $(A+B)^2 \leq 2A^2 + 2B^2$.

Since for the function $f_1(x)$ in the $C_0^\infty(G)$ class the estimate (3.2.9) holds by virtue of Lemma 3.4, we obtain from (3.2.35) and (3.2.36), invoking $(A+B)^2 \leq 2A^2 + 2B^2$, that

$$\int_{G_R}\int_0^{\Lambda} \lambda^{s-1/2}\left|\overset{s}{\sigma}{}'_\lambda(x,f) - \overset{s}{\sigma}{}''_\lambda(x,f)\right|^2 dx d\lambda$$

$$\le 8\|f\|^2_{L_2(G)} + 2\int_{G_R}\int_0^{\Lambda} \lambda^{s-1/2}\left|\overset{s}{\sigma}{}'_\lambda(x,f-f_1) - \overset{s}{\sigma}{}''_\lambda(x,f-f_1)\right|^2 dx d\lambda$$

$$\le 8C\|f\|^2_{L_2(G)} + 4\int_{G_R}\int_0^{\Lambda} \lambda^{s-1/2}\left|\overset{s}{\sigma}{}'_\lambda(x,f-f_1)\right|^2 dx d\lambda \tag{3.2.37}$$

$$+4\int_{G_R}\int_0^{\Lambda} \lambda^{s-1/2}\left|\overset{s}{\sigma}{}''_\lambda(x,f-f_1)\right|^2 dx d\lambda$$

(C is the constant in (3.2.9)).

Both integrals on the right-hand side of (3.2.27) are estimated in a similar manner; therefore we shall restrict ourselves to estimating the first of them.

Making use of expression (3.2.1) for the Riesz mean and of the Parseval equality, we obtain that

$$\int_{G_R}\int_0^{\Lambda} \lambda^{s-1/2}\left|\overset{s}{\sigma}{}'_\lambda(x,f-f_1)\right|^2 dx d\lambda$$

$$= \int_0^{\Lambda} \lambda^{s-1/2}\left\{\sum_{i=1}^{m'}\int_0^{\lambda}\left[(\widehat{f-f_1})'_i(t)\right]^2\left(1-\frac{t}{\lambda}\right)^{2s} d\rho'(t)\right\} d\lambda \tag{3.2.38}$$

(on the condition that the expansion on the right-hand side of (3.2.38) is convergent).

Reversing the order of integration with respect to t and λ on the right-hand side of (3.2.38), we obtain that the right-hand side of (3.2.38) is

$$\sum_{i=1}^{m'}\int_0^{\Lambda}\left[(\widehat{f-f_1})'_i(t)\right]^2\left\{\int_t^{\Lambda}\lambda^{-s-1/2}(\lambda-t)^{2s}d\lambda\right\} d\rho'(t). \tag{3.2.39}$$

With a view to computing the integral within the braces in (3.2.39), we consider the following cases separately: (1) $s \ge 0$; (2) $-1/2 < s < 0$.

In the former case the integral between the braces is majorized as

$$\int_t^\Lambda \lambda^{-1/2}(\lambda - t)^s d\lambda \leq \int_t^\Lambda \lambda^{s-1/2} d\lambda \leq \int_0^\Lambda \lambda^{s-1/2} d\lambda = \frac{\Lambda^{s+1/2}}{s+1/2}.$$

In the case $-1/2 < s < 0$, the integral between the braces in (3.2.39) is converted, with the aid of the change of variables $\mu = \lambda - t$, to the form

$$\int_0^{\Lambda - t} (\mu + t)^{-s-1/2} \mu^{2s} d\mu \leq \int_0^{\Lambda - t} \mu^{-s-1/2} \mu^{2s} d\mu \leq \int_0^\Lambda \mu^{s-1/2} d\mu = \frac{\Lambda^{s+1/2}}{s+1/2}.$$

Thus, in both cases the integral between the braces in (3.2.39) does not exceed $\Lambda^{s+1/2}(s+1/2)^{-1}$. Therefore, from (3.2.38), (3.2.39), from the Parseval inequality, and from inequality (3.2.34), we obtain the inequality

$$\int_{G_R} \int_0^\Lambda \lambda^{s-1/2} \left| \overset{s}{\sigma}{}'_\lambda(x, f - f_1) \right|^2 dx d\lambda$$

$$\leq \frac{\Lambda^{s+1/2}}{s+1/2} \|f - f_1\|_1 \|^2_{L_2(G)} \leq (s+1/2)^{-1} \|f\|_{L_2(G)}.$$

In much the same manner one arrives at the inequality

$$\int_{G_R} \int_0^\Lambda \lambda^{s-1/2} \left| \overset{s}{\sigma}{}''_\lambda(x, f - f_1) \right|^2 dx d\lambda \leq (s+1/2)^{-1} \|f\|_{L_2(G)}.$$

The last two inequalities and (3.2.37) lend support to the validity of the desired estimate (3.2.33) with the constant C_2 equal to $8C + 8(s+1/2)^{-1}$.

The proof of Lemma 3.6 is complete.

LEMMA 3.7. *Given any $s > 0$, for an arbitrary N-dimensional domain G and an arbitrary function $f(x)$ in the class $L_2(G)$ there is a numerical sequence $\{\bar{\lambda}_n\}$ such that $(n-1)^2 \leq \bar{\lambda}_n \leq n^2$ holds for all numbers n, beginning with a certain n, and for the Riesz means (3.2.1) and (3.2.2) of two arbitrary self-adjoint nonnegative extensions $\widehat{A}'$ and $\widehat{A}''$ of the Laplace operator in domain G for almost all points of domain G the following estimate holds:*

$$\left| \overset{s\,\prime}{\sigma}_{\bar{\lambda}_n}(x, f) - \overset{s\,\prime\prime}{\sigma}_{\bar{\lambda}_n}(x, f) \right| = o(\bar{\lambda}_n^{-s/2}). \tag{3.2.40}$$

PROOF: By virtue of Lemma 3.6, for any function $f(x)$ in $L_2(G)$ and for an arbitrary sufficiently small $R > 0$ the "double" integral on the left-hand side of

(3.2.9) is convergent. Representing this integral as the sum of integrals

$$\int\limits_{G_R}\int\limits_0^\infty \lambda^{s-1/2}\left|\overset{s}{\sigma}{}'_\lambda(x,f)-\overset{s}{\sigma}{}''_\lambda(x,f)\right|^2 dx\,d\lambda$$
$$=\sum_{n=1}^\infty \int\limits_{(n-1)^2}^{n^2}\lambda^s\left\{\int\limits_{G_R}\left[\overset{s}{\sigma}{}'_\lambda(x,f)-\overset{s}{\sigma}{}''_\lambda(x,f)\right]^2 dx\right\}\frac{d\lambda}{\sqrt{\lambda}}, \tag{3.2.41}$$

we apply the mean-value formula to each of the integrals under the summation sign on the right-hand side (3.2.41). A condition for applicability of the mean-value formula is the continuity in λ on the segments $[(n-1)^2, n^2]$ of the function within the braces on the right-hand side of (3.2.41). We denote this function by $\Phi(\lambda)$.

It suffices to show that for any fixed $\lambda \geq 1$ the difference $\Phi(\lambda+\Delta\lambda)-\Phi(\lambda)$ tends to zero as $\Delta\lambda \to 0$.

Note that

$$|\Phi(\lambda+\Delta\lambda)-\Phi(\lambda)| = \Big|\int\limits_{G_R}\left[\overset{s}{\sigma}{}'_{\lambda+\Delta\lambda}(x,f)-\overset{s}{\sigma}{}''_{\lambda+\Delta\lambda}(x,f)\right]^2 dx$$
$$-\int\limits_{G_R}\left[\overset{s}{\sigma}{}'_\lambda(x,f)-\overset{s}{\sigma}{}''_\lambda(x,f)\right]^2 dx\Big|.$$

By virtue of the triangle inequality and the Parseval equality for any $s \geq 0$, we have

$$\left\{\int\limits_{G_R}\left[\overset{s}{\sigma}{}'_{\lambda+\Delta\lambda}(x,f)-\overset{s}{\sigma}{}''_{\lambda+\Delta\lambda}(x,f)\right]^2 dx\right\}^{1/2}+\left\{\int\limits_{G_R}\left[\overset{s}{\sigma}{}'_\lambda(x,f)-\overset{s}{\sigma}{}''_\lambda(x,f)\right]^2 dx\right\}^{1/2}$$
$$\leq \left\{\int\limits_G\left[\overset{s}{\sigma}{}'_{\lambda+\Delta\lambda}(x,f)\right]^2 dx\right\}^{1/2}+\left\{\int\limits_G\left[\overset{s}{\sigma}{}''_{\lambda+\Delta\lambda}(x,f)\right]^2 dx\right\}^{1/2}$$
$$+\left\{\int\limits_G\left[\overset{s}{\sigma}{}'_\lambda(x,f)\right]^2 dx\right\}^{1/2}+\left\{\int\limits_G\left[\overset{s}{\sigma}{}''_\lambda(x,f)\right]^2 dx\right\}^{1/2} \leq 4\|f\|_{L_2(G)}.$$

Consequently, the following inequality holds:

$$|\Phi(\lambda+\Delta\lambda)-\Phi(\lambda)| \leq 4|f|| \left| \left\{ \int\limits_{G_R} \left[\overset{s}{\sigma}{}'_{\lambda+\Delta\lambda}(x,f) - \overset{s}{\sigma}{}''_{\lambda+\Delta\lambda}(x,f) \right]^2 dx \right\}^2 \right.$$

$$\left. - \left\{ \int\limits_{G_R} \left[\overset{s}{\sigma}{}'_{\lambda}(x,f) - \overset{s}{\sigma}{}''_{\lambda}(x,f) \right]^2 dx \right\}^{1/2} \right| .$$

Making use of the triangle inequality on the right-hand side, we obtain for the norm in $L_2(G)$ that

$$|\Phi(\lambda+\Delta\lambda)-\Phi(\lambda)|$$

$$\leq 4|f|| \left\{ \int\limits_{G_R} \left[\overset{s}{\sigma}{}'_{\lambda+\Delta\lambda}(x,f) - \overset{s}{\sigma}{}''_{\lambda+\Delta\lambda}(x,f) - \overset{s}{\sigma}{}'_{\lambda}(x,f) + \overset{s}{\sigma}{}''_{\lambda}(x,f) \right]^2 dx \right\}^{1/2}$$

and the more so

$$\begin{aligned} |\Phi(\lambda+\Delta\lambda)-\Phi(\lambda)| &\leq 4\|f\| \left\{ \int\limits_{G} \left[\overset{s}{\sigma}{}'_{\lambda+\Delta\lambda}(x,f) - \overset{s}{\sigma}{}'_{\lambda}(x,f) \right]^2 dx \right\}^{1/2} \\ &+4\|f\| \left\{ \int\limits_{G} \left[\overset{s}{\sigma}{}'_{\lambda+\Delta\lambda}(x,f) - \overset{s}{\sigma}{}''_{\lambda}(x,f) \right]^2 dx \right\}^{1/2} \end{aligned} \tag{3.2.42}$$

It suffices to prove that for any fixed $\lambda \geq 1$, either of the two expressions enclosed within the braces in (3.2.42) tends to zero as $\Delta\lambda \to 0$. The proof will be given for the former expression only; the proof for the latter expression is analogous and will be an easy exercise for the reader. It suffices to note that, by virtue of the Parseval equality,

$$\int\limits_G \left[\overset{s}{\sigma}'_{\lambda+\Delta\lambda}(x,f) - \overset{s}{\sigma}'_{\lambda}(x,f)\right]^2 dx$$

$$= \sum_{i=1}^{m'} \int\limits_0^{\lambda-2|\Delta\lambda|} \left|\widehat{f}_i'(t)\right|^2 \left[\left(1-\frac{t}{\lambda+\Delta\lambda}\right)^s - \left(1-\frac{t}{\lambda}\right)^s\right]^2 d\rho'(t)$$

$$+ \sum_{i=1}^{m'} \int\limits_{\lambda-2|\Delta\lambda|}^{\lambda+\Delta\lambda} \left|\widehat{f}_i'(t)\right|^2 \left(1-\frac{t}{\lambda+\Delta\lambda}\right)^{2s} d\rho'(t)$$

$$- \sum_{i=1}^{m'} \int\limits_{\lambda-2|\Delta\lambda|}^{\lambda} \left|\widehat{f}_i'(t)\right|^2 \left(1-\frac{t}{\lambda}\right)^{2s} d\rho'(t),$$

and to take into account that for all t in the segment $[0, \lambda - 2|\Delta\lambda|]$

$$\left[\left(1-\frac{t}{\lambda+\Delta\lambda}\right)^s - \left(1-\frac{t}{\lambda}\right)^s\right] = O\left(\frac{|\Delta\lambda|^s}{\lambda^{s+1}}\right),$$

for all t in the segment $[\lambda - 2|\Delta\lambda|, \lambda + \Delta\lambda]$

$$\left(1-\frac{t}{\lambda+\Delta\lambda}\right)^s = O\left(\frac{|\Delta\lambda|^s}{\lambda^s}\right),$$

and, finally for all t in the segment $[\lambda - 2|\Delta\lambda|, \lambda]$

$$\left(1-\frac{t}{\lambda}\right)^s = O\left(\frac{|\Delta\lambda|^s}{\lambda^s}\right).$$

This completes the proof of continuity in λ on each segment $[(n-1)^2, n^2]$ for $n > 1$ for the function between the braces in (3.2.41).

Applying the mean-value formula to each integral under the sign of summation on the right-hand side of (3.2.41), we come to the conclusion that there is a value of

$\bar{\lambda}_n$ subject to $(n-1)^2 \leq \bar{\lambda}_n \leq n^2$ such that

$$\int\limits_{(n-1)^2}^{n^2} \lambda^s \left\{ \int\limits_{G_R} \left[\overset{s}{\sigma}{}'_\lambda(x,f) - \overset{s}{\sigma}{}''_\lambda(x,f) \right]^2 dx \right\} \frac{d\lambda}{\sqrt{\lambda}}$$

$$= \bar{\lambda}_n^s \int\limits_{G_R} \left[\overset{s}{\sigma}{}'_{\bar{\lambda}_n}(x,f) - \overset{s}{\sigma}{}''_{\bar{\lambda}_n}(x,f) \right]^2 dx \int\limits_{(n-1)^2}^{n^2} \frac{d\lambda}{\sqrt{\lambda}} \tag{3.2.43}$$

$$= 2 \int\limits_{G_R} \left[\bar{\lambda}_n^{s/2} \left[\overset{s}{\sigma}{}'_{\bar{\lambda}_n}(x,f) - \overset{s}{\sigma}{}''_{\bar{\lambda}_n}(x,f) \right] \right\}^2 dx.$$

From (3.2.9), (3.2.41), and (3.2.43) we infer that

$$\sum_{i=2}^{\infty} \int\limits_{G_R} \left[\bar{\lambda}_n^{s/2} \left(\overset{s}{\sigma}{}'_{\bar{\lambda}_n}(x,f) - \overset{s}{\sigma}{}''_{\bar{\lambda}_n}(x,f) \right) \right]^2 dx < \infty.$$

It follows from the last inequality and from the familiar theorem of B. Levi that the series

$$\sum_{n=2}^{\infty} \left[\bar{\lambda}_n^{s/2} \left(\overset{s}{\sigma}{}'_{\bar{\lambda}_n}(x,f) - \overset{s}{\sigma}{}''_{\bar{\lambda}_n}(x,f) \right) \right]^2 \tag{3.2.44}$$

is convergent almost everywhere on the set G_R and, in view of $R > 0$ being arbitrary, also almost everywhere in the domain G. The validity of estimate (3.2.40) for almost all points x in the domain G is thus a sufficient condition for the convergence of series (3.2.44) almost everywhere in domain G.

This completes the proof of Lemma 3.7.

3.2.3. Direct Proof of Theorem 3.4

To begin, we give a proof of the first assertion of the theorem on the validity of estimate (3.2.3) for $s > 0$, for all sufficiently large λ, and for almost all points x of domain G.

We make use of the known representation of the Riesz means of order s via the

Riesz means of lower order δ[19)]

$$\lambda^s \overset{s}{\sigma}_\lambda(x,f) = C(s,\delta) \int_0^\lambda (\lambda - t)^{s-\delta-1} t^\delta \overset{\delta}{\sigma}_t(x,f)\,dt, \tag{3.2.45}$$

where $C(s,\delta) = \Gamma(s+1)[\Gamma(s-\delta)\Gamma(\delta+1)]^{-1}$, $s > \delta > -1$.

Representation (3.2.45) implies, in particular, that for $s > 1/2$ the representation

$$\frac{d}{d\lambda}\left[\lambda^s \overset{s}{\sigma}_\lambda(x,f)\right] = \lambda^{s-1} \overset{s-1}{\sigma}_\lambda(x,f) \tag{3.2.46}$$

holds.

Let λ and $\bar\lambda$ be two arbitrary values of the variable λ such that $\lambda \geq \bar\lambda \geq 1$ and $\lambda - \bar\lambda \leq C\sqrt{\bar\lambda}$ for a certain fixed constant $C \geq 1$.

By virtue of Lemma 3.7, to prove estimate (3.2.3) for almost all points x of the set G_R, it suffices to show that for the two arbitrary values λ and $\bar\lambda$ the estimate[20)]

$$\lambda^s\left[\overset{s}{\sigma}{}'_\lambda(x,f) - \overset{s}{\sigma}{}''_\lambda(x,f)\right] - \bar\lambda^s\left[\overset{s}{\sigma}{}'_{\bar\lambda}(x,f) - \overset{s}{\sigma}{}''_{\bar\lambda}(x,f)\right] = o(\lambda^{s/2}) \tag{3.2.47}$$

holds almost everywhere in G_R.

In establishing estimate (3.2.47), we consider in turn two cases: (a) $s > 1/2$; (b) $0 < s \leq 1/2$.

In the case $s > 1/2$, we integrate the difference of relations (3.2.46), written successively for extensions $\widehat{A}'$ and $\widehat{A}''$, with respect to the variable λ between the limits from $\bar\lambda$ to λ, where $\lambda \geq \bar\lambda$, $\lambda - \bar\lambda \leq C\sqrt{\bar\lambda}$. The result is

$$\lambda^s\left[\overset{s}{\sigma}{}'_\lambda(x,f) - \overset{s}{\sigma}{}''_\lambda(x,f)\right] - \bar\lambda^s\left[\overset{s}{\sigma}{}'_{\bar\lambda}(x,f) - \overset{s}{\sigma}{}''_{\bar\lambda}(x,f)\right]$$

$$= \int_{\bar\lambda}^{\lambda} t^{s-1}\left[\overset{s-1}{\sigma}{}'_t(x,f) - \overset{s-1}{\sigma}{}''_t(x,f)\right] dt.$$

Operating in terms of absolute values and applying the Cauchy–Buniakowski inequal-

---

19) This relation is easily checked: it suffices to substitute $t^\delta \overset{\delta}{\sigma}_t = \int_0^\beta (t-\tau)^\delta S_\tau\,d\tau$ into its right-hand side and to reverse the order of integration with respect to t and τ, keeping in mind that $\lambda^s \overset{s}{\sigma}_\lambda \int_0^\lambda (\lambda - \tau)^s S_\tau\,d\tau$.

20) Indeed, if the estimate (3.2.47) holds for the two arbitrary values of λ and $\bar\lambda$, then for any $\lambda \geq 2$ there is an $n \geq 2$ such that $n^2 \leq \lambda \leq (n+1)^2$, and for this n there is a $\bar\lambda_n$ such that $(n-1)^2 \leq \bar\lambda \leq n^2$, implying thereby the validity of estimate (3.2.40) (note that $\lambda \geq \bar\lambda_n \geq 1$ and $\lambda - \bar\lambda \leq C\sqrt{\bar\lambda}$). A comparison of (3.2.40) with (3.2.47) taken at $\bar\lambda = \bar\lambda_n$ allows us to establish the validity of estimate (3.2.3).

ity, we obtain

$$\left|\lambda^s\left[\overset{s}{\sigma}{}'_\lambda(x,f)-\overset{s}{\sigma}{}''_\lambda(x,f)\right]-\bar\lambda^s\left[\overset{s}{\sigma}{}'_{\bar\lambda}(x,f)-\overset{s}{\sigma}{}''_{\bar\lambda}(x,f)\right]\right|$$

$$\leq\left\{\int_{\bar\lambda}^{\lambda}t^{s-3/2}\left[\overset{s-1}{\sigma}{}'_t(x,f)-\overset{s-1}{\sigma}{}''_t(x,f)\right]^2dt\right\}^{1/2}\left\{\int_{\bar\lambda}^{\lambda}t^{s-1/2}dt\right\}^{1/2}. \tag{3.2.48}$$

One observes that, by virtue of the inequality $\lambda-\bar\lambda\leq C\sqrt{\lambda}$, the estimate

$$\left\{\int_{\bar\lambda}^{\lambda}t^{s-1/2}dt\right\}^{1/2}=O(\lambda^{s/2}) \tag{3.2.49}$$

holds, whereas the assumption $s>1/2$ and the convergence of integral (3.2.9) for any $s>-1/2$ imply that

$$\left\{\int_{\bar\lambda}^{\lambda}t^{s-3/2}\left[\overset{s-1}{\sigma}{}'_t(x,f)-\overset{s-1}{\sigma}{}''_t(x,f)\right]^2dt\right\}^{1/2}=o(1) \tag{3.2.50}$$

holds almost everywhere on G_R.

Relations (3.2.48)–(3.2.50) imply that the estimate (3.2.47) holds for $s>1/2$.

To prove the estimate (3.2.3), we merely have to establish the validity of estimate (3.2.47) for $0<s<1/2$.

Writing relation (3.2.45) for the extensions $\widehat{A}'$ and $\widehat{A}''$ and for the quantities λ and $\bar\lambda$ and assuming $\delta=s-1/2$, we obtain

$$\left[\lambda^s\left(\overset{s}{\sigma}{}'_\lambda(x,f)-\overset{s}{\sigma}{}''_\lambda(x,f)\right)\right]-\left[\bar\lambda^s\left(\overset{s}{\sigma}{}'_{\bar\lambda}(x,f)-\overset{s}{\sigma}{}''_{\bar\lambda}(x,f)\right)\right]$$

$$=C\left(s,s-\frac{1}{2}\right)\left\{\int_0^{\lambda}(\lambda-t)^{-1/2}t^{s-1/2}\left[\overset{s-1/2}{\sigma}{}'_t(x,f)-\overset{s-1/2}{\sigma}{}''_t(x,f)\right]dt\right. \tag{3.2.51}$$

$$\left.-\int_0^{\bar\lambda}(\bar\lambda-t)^{-1/2}t^{s-1/2}\left[\overset{s-1/2}{\sigma}{}'_t(x,f)-\overset{s-1/2}{\sigma}{}''_t(x,f)\right]dt\right\}.$$

Considering that $0\leq\lambda-\bar\lambda\leq C\sqrt{\lambda}$, the equality (3.2.51) is rewritten in the following manner:

$$\left[\lambda^s\left(\overset{s}{\sigma}{}'_\lambda(x,f)-\overset{s}{\sigma}{}''_\lambda(x,f)\right)\right]-\left[\bar\lambda^s\left(\overset{s}{\sigma}{}'_{\bar\lambda}(x,f)-\overset{s}{\sigma}{}''_{\bar\lambda}(x,f)\right)\right]$$

$$=C\left(s,s-\tfrac{1}{2}\right)[I_1+I_2+I_3], \tag{3.2.52}$$

where

$$I_1 = \int\limits_0^{\lambda-2(\lambda-\bar\lambda)} \left[(\lambda-t)^{-1/2} - (\bar\lambda - t)^{-1/2}\right] t^{s-1/2} \left[{}^{s-1/2}\sigma'_t(x,f) - {}^{s-1/2}\sigma''_t(x,f)\right] dt, \tag{3.2.53}$$

$$I_2 = \int\limits_{\lambda-2(\lambda-\bar\lambda)}^{\lambda} (\lambda-t)^{-1/2} t^{s-1/2} \left[{}^{s-1/2}\sigma'_t(x,f) - {}^{s-1/2}\sigma''_t(x,f)\right] dt, \tag{3.2.54}$$

$$I_3 = \int\limits_{\lambda-2(\lambda-\bar\lambda)}^{\bar\lambda} (\bar\lambda-t)^{-1/2} t^{s-1/2} \left[{}^{s-1/2}\sigma'_t(x,f) - {}^{s-1/2}\sigma''_t(x,f)\right] dt. \tag{3.2.55}$$

It suffices to show that almost everywhere on the set G_R each of the integrals I_1, I_2, and I_3 equals $o(\lambda^{s/2})$.

With the aid of the Cauchy–Buniakowski inequality and the trivial inequality

$$[(\lambda-t)^{-1/2} - (\bar\lambda-t)^{-1/2}]^2 \le (\bar\lambda-t)^{-1} - (\lambda-t)^{-1},$$

holding for all $0 \le t \le \bar\lambda \le \lambda$, we obtain

$$|I_1| \le \left\{ \int\limits_0^{\lambda-2(\lambda-\bar\lambda)} [(\bar\lambda-t)^{-1} - (\lambda-t)^{-1}] dt \right\}^{1/2} \times \left\{ \int\limits_0^{\lambda} t^{2s-1} \left[{}^{s-1/2}\sigma'_t(x,f) - {}^{s-1/2}\sigma''_t(x,f)\right]^2 dt \right\}^{1/2}. \tag{3.2.56}$$

Next, by virtue of the condition $0 \le \lambda - \bar\lambda \le C\sqrt{\lambda}$, we obtain

$$\int\limits_0^{\lambda-2(\lambda-\bar\lambda)} [(\bar\lambda-t)^{-1} - (\lambda-t)^{-1}] dt = \left\{ \ln \frac{\lambda-t}{\bar\lambda-t} \right\} \Bigg|_{t=0}^{t=\lambda-2(\lambda-\bar\lambda)} = \ln 2 + \ln \frac{\bar\lambda}{\lambda} = O(1). \tag{3.2.57}$$

We partition the latter integral on the right-hand side of (3.2.56) into the sum of two integrals:

$$\int_0^\lambda t^{2s-1}\left[{}^{s-1/2}\sigma'_t(x,f) - {}^{s-1/2}\sigma''_t(x,f)\right]^2 dt$$

$$= \int_0^{\sqrt{\lambda}} t^{2s-1}\left[{}^{s-1/2}\sigma'_t(x,f) - {}^{s-1/2}\sigma''_t(x,f)\right]^2 dt$$

$$+ \int_{\sqrt{\lambda}}^{\lambda} t^{2s-1}\left[{}^{s-1/2}\sigma'_t(x,f) - {}^{s-1/2}\sigma''_t(x,f)\right]^2 dt.$$

We infer from the last equality that

$$\int_0^\lambda t^{2s-1}\left[{}^{s-1/2}\sigma'_t(x,f) - {}^{s-1/2}\sigma''_t(x,f)\right]^2 dt$$

$$\leq (\sqrt{\lambda})^s \int_0^{\sqrt{\lambda}} t^{s-1}\left[{}^{s-1/2}\sigma'_t(x,f) - {}^{s-1/2}\sigma''_t(x,f)\right]^2 dt$$

$$+ \lambda^s \int_{\sqrt{\lambda}}^{\lambda} t^{s-1}\left[{}^{s-1/2}\sigma'_t(x,f) - {}^{s-1/2}\sigma''_t(x,f)\right]^2 dt,$$

and, therefore, by virtue of estimate (3.2.9) almost everywhere on the set G_R we have

$$\int_0^\lambda t^{2s-1}\left[{}^{s-1/2}\sigma'_t(x,f) - {}^{s-1/2}\sigma''_t(x,f)\right]^2 dt = o(\lambda^s). \tag{3.2.58}$$

It follows from relations (3.2.56)–(3.2.58) that

$$I_1 = o(\lambda^{s/2})$$

is valid for almost all points x in G_R.

It remains to show that each of the integrals (3.2.54) and (3.2.55) equals $o(\lambda^{s/2})$ almost everywhere on the set G_R. We restrict ourselves to consideration of the integral (3.2.54); the same reasoning applies equally to the integral (3.2.55).

Taking into account that $s > 0$ and replacing s by $s - 1/2$ and δ by $s/2 - 1/2$ in formula (3.2.45) for the extensions $\widehat{A}'$ and $\widehat{A}''$, we obtain from (3.2.45)

$$t^{s-1/2}\left[{}^{s-1/2}\sigma'_t(x,f) - {}^{s-1/2}\sigma''_t(x,f)\right]$$

$$= C\left(s-\frac{1}{2},\frac{s}{2}-\frac{1}{2}\right)\int\limits_0^t (t-u)^{s/2-1}u^{s/2-1/2}\left[{}^{s/2-1/2}\sigma'_u(x,f) - {}^{s/2-1/2}\sigma''_u(x,f)\right]du. \tag{3.2.59}$$

Now, we make use of the fact that for the so-called fractional integral

$$f_\alpha(t) = \int\limits_0^t (t-u)^{\alpha-1}f(u)du$$

for any $0 < \alpha < 1/2$, $1/p = 1/2 - \alpha$ and for any function $f(x)$ in the class $L_2[0,\lambda]$ the following estimate holds:

$$\|f_\alpha\|_{L_p[0,\lambda]} \le C_\alpha \|f\|_{L_2[0,\lambda]} \tag{3.2.60}$$

with a constant C_α dependent on α only.[21]

Applying the estimate (3.2.60) taken at $\alpha = s/2$ to the fractional integral on the right-hand side of (3.2.59),[22] we obtain the inequality

$$\left\| t^{s-1/2}\left[{}^{s-1/2}\sigma'_t(x,f) - {}^{s-1/2}\sigma''_t(x,f)\right]\right\|_{L_p[0,\lambda]}$$

$$\le C\left\{\int\limits_0^\lambda u^{s-1}\left[{}^{s/2-1/2}\sigma'_u(x,f) - {}^{s/2-1/2}\sigma''_u(x,f)\right]^2 du\right\}^{1/2}, \tag{3.2.61}$$

where $1/p = (1-s)/2$.

Extending the argumentation that we have applied to the latter integral on the right-hand side of (3.2.56) to the integral on the right-hand side of (3.2.61), we obtain the relation

$$\left\| t^{s-1/2}\left[{}^{s-1/2}\sigma'_t(x,f) - {}^{s-1/2}\sigma''_t(x,f)\right]\right\|_{L_p[0,\lambda]} = o(\lambda^{s/4}), \tag{3.2.62}$$

which holds for almost all points x in the set G_R at $1/p = (1-s)/2$.

Now, to estimate integral (3.2.54), it suffices to apply the Hölder inequality to it:

$$|I_2| \le \left\{\int\limits_{\lambda-2(\lambda-\bar{\lambda})}^{\lambda} (\lambda-t)^{-q/2}dt\right\}^{1/q} \left\| t^{s-1/2}\left[{}^{s-1/2}\sigma'_t(x,f) - {}^{s-1/2}\sigma''_t(x,f)\right]\right\|_{L_p[0,\lambda]}$$

---

21) See, for example, the monograph of G. G. Hardy, J. E. Littlewood, and G. Polya [23, p. 348].

22) Since $0 < s \le 1/2$, one has $\alpha \le 1/4$, and the condition $\alpha < 1/2$ is thus satisfied.

at $1/q = (1+s)/2$, $1/p = (1-s)/2$, and to make use of estimate (3.2.62) and of the simple estimate

$$\left\{ \int_{\lambda-2(\lambda-\bar\lambda)}^{\lambda} (\lambda-t)^{-q/2} dt \right\}^{1/q} = \left\{ \int_0^{2(\lambda-\bar\lambda)} u^{-1/(1+s)} du \right\}^{(1+s)/2}$$

$$\leq \left\{ \int_0^{2C\sqrt{\lambda}} u^{-1/(1+s)} du \right\}^{(1+s)/2} = O(\lambda^{s/4}).$$

We obtain that $I_2 = o(\lambda^{s/2})$ for almost all points x in the set G_R.

Insofar as the same reasoning applies to integral (3.2.55), the proof of estimate (3.2.3) for any $s > 0$ is thus complete.

What we have left to do is to prove the second assertion of Theorem 3.4 stating that for an arbitrary function $f(x)$ in $L_2(G)$ becoming zero almost everywhere in a certain domain D interior to G, the estimates (3.2.3′) hold for $s > 0$ and for all sufficiently large $\lambda > 0$ almost everywhere in D.

For the sake of definiteness we shall focus on the first estimate (3.2.3′), that is, we shall consider the extension $\widehat{A}'$.

For an arbitrary $R > 0$ we fix an arbitrary point x in the set D_R and apply to this point x the same argumentation that has been used in proving Lemma 3.3.

In doing so, we arrive at relation (3.2.17) in which now two terms have become zero: the term on the left-hand side and the first term on the right-hand side. In other words, we obtain that the relation

$$\overset{s}{\sigma}{}'_\lambda(x,f) = 2^s \Gamma(s+1) \lambda^{-s/2} \sum_{i=1}^{m'} \int_0^\infty \widehat{f}'_i(t) u'_i(x,t) I_t^\lambda(R) d\rho'(t)$$

holds for any point x in the set D_R. If so, following closely the line of reasoning of Lemma 3.3, we obtain that the estimate

$$\int_{D_R} \int_0^\infty \lambda^{s-1/2} \left[\overset{s}{\sigma}{}'_\lambda(x,f)\right]^2 dx d\lambda \leq C \|f\|_{L_2(G)} \tag{3.2.63}$$

is valid for all $s > -1/2$.

Based on estimate (3.2.63) and invoking the arguments that have been used in proving Lemma 3.7, we establish that for $s > 0$ for the function of interest there is a numerical sequence $\{\bar\lambda_n\}$ such that for all numbers n, beginning with a certain n, the condition $(n-1)^2 \leq \bar\lambda_n \leq n^2$ holds, and for almost all points x in domain D the estimate

$$\left|\overset{s}{\sigma}{}'_{\bar\lambda_n}(x,f)\right| = o(\bar\lambda_n^{-s/2}) \tag{3.2.64}$$

is valid.

With reference to estimate (3.2.64) and following closely the line of reasoning adopted in proving the first assertion of Theorem 3.4, we come to the validity of the first estimate (3.2.3′) for almost all points in domain D. No special proof is needed to establish the validity of the second estimate (3.2.3′).

The proof of Theorem 3.4 is thus complete.

COMMENTS ON CHAPTER 3

The main results of Section 3.1 were first reported by Il'in [33], [34], [35].

The results of Section 3.2 concerned with the generalized principle of localization for the Riesz means of spectral decompositions were reported by Il'in in his earlier works [31], [32].

The results on the generalized approach to the problem of equiconvergence of the Riesz means (Section 3.2) were reported in 1988 in a paper by Il'in [40].

The generalized principle of localization of the Riesz means was extended to the Laplace–Beltrami operator on a harmonic Riemannian manifold in the papers of Il'yushina [48], [49] in 1971.

Chapter 4

Self-Adjoint Nonnegative Extensions of an Elliptic Operator of Second Order

Our intention in this chapter is to show that the theorems on the exact conditions of uniform convergence and localization of spectral decompositions that have been established by us in Chapter 2 for an arbitrary self-adjoint nonnegative extension of the Laplace operator remain valid also for arbitrary self-adjoint nonnegative extensions of a general elliptic operator of second order L.

The formalism that we employ in the present chapter resembles that expounded in Chapter 2.

By way of example, in proving a negative-type theorem for the Riesz means of a "small" nonnegative order s, the technique differs in essence from the case of the Laplace operator in the use, instead of the mean-value formula for a regular solution of the equation $\Delta u + \lambda u = O$, of the mean-value formula for a regular solution of the elliptic equation wit spectral parameter $-Lu + \lambda \rho_0(x) u = O$ in the specific form established by E. I. Moiseev. This necessitates the study of an additional summand (which becomes zero in the case of the Laplace operator) in the expression for the Riesz means.

The occurrence of a strictly positive weighting factor $\rho(x)$ in the spectral parameter equation allows us to consider the elliptic operator on an arbitrary Riemannian manifold. In particular, our scheme involves the Beltrami–Laplace operator on a closed manifold in an arbitrary (not necessarily symmetric or harmonic) Riemannian space.

The present chapter is composed of three sections. In Section 4.1, ancillary (essentially self-contained) propositions are established, bearing upon the fundamental functions generated by an arbitrary ordered spectral representation of the space L_2 with weight $\rho_0(x)$ with respect to an arbitrary self-adjoint nonnegative extension of the elliptic operator with the same weight. In Section 4.2, we establish a negative-type theorem for each individual self-adjoint nonnegative extension of an elliptic operator of second order L with the aforementioned weight; in Section 4.3, a main positive-type theorem is proved.

4.1. ANCILLARY PROPOSITIONS ABOUT FUNDAMENTAL FUNCTIONS

4.1.1. Ordered Weighted Spectral Representations of the Space L_2

Assume that G is an arbitrary domain in the space E_N and L is an arbitrary elliptic, formally self-adjoint nonnegative operator of second order defined in domain G as

$$Lu = -\sum_{i,j=1}^{N} \frac{\partial}{\partial x_i}\left[a_{ij}(x)\frac{\partial u}{\partial x_j}\right] + c(x)u. \tag{4.1.1}$$

We assume that the coefficients $a_{ij}(x)$ and $c(x)$ in (4.1.1) are definite and infinitely differentiable in domain G, and the coefficients $a_{ij}(x)$ satisfy the condition of uniform ellipticity in this domain, that is, the relations

$$a_{ij}(x) = a_{ji}(x),$$

$$\sum_{i,j=1}^{N} a_{ij}(x)\xi_i\xi_j \geq \alpha \sum \xi_i^2 \qquad \text{for} \qquad \alpha > 0$$

hold for all points x of domain G and for all real numbers $\xi_1, \xi_2, \ldots, \xi_N$, the coefficient $c(x)$ being nonnegative in this domain.

Further, assume that $\rho_0(x)$ is an arbitrary function, bounded from below by a constant and infinitely differentiable in G; this function will be referred to as the weight.

Finally, assume that $\widehat{A}$ is an arbitrary self-adjoint nonnegative extension of operator (4.1.1) in domain G with weight $\rho_0(x)$ which is defined in the same manner as in Section 2.1.1, however, with the weighting factor $\rho_0(x)$ introduced in all scalar products.[1)]

In what follows we shall make use of the symbol $L_2(G, \rho_0(x))$ to denote the space L_2 in which the norm is introduced with the weighting factor $\rho_0(x)$ added. By

---

[1)] That is, this time the scalar product is defined as

$$(u, v) = \int_G u(x)v(x)\rho_0(x)\,dx.$$

analogy with what was said in Section 2.1.2, we introduce the concept of an ordered spectral representation of the space $L_2(G, \rho_0(x))$ with respect to an arbitrary self-adjoint nonnegative extension $\widehat{A}$ of the operator (4.1.1) with weight $\rho_0(x)$, and we establish the Gårding–Browder–Mautner theorem. In our case, the involvement of the infinitely differentiable weighting factor $\rho_0(x)$, bounded from below by a positive constant, requires no change in the proof of this theorem as outlined in the familiar monograph of Dunford and Schwartz [16, pp. 875–876].

We reformulate the Gårding–Browder–Mautner theorem in a more convenient form.

For each self-adjoint nonnegative extension $\widehat{A}$ of the elliptic operator (4.1.1) in domain G with weight $\rho_0(x)$, there exists at least one ordered representation of the space $L_2(G, \rho_0(x))$ with a spectral measure $\rho(\lambda)$, with sets of multiplicity e_i, with fundamental functions $u_i(x, \lambda)$ $(i = 1, 2, \dots, \widehat{m})$ of multiplicity $\widehat{m} \le \infty$ such that the following requirements are fulfilled:

(1) the fundamental functions $u_i(x, \lambda)$ are measurable with respect to the product of the Lebesgue measure of domain G with weight $\rho_0(x)$ and the spectral measure $\rho(\lambda)$, they become zero on the complements of the sets e_i, they belong to the class $C^\infty(G)$ for each fixed $\lambda \ge O$, and they satisfy, inside G, the elliptic differential equation

$$Lu_i(x, \lambda) = \lambda \rho_0(x, \lambda) u_i(x, \lambda); \tag{4.1.2}$$

(2) the Fourier image

$$\widehat{f}_i(\lambda) = \int_G f(y) u_i(y, \lambda) \rho_0(y) dy \qquad (i = 1, 2, \dots, \widehat{m}) \tag{4.1.3}$$

is defined on the set e_i for each function $f(x)$ in the class $L_2(G, \rho_0(x))$ as an element of the space L_2 with respect to the variable λ with measure $\rho(\lambda)$ such that the spectral decomposition $E_\lambda f(x)$ of each function $f(x)$ in the class $L_2(G, \rho_0(x))$ takes the form

$$E_\lambda f(x) = \sum_{i=1}^{\widehat{m}} \int_0^\lambda \widehat{f}_i(\lambda) u_i(x, t) d\rho(t) \tag{4.1.4}$$

and converges to $f(x)$ in the metric $L_2(G, \rho_0(x))$ as $\lambda \to \infty$;

(3) for any two functions $f(x)$ and $g(x)$ in the $L_2(G, \rho_0(x))$ class the Parseval equality holds,

$$\int_G f(x) g(x) \rho_0(x) dx = \sum_{i=1}^{\widehat{m}} \int_0^\infty \widehat{f}_i(t) \widehat{g}_i(t) d\rho(t), \tag{4.1.5}$$

and, in particular, for any function $f(x)$ in the class $L_2(G, \rho_0(x))$,

$$\int_G f^2(x) \rho_0(x)\, dx = \sum_{i=1}^{\widehat{m}} \int_0^\infty \widehat{f}_i^2(t) d\rho(t). \tag{4.1.6}$$

4.1.2. Moiseev Mean-Value Formula

Each fundamental function $u_i(x,\lambda)$, for any fixed $\lambda \geq 0$, is, in compliance with the Garding-Browder-Mautner theorem, a solution of Eq. (4.1.2), regular in domain G, with the elliptic operator L defined by (4.1.1) and with the infinitely differentiable and strictly positive weight function $\rho_0(x)$.

In this section, we establish the mean-value formula for any regular solution $u_i(x,\lambda)$ of Eq. (4.1.2) following the line of reasoning by Moiseev [62].

Assume that the function $u(x,\lambda)$ is a solution of the equation

$$-Lu + \lambda\rho_0(x)u = 0,$$

regular in domain G, for a certain fixed $\lambda \geq 0$ on the condition that Lu is an operator of the form (4.1.1), uniformly elliptic in domain G, with coefficients $a_{ij}(x)$ and $c(x)$ infinitely differentiable in domain G such that $c(x) \geq 0$ everywhere in G, and the weight function $\rho_0(x)$ is bounded from below by a positive constant and is infinitely differentiable in domain G.

We fix an arbitrary point x of the open domain G and introduce, within a sufficiently small vicinity of this point, a nonorthogonal (generally speaking) system of coordinates $r, \varphi_1, \varphi_2, \dots, \varphi_{N-1}$; here r is understood to mean the geodetic distance between a variable point y and a fixed point x measured along the rays of the equation

$$\sum_{i,j=1}^{N} a_{ij}(x)\frac{\partial r}{\partial x_i}\frac{\partial r}{\partial x_j} = \rho_0(x), \tag{4.1.7}$$

and $\varphi_1, \varphi_2, \dots, \varphi_{N-1}$ are "angular" coordinates varied within the respective ranges $0 \leq \varphi_1 \leq \pi, \dots, 0 \leq \varphi_{N-2} \leq \pi$, $0 \leq \varphi_{N-1} \leq 2\pi$.

The element of volume in the adopted system of coordinates is written as

$$dy = J(x, r, \varphi_1, \dots, \varphi_{N-1})dr\, d\varphi_1 \dots d\varphi_{N-1},$$

where $J(x, r, \varphi_1, \dots, \varphi_{N-1})$ is used to denote a Jacobian.

For an arbitrary function f, defined in a vicinity of the point x, and for a sufficiently small $r \geq 0$ we denote the integral

$$r^{1-N}\int_0^{\pi}\dots\int_0^{\pi}\int_0^{2\pi} f(r,\varphi_1,\dots,\varphi_{N-1})\, J(x,r,\varphi_1,\dots,\varphi_{N-1})\, d\varphi_1\dots, d\varphi_{N-1}$$

by the symbol

$$\int_{\omega} f(x + r\omega)\, d\omega,$$

that is, we set by definition that

$$\int_{\omega} f(x+r\omega)\,d\omega$$

$$= r^{1-N}\int_0^{\pi}\dots\int_0^{\pi}\int_0^{2\pi} f(r,\varphi_1,\dots,\varphi_{N-1})\,J(x,r,\varphi_1,\dots,\varphi_{N-1})\,d\varphi_1\dots,d\varphi_{N-1}. \tag{4.1.8}$$

It follows immediately from (4.1.8) that, given a sufficiently small $R > 0$,

$$\int_0^R r^{N-1}\left(\int_{\omega} f(x+r\omega)d\omega\right)dr = \int_{r_{xy}\le R} f(y)dy, \tag{4.1.9}$$

where r_{xy} is the geodetic distance between points x and y as measured along the rays of Eq. (4.1.7).

The main result of E. I. Moiseev is his proof of the existence of two smooth functions $p(x,y)$ and $q(x,y)$ which are defined for all y within a sufficiently small vicinity of the point x and which satisfy the conditions $0 < a \le p(x,y) \le b < \infty$, $-\infty < a_1 \le q(x,y) \le b_1 < \infty$. These two functions are such that they guarantee, for any regular solution $u(x,\lambda)$ of the equation $-Lu + \lambda\rho_0(x)u = 0$ with an arbitrary $\lambda \ge 0$ and any sufficiently small $r > 0$, the validity of the equality

$$\int_{\omega} u(x+r\omega,\lambda)\rho_0(x+r\omega)p(x,x+r\omega)d\omega$$

$$= (2\pi)^{N/2}u(x,\lambda)\rho_0(x)(r\sqrt{\lambda})^{-\nu}J_\nu(r\sqrt{\lambda}) \tag{4.1.10}$$

$$+\rho_0(x)r^{-\nu}\int_0^r w_\nu(\tau\sqrt{\lambda},r\sqrt{\lambda})\tau^{\nu}\left(\int_{\omega} u(x+\tau\omega,\lambda)q(x,x+\tau\omega)d\omega\right)d\tau,$$

where $\nu = (N-2)/2$, $w_\nu(a,b) = J_\nu(a)Y_\nu(b) - J_\nu(b)Y_\nu(a)$, $J_\nu(x)$ being a Bessel function and $Y_\nu(x)$ a Neyman function.

Equality (4.1.10) will be referred to as the *Moiseev mean-value formula*.

Specifically, when an elliptic operator of second order L is a Laplace operator ($L = -\Delta$), the function $q(x,y)$ in (4.1.10) becomes identically zero, and the second term on the right-hand side of (4.1.10) vanishes.

One will meet with little difficulty in ensuring that the function $p(x,y)$ in (4.1.10) uniformly with respect to x on each compact set of domain G satisfies, for $\omega_N =$

$2(\pi)^{N/2}\left[\Gamma(N/2)\right]^{-1}$, the following relation:

$$\frac{1}{\omega_N}\int\limits_{\omega} p(x+r\omega, x)d\omega = 1 + O(r),$$

which holds for all sufficiently small $r > 0$.

Indeed, since the derivative of the function $p(x,y)$ with respect to each variable is uniformly bounded on each compact set of domain $G \times G$, it suffices to show that the relation

$$\frac{1}{\omega_N}\int\limits_{\omega} p(x, x+r\omega)d\omega = 1 + O(r)$$

holds uniformly with respect to x for all sufficiently small $r > 0$ on each compact set of domain G.

This is easily done by applying the mean-value formula

$$\int\limits_{\omega} u(x+r\omega,\lambda)\rho_0(x+r\omega)p(x,x+r\omega)d\omega$$
$$= u(x+r\omega^*,\lambda)\rho_0(x+r\omega^*)\int\limits_{\omega} p(x,x+r\omega)d\omega$$

to the integral on the left-hand side of (4.1.10) and then passing to the limit in (4.1.10) as $r \to 0+0$. .

We obtain in such a manner that

$$\lim_{r\to 0+0}\frac{1}{\omega_N}\int\limits_{\omega} p(x, x+r\omega)d\omega = 1.$$

Next, the relation $\frac{1}{\omega_N}\int\limits_{\omega} p(x,x+r\omega)d\omega = 1 + O(r)$ is derived by applying the Lagrange formula with allowance made for the boundedness (uniform on any compact set) of the derivative $\partial/\partial r\, p(x, x + +r\omega)$.

The mean-value formula (4.1.10) enables one to obtain a specialized representation for the Fourier image (4.1.3) of a function $F(x,y)$ which is the product of the function $p(x,y)$ and an arbitrary function f dependent only on the geodetic distance $r = r_{xy}$ between a variable point y and a fixed point x and distinct from zero only for sufficiently small r_{xy}.

Assume that x is an arbitrary fixed point in the open domain G, R is a positive sufficiently small number (in all cases, smaller than the geodetic distance between the point x and the boundary of domain G), $r = r_{xy}$ is the geodetic distance between a variable point y and a fixed point x, $f(x)$ is an arbitrary function becoming zero at $r \geq R$; the function $F(x,y)$ takes the form

$$F(x,y) = \begin{cases} p(x,y)f(r_{xy}) & \text{for} \quad r \leq R, \\ 0 & \text{for} \quad r > R. \end{cases} \tag{4.1.11}$$

The Fourier image $\widehat{F}_i(x,\lambda)$ of function (4.1.11) corresponding to an arbitrary ordered spectral representation of the space $L_2(G,\rho_0(x))$ with respect to the self-adjoint nonnegative extension $\widehat{A}$ of the elliptic operator L with weight $\rho_0(x)$ takes the form

$$\begin{aligned} \widehat{F}_i(x,\lambda) &= (2\pi)^{N/2}\rho_0(x)u_i(x,\lambda)\lambda^{(2-N)/4} \\ &\quad \times \int_0^R f(r)r^{N/2}J_{(N-2)/2}(r\sqrt{\lambda})dr + \widetilde{F}_i(x,\lambda) \\ &\quad (i = 1,2,\dots,\widehat{m}), \end{aligned} \tag{4.1.12}$$

where the "additional" term $\widetilde{F}_i(x,\lambda)$ is defined as

$$\begin{aligned} \widetilde{F}_i(x,\lambda) &= \rho_0(x)\int_0^R r^{N/2}f(r) \\ &\times \left[\int_0^r \tau^{\nu}w_{\nu}(\tau\sqrt{\lambda},r\sqrt{\lambda})\left(\int_{\omega} u_i(x+\tau\omega,\lambda)q(x,x+\tau\omega)d\omega\right)d\tau\right]dr \\ &\quad (i = 1,2,\dots,\widehat{m}). \end{aligned} \tag{4.1.13}$$

Indeed, in order to obtain the Fourier image of function (4.1.11) through representations (4.1.12), (4.1.13), it suffices to observe that, in accordance with (4.1.3), the Fourier image $\widehat{F}_i(x,\lambda)$ is

$$\begin{aligned} \widehat{F}_i(x,\lambda) & \int_G F(x,y)u_i(y,\lambda)\rho_0(y)dy \\ &= \int_0^R f(r)r^{N-1}\left(\int_{\omega} u_i(x+r\omega,\lambda)\rho_0(x+r\omega)p(x,x+r\omega)d\omega\right)dr, \end{aligned}$$

and then to make use of the Moiseev mean-value formula (4.1.10).

Note that for the particular case of the Laplace operator, the additional term (4.1.13) in (4.1.12) becomes zero.

4.1.3. Estimation of the Integral of the Squares of Fundamental Functions

This section is mainly concerned with the establishment of a proposition which is a generalization of Lemma 2.1 (Section 2.3.2).

LEMMA 4.1. *For any real $\mu \geq 0$ uniformly with respect to x in each subdomain G' strictly interior to domain G, the following estimate holds:*

$$\sum_{i=1}^{\widehat{m}} \int\limits_{\mu \leq \sqrt{\lambda} \leq \mu+1} u_i^2(x,\lambda)\,d\rho(\lambda) = O\left[(\mu+1)^{N-1}\right]. \tag{4.1.14}$$

PROOF: 1^0. We start by establishing the "rough" estimate

$$\sum_{i=1}^{\widehat{m}} \int\limits_{\mu \leq \sqrt{\lambda} \leq \mu+1} u_i^2(x,\lambda)\,d\rho(\lambda) = O\left[(\mu+1)^{N+4}\right], \tag{4.1.15}$$

likewise valid for any $\mu \geq 0$ and uniform in x within each subdomain G' strictly interior to domain G.

Instead of the rough estimate (4.1.15), we shall establish a stronger result from which this estimate immediately follows: we wish to prove that for an arbitrary subdomain G' strictly interior to domain G, there exists a constant C such that for all x from G' the following relation holds:

$$\sum_{i=1}^{\widehat{m}} \int\limits_0^\infty \frac{u_i^2(x,\lambda)}{(1+\lambda)^{N/2+2}}\,d\rho(\lambda) \leq C. \tag{4.1.16}$$

Let L be the elliptic operator (4.1.1), and E be a unit operator. Now, we define the elliptic operator

$$\widetilde{L} = \frac{1}{\rho_0(x)} L + E. \tag{4.1.17}$$

We use the symbol $K_1(x,y)$ to denote the main fundamental solution for the operator $\widetilde{L}$,[2)] and the symbol $K_s(x,y)$ to denote the iteration of order s for the kernel $K_1(x,y)$.

We fix a subdomain G' strictly interior to domain G and an arbitrary point x in subdomain G'; next, we fix an R smaller in value than the distance between G' and the boundary of G, and define a "cutting" function $\eta(x,y)$ which is unity at $|x-y| \leq R/2$, zero at $|x-y| \geq R$, and which is infinitely differentiable over the entire E^N.

Further, we assume that $s = [N/4]+1$ (where $[N/4]$ denotes the integer part of $N/4$), and we consider the function $v(x,y) = K_s(x,y)\eta(x,y)$.

Since $s = [N/4]+1$, the iterated kernel $K_s(x,y)$ has a square-integrable singularity, and therefore for any fixed point x the function $[\rho_0(y)]^{-1}v(x,y)$ (as a function of y)

---

2) Concerning the definition and existence of the main fundamental solution for an arbitrary elliptic operator of second order, the reader is referred to the monograph of Miranda [60, p. 69].

belongs to the class $L(G, \rho_0(x,y))$, and the Parseval equality (see equality (4.1.6)) holds for this function:

$$\int_G v^2(x,y)[\rho_0(y)]^{-1}dy = \sum_{i=1}^{\widehat{m}} \int_0^\infty \left| \int_G v(x,y)u_i(y,\lambda)dy \right|^2 d\rho(\lambda). \tag{4.1.18}$$

We observe now that each fundamental function $u_i(y,\lambda)$ is a regular solution of the equation $\widetilde{L}u_i(y,\lambda) = (\lambda+1)u_i(y,\lambda)$. Therefore, applying successively the second Green's formula to a pair of functions, one of which is $u_i(y,\lambda)$ and the other of which is, respectively, $v(x,y)$, $\widetilde{L}v(x,y)$, ..., $\widetilde{L}^{s-1}v(x,y)$, we arrive at the relations[3)]

$$\int u_i(y,\lambda)\widetilde{L}v(x,y)dy = \int v(x,y)\widetilde{L}u_i(y,\lambda)dy$$
$$= (\lambda+1)\int v(x,y)u_i(y,\lambda)dy,$$

$$\int u_i(y,\lambda)\widetilde{L}^2v(x,y)dy = \int \widetilde{L}u_i(y,\lambda)\widetilde{L}v(x,y)dy$$
$$= (\lambda+1)\int u_i(y,\lambda)\widetilde{L}v(x,y)dy,$$

. . . .

$$\int u_i(y,\lambda)\widetilde{L}^{s-1}v(x,y)dy = \int \widetilde{L}u_i(y,\lambda)\widetilde{L}^{s-2}v(x,y)dy$$
$$= (\lambda+1)\int v(x,y)u_i(y,\lambda)dy,$$

The following equality stems from these relations:

$$\int u_i(y,\lambda)\widetilde{L}^{s-1}v(x,y)dy = (\lambda+1)^{s-1}\int v(x,y)u_i(y,\lambda)dy. \tag{4.1.19}$$

Now we take into account that

$$\widetilde{L}^{s-1}v(x,y) = \begin{cases} \widetilde{L}^{s-1}K_s(x,y) = K_1(x,y) & \text{for } |x-y| \le R/2, \\ 0 & \text{for } |x-y| \ge R, \\ \in C^\infty & \text{for } |x-y| > R/2. \end{cases}$$

Having applied Green's formula once again to the regular solution of $u_i(y,\lambda)$ and to the function $\widetilde{L}^{s-1}v(x,y)$ exhibiting a singularity of the fundamental solution, we obtain the relation

$$\int u_i(y,\lambda)\widetilde{L}^s v(x,y)dy = u_i(x,\lambda) + (\lambda+1)\int u_i(y,\lambda)\widetilde{L}^{s-1}v(x,y)dy. \tag{4.1.20}$$

---

3) The integrals may be assumed as taken over the entire E^N; in fact, the integration is carried out only with respect to the radius R of a ball centered at point x.

Relations (4.1.19) and (4.1.20) imply the inequality

$$\frac{|u_i(x,\lambda)}{(\lambda+1)^s} \le \left|\int_G v(x,y)u_i(y,\lambda)dy\right| + \frac{1}{(\lambda+1)^s}\left|\int_G u_i(y,\lambda)\widetilde{L}^s v(x,y)dy\right|,$$

which, in turn, implies that

$$\sum_{i=1}^{\widetilde{m}}\int_0^\infty \frac{|u_i(x,\lambda)|^2}{(\lambda+1)^{2s}}\,d\rho(\lambda)$$

$$\le 2\sum_{i=1}^{\widehat{m}}\int_0^\infty \left|\int_G v(x,y)u_i(y,\lambda)dy\right|^2 d\rho(\lambda) \tag{4.1.21}$$

$$+2\sum_{i=1}^{\widehat{m}}\int_0^\infty \frac{1}{(\lambda+1)^{2s}}\left|\int_G u_i(y,\lambda)\widetilde{L}^s v(x,y)dy\right|^2 d\rho(\lambda).$$

The boundedness of the former sum on the right-hand side of (4.1.21) stems from the Parseval equality (4.1.18), and the boundedness of the latter sum follows from the inequality $(\lambda+1)^{-2s} \le 1$, from the Parseval equality, and from the fact that the function $[\rho_0(y)]^{-1}\widetilde{L}^s v(x,y)$ becomes zero at $|x-y| \le R/2$ and $|x-y| \ge R$ and belongs to the class $L_2(G,\rho_0(y))$.

Thus, we have provided evidence that the left-hand side of (4.1.21) is bounded for any x in subdomain G'.

Inasmuch as $2s = 2[N/4]+2 \le N/2+2$, the above reasoning leads to inequality (4.1.16) and, consequently, to the rough estimate (4.1.15).

This completes the proof of the first part of Lemma 4.1.

2^0. We now proceed with a proof of the exact estimate (4.1.14).[4)]

We fix an arbitrary subdomain G' strictly interior to domain G, we also fix an arbitrary point x of subdomain G' and an arbitrary sufficiently small $R > 0$, *a fortiori* smaller than the distance between the subdomain G' and the boundary of G. We denote by the symbol $r = |x-y|$ the geodetic distance between a variable point y and a fixed point x measured along the rays of Eq. (4.1.7). Next, we compute the Fourier image for the function $p(x,y)w(r)$, where $p(x,y)$ is the function in the Moiseev mean-value formula (4.1.10), and the $w(r)$ takes the form

$$w(r) = \begin{cases} (2\pi)^{-N/2}\mu^{N/2}r^{-\nu}J_\nu(\mu r)[\rho_0(x)]^{-1} & \text{for} \quad r = |x-y| \in [R/2, R] \\ 0 & \text{for other} \quad r = |x-y|, \end{cases} \tag{4.1.22}$$

[4)] We have already seen in Chapter 1 that this estimate is accurate to within an order of magnitude with respect to μ for the special case of the Laplace operator.

where, by convention, $\nu = (N-2)/2$.

One observes that it suffices to prove estimate (4.1.14) for values of $\mu \geq \mu_0$, where μ_0 is a sufficiently large fixed positive number, since for the values of $\mu \leq \mu_0$ the estimate (4.1.14) stems from the rough estimate (4.1.15) that has been proved above. Therefore, in what follows we assume that the μ in (4.1.22) is in excess of a sufficiently large fixed number μ_0.

According to relations (4.1.12) and (4.1.13), the Fourier image $\widehat{w}_i(x,\lambda)$ of function (4.1.22) with respect to the fundamental function $u_i(y,\lambda)$ of an arbitrary ordered spectral representation of the space $L_2(G,\rho_0(y))$, associated with the extension $\widehat{A}$ of the operator (4.1.1) with weight $\rho_0(y)$, takes the form

$$\widehat{w}_i(x,\lambda) = \bar{w}_i(x,\lambda) + \widetilde{w}_i(x,\lambda), \tag{4.1.23}$$

where

$$\bar{w}_i(x,\lambda) = \mu^{N/2}\lambda^{-\nu/2}u_i(x,\lambda)\int\limits_{R/2}^{R} rJ_\nu(r\mu)J_\nu(r\sqrt{\lambda})dr, \tag{4.1.24}$$

and the additional term $\widetilde{w}_i(x,\lambda)$ is

$$\widetilde{w}_i(x,\lambda) = \mu^{N/2}(2\pi)^{-N/2}\int\limits_{R/2}^{R} rJ_\nu(r\mu)\left[\int\limits_0^r \tau^\nu w_\nu(\tau\sqrt{\lambda}, r\sqrt{\lambda}) \times \left(\int\limits_\omega u_i(x+\tau\omega,\lambda)q(x,x+\tau\omega)d\omega\right)d\tau\right]dr. \tag{4.1.25}$$

We have shown (see Sections 1.1.2 and 2.3.2) that, given sufficiently large positive μ_0, for all λ subject to $\mu_0 < \mu \leq \sqrt{\lambda} \leq \mu + 1$, there is a constant $\alpha > 0$ such that the inequality

$$|\bar{w}_i(x,\lambda)| \geq \alpha|u_i(x,\lambda)| \tag{4.1.26}$$

holds for $\bar{w}_i(x,\lambda)$ as defined by (4.1.24).

The validity of estimate

$$\int\limits_G [w(|x-y|)]^2\rho_0(y)dy = O(\mu^{N-1}) \tag{4.1.27}$$

for function (4.1.22), uniform with respect to x in subdomain G', is proved in much the same manner.

We apply the Parseval equality (4.1.6) to function (4.1.22) in such a manner that there remains on the right-hand side of (4.1.6) only the integral taken for the values of λ subject to the condition $\mu_0 < \mu \le \sqrt{\lambda} \le \mu + 1$; next, making use of representation (4.1.23), we obtain

$$\sum_{i=1}^{\widehat{m}} \int\limits_{\mu \le \sqrt{\lambda} \le \mu+1} |\bar{w}_i(x,\lambda)|^2 \, d\rho(\lambda)$$
$$\le 2 \int\limits_G [w(|x-y|)]^2 \rho_0(y) dy + 2 \sum_{i=1}^{\widehat{m}} \int\limits_{\mu \le \sqrt{\lambda} \le \mu+1} |\widetilde{w}_i(x,\lambda)|^2 \, d\rho(\lambda). \tag{4.1.28}$$

A comparison of inequality (4.1.28) with relations (4.1.26) and (4.1.27) leads to the estimate

$$\sum_{i=1}^{\widehat{m}} \int\limits_{\mu \le \sqrt{\lambda} \le \mu+1} u_i^2(x,\lambda) d\rho(\lambda)$$
$$\le O(\mu^{N-1}) + \frac{2}{\alpha^2} \sum_{i=1}^{\widehat{m}} \int\limits_{\mu \le \sqrt{\lambda} \le \mu+1} |\widetilde{w}_i(x,\lambda)|^2 \, d\rho(\lambda). \tag{4.1.29}$$

We now prove, with reference to (4.1.29), that *if the estimate*

$$\sum_{i=1}^{\widehat{m}} \int\limits_{\mu \le \sqrt{\lambda} \le \mu+1} u_i^2(x,\lambda) d\rho(\lambda) = O(\mu^s) \tag{4.1.30}$$

holds uniformly with respect to x in an arbitrary subdomain G' strictly interior to domain G for the quantity on the left-hand side of (4.1.29) and for a certain $s \ge N-1$, then the estimate

$$\sum_{i=1}^{\widehat{m}} \int\limits_{\mu \le \sqrt{\lambda} \le \mu+1} u_i^2(x,\lambda) d\rho(\lambda) = O(\mu^{\max(N-1, s-2)}) \tag{4.1.31}$$

also holds for the same quantity uniformly with respect to x in an arbitrary subdomain G'' strictly interior to subdomain G'.

Note that the formulated assertion, once proved, lends immediate support to the estimate (4.1.14) (uniform with respect to x in a strictly interior subdomain G') lying at the heart of Lemma 4.1. Indeed, by virtue of the previously established rough estimate (4.1.15), the estimate (4.1.30) holds at $s = N + 4$. If so, the estimate (4.1.31) also holds at $s = N + 4$, that is, the estimate (4.1.30) holds at $s = N + 2$.

This ensures, by virtue of (4.1.31), the validity of estimate (4.1.30) at $s = N$. Applying the formulated assertion once again at $s = N$, we obtain the desired estimate (4.1.4) from (4.1.30).

Briefly, the formulated assertion enables one to obtain the exact estimate (4.1.14) from the rough estimate (4.1.15) in three steps. The fact that in each successive step we take an arbitrary subdomain strictly interior not to the main domain G, but rather to an arbitrary strictly interior subdomain chosen at the preceding step, incurs no loss of generality inasmuch as all strictly interior subdomains are chosen arbitrarily.

Thus, to prove Lemma 4.1, we must show that the assertion that has been formulated above is correct.

To this end it suffices to show, by virtue of inequality (4.1.29), that if the estimate (4.1.30) is correct uniformly with respect to x in every subdomain G' strictly interior to domain G for a certain $s \geq N - 1$, then the estimate

$$\sum_{i=1}^{\widehat{m}} \int\limits_{\mu \leq \sqrt{\lambda} \leq \mu+1} |\widetilde{w}_i(x,\lambda)|^2 \, d\rho(\lambda) = O(\mu^{N-1}) + O(\mu^{s-2}) \tag{4.1.32}$$

holds uniformly with respect to x in every subdomain G'' strictly interior to domain G'.

We write the left-hand side of (4.1.32), with reference to equality (4.1.25), as

$$\sum_{i=1}^{\widehat{m}} \int\limits_{\mu \leq \sqrt{\lambda} \leq \mu+1} |\widetilde{w}_i(x,\lambda)|^2 \, d\rho(\lambda)$$

$$= \left(\frac{\mu}{2\pi}\right)^N \sum_{i=1}^{\widehat{m}} \int\limits_{\mu \leq \sqrt{\lambda} \leq \mu+1} \left\{ \int\limits_{R/2}^{R} r J_\nu(r\mu) \left[\int\limits_0^r \tau^\nu w_\nu(\tau\sqrt{\lambda}, r\sqrt{\lambda}) \right.\right. \tag{4.1.33}$$

$$\left.\left. \times \left(\int\limits_\omega u_i(x+\tau\omega,\lambda) q(x, x+\tau\omega) d\omega \right) d\tau \right] dr \right\}^2 d\rho(\lambda).$$

We partition the integral with respect to the variable τ, enclosed within the square brackets on the right-hand side of (4.1.33), into the sum of two integrals:

$$\int_0^r = \int_0^{1/\mu} + \int_{1/\mu}^r .$$

At the next step, we arrive at the inequality

$$\sum_{i=1}^{\widehat{m}} \int_{\mu\le\sqrt{\lambda}\le\mu+1} |\widetilde{w}_i(x,\lambda)|^2\, d\rho(\lambda)$$

$$\le 2\left(\frac{\mu}{2\pi}\right)^N \sum_{i=1}^{\widehat{m}} \int_{\mu\le\sqrt{\lambda}\le\mu+1} \left\{\int_{R/2}^{R} rJ_\nu(r\mu)\left[\int_0^{1/\mu} \tau^\nu w_\nu(\tau\sqrt{\lambda}, r\sqrt{\lambda})\right.\right.$$

$$\left.\left.\times\left(\int_\omega u_i(x+\tau\omega,\lambda)q(x,x+\tau\omega)d\omega\right)d\tau\right]dr\right\}^2 d\rho(\lambda) \tag{4.1.34}$$

$$+2\left(\frac{\mu}{2\pi}\right)^N \sum_{i=1}^{\widehat{m}} \int_{\mu\le\sqrt{\lambda}\le\mu+1} \left\{\int_{R/2}^{R} rJ_\nu(r\mu)\left[\int_{1/\mu}^{r} \tau^\nu w_\nu(\tau\sqrt{\lambda}, r\sqrt{\lambda})\right.\right.$$

$$\left.\left.\times\left(\int_\omega u_i(x+\tau\omega,\lambda)q(x,x+\tau\omega)d\omega\right)d\tau\right]dr\right\}^2 d\rho(\lambda).$$

Assume that G' is an arbitrary subdomain strictly interior to domain G, and G'' is an arbitrary subdomain strictly interior to domain G'. We fix a positive R smaller than the distance between the domain G'' and the boundary of domain G'.

It suffices to show that if for a certain $s \ge N-1$ the estimate (4.1.30) holds uniformly with respect to x in subdomain G', then uniformly with respect to x in subdomain G'' the former term on the right-hand side of (4.1.34) has the order of $O(\mu^{s-2})$, and the latter term has the order of $O(\mu^{N-1})$.

First, we estimate the former term on the right-hand side of (4.1.34), denoting it for short by $\mathcal{K}_1$.

Since, by virtue of the Cauchy–Buniakowski inequality (also taking into account that $|q(x,y)| \le C < \infty$),

$$\left\{ \int_{R/2}^{R} r J_\nu(r\mu) \left[\int_0^{1/\mu} \tau^\nu w_\nu(\tau\sqrt{\lambda}, r\sqrt{\lambda}) \times \left(\int_\omega u_i(x+\tau\omega,\lambda) q(x,x+\tau\omega) d\omega \right) d\tau \right] dr \right\}^2$$

$$\le \frac{R}{2} \int_{R/2}^{R} r^2 J_\nu^2(r\mu) \left[\int_0^{1/\mu} \tau^\nu w_\nu(\tau\sqrt{\lambda}, r\sqrt{\lambda}) \times \left(\int_\omega u_i(x+\tau\omega,\lambda) q(x,x+\tau\omega) d\omega \right) d\tau \right]^2 dr,$$

$$\left[\int_0^{1/\mu} \tau^\nu w_\nu(\tau\sqrt{\lambda}, r\sqrt{\lambda}) \left(\int_\omega u_i(x+\tau\omega,\lambda) q(x,x+\tau\omega) d\omega \right) d\tau \right]^2$$
$$\le \frac{1}{\mu} \int_0^{1/\mu} \tau^{2\nu} w_\nu^2(\tau\sqrt{\lambda}, r\sqrt{\lambda}) \left(\int_\omega u_i(x+\tau\omega,\lambda) q(x,x+\tau\omega) d\omega \right)^2 d\tau,$$

$$\left(\int_\omega u_i(x+\tau\omega,\lambda) q(x,x+\tau\omega) d\omega \right)^2 \le C_1 \int_\omega u_i^2(x+\tau\omega,\lambda) d\omega,$$

then, allowing for

$$w_\nu^2(\tau\sqrt{\lambda}, r\sqrt{\lambda}) \le 2J_\nu^2(\tau\sqrt{\lambda})Y_\nu^2(r\sqrt{\lambda}) + 2J_\nu^2(r\sqrt{\lambda})Y_\nu^2(\tau\sqrt{\lambda}),$$

we obtain the following estimate[5] for the former term $\mathcal{K}$:

$$\mathcal{K}_1 \le C_2\mu^{N-1}\sum_{i=1}^{\widehat{m}} \int_{\mu\le\sqrt{\lambda}\le\mu+1} \left\{ \int_{R/2}^{R} r^2 J_\nu^2(r\mu) J_\nu^2(r\sqrt{\lambda}) \right.$$

$$\times \left[\int_0^{1/\mu} \tau^{2\nu} Y_\nu^2(\tau\sqrt{\lambda}) \left(\int_\omega u_i^2(x+\tau\omega,\lambda)\,d\omega \right) d\tau \right] dr \Bigg\} d\rho(\lambda)$$

$$+C_2\mu^{N-1}\sum_{i=1}^{\widehat{m}} \int_{\mu\le\sqrt{\lambda}\le\mu+1} \left\{ \int_{R/2}^{R} r^2 J_\nu^2(r\mu) Y_\nu^2(r\sqrt{\lambda}) \right. \tag{4.1.35}$$

$$\times \left[\int_0^{1/\mu} \tau^{2\nu} J_\nu^2(\tau\sqrt{\lambda}) \left(\int_\omega u_i^2(x+\tau\omega,\lambda)\,d\omega \right) d\tau \right] dr \Bigg\} d\rho(\lambda).$$

Recalling that $\mu > \mu_0$ (where μ_0 is a sufficiently large positive number), the estimates

$$J_\nu^2(r\mu) \le \frac{C_3}{r\mu}, \quad J_\nu^2(r\sqrt{\lambda}) \le \frac{C_4}{r\mu}, \quad Y_\nu^2(r\sqrt{\lambda}) \le \frac{C_5}{r\mu} \tag{4.1.36}$$

hold for all $\mu \le \sqrt{\lambda} \le \mu+1$ and for all r in the segment $R/2 \le r \le R$, whereas the estimates

$$\begin{cases} J_\nu^2(\tau\sqrt{\lambda}) \le C_6(\tau\mu)^{2\nu}, \\ Y_\nu^2(\tau\sqrt{\lambda}) \le \begin{cases} C_7(\tau\mu)^{-2\nu} & \text{for } N\ne 2, \text{ that is, for } \nu\ne 0, \\ C_8(\tau\mu)^{-1/2} & \text{for } N=2, \text{ that is, for } \nu = 0, \end{cases} \end{cases} \tag{4.1.37}$$

hold for all τ in the segment $0 \le \tau \le 1/\mu$.

Invoking (4.1.36) and (4.1.37) for estimating the right-hand side of (4.1.35), we

---

[5] We make use of the symbols C_1, C_2, ... to denote constants that are dependent on N, $\sup_{x\in G}|q(x,y)|$ and on the fixed subdomains G' and G'' only.

obtain

$$\mathcal{K}_1 \le C_9\mu^{N-3}$$

$$\times \int_0^{1/\mu} \left\{ \int_\omega \left[\sum_{i=1}^{\widehat{m}} \int_{\mu \le \sqrt{\lambda} \le \mu+1} u_i^2(x+\tau\omega,\lambda)d\rho(\lambda) \right] d\omega \right\} \tau^{-1/2}\mu^{-2\nu-1/2}d\tau \tag{4.1.38}$$

$$+C_{10}\mu^{N-3} \int_0^{1/\mu} \left\{ \int_\omega \left[\sum_{i=1}^{\widehat{m}} \int_{\mu \le \sqrt{\lambda} \le \mu+1} u_i^2(x+\tau\omega,\lambda)d\rho(\lambda) \right] d\omega \right\} (\tau^2\mu)^{2\nu}d\tau.$$

Taking into account that $2\nu = N - 2$ and that for any point $x \in G''$ the value of $x + \tau\omega$ falls into subdomain G', and invoking (4.1.30) for estimating the terms on the right-hand side of (4.1.38) within the square brackets, we infer that for all $x \in G''$ the right-hand side of (4.1.38) does not exceed the value of $C_{11}\mu^{s-2}$.

We have established thereby that the former term on the right-hand side of (4.1.34) has the order of $O(\mu^{s-2})$ uniformly with respect to $x \in G''$.

It remains to prove that the latter term on the right-hand side of (4.1.34) (now denoted for brevity by $\mathcal{K}_2$) has the order of $O(\mu^{N-1})$ uniformly with respect to $x \in G''$.

By virtue of the Cauchy–Buniakowski inequality the latter term $\mathcal{K}_2$ is majorized as shown below:

$$\mathcal{K}_2 \le R\left(\frac{\mu}{2\pi}\right)^N \sum_{i=1}^{\widehat{m}} \int_{\mu \le \sqrt{\lambda} \le \mu+1} \left\{ \int_{R/2}^{R} r^2 J_\nu^2(r\mu) \left[\int_{1/\mu}^{r} \tau^\nu w_\nu(\tau\sqrt{\lambda}, r\sqrt{\lambda}) \right.\right.$$

$$\left.\left. \times \left(\int_\omega u_i(x+\tau\omega,\lambda)q(x,x+\tau\omega)d\omega \right) d\tau \right]^2 dr \right\} d\rho(\lambda). \tag{4.1.39}$$

Let us estimate the square of the term within the square brackets on the right-hand side of (4.1.39).

We note that, by virtue of the equality

$$w_\nu(\tau\sqrt{\lambda}, r\sqrt{\lambda}) = J_\nu(r\sqrt{\lambda})Y_\nu(\tau\sqrt{\lambda}) - J_\nu(\tau\sqrt{\lambda})Y_\nu(r\sqrt{\lambda}),$$

the following estimate holds:

$$\left[\int\limits_{1/\mu}^{r} \tau^{\nu} w_{\nu}(\tau\sqrt{\lambda}, r\sqrt{\lambda}) \left(\int\limits_{\omega} u_i(x+\tau\omega,\lambda) q(x, x+\tau\omega) d\omega\right) d\tau\right]^2$$

$$\leq 2J_{\nu}^2(r\sqrt{\lambda}) \left[\int\limits_{1/\mu}^{r} \tau^{\nu} Y_{\nu}(\tau\sqrt{\lambda}) \left(\int\limits_{\omega} u_i(x+\tau\omega,\lambda) q(x, x+\tau\omega) d\omega\right) d\tau\right]^2 \tag{4.1.40}$$

$$+ 2Y_{\nu}^2(r\sqrt{\lambda}) \left[\int\limits_{1/\mu}^{r} \tau^{\nu} J_{\nu}(\tau\sqrt{\lambda})\right.$$

$$\left.\times \left(\int\limits_{\omega} u_i(x+\tau\omega,\lambda) q(x, x+\tau\omega) d\omega\right) d\tau\right]^2 .$$

Both terms on the right-hand side of (4.1.40) are subject to a similar estimation; therefore, we confine ourselves to estimating the former of the two.

In this term, we take by parts the integral enclosed within the square brackets, assuming

$$u(\tau) = \tau^{-1/4} Y_{\nu}(\tau\sqrt{\lambda}),$$

$$v(\tau) = \int\limits_{1/\mu}^{\tau} \rho^{\nu+1/4} \left(\int\limits_{\omega} u_i(x+\rho\omega,\lambda) q(x, x+\rho\omega) d\omega\right) d\rho.$$

We arrive at the equality

$$\int\limits_{1/\mu}^{r} \tau^{\nu} Y_{\nu}(\tau\sqrt{\lambda}) \left(\int\limits_{\omega} u_i(x+\tau\omega,\lambda) q(x, x+\tau\omega) d\omega\right) d\tau$$

$$= r^{-1/4} Y_{\nu}(r\sqrt{\lambda}) \int\limits_{1/\mu}^{r} \rho^{\nu+1/4} \left(\int\limits_{\omega} u_i(x+\rho\omega,\lambda) q(x, x+\rho\omega) d\omega\right) d\rho$$

$$- \int\limits_{1/\mu}^{r} \frac{d}{d\tau}\left[\tau^{-1/4} Y_{\nu}(\tau\sqrt{\lambda})\right] \int\limits_{1/\mu}^{\tau} \rho^{\nu+1/4} \left(\int\limits_{\omega} u_i(x+\rho\omega,\lambda) q(x, x+\rho\omega) d\omega\right) d\rho.$$

From this equality, with reference to $(A+B)^2 \le 2A^2 + 2B^2$, we obtain

$$\left[\int_{1/\mu}^{r} \tau^{\nu} Y_\nu(\tau\sqrt{\lambda}) \left(\int_{\omega} u_i(x+\tau\omega,\lambda) q(x, x+\tau\omega) d\omega\right) d\tau\right]^2$$

$$\le 2r^{-1/2} Y_\nu^2(r\sqrt{\lambda}) \left[\int_{1/\mu}^{r} \rho^{\nu+1/4} \left(\int_{\omega} u_i(x+\rho\omega,\lambda) q(x,x+\rho\omega) d\omega\right) d\rho\right]^2$$

$$+2\int_{1/\mu}^{r} \tau^{3/4} \left\{\frac{d}{d\tau}\left[\tau^{-1/4} Y_\nu(\tau\sqrt{\lambda})\right]\right\}^2 d\tau \tag{4.1.41}$$

$$\times \int_{1/\mu}^{r} \tau^{-3/4} \left[\int_{1/\mu}^{\tau} \rho^{\nu+1/4} \left(\int_{\omega} u_i(x+\rho\omega,\lambda) q(x,x+\rho\omega) d\omega\right) d\rho\right]^2 d\tau.$$

We note now, by virtue of relation (4.1.9), that

$$\int_{1/\mu}^{\tau} \rho^{\nu+1/4} \left(\int_{\omega} u_i(x+\rho\omega,\lambda) q(x,x+\rho\omega) d\omega\right) d\rho$$

$$= \int_{1/\mu \le r_{xy} \le \tau} \frac{u_i(y,\lambda) q(x,y)}{\rho_0(y) r_{xy}^{N/2-1/4}} \rho_0(y) dy, \tag{4.1.42}$$

where the symbol r_{xy} denotes the geodetic distance between a fixed point x and a variable point y.

Next, we note that, by virtue of the recurrence relation $(t^\nu Y_\nu(t))' = t^\nu Y_{\nu-1}(t)$, the equation

$$\frac{d}{d\tau}\left[\tau^{-1/4} Y_\nu(\tau\sqrt{\lambda})\right] = \frac{d}{d\tau}\left\{\tau^{-1/4-\nu}\left[\tau^\nu Y_\nu(\tau\sqrt{\lambda})\right]\right\}$$

$$= \tau^{-1/4-\nu}\tau^\nu \sqrt{\lambda} Y_{\nu-1}(\tau\sqrt{\lambda}) - \left(\frac{1}{4}+\nu\right) \tau^{-5/4-\nu}\tau^\nu Y_\nu(\tau\sqrt{\lambda})$$

$$= \tau^{-1/4}\sqrt{\lambda} Y_{\nu-1}(\tau\sqrt{\lambda}) - \left(\frac{1}{4}+\nu\right) \tau^{-5/4} Y_\nu(\tau\sqrt{\lambda}) - \left(\frac{1}{4}+\nu\right) \tau^{-5/4} Y_\nu(\tau\sqrt{\lambda})$$

holds and, consequently,

$$\tau^{3/4}\left\{\frac{d}{d\tau}\left[\tau^{-1/4}Y_\nu(\tau\sqrt{\lambda})\right]\right\}^2$$
$$\leq 2\tau^{1/4}(\mu+1)^2Y_{\nu-1}^2(\tau\sqrt{\lambda})+2\left(\frac{1}{4}+\nu\right)^2\tau^{-7/4}Y_\nu^2(\tau\sqrt{\lambda}).$$

From the last inequality and from the third estimate (4.1.36) we obtain

$$\tau^{3/4}\left\{\frac{d}{d\tau}\left[\tau^{-1/4}Y_\nu(\tau\sqrt{\lambda})\right]\right\}^2 \leq C_{11}(\tau^{-3/4}\mu+\tau^{-11/4}\mu^{-1})$$

and therefore

$$\int\limits_{1/\mu}^{r}\tau^{3/4}\left\{\frac{d}{d\tau}\left[\tau^{-1/4}Y_\nu(\tau\sqrt{\lambda})\right]\right\}^2 d\tau \leq C_{12}(r^{1/4}\mu+\mu^{3/4}) \leq C_{13}\mu r^{1/4}. \qquad (4.1.43)$$

From relations (4.1.41)–(4.1.43) and the third relation (4.1.36) it follows that

$$\left[\int\limits_{1/\mu}^{r}\tau^\nu Y_\nu(\tau\sqrt{\lambda})\left(\int\limits_\omega u_i(x+\tau\omega,\lambda)q(x,x+\tau\omega)d\omega\right)d\tau\right]^2$$
$$\leq 2r^{-3/2}\mu^{-1/2}\left[\int\limits_{1/\mu\leq r_{xy}\leq r}\frac{u_i(y,\lambda)q(x,y)}{\rho_0(y)r_{xy}^{N/2-1/4}}\rho_0(y)dy\right]^2 \qquad (4.1.44)$$
$$+2C_{13}\mu r^{1/4}\int\limits_{1/\mu}^{r}\tau^{-3/4}\left[\int\limits_{1/\mu\leq r_{xy}\leq\tau}\frac{u_i(y,\lambda)q(x,y)}{\rho_0(y)r_{xy}^{N/2-1/4}}\rho_0(y)dy\right]^2 d\tau.$$

A similar estimation applies to the second integral on the right-hand side of (4.1.40), enclosed within square brackets. The following estimate holds for the square of this integral:

$$\left[\int_{1/\mu}^{r} \tau^{\nu} J_{\nu}(\tau\sqrt{\lambda}) \left(\int_{\omega} u_i(x+\tau\omega,\lambda) q(x, x+\tau\omega) d\omega\right) d\tau\right]^2$$

$$\leq 2r^{-3/2}\mu^{-1/2} \left[\int_{1/\mu \leq r_{xy} \leq r} \frac{u_i(y,\lambda)q(x,y)}{\rho_0(y) r_{xy}^{N/2-1/4}} \rho_0(y) dy\right]^2 \tag{4.1.45}$$

$$+ 2C_{14}\mu\, r^{1/4} \int_{1/\mu}^{r} \tau^{-3/4} \left[\int_{1/\mu \leq r_{xy} \leq \tau} \frac{u_i(y,\lambda)q(x,y)}{\rho_0(y) r_{xy}^{N/2-1/4}} \rho_0(y) dy\right]^2 d\tau.$$

Applying (4.1.44), (4.1.45) and the second and third estimates (4.1.36) to the right-hand side of (4.1.40), we arrive at the inequality

$$\left[\int_{1/\mu}^{r} \tau^{\nu} w_{\nu}(\tau\sqrt{\lambda}, r\sqrt{\lambda}) \left(\int_{\omega} u_i(x+\tau\omega,\lambda) q(x, x+\tau\omega) d\omega\right) d\tau\right]^2$$

$$\leq C_{15} r^{-5/2}\mu^{-3/2} \left[\int_{1/\mu \leq r_{xy} \leq r} \frac{u_i(y,\lambda)q(x,y)}{\rho_0(y) r_{xy}^{N/2-1/4}} \rho_0(y) dy\right]^2 \tag{4.1.46}$$

$$+ C_{16} r^{-3/4} \int_{1/\mu}^{r} \tau^{-3/4} \left[\int_{1/\mu \leq r_{xy} \leq \tau} \frac{u_i(y,\lambda)q(x,y)}{\rho_0(y) r_{xy}^{N/2-1/4}} \rho_0(y) dy\right]^2 d\tau.$$

Substituting (4.1.46) into (4.1.39) and making use of the first estimate in (4.1.36), we obtain

$$\mathcal{K}_2 \le C_{17}\mu^{N-5/2} \int_{R/2}^{R} r^{-3/2}$$

$$\times \left\{ \sum_{i=1}^{\widehat{m}} \int_{\mu \le \sqrt{\lambda} \le \mu+1} \left[\int_{1/\mu \le r_{xy} \le r} \frac{u_i(y,\lambda) q(x,y)}{\rho_0(y) r_{xy}^{N/2-1/4}} \rho_0(y) dy \right]^2 d\rho(\lambda) \right\} dr \tag{4.1.47}$$

$$+ C_{18}\mu^{N-1} \int_{R/2}^{R} r^{1/4} \left\{ \int_{1/\mu}^{r} \tau^{-3/4} \left[\sum_{i=1}^{\widehat{m}} \int_{\mu \le \sqrt{\lambda} \le \mu+1} \left(\int_{1/\mu \le r_{xy} \le \tau} \frac{u_i(y,\lambda) q(x,y)}{\rho_0(y) r_{xy}^{N/2-1/4}} \right.\right.\right.$$

$$\left.\left.\left. \times \rho_0(y) dy \right)^2 d\rho(\lambda) \right] d\tau \right\} dr.$$

Since for any point x of subdomain G'' and for any $\tau \le r \le R$ the function

$$f(y) = \begin{cases} \dfrac{q(x,y)}{\rho_0(y) r_{xy}^{N/2-1/4}} & \text{for} \quad 1/\mu \le r_{xy} \le \tau, \\ 0 & \text{for all other values of} \quad y \end{cases}$$

belongs to the class $L_2(G, \rho_0(y))$ and its norm in this class for all $x \in G''$ is bounded by the same constant C_{19}, the Parseval equality for this function

$$\sum_{i=1}^{\widehat{m}} \int_0^{\infty} \left[\int_{1/\mu \le r_{xy} \le \tau} \frac{u_i(y,\lambda) q(x,y)}{\rho_0(y) r_{xy}^{N/2-1/4}} \rho_0(y) dy \right]^2 d\rho(\lambda) = \int_G f^2(y) \rho_0(y) dy \le C_{19}$$

and the inequality (4.1.47) imply the estimate

$$\mathcal{K}_2 \le C_{20}\mu^{N-5/2} \int_{R/2}^{R} r^{-3/2} dr + C_{12}\mu^{N-1} \int_{R/2}^{R} r^{1/2} dr \le C_{22}\mu^{N-1},$$

uniform with respect to x in subdomain G''.

This proves that the latter term on the right-hand side of (4.1.34) has the order of $O(\mu^{N-1})$ uniformly with respect to x in subdomain G''.

The proof of Lemma 4.1 is thus complete.

A moment's consideration shows that two simple corollaries, similar to the case of the Laplace operator, are to be inferred from Lemma 4.1.

COROLLARY 1. *For any $\mu \geq 1$ and for any t in the segment $1 \leq t \leq \mu$ uniformly with respect to x within every subdomain G' strictly interior to domain G the following estimate holds:*

$$\sum_{i=1}^{\widehat{m}} \int_{|\sqrt{\lambda}-\mu|\leq t} u_i^2(x,\lambda)d\rho(\lambda) = t\,O(\mu^{N-1}). \tag{4.1.48}$$

COROLLARY 2. *For any $\delta > 0$ and for all $\lambda \geq 1$ uniformly with respect to x within every subdomain G' strictly interior to domain G the following estimates hold:*

$$\sum_{i=1}^{\widehat{m}} \int_{1}^{\lambda} u_i^2(x,t)t^{\delta-N/2}d\rho(t) = O(\lambda^{\delta}), \tag{4.1.49}$$

$$\sum_{i=1}^{\widehat{m}} \int_{\lambda}^{\infty} u_i^2(x,t)t^{-\delta-N/2}d\rho(t) = O(\lambda^{-\delta}). \tag{4.1.50}$$

In particular, for any $\delta > 0$ one may assert that the term

$$\sum_{i=1}^{\widehat{m}} \int_{1}^{\infty} u_i^2(x,t)t^{-\delta-N/2}d\rho(t) \tag{4.1.51}$$

is uniformly bounded with respect to x in G'.

To prove these corollaries one merely has to go back to the argumentation expounded at the end of Section 2.3.2, with a specific reference to estimate (4.1.14) irrespective of the form of the elliptic operator.

4.1.4. Fractional Powers of Self-Adjoint Extensions of Elliptic Operators

Let $\widehat{A}$ be an arbitrary self-adjoint nonnegative extension of elliptic operator (4.1.1). We assume, with no loss of generality, that this extension is strictly positive,[6] that is, we assume that there exists a constant $\mu_0 \geq 0$ such that for all $u(x)$ from $C_0^\infty(G)$ the inequality $(\widehat{A}u, u) \geq \mu_0(u,u)$ holds.

[6] This can always be ensured by appropriately changing the coefficient $c(x)$ of operator (4.1.1).

We denote by $\{E_\lambda\}$ a spectral family of the extension $\widehat{A}$ and consider, for any real $\alpha > 0$, an operator $\widehat{A}^{-\alpha}$ defined as

$$\widehat{A}^{-\alpha} = \int_{\mu_0}^{\infty} \lambda^{-\alpha} dE_\lambda. \tag{4.1.52}$$

One infers from the above cited work of L. Gårding (see Chapter 2) that the operator (4.1.52), for the case where one of the points x or y lies inside G, is an integral operator with kernel $T_\alpha(x, y)$; it appears reasonable to call $T_\alpha(x, y)$ *fractional kernel* (or, more exactly, the *kernel of order* α).

However, for the general self-adjoint elliptic operator of second order with variable coefficients, one can hardly expect it to be a feasible task to explore in detail the structure of a fractional kernel, as we have done for the Laplace operator (see Section 2.3.3).

Therefore, we restrict ourselves to expounding results relevant to the spectral theory, specifically, to the establishment of negative- and positive-type theorems on localization and convergence of spectral decompositions.

Let n be any number, $\mathcal{D}(\widehat{A}^n)$ be the domain of definition for the operator $\widehat{A}^n$, and let the symbol $\widetilde{\Delta}(\widehat{A})$ denote the class

$$\widetilde{\Delta}(\widehat{A}) = \bigcap_{n=1}^{\infty} \mathcal{D}(\widehat{A}^n), \tag{4.1.53}$$

which is the intersection of domains of definition for all positive integer powers of operator $\widehat{A}$.

In exploring the convergence and localization of spectral decompositions, one can ignore functions in the class (4.1.53), since the spectral decompositions of such functions trivially converge absolutely and uniformly on any compact set of the main domain G.

Otherwise stated, we can study the spectral decompositions of functions that belong to one or another class of differentiable functions to within a summand belonging to the class (4.1.53).

We shall make use of the symbol $\overset{0}{H}{}_p^\alpha(\Omega)$ for an arbitrary domain Ω to denote the class of functions obtained by closing, in the norm of the Nikol'skii class $H_p^\alpha(\Omega)$, a set of functions from $C_0^\infty(\Omega)$.

We shall need the following two assertions.

ASSERTION 1. *If D is a bounded domain whose closure is contained in an arbitrary N-dimensional domain G of interest, and Ω is a domain whose closure is contained in the open domain D, then for any real $\alpha > 0$, $1 \le p < \infty$ and for any function h, finite in domain Ω and belonging to the class $L_p(G)$, there is a function f, finite in D and belonging to the Nikol'skii class $H_p^{2\alpha}(G)$, such that $f - \widehat{A}^{-\alpha} h \in \widetilde{\Delta}(\widehat{A})$.*

ASSERTION 2. *Let D, Ω, and G have the same meanings as those in Assertion 1. Then for any $\alpha > 0$, $s > 0$, and $1 \leq p \leq \infty$ the operator $\widehat{A}^{-\alpha}$ implements isomorphism of the Nikol'skii classes H_p^s and $\overset{0}{H}{}_p^{s+2\alpha}$ modulo $\widetilde{\Delta}(\widehat{A})$. This signifies that:*

(1) for any function h in the class $H_p^s(\Omega)$ there exists a function f in the class $\overset{0}{H}{}_p^{s+2\alpha}(D)$ such that $f - \widehat{A}^{-\alpha}h \in \widetilde{\Delta}(\widehat{A})$;

(2) for any function f in the class $\overset{0}{H}{}_p^{s+2\alpha}(\Omega)$ there exists a function h in the class $H_p^s(D)$ such that $f - \widehat{A}^{-\alpha}h \in \widetilde{\Delta}(\widehat{A})$.

These assertions have been proved by Sh. A. Alimov [3] both for an elliptic operator of second order and for an elliptic operator of order m.

4.2. THEOREMS OF NEGATIVE TYPE

In this section we are concerned with negative-type theorems for an arbitrary self-adjoint nonnegative extension $\widehat{A}$ of the elliptic operator of second order (4.1.1) in an arbitrary N-dimensional domain G with weight $\rho_0(x)$. Earlier, analogous Theorems 2.1 and 2.2 have been established by us for the Laplace operator (see Chapter 2).

Assume that the coefficients of operator (4.1.1) and the weight function $\rho_0(x)$ satisfy the conditions formulated in Section 4.1.1.

Here we are concerned with the following two propositions.

THEOREM 4.1. *(On conditions that are not sufficient for guaranteeing the localization of the Riesz means of order s in the Zygmund–Hölder classes). Assume that $N \geq 2$, $0 \leq s < (N-1)/2$, G is an arbitrary domain in the space E^N, $\widehat{A}$ is an arbitrary self-adjoint nonnegative extension of the elliptic operator (4.1.1) in domain G with weight $\rho_0(x)$, x_0 is a fixed interior point of domain G, α is a fixed real number subject to $0 < \alpha < (N-1)/2 - s$. If so, there exists a function $f(x)$ satisfying the following conditions: (1) $f(x)$ is finite in domain G and becomes zero in a certain vicinity D of the point x_0; (2) $f(x)$ belongs to the Zygmund–Hölder class $C^\alpha(G)$; (3) the Riesz means $\sigma_\lambda^s(x_0, f)$ of order s and size λ of a spectral decomposition of function $f(x)$ are infinite as $\lambda \to \infty$ at point x_0.*

THEOREM 4.2. *(On conditions that are not sufficient for guaranteeing the localization of the Riesz means of order s in the Sobolev–Liouville, Nikol'skii, and Besov classes). Assume that $N \geq 2$, $0 \leq s < (N-1)/2$, G is an arbitrary domain in the space E^N, $\widehat{A}$ is an arbitrary self-adjoint nonnegative extension of the elliptic operator (4.1.1) in domain G with weight $\rho_0(x)$, x_0 is a fixed interior point of domain G, α is a fixed real number subject to $0 < \alpha < (N-1)/2 - s$. If so, there exists a function $f(x)$ satisfying the following conditions: (1) $f(x)$ is finite in domain G and becomes zero in a certain vicinity D of the point x_0; (2) $f(x)$ belongs to each of the Sobolev–Liouville $L_p^\alpha(G)$, Nikol'skii $H_p^\alpha(G)$, and Besov $B_{p,\theta}^\alpha(G)$ classes with order of differentiability α, with any degree of summability $p \geq 1$ and (in the case of a Besov class) with any*

$\theta \geq 1$; (3) the Riesz means $\sigma_\lambda^s(x,f)$ of order s and size λ of a spectral decomposition of function $f(x)$ are infinite as $\lambda \to \infty$ at point x_0.

Following the same line of reasoning as in Section 2.2.1, one can easily show that Theorem 4.2 is a simple consequence of Theorem 4.1.

With this in mind, we shall focus on proving Theorem 4.1 only.

We divide Section 4.2 into four minor sections. In Section 4.2.1, a proof is given of the central lemma on the lower-bound estimate for the Lebesgue function of the Riesz means of a small nonnegative order; in Section 4.2.2, a proof is given of the main inequality used in proving the lemma in Section 4.2.1; in Section 4.2.3, the proof of the central lemma on the lower-bound estimate of the Lebesgue function is extended to the Riesz means of any nonnegative order; finally, Section 4.4.2 is concerned with a direct proof of Theorem 4.1.

4.2.1. Proof of the Central Lemma for the Riesz Means of a Small Nonnegative Order

LEMMA 4.2. *(On the lower-bound estimate of the Lebesgue function of the Riesz means of order $0 \leq s < 1/2$). Assume that G is an arbitrary domain in the space E^N, $\widehat{A}$ is an arbitrary self-adjoint nonnegative extension of operator (4.1.1) in domain G with weight $\rho_0(x)$; taken with respect to $\widehat{A}$ is an arbitrary ordered spectral representation of the space*[7] *$L_2(G,\rho_0(x))$ with a measure $\rho(\lambda)$, with sets of multiplicity e_i, with fundamental functions $u_i(x,\lambda)$ $(i = 1,2,\dots,\widehat{m})$ of multiplicity $\widehat{m} \leq \infty$; x_0 is a fixed point of the open domain G; r_{x_0y} is the geodetic distance between the point x_0 and a variable point y measured along the rays of Eq. (4.1.7) emanating from the point x_0; E is a ring layer of the form $E = \{R^4/4 \leq r_{x_0y} \leq R^4\}$ entirely contained in domain G. If so, for any s in the half-interval $0 \leq s < 1/2$ we can fix a positive number R so small that for an $\alpha_0 > 0$ and for all sufficiently large $\lambda > 0$ the following inequality holds:*

$$\int_E \left| \sum_{i=1}^{\widehat{m}} \int_0^\lambda u_i(x_0,t)u_i(y,t)\left(1-\frac{t}{\lambda}\right)^s d\rho(t) \right| \rho_0(y)dy \geq \alpha_0 \lambda^{(N-1)/4-s/2}. \tag{4.2.1}$$

REMARK. Denoting the Riesz means of order s for the spectral function of extension $\widehat{A}$ by the symbol $\theta^s(x,y,\lambda)$, the estimate (4.2.1) can be rewritten in the form

$$\int_E |\theta^s(x_0,y,\lambda)|\, \rho_0(y)dy \geq \alpha_0 y^{(N-1)/4-s/2}. \tag{4.2.1'}$$

[7] See Section 4.1.1.

The more so, it follows from inequality (4.2.1′) that with the ring layer replaced by the entire domain G, the inequality

$$\int_G |\theta^s(x_0, y, \lambda)|\, \rho_0(y)dy \geq \alpha_0 y^{(N-1)/1-s/2}$$

holds, which is in fact a lower-bound estimate for the Lebesgue function of the Riesz means of order s.

PROOF OF LEMMA 4.2. We fix an arbitrary point x_0 in the open domain G and define a family of rays of Eq. (4.1.7) in the vicinity of the point x_0. We assume that the positive number R is small enough to enable the placement of a geodetic ball of radius R centered at x_0 within the domain G. Now we consider the function

$$V^\lambda(x_0, y) = p(x_0, y)\overset{\lambda}{v}(r_{x_0y}),$$

where r_{x_0y} is the geodetic distance between the point x_0 and a variable point y, $p(x, y)$ is the function defined in the Moiseev mean-value formula (4.1.10) (see Section 4.1.2), and the function $\overset{\lambda}{v}(r_{x_0y})$ is, to within a factor $\rho_0(x_0)$, the same as that for the Laplace operator (see function (2.4.5), Section 2.4.1):

$$\overset{\lambda}{v}(r_{x_0y}) = \begin{cases} \dfrac{\Gamma(s+1)2^s(2\pi)^{-N/2}}{\rho_0(x_0)} \lambda^{N/4-s/2} r_{x_0y}^{-(N/2+s)} J_{N/2+s}(r_{x_0y}\sqrt{\lambda}) & \text{for} \quad r_{x_0y} \leq R, \\ 0 & \text{for} \quad r_{x_0y} > R. \end{cases} \tag{4.2.2}$$

By virtue of relations (4.1.11)–(4.1.13), the Fourier image $\widehat{V}_i^\lambda(x_0, t)$ of a function $V^\lambda(x_0, y)$ with respect to a fundamental function $u_i(y, t)$ is

$$\widehat{V}_i^\lambda(x_0, t) = (2\pi)^{N/2}\rho_0(x_0)u_i(x_0, t)t^{(2-N)/4} \int_0^R \overset{\lambda}{v}(r)r^{N/2} J_{N/2-1}(r\sqrt{t})dr + \widetilde{V}_i^\lambda(x_0, t) \tag{4.2.3}$$

$$(i = 1, 2, \ldots, \widehat{m}),$$

where the additional term $\widetilde{V}_i^\lambda(x_0,t)$ takes the form

$$\widetilde{V}_i^\lambda(x_0,t) = \rho_0(x_0)\int_0^R r^{N/2}\overset{\lambda}{v}(r)\left[\int_0^r \tau^\nu w_\nu(\tau\sqrt{t}, r\sqrt{t}) \times\left(\int_\omega u_i(x_0+\tau\omega,\lambda)q(x_0,x_0+\tau\omega)d\omega\right)d\tau\right]dr. \tag{4.2.4}$$

One will recall that $\nu = (N-2)/2$, $w_\nu(a,b) = J_\nu(a)Y_\nu(b) - J_\nu(b)Y(a)$.

Substituting the expression (4.2.2) for function $\overset{\lambda}{v}(r)$ into the right-hand of (4.2.3) and (4.2.4), the Fourier image $\widehat{V}_i^\lambda(x_0,t)$ can be rewritten as

$$\widehat{V}_i^\lambda(x_0,t) = \Gamma(s+1)2^s\lambda^{N/4-s/2}t^{(2-N)/4}u_i(x_0,t) \times\int_0^R J_{N/2+s}(r\sqrt{\lambda})J_{N/2-1}(r\sqrt{t})r^{-s}dr + \widetilde{V}_i^\lambda(x_0,t) \tag{4.2.5}$$

$$(i = 1, 2, \dots, \widehat{m}),$$

where the additional term $\widetilde{V}_i^\lambda(x_0,t)$ is

$$\widetilde{V}_i^\lambda(x_0,t) = \Gamma(s+1)2^s(2\pi)^{-N/2}\lambda^{N/4-s/2}\int_0^R r^{-s}J_{N/2+s}(r\sqrt{\lambda}) \times\left[\int_0^r \tau^\nu w_\nu(\tau\sqrt{t}, r\sqrt{t})\left(\int_\omega u_i(x_0+\tau\omega,\lambda)q(x_0,x_0+\tau\omega)d\omega\right)d\tau\right]dr. \tag{4.2.6}$$

The right-hand side of (4.2.5), without the additional term $\widetilde{V}_i^\lambda(x_0,t)$, looks the same as that in the case of the Laplace operator (see Section 2.4.1). Therefore, carrying out the transformation $\int_0^R = \int_o^\infty - \int_R^\infty$, we rewrite the Fourier image $\widehat{V}_i^\lambda(x_0,t)$ as

$$\widehat{V}_i^\lambda(x_0,t) = \delta_t^\lambda u_i(x_0,t)\left(1-\frac{t}{\lambda}\right)^s - \Gamma(s+1)2^s\lambda^{N/4-s/2}t^{(2-N)/4}u_i(x_0,t)I_t^\lambda(R) + \widetilde{V}_i^\lambda(x_0,t), \tag{4.2.7}$$

where $\widetilde{V}_i^\lambda(x_0,t)$ is the additional term as defined by relation (4.2.6), and the quantities δ_t^λ and $I_t^\lambda(R)$ have the same meaning as those in Section 2.4.1, that is,

$$\delta_t^\lambda = \begin{cases} 1 & \text{for} \quad t > \lambda, \\ 0 & \text{for} \quad t \le \lambda, \end{cases}$$

$$I_t^\lambda(R) = \int\limits_R^\infty J_{N/2+s}(r\sqrt{\lambda})J_{N/2-1}(r\sqrt{t})r^{-s}dr. \tag{4.2.8}$$

Note that, given $s = 0$ and $t = \lambda$, we take $\delta_t^\lambda = 1/2$ and $(1 - t/\lambda)^s = 1$.

In much the same manner as in Section 2.4.1, we multiply (4.2.7) by the fundamental function $u_i(y,t)$ and subject the equality thus obtained, first, to summation over all the numbers i from 1 to $\widehat{m}$ and, next, to integration over the spectral measure $\rho(t)$ between the limits from 0 to Λ with respect to t; here $\Lambda = \Lambda(\lambda)$ is a sufficiently large number whose choice will be specified below. As a result, we obtain the following equality:

$$\sum_{i=1}^{\widehat{m}} \int\limits_0^\Lambda \widehat{V}_i^\lambda(x_0,t)u_i(y,t)d\rho(t) = \sum_{i=1}^{\widehat{m}} \int\limits_0^\lambda u_i(x_0,t)u_i(y,t)\left(1 - \frac{t}{\lambda}\right)^s d\rho(t)$$

$$- \Gamma(s+1)2^s\lambda^{N/4-s/2} \sum_{i=1}^{\widehat{m}} \int\limits_0^\Lambda u_i(x_0,t)u_i(y,t)t^{(2-N)/4} I_t^\lambda(R)d\rho(t) \tag{4.2.9}$$

$$+ \sum_{i=1}^{\widehat{m}} \int\limits_0^\Lambda \widetilde{V}_i^\lambda(x_0,t)u_i(y,t)d\rho(t).$$

(Here we take into account relation (4.2.8) and the fact that the integral on the left-hand side and the first two integrals under the summation sign on the right-hand side of (4.2.9) are absolutely convergent (for the proof of this see Section 2.4.1). It is to be inferred, therefore, that the third integral under the summation sign on the right-hand side of (4.2.9) is also absolutely convergent.)

The following inequality stems from (4.2.9):

$$\left|\sum_{i=1}^{\widehat{m}}\int_0^{\lambda} u_i(x_0,t)u_i(y,t)\left(1-\frac{t}{\lambda}\right)^s d\rho(t)\right| \geq \left|\sum_{i=1}^{\widehat{m}}\int_0^{\Lambda} \widehat{V}_i^{\lambda}(x_0,t)u_i(y,t)d\rho(t)\right|$$

$$-\Gamma(s+1)2^s\lambda^{N/4-s/2}\left|\sum_{i=1}^{\widehat{m}}\int_0^{\Lambda} u_i(x_0,t)u_i(y,t)t^{(2-N)/4}I_t^{\lambda}(R)d\rho(t)\right|$$

$$-\left|\sum_{i=1}^{\widehat{m}}\int_0^{\Lambda} \widetilde{V}_i^{\lambda}(x_0,t)u_i(y,t)d\rho(t)\right|.$$

The integration of this inequality with weight $\rho_0(y)$ on the coordinates of point y over the ring layer $E = \{R^4/4 \leq r_{x_0y} \leq R^4\}$ yields

$$\int_E \left|\sum_{i=1}^{\widehat{m}}\int_0^{\lambda} u_i(x_0,t)u_i(y,t)\left(1-\frac{t}{\lambda}\right)^s d\rho(t)\right| \rho_0(y)dy$$

$$\geq \int_E \left|\sum_{i=1}^{\widehat{m}}\int_0^{\Lambda} \widehat{V}_i^{\lambda}(x_0,t)u_i(y,t)d\rho(t)\right| \rho_0(y)dy$$

$$-\Gamma(s+1)2^s\lambda^{N/4-s/2}\int_E \left|\sum_{i=1}^{\widehat{m}}\int_0^{\Lambda} u_i(x_0,t)u_i(y,t)t^{(2-N)/4}I_t^{\lambda}(R)d\rho(t)\right| \rho_0(y)dy$$

$$-\int_E \left|\sum_{i=1}^{\widehat{m}}\int_0^{\Lambda} \widetilde{V}_i^{\lambda}(x_0,t)u_i(y,t)d\rho(t)\right| \rho_0(y)dy. \tag{4.2.10}$$

Repeating word for word the arguments that have been used in proving Lemma 2.3 (see Section 2.4.1), based only on the Parseval equality, on estimates (4.1.48)–(4.1.50), and on the estimate of (4.2.8) (see Section 2.4.2), we establish that there exists a positive constant β, dependent on N, s and on the positive infimum of the weight function $\rho_0(x)$, such that for a sufficiently small $R > 0$, for any $\lambda > 0$, and for all sufficiently large $\Lambda(\lambda, R)$ the following inequalities hold[8]):

$$\int_E \left| \sum_{i=1}^{\widehat{m}} \int_0^\Lambda \widehat{V}_i^\lambda(x_0,t) u_i(y,t) d\rho(t) \right| \rho_0(y) dy \geq 2\beta R^{2N-2-4s} \lambda^{(N-1)/4-s/2}, \tag{4.2.11}$$

$$\Gamma(s+1)2^s \int_E \left| \sum_{i=1}^{\widehat{m}} \int_0^\Lambda u_i(x_0,t) u_i(y,t) t^{(2-N)/4} I_t^\lambda(R) d\rho(t) \right| \rho_0(y) dy \leq \beta R^{2N-2-4s}. \tag{4.2.12}$$

One will infer from inequalities (4.2.10)–(4.2.12) that in order to prove Lemma 4.2 it suffices to show that the inequality

$$\int_E \left| \sum_{i=1}^{\widehat{m}} \int_0^\Lambda \widetilde{V}_i^\lambda(x_0,t) u_i(y,t) d\rho(t) \right| \rho_0(y) dy \leq \frac{\beta}{2} R^{2N-2-4s} \lambda^{(N-1)/4-s/2} \tag{4.2.13}$$

holds for a sufficiently small $R > 0$ and for any $\lambda > 0$, given sufficiently large $\Lambda(\lambda, R)$.

We now proceed with a proof of inequality (4.2.13). We make use of $\widetilde{V}^\lambda(x_0, y)$ to denote a function whose Fourier image in the expansion with respect to fundamental functions $u_i(y,t)$ is $\widetilde{V}_i^\lambda(x_0,t)$ $(i = 1, 2, \dots, \widehat{m})$.

In order to show the existence of such a function as an element of the space $L_2(G, \rho_0(y))$, it suffices to establish

$$\sum_{i=1}^{\widehat{m}} \int_0^\infty \left| \widetilde{V}_i^\lambda(x_0,t) \right|^2 d\rho(t) < \infty$$

for each fixed $\lambda > 0$.

We wish to prove a stronger inequality

$$\sum_{i=1}^{\widehat{m}} \int_0^\infty \left| \widetilde{V}_i^\lambda(x_0,t) \right|^2 d\rho(t) \leq C \lambda^{(N-1)/2-s}, \tag{4.2.14}$$

where C is a constant independent of either λ or R.

First, we make certain that inequality (4.2.14) provides a route to proving Lemma 4.2, and after that we shall turn our attention to a proof of the inequality in question.

---

[8]) The involvement of the weighting factor $\rho_0(y)$ in this particular case does not violate the logic of the reasoning, considering that this factor is bounded on both sides by positive constants.

Thus, we assume that the inequality has already been proved. We note that by the definition of a spectral decomposition the left-hand side of (4.2.13) can be written in the form

$$\int_E \left| E_\Lambda(y, \widetilde{V}^\lambda(x_0, y)) \right| \rho_0(y) dy, \tag{4.2.15}$$

where $E_\Lambda(y, \widetilde{V}^\lambda(x_0, y))$ signifies the spectral decomposition of a function $\widetilde{V}^\lambda(x_0, y)$ taken at a point y (the point x_0 is assumed to be a fixed one).

Since the function $\widetilde{V}^\lambda(x_0, y)$ belongs to the class $L_2(G, \rho_0(y))$, the quantity (4.2.15) tends to $\int_E \left|\widetilde{V}^\lambda(x_0, y)\right| \rho_0(y)dy$ as $\Lambda \to \infty$.

Therefore, for every $\lambda > 0$ and every $R > 0$ the number Λ may be taken arbitrarily large such that the quantity (4.2.15) becomes different from $\int_E \left|\widetilde{V}^\lambda(x_0, y)\right| \rho_0(y)dy$ in absolute value by less than $\frac{\beta}{4} R^{2N-2-4s} \lambda^{(N-1)/4-s/2}$.

With Λ thus chosen, to establish inequality (4.2.13) for a sufficiently small $R > 0$ one only needs to show that

$$\int_E \left|\widetilde{V}^\lambda(x_0, y)\right| \rho_0(y) < \frac{\beta}{4} R^{2N-2-4s} \lambda^{(N-1)/4-s/2}. \tag{4.2.16}$$

To this end, we apply the Cauchy–Buniakowski inequality to the left-hand side of (4.2.16) and then we make use of the Parseval equation (4.1.6) for the function $\widetilde{V}^\lambda(x, y)$, inequality (4.2.14), and the trivial estimate $\int_E \rho_0(y)dy = O(R^{4N})$.

We thus obtain

$$\begin{aligned}
&\int_E \left|\widetilde{V}^\lambda(x_0, y)\right| \rho_0(y)dy \\
&\le \left[\int_G \left|\widetilde{V}^\lambda(x_0, y)\right|^2 \rho_0(y)dy\right]^{1/2} \left[\int_E \rho_0(y)dy\right]^{1/2} \\
&= O(R^{2N}) \left[\sum_{i=1}^{\widehat{m}} \int_0^\infty \left|\widetilde{V}_i^\lambda(x_0, t)\right|^2 d\rho(t)\right]^{1/2} \\
&= O(R^{2N} \lambda^{(N-1)/4-s/2}) = \lambda^{(N-1)/4-s/2} o(R^{2N-2-4s}).
\end{aligned}$$

It follows therefrom that for any sufficiently small $R > 0$

$$\int_E \left|\widetilde{V}^\lambda(x_0, y)\right| \rho_0 dy < \frac{\beta}{4} R^{2N-2-4s} \lambda^{(N-1)/4-s/2}.$$

We have thereby shown the validity of inequality (4.2.16) (and consequently of inequality (4.2.13)) provided that inequality (4.2.14) holds.

To complete the proof of Lemma 4.2 all we have left to do is to establish inequality (4.2.14).

4.2.2. Proof of the Main Inequality (4.2.14)

By virtue of relation (4.2.6), to prove inequality (4.2.14) it suffices to establish the existence of a constant A, independent of R and λ, such that

$$I(R,\lambda) = \sum_{i=1}^{\widehat{m}} \int_0^\infty \Bigg\{ \int_0^R r^{-s} J_{N/2+s}(r\sqrt{\lambda}) \Bigg[\int_0^r \tau^\nu w_\nu(\tau\sqrt{t}, r\sqrt{t}) \times \left(\int_\omega u_i(x_0+\tau\omega, t) q(x_0, x_0+\tau\omega) d\omega \right) d\tau \Bigg] dr \Bigg\}^2 d\rho(t) \le \frac{A}{\sqrt{\lambda}}. \tag{4.2.17}$$

Reversing the order of integration with respect to r and τ in the expression within the braces in (4.2.17), we obtain

$$\begin{aligned} &\int_0^R r^{-s} J_{N/2+s}(r\sqrt{\lambda}) \Bigg[\int_0^r \tau^\nu w_\nu(\tau\sqrt{t}, r\sqrt{t}) \times \left(\int_\omega u_i(x_0+\tau\omega, t) q(x_0, x_0+\tau\omega) d\omega \right) d\tau \Bigg] dr \\ &= \int_0^R \tau^\nu \left(\int_\omega u_i(x_0+\tau\omega, t) q(x_0, x_0+\tau\omega) d\omega \right) \times \left[\int_\tau^R r^{-s} J_{N/2+s}(r\sqrt{\lambda}) w_\nu(\tau\sqrt{t}, r\sqrt{t}) dr \right] d\tau. \end{aligned} \tag{4.2.18}$$

We substitute (4.2.18) into (4.2.17) and carry out integration by parts with respect to τ, assuming

$$u(\tau) = \tau^{-\delta} \int_\tau^R r^{-s} J_{N/2+s}(r\sqrt{\lambda}) w_\nu(\tau\sqrt{t}, r\sqrt{t}) dr,$$

$$v(r) = \int_0^\tau \tau_1^{\nu+\delta} \left(\int_\omega u_i(x_0 + \tau_1\omega, t) q(x_0, x_0 + \tau_1\omega) d\omega \right) d\tau_1,$$

where $\delta = 1/6 - s/3$. Note that, since $0 \le s < 1/2$ by the condition of Lemma 4.2, the δ belongs to the half-interval $0 < \delta \le 1/6$.

Taking into account that both substitutions $[u(\tau)v(\tau)]|_{\tau=0}^{\tau=R}$ become zero and that, by virtue of the relation $w_\nu(\tau\sqrt{t}, r\sqrt{t}) = J_\nu(\tau\sqrt{t})Y_\nu(r\sqrt{t}) - J_\nu(r\sqrt{t})Y_\nu(\tau\sqrt{t})$, the equality $w_\nu(\tau\sqrt{t}, \tau\sqrt{t}) = O$ holds, and with reference to relation (4.1.9) (see Section 4.1.2), we obtain that the left-hand side of (4.2.17) is defined by the equality

$$I(R,\lambda) = \sum_{i=1}^{\widehat{m}} \int_0^\infty \Bigg\{ \int_0^R \Bigg[\int_{r_{x_0 y} \le \tau} \frac{u_i(y,t) q(x_0, y)}{\rho_0(y) r_{x_0 y}^{N/2-\delta}} dy \Bigg]$$

$$\times \left[\int_\tau^R r^{-s} J_{N/2+s}(r\sqrt{\lambda}) \frac{d}{d\tau} (\tau^{-\delta} w_\nu(\tau\sqrt{t}, r\sqrt{t})) dr \right] d\tau \Bigg\}^2 d\rho(t). \tag{4.2.19}$$

Establishing the estimate $I(R,\lambda) = O(\lambda^{-1/2})$ for (4.2.19) implies that inequality (4.2.14) has been proved.

We now proceed with a proof of this estimate. It may be assumed, with no loss of generality, that $R < 1$. We partition the integral with respect to t between the limits from 0 to ∞ into the sum of two integrals, one from 0 to $1/R^2$ and the other one from $1/R^2$ to ∞. For the latter integral we partition the inner integral with respect to the variable τ between the limits from 0 to R into the sum of two integrals, one from 0 to $1/\sqrt{t}$ and the other one from $1/\sqrt{t}$ to R. For the integral with respect to the variable t taken between the limits from 0 to $1/R^2$ we partition the integration with respect to the variable r between the limits from τ to R into the sum of two integrations between the limits from τ to $1/\sqrt{t}$ and from $1/\sqrt{t}$ to R.

Allowing for the trivial equality

$$\begin{aligned} &\frac{d}{d\tau}\left[\tau^{-\delta} w_\nu(\tau\sqrt{t}, r\sqrt{t})\right] \\ &= J_\nu(r\sqrt{t}) \left[\sqrt{t}\,\tau^{-\delta} Y_{\nu+1}(\tau\sqrt{t}) - (\nu-\delta)\tau^{-\delta-1} Y_\nu(\tau\sqrt{t})\right] \\ &\quad - Y_\nu(r\sqrt{t}) \left[\sqrt{t}\,\tau^{-\delta} J_{\nu+1}(\tau\sqrt{t}) - (\nu-\delta)\tau^{-\delta-1} J_\nu(\tau\sqrt{t})\right], \end{aligned} \tag{4.2.20}$$

we can reduce the proof of the estimate $I(R,\lambda) = O(\lambda^{-1/2})$ for the quantity (4.2.19) to establishing that each of the five quantities

$$I_1(R,\lambda) = \sum_{i=1}^{\widehat{m}} \int\limits_{1/R^2}^{\infty} \left\{ \int\limits_{1/\sqrt{t}}^{R} \left[\int\limits_{r_{x_0 y} \le \tau} \frac{u_i(y,t)q(x_0,y)}{\rho_0(y) r_{x_0 y}^{N/2-\delta}} \rho_0(y)dy \right] \right.$$
$$\left. \times \left[\int\limits_{\tau}^{R} r^{-s} J_{N/2+s}(r\sqrt{\lambda}) \frac{d}{d\tau}(\tau^{-\delta} w_\nu(\tau\sqrt{t}, r\sqrt{t}))dr \right] d\tau \right\}^2 d\rho(t), \tag{4.2.21}$$

$$I_2(R,\lambda) = \sum_{i=1}^{\widehat{m}} \int\limits_{1/R^2}^{\infty} \left\{ \int\limits_{0}^{1/\sqrt{t}} \left[\int\limits_{r_{x_0 y} \le \tau} \frac{u_i(y,t)q(x_0,y)}{\rho_0(y) r_{x_0 y}^{N/2-\delta}} \rho_0(y)dy \right] \right.$$
$$\times \left[\sqrt{t}\,\tau^{-\delta} Y_{\nu+1}(\tau\sqrt{t}) - (\nu-\delta)\tau^{-\delta-1} Y_\nu(\tau\sqrt{t}) \right] \tag{4.2.22}$$
$$\left. \times \left[\int\limits_{\tau}^{R} r^{-s} J_{N/2+s}(r\sqrt{\lambda}) J_\nu(r\sqrt{t})dr \right] d\tau \right\}^2 d\rho(t),$$

$$I_3(R,\lambda) = \sum_{i=1}^{\widehat{m}} \int\limits_{1/R^2}^{\infty} \left\{ \int\limits_{0}^{1/\sqrt{t}} \left[\int\limits_{r_{x_0 y} \le \tau} \frac{u_i(y,t)q(x_0,y)}{\rho_0(y) r_{x_0 y}^{N/2-\delta}} \rho_0(y)dy \right] \right.$$
$$\times \left[(\nu-\delta)\tau^{-\delta-1} J_\nu(\tau\sqrt{t}) - \sqrt{t}\,\tau^{-\delta} J_{\nu+1}(\tau\sqrt{t}) \right] \tag{4.2.23}$$
$$\left. \times \left[\int\limits_{\tau}^{1/\sqrt{t}} r^{-s} J_{N/2+s}(r\sqrt{\lambda}) Y_\nu(r\sqrt{t})dr \right] d\tau \right\}^2 d\rho(t),$$

$$I_4(R,\lambda) = \sum_{i=1}^{\widehat{m}} \int\limits_{1/R^2}^{\infty} \left\{ \int\limits_{0}^{1/\sqrt{t}} \left[\int\limits_{r_{x_0 y} \le \tau} \frac{u_i(y,t)q(x_0,y)}{\rho_0(y) r_{x_0 y}^{N/2-\delta}} \rho_0(y)dy \right] \right.$$
$$\times \left[(\nu-\delta)\tau^{-\delta-1} J_\nu(\tau\sqrt{t}) - \sqrt{t}\,\tau^{-\delta} J_{\nu+1}(\tau\sqrt{t}) \right] \tag{4.2.24}$$
$$\left. \times \left[\int\limits_{1/\sqrt{t}}^{R} r^{-s} J_{N/2+s}(r\sqrt{\lambda}) Y_\nu(r\sqrt{t})dr \right] d\tau \right\}^2 d\rho(t),$$

$$I_5(R,\lambda) = \sum_{i=1}^{\widehat{m}} \int_0^{1/R^2} \left\{ \int_0^R \left[\int_{r_{x_0y}\le\tau} \frac{u_i(y,t)q(x_0,y)}{\rho_0(y)r_{x_0y}^{N/2-\delta}}\rho_0(y)dy \right] \right.$$
$$\left. \times \left[\int_\tau^R r^{-s}J_{N/2+s}(r\sqrt{\lambda})\frac{d}{d\tau}(\tau^{-\delta}w_\nu(\tau\sqrt{t},r\sqrt{t}))dr \right] d\tau \right\}^2 d\rho(t) \tag{4.2.25}$$

has the order of $O(\lambda^{-1/2})$, the estimation of the O-terms being independent of either λ or R.

We start by establishing the estimate $I_1(R,\lambda) = O(\lambda^{-1/2})$. It follows from expression (4.2.20) that the estimate[9)]

$$\left| \frac{d}{d\tau}\left[\tau^{-\delta}w_\nu(\tau\sqrt{t},r\sqrt{t})\right] \right| \le C_1 r^{-1/2}\tau^{-1/2-\delta} \tag{4.2.26}$$

is valid for $r \ge \tau \ge 1/\sqrt{t}$.

Making use of this estimate and of the familiar inequality

$$\left| J_{N/2+s}(r\sqrt{\lambda}) \right| \le C_2 r^{-1/2}\lambda^{-1/4}, \tag{4.2.27}$$

we obtain

$$\left| \int_\tau^R r^{-s}J_{N/2+s}(r\sqrt{\lambda})\frac{d}{d\tau}\left[\tau^{-\delta}w_\nu(\tau\sqrt{t},r\sqrt{t})\right] dr \right| \le C_3\lambda^{-1/4}\tau^{-1/2-s-2\delta}. \tag{4.2.28}$$

(Note that on the right-hand side of (4.2.28) at $0 < s < 1/2$ the exponent of τ may be taken equal to $-1/2-s-\delta$. Here we have chosen an exponent $-1/2-s-2\delta$ to account for the case $s = 0$.)

Now, applying the Cauchy–Buniakowski inequality to the integral within the braces in (4.2.21) and making use of estimate (4.2.28), we obtain

---

9) In what follows we make use of the symbols C_i $(i = 1, 2, \ldots)$ to denote constants which are dependent on N and $0 \le s < 1/2$ only.

$$|I_1(R,\lambda)| \leq C_4\lambda^{-1/2}\sum_{i=1}^{\widehat{m}}\int\limits_{1/R^2}^{\infty}\left[\int\limits_0^R \frac{d\tau}{\tau^{1-\delta}}\right]$$

$$\times\left[\int\limits_0^R\left(\int\limits_{r_{x_0 y}\leq\tau}\frac{u_i(y,t)q(x_0,y)}{\rho_0(y)r_{x_0y}^{N/2-\delta}}\rho_0(y)dy\right)^2\tau^{-2s-5\delta}\right]d\rho(t)$$

$$\leq C_5\lambda^{-1/2}\int\limits_0^R\tau^{-2s-5\delta}\left\{\sum_{i=1}^{\widehat{m}}\int\limits_0^{\infty}\left(\int\limits_{r_{x_0y}\leq\tau}\frac{u_i(y,t)q(x_0,y)}{\rho_0(y)r_{x_0y}^{N/2-\delta}}\rho_0(y)dy\right)^2 d\rho(t)\right\}d\tau.$$

The term within the braces on the right-hand side of this inequality, is, by virtue of the Parseval equality (4.1.6),

$$\int\limits_{r_{x_0y}\leq\tau}\frac{q^2(x_0,y)}{\rho_0^2(y)r_{x_0y}^{N-2\delta}}\rho_0(y)dy \leq C_6,$$

which, in view of the equality $-2s-5\delta=-1+\delta$, leads to the estimate

$$|I_1(R,\lambda)| \leq C_7\lambda^{-1/2}\int\limits_0^R\tau^{-1+\delta}d\tau \leq C_8\lambda^{-1/2}.$$

This completes the proof of the estimate $I_1(R,\lambda)=O(\lambda^{-1/2})$.

In establishing the estimate $I_2(R,\lambda)=O(\lambda^{-1/2})$, we take into account that $\tau\leq 1/\sqrt{t}$ and, consequently,

$$\begin{aligned}&\left|\sqrt{t}\,\tau^{-\delta}Y_{\nu+1}(\tau\sqrt{t})-(\nu-\delta)\tau^{-\delta-1}Y_\nu(\tau\sqrt{t})\right|\\ &\qquad\leq C_9\tau^{-\delta-\nu-1}(\sqrt{t})^{-\nu}\leq C_9\tau^{-2\delta-\nu-1}(\sqrt{t})^{-\nu-\delta}.\end{aligned} \tag{4.2.29}$$

Moreover, by virtue of the inequality $|J_\nu(r\sqrt{t})|\leq C_{10}r^{-1/2}t^{-1/4}$ and estimate

(4.2.27), we arrive at the inequality

$$\left|\int_{\tau}^{R} r^{-s} J_{N/2+s}(r\sqrt{\lambda}) J_{\nu}(r\sqrt{t})dr\right|$$
$$\le C_{11}\lambda^{-1/4}t^{-1/4}\int_{\tau}^{R} r^{-1-s-\delta}dr \le C_{12}\lambda^{-1/4}(\sqrt{t})^{-1/2}\tau^{-s-\delta}, \tag{4.2.30}$$

(with allowance made for $r \le R < 1$ and, consequently, $r^{-1-s} \le r^{-1-s-\delta}$).

Multiplying inequalities (4.2.9) and (4.2.30) term by term, we come to the estimate

$$\Big|\sqrt{t}\tau^{-\delta}Y_{\nu+1}(\tau\sqrt{t})$$
$$-(\nu-\delta)\tau^{-\delta-1}Y_{\nu}(\tau\sqrt{t})\Big|\left|\int_{\tau}^{R} r^{-s} J_{N/2+s}(r\sqrt{\lambda}) J_{\nu}(r\sqrt{t})dr\right| \tag{4.2.31}$$
$$\le C_{12}\lambda^{-1/4}(\sqrt{t})^{-1/2-\nu-\delta}\tau^{-s-3\delta-\nu-1}.$$

Now, applying the Cauchy–Buniakowski inequality to the integral within the braces in (4.2.22) and making use of estimate (4.2.31) and the relation $-2s-6\delta=-1$, we obtain

$$|I_2(R,\lambda)| \le C_{13}^2\lambda^{-1/2}\sum_{i=1}^{\widehat{m}}\int_{1/R^2}^{\infty}\left[\int_0^{1/\sqrt{t}} d\tau\right]$$
$$\times\left[\int_0^{1/\sqrt{t}}\left(\int_{r_{x_0y}\le\tau}\frac{u_i(y,t)q(x_0,y)}{\rho_0(y)r_{x_0y}^{N/2-\delta}}\rho_0(y)dy\right)^2\tau^{-2s-6\delta-2\nu-2}\right]t^{-\nu-1/2-\delta}d\rho(t)$$
$$\le C_{13}^2\lambda^{-1/2}\sum_{i=1}^{\widehat{m}}\int_{1/R^2}^{\infty}t^{-\nu-1-\delta}\left\{\int_0^{1/\sqrt{t}}\left[\int_{r_{x_0y}\le\tau}u_i^2(y,t)dy\right]\right.$$
$$\left.\times\left[\int_{r_{x_0y}\le\tau}\frac{q^2(x_0,y)}{r_{x_0y}^{N-2\delta}}dy\right]\tau^{-3-2\nu}d\tau\right\}d\rho(t).$$

Further, taking into account $\nu + 1 = N/2$ and

$$\int\limits_{r_{x_0 y} \le \tau} \frac{q^2(x_0, y)}{r_{x_0 y}^{N-2\delta}} dy = O(\tau^{2\delta}),$$

we arrive at the inequality

$$|I_2(R, \lambda)|$$

$$\le C_4 \lambda^{-1/2} \int\limits_0^R \left\{ \int\limits_{r_{x_0 y} \le \tau} \left[\sum_{i=1}^{\widehat{m}} \int\limits_{1/R^2}^{\infty} t^{-N/2-\delta} u_i^2(y, t) d\rho(t) \right] dy \right\} \tau^{-3+2\delta-2\nu} d\tau.$$

Next, we recall that, by virtue of estimate (4.1.50) (see Section 4.1.3), the term within the brackets in the last inequality is uniformly bounded (for all positive $R < 1$ and for any fixed point x_0) and that $\int\limits_{r_{x_0 y} \le \tau} dy = O(\tau^{2\nu+2})$.

We thereby obtain

$$|I_2(R, \lambda)| \le C_{15} \lambda^{-1/2} \int\limits_0^R \tau^{-1+2\delta} d\tau \le C_{16} \lambda^{-1/2}.$$

This completes the derivation of the estimate $I_2(R, \lambda) = O(\lambda^{-1/2})$.

To estimate $I_3(R, \lambda)$, we take into account $\tau \le r \le 1/\sqrt{t}$. Making use of the estimates

$$|Y_\nu(r\sqrt{t})| \le C_{17}(r\sqrt{t})^{-\nu-\delta-1/2},$$
$$|J_\nu(\tau\sqrt{t})| \le C_{18}(\tau\sqrt{t})^{-1/2}$$

and invoking inequality (4.2.27), we obtain

$$|(\nu - \delta)\tau^{-\delta-1} J_\nu(\tau\sqrt{t}) - \sqrt{t}\, \tau^{-\delta} J_{\nu+1}(\tau\sqrt{t})| \le C_{19} \tau^{-1-\delta}, \tag{4.2.32}$$

$$\left| \int\limits_\tau^{1/\sqrt{t}} r^{-s} J_{N/2+s}(r\sqrt{\lambda}) Y_\nu(r\sqrt{t}) dr \right|$$

$$\le C_{20} \lambda^{-1/4} (\sqrt{t})^{-\nu-\delta-1/2} \int\limits_\tau^{1/\sqrt{t}} r^{-s-\nu-1-\delta} dr \tag{4.2.33}$$

$$\le C_{21} \lambda^{-1/4} (\sqrt{t})^{-\nu-\delta-1/2} \tau^{-s-\nu} \le C_{21} \lambda^{-1/4} (\sqrt{t})^{-\nu-\delta-1/2} \tau^{-s-\nu-2\delta}.$$

Multiplying inequalities (4.2.32) and (4.2.33) term by term brings us to the estimate

$$\begin{aligned} &|(\nu-\delta)\tau^{-\delta-1}J_\nu(\tau\sqrt{t}) \\ &\quad -\sqrt{t}\tau^{-\delta}J_{\nu+1}(\tau\sqrt{t})|\left|\int_\tau^{1/\sqrt{t}} r^{-s}J_{N/2+s}(r\sqrt{\lambda})Y_\nu(r\sqrt{t})dr\right| \\ &\le C_{22}\lambda^{-1/4}(\sqrt{t})^{-\nu-\delta-1/2}\tau^{-s-\nu-1-3\delta}. \end{aligned} \tag{4.2.34}$$

The right-hand side of (4.2.34) in order of magnitude is the same as the right-hand side of (4.2.31).

Therefore, to obtain an estimate of $I_3(R,\lambda) = O(\lambda^{-1/2})$ all we have to do is to go back, starting from estimate (4.2.34), to the argumentation that was applied above in obtaining the estimate $I_2(R,\lambda) = O(\lambda^{-1/2})$ with reference to estimate (4.2.31).

Thus, the estimate $I_3(R,\lambda) = O(\lambda^{-1/2})$ may be regarded as completed.

Let us now proceed to establish the estimate $I_4(R,\lambda) = O(\lambda^{-1/2})$. In this particular case $r > 1/\sqrt{t} \ge \tau$ and, based on inequality (4.2.27) and on the estimate $|Y_\nu(r\sqrt{t})| \le C_{23}(r\sqrt{t})^{-1/2}$, we obtain

$$\begin{aligned} &\left|\int_{1/\sqrt{t}}^{R} r^{-s}J_{N/2+s}(r\sqrt{\lambda})Y_\nu(r\sqrt{t})dr\right| \\ &\le C_{24}\lambda^{-1/4}(\sqrt{t})^{-1/2}\int_{1/\sqrt{t}}^{R} r^{-s-1}dr \le C_{24}\lambda^{-1/4}(\sqrt{t})^{-1/2}\int_{1/\sqrt{t}}^{R} r^{-s-1-\delta}dr \\ &\le C_{25}\lambda^{-1/4}(\sqrt{t})^{-1/2}(\sqrt{t})^{s+\delta} \le C_{25}\lambda^{-1/4}(\sqrt{t})^{-1/2}\tau^{-s-\delta}. \end{aligned}$$

From this inequality and from inequality (4.2.32) we obtain, with reference to $1 \le \tau^{-\nu-\delta}(\sqrt{t})^{-\nu-\delta}$,

$$\begin{aligned} &|(\nu-\delta)\tau^{-\delta-1}J_\nu(\tau\sqrt{t}) \\ &\quad -\sqrt{t}\tau^{-\delta}J_{\nu+1}(\tau\sqrt{t})|\left|\int_{1/\sqrt{t}}^{R} r^{-s}J_{N/2+s}(r\sqrt{\lambda})Y_\nu(r\sqrt{t})dr\right| \\ &\le C_{26}\lambda^{-1/4}(\sqrt{t})^{-\nu-\delta-1/2}\tau^{-s-\nu-1-3\delta}. \end{aligned} \tag{4.2.35}$$

Estimate (4.2.35) is identical to estimate (4.2.31). Therefore, the estimate $I_4(R,\lambda) = O(\lambda^{-1/2})$ is obtained from (4.2.35) in much the same manner as the estimate $I_2(R,\lambda) = O(\lambda^{-1/2})$ is obtained from (4.2.31).

The proof of the estimate $I_4(R,\lambda) = O(\lambda^{-1/2})$ is thus complete.

To establish the main estimate (4.2.14) and to prove Lemma 4.2 all we have left to do is to show that $I_5(R,\lambda) = O(\lambda^{-1/2})$. Now proceed to just this point.

In this particular case, $\tau \le r \le R \le 1/\sqrt{t}$. Inasmuch as for $N > 2$ (that is, for $\nu > 0$) the inequalities $|Y_\nu(\rho)| < C_{27}\rho^{-\nu}$, $|J_\nu(\rho)| \le C_{28}\rho^\nu$ hold for any positive $\rho \le 1$, and for $N = 2$ (that is, for $\nu = 0$) these two inequalities hold for $J_{\nu+1}(\rho)$ and $Y_{\nu+1}(\rho)$, and keeping in mind the estimates $Y_0(\rho) = (2/\pi)(\ln \rho/2) + O(1)$ and $J_0(\rho) = 1 + O(\rho^2)$, we arrive at the following estimate for (4.2.20):

(a) for $N > 2$ (that is, for $\nu > 0$)

$$\left|\frac{d}{d\tau}\left[\tau^{-\delta} w_\nu(\tau\sqrt{t}, r\sqrt{t})\right]\right| \le C_{29}\left[r^\nu \tau^{-\delta-\nu-1} + r^{-\nu}\tau^{-\delta+\nu-1}\right];$$

(b) for $N = 2$ (that is, for $\nu > 0$)

$$\left|\frac{d}{d\tau}\left[\tau^{-\delta} w_\nu(\tau\sqrt{t}, r\sqrt{t})\right]\right| \le C_{30}\left[r^\nu \tau^{-2\delta-\nu-1} + r^{-\nu}\tau^{-2\delta+\nu-1}\right]. \tag{4.2.36}$$

As is seen, estimate (4.2.36) holds for any $N \ge 2$. Using this estimate and inequality (4.2.27) and taking into account that $r^{-\nu}\tau^\nu \le 1$, we obtain

$$\begin{aligned}
&\left|\int_\tau^R r^{-s} J_{N/2+s}(r\sqrt{\lambda})\frac{d}{d\tau}\left[\tau^{-\delta} w_\nu(\tau\sqrt{t}, r\sqrt{t})\right] dr\right| \\
&\le C_{31}\lambda^{-1/4}\int_\tau^R \left[r^\nu\tau^{-2\delta-\nu-1} + r^{-\nu}\tau^{-2\delta+\nu-1}\right] r^{-s-1/2} dr \\
&\le C_{32}\lambda^{-1/4}\left[\tau^{-2\delta-\nu-1}R^{\nu-s+1/2} + \tau^{-2\delta-1}R^{-s+1/2}\right].
\end{aligned} \tag{4.2.37}$$

Applying the Cauchy–Buniakowski inequality to the integral within the braces in

(4.2.25), we obtain

$$|I_5(R,\lambda)| \leq \sum_{i=1}^{\widehat{m}} \int_0^{1/R^2} \left\{ \left[\int_0^R d\tau \right] \left[\int_0^R \left(\int_{r_{x_0 y} \leq \tau} \frac{u_i^2(y,t) q^2(x_0,y)}{r_{x_0 y}^{N-2\delta}} dy \right) \right. \right.$$

$$\left. \left. \times \left(\int_{r_{x_0 y} \leq \tau} dy \right) \left(\int_\tau^R r^{-s} J_{N/2+s}(r\sqrt{\lambda}) \frac{d}{d\tau}(\tau^{-\delta} w_\nu(\tau\sqrt{t}, r\sqrt{t}))dr \right)^2 d\tau \right] \right\} d\rho(t).$$

Next, taking into account that $\int\limits_{r_{x_0 y} \leq \tau} = O(\tau^N) = O(\tau^{2\nu+2})$ and making use of inequality (4.2.37), we obtain

$$|I_5(R,\lambda)| \leq C_{33} \lambda^{-1/2} R \int_0^R \left\{ \int_{r_{x_0 y} \leq \tau} \left[\sum_{i=1}^{\widehat{m}} \int_0^{1/R^2} u_i^2(y,t) d\rho(t) \right] \right. \tag{4.2.38}$$

$$\left. \times \frac{q^2(x_0,y)}{r_{x_0 y}^{N-2\delta}} dy \right\} (R^{2\nu-2s+1} \tau^{-4\delta} + R^{-2s+1} \tau^{2\nu-4\delta}) d\tau.$$

We note, finally, that the expression within the brackets in (4.2.38) equals $O(R^{-N})$ by virtue of estimate (4.1.48) taken at $t = \mu$ (for all $R < 1$ and for any fixed point x_0).

It follows therefrom that

$$|I_5(R,\lambda)| \leq C_{34} \lambda^{-1/2} R^{1-N} \int_0^R \left(\int_{r_{x_0 y} \leq \tau} \frac{q^2(x_0,y)}{r_{x_0 y}^{N-2\delta}} dy \right) \tag{4.2.39}$$

$$\times (R^{N-1-2s} \tau^{-4\delta} + R^{-2s+1} \tau^{N-2-4\delta}) d\tau.$$

Recalling that $R < 1$ and taking into account

$$\int_{r_{x_0 y} \leq \tau} \frac{q^2(x_0,y)}{r_{x_0 y}^{N-2\delta}} dy = O(\tau^{2\delta}),$$

we finally obtain from (4.2.39) that

$$|I_5(R,\lambda)| \leq C_{35} \lambda^{-1/2} R^{1-2s-2\delta} = C_{35} \lambda^{-1/2} R^{4\delta} \leq C_{35} \lambda^{-1/2}.$$

This completes the proof of the estimate $I_5(R,\lambda) = O(\lambda^{-1/2})$.

Hence, the main inequality (4.2.14) has been proved, and the proof of Lemma 4.2 is thus complete.

It follows from the above reasoning that for all x_0 that belong to an arbitrary fixed compact set of domain G, the inequality (4.2.14) holds with the same constant C on its right-hand side.

4.2.3. On the Exactness (to within an Order of Magnitude) of the Main Estimate for the Central Lemma

Naturally the question arises: Is the main lower-bound estimate (4.2.1) exact to within an order of magnitude with respect to λ? The following assertion answers this question in the affirmative.

LEMMA 4.3. *Assume that all the notations of Lemma 4.2 are retained, all the conditions of the lemma are fulfilled, and a positive number R has been fixed in a similar manner for any s in the interval of $0 \le s < 1/2$. If so, there is a positive constant α_1 such that for all sufficiently large $\lambda > 0$ the upper-bound estimate*

$$\int_E \left| \sum_{i=1}^{\widehat{m}} \int_0^{\lambda} u_i(x_0,t)u_i(y,t)\left(1-\frac{t}{\lambda}\right)^s d\rho(t) \right| \rho_0(y)dy \le \alpha_1 \lambda^{(N-1)/4-s/2} \tag{4.2.40}$$

holds, where the λ on the right-hand side is of the same power of $(N-1)/4-s/2$ as that on the right-hand side of estimate (4.2.1).

PROOF OF LEMMA 4.3: From equality (4.2.9), established in proving Lemma 4.2 (see Section 4.2.1), one infers that

$$\begin{aligned} &\left| \sum_{i=1}^{\widehat{m}} \int_0^{\lambda} u_i(x_0,t)u_i(y,t)\left(1-\frac{t}{\lambda}\right)^s d\rho(t) \right| \\ &\le \left| \sum_{i=1}^{\widehat{m}} \int_0^{\Lambda} \widehat{V}_i^{\lambda}(x_0,t)u_i(y,t)d\rho(t) \right| \\ &\quad + \Gamma(s+1)2^s\lambda^{N/4-s/2} \left| \sum_{i=1}^{\widehat{m}} \int_0^{\Lambda} u_i(x_0,t)u_i(y,t)t^{(2-N)/4} I_t^{\lambda}(R)d\rho(t) \right| \\ &\quad + \left| \sum_{i=1}^{\widehat{m}} \int_0^{\Lambda} \widetilde{V}_i^{\lambda}(x_0,t)u_i(y,t)d\rho(t) \right|. \end{aligned}$$

Integration of this inequality with weight $\rho_0(y)$ over the ring layer $E = |R^4/4 \leq r_{x_0y} \leq R^4|$ yields

$$\int_E \left| \sum_{i=1}^{\widehat{m}} \int_0^{\lambda} u_i(x_0,t)u_i(y,t)\left(1-\frac{t}{\lambda}\right)^s d\rho(t)\right| \rho_0(y)dy$$

$$\leq \int_E \left| \sum_{i=1}^{\widehat{m}} \int_0^{\Lambda} \widehat{V}_i^{\lambda}(x_0,t)u_i(y,t)d\rho(t)\right| \rho_0(y)dy$$

$$+\Gamma(s+1)2^s\lambda^{N/4-s/2} \int_E \left| \sum_{i=1}^{\widehat{m}} \int_0^{\Lambda} u_i(x_0,t)u_i(y,t)t^{(2-N)/4} I_t^{\lambda}(R)d\rho(t)\right| \rho_0(y)dy$$

$$+\int_E \left| \sum_{i=1}^{\widehat{m}} \int_0^{\Lambda} \widetilde{V}_i^{\lambda}(x_0,t)u_i(y,t)d\rho(t)\right| \rho_0(y)dy.$$

(4.2.41)

It follows from inequality (4.2.41) and from the upper-bound estimates (4.2.12) and (4.2.13) for sufficiently large Λ that in order to establish the desired estimate (4.2.40), it suffices to establish the estimate

$$\int_E \left| \sum_{i=1}^{\widehat{m}} \int_0^{\Lambda} \widehat{V}_i^{\lambda}(x_0,t)u_i(y,t)d\rho(t)\right| \rho_0(y)dy = O(\lambda^{(N-1)/4-s/2}) \qquad (4.2.42)$$

for all sufficiently large Λ.

Since, for any fixed $\lambda > 0$, the function $V^{\lambda}(x_0,y)$ belongs to the class $L_2(G,\rho_0(y))$, the spectral decomposition of this function converges in the $L_2(G,\rho_0(y))$ metric by virtue of the Gårding–Browder–Mautner theorem (see Section 4.1.1); consequently, by virtue of the Cauchy–Buniakowski inequality, it also converges in the $L_1(E,\rho_0(y))$ metric.

Therefore, to prove the validity of estimate (4.2.42) for all sufficiently large Λ, it will be enough to establish the estimate

$$\int_E \left|V^{\lambda}(x_0,y)\right| \rho_0(y)dy = O(\lambda^{(N-1)/4-s/2}). \qquad (4.2.43)$$

Inasmuch as $V^{\lambda}(x_0,y) = p(x_0,y)\overset{\lambda}{v}(r_{x_0y})$, where $p(x_0,y)$ is a function bounded on both sides (see the Moiseev mean-value formula, Section 4.1.2) and $\overset{\lambda}{v}(r_{x_0y})$ is a function defined by relation (4.2.2), it suffices in establishing estimate (4.2.43) to make use of the familiar estimate for a Bessel function,

$$\left|J_{N/2+s}(r\sqrt{\lambda})\right| = O(r^{-1/2}\lambda^{-1/4}).$$

The validity of estimate (4.2.43) follows from the above Bessel function estimate and from the fact that for any $N \geq 1$

$$\int\limits_{R^4/4 \leq r_{x_0 y} \leq R^4} r_{x_0 y}^{-(N+1)/2} dy \leq C = \text{const}.$$

This completes the proof of Lemma 4.3.

4.2.4. Lemma on the Riesz Means of any Nonnegative Order

LEMMA 4.4. *Assume that all the notations of Lemma 4.2 are retained, and all the conditions of the lemma are fulfilled. If so, for any $s \geq 0$ and any $\delta > 0$ a positive number R can be fixed so small that the expression*

$$\lambda^{-(N-1)/4+s/2+\delta} \int\limits_E \left| \sum_{i=1}^{\widehat{m}} \int\limits_0^{\lambda} u_i(x_0,t) u_i(y,t) \left(1 - \frac{t}{\lambda}\right)^s d\rho(t) \right| \rho_0(y) dy \qquad (4.2.44)$$

is an unbounded function of λ on the half-line $\lambda \geq 1$.

PROOF: We note, first of all, that at $0 \leq s < 1/2$ and for any $\delta > 0$ the statement of Lemma 4.4 immediately follows from Lemma 4.2; consequently, under the specified conditions of $0 \leq s < 1/2$ and any $\delta > 0$, Lemma 4.4 may regarded as proved.

We therefore have to prove that Lemma 4.4 holds also at $s \geq 1/2$, for any $\delta > 0$.

Since, for any s_0 in the interval $0 < s_0 < 1/2$ and for any $\delta > 0$, the assertion of Lemma 4.4 on the unboundedness of (4.2.44) on the half-line $\lambda \geq 1$ holds, there exists a function $h(y)$ (by virtue of the above-mentioned resonance-type theorem[10]) in the class $L_\infty(E)$ for any s_0 and for any $0 < \delta < 1$ such that the expression

$$\lambda^{-(N-1)/4+s_0/2+\delta} \int\limits_E \left[\sum_{i=1}^{\widehat{m}} \int\limits_0^{\lambda} u_i(x_0,t) u_i(y,t) \left(1 - \frac{t}{\lambda}\right)^{s_0} d\rho(t) \right] \rho_0(y) h(y) dy \quad (4.2.45)$$

is unbounded in value on the half-line $\lambda \geq 1$.

With reference to (4.2.25) and keeping in mind that

$$\lambda^{-(N-1)/4} \int\limits_E \left[\sum_{i=1}^{\widehat{m}} \int\limits_0^{\lambda} u_i(x_0,t) u_i(y,t) d\rho(t) \right] \rho_0(y) h(y) dy \qquad (4.2.46)$$

[10] See, for example, Kaczmarz and Steinhaus [51, p. 31, Assertion 3].

is bounded on this half-line for any function $h(y)$ in the class $L_\infty(E)$ (see Lemma 4.3, the case $s = 0$), we shall convince ourselves now that the expression

$$\lambda^{-(N-1)/4+s/2+\delta} \int_E \left[\sum_{i=1}^{\widehat{m}} \int_0^\lambda u_i(x_0,t)u_i(y,t) \left(1 - \frac{t}{\lambda}\right)^s d\rho(t) \right] \rho_0(y)h(y)dy \quad (4.2.47)$$

is also unbounded in value on the half-line $\lambda \geq 1$ for any $s > s_0$, for any δ in the interval of $0 < \delta < 1$, and for a function $h(y)$ in the class $L_\infty(E)$.

This having been done, the proof of Lemma 4.4 is complete, since the unboundedness of (4.2.47) on the half-line $\lambda \geq 1$ implies also the unboundedness of (4.2.44) on $\lambda \geq 1$ for any $s > s_0$ and for any δ in the interval $0 < \delta < 1$ (and, consequently, for any positive δ).

For convenience, we adopt the following notation:

$$\sigma_\lambda^s(h) = \sum_{i=1}^{\widehat{m}} \int_0^\lambda u_i(x_0,t) \left[\int_E u_i(y,t)\rho_0(y)h(y)dy \right] \left(1 - \frac{t}{\lambda}\right)^s d\rho(t). \quad (4.2.48)$$

Expressions (4.2.45), (4.2.46), and (4.2.47) can now be rewritten as

$$\lambda^{-(N-1)/4+s_0/2+\delta}\sigma_\lambda^{s_0}(h), \quad (4.2.45')$$

$$\lambda^{-(N-1)/4}\sigma_\lambda^0(h), \quad (4.2.46')$$

$$\lambda^{-(N-1)/4+s/2+\delta}\sigma_\lambda^s(h). \quad (4.2.47')$$

To prove Lemma 4.4, it will suffice to establish the validity of the following assertion:

If there exists a constant C such that for all $h(y)$ in $L_\infty(E)$

$$\left|\lambda^{-(N-1)/4}\sigma_\lambda^0(h)\right| \leq C, \quad (4.2.49)$$

and if there exists a function $h(y)$ in the class $L_\infty(E)$ such that for a certain $s_0 > 0$ and for any δ in the interval $0 < \delta < 1$

$$\overline{\lim_{\lambda\to+\infty}} \left|\lambda^{-(N-1)/4+s_0/2+\delta}\sigma_\lambda^{s_0}(h)\right| = +\infty, \quad (4.2.50)$$

then for any $s > s_0$ and for any δ in the interval $0 < \delta < 1$ there exists a function $h(y)$ in the class $L_\infty(E)$ such that

$$\overline{\lim_{\lambda\to\infty}} \left|\lambda^{-(N-1)/4+s/2+\delta}\sigma_\lambda^{s}(h)\right| = +\infty. \quad (4.2.51)$$

The proof of this assertion will be carried out by contradiction, as has been suggested by Sh. A. Alimov.

Assume for a certain $s > s_0$, for a certain δ in the interval $o < \delta < 1$, and for any function $h(y)$ in the class $L_\infty(E)$ that there exists a constant C_1 such that

$$\left|\lambda^{-(N-1)/4+s/2+\delta}\sigma_\lambda^s(h)\right| \le C_1. \tag{4.2.52}$$

We fix a positive ε subject to

$$0 < \varepsilon < \frac{2\delta s_0}{s+2\delta}; \tag{4.2.53}$$

next, we consider the function of complex argument z

$$\Phi(z) = \lambda^{-(N-1)/4+(s/2+\delta)z}\sigma_\lambda^{sz+\varepsilon}(h), \tag{4.2.54}$$

defined within the band $0 \le \operatorname{Re} z \le 1$.

Let us ensure that for any function $h(y)$ in the class $L_\infty(E)$ the function (4.2.54) is bounded at the boundary of the band $0 \le \operatorname{Re} z \le 1$, that is, for any $h(y)$ in $L_\infty(E)$ there exist constants C_2 and C_3 such that the conditions

$$|\Phi(0+i\eta)| \le C_2, \tag{4.2.55}$$

$$|\Phi(1+i\eta)| \le C_3 \tag{4.2.56}$$

hold for all real η.

We make use of the well-known relation for the Riesz means of two arbitrary complex orders α and β such that $\operatorname{Re}\beta > \operatorname{Re}\alpha \ge 0$:

$$\lambda^\beta\sigma_\lambda^\beta(h) \quad = \Gamma(\beta+1)\left[\Gamma(\beta-\alpha)\Gamma(\alpha+1)\right]^{-1}\int_0^\lambda (\lambda-t)^{\beta-\alpha-1}t^\alpha\sigma_t^\alpha(h)dt. \tag{4.2.57}$$

In order to establish (4.2.55), we make use of inequality (4.2.49), definition (4.2.54), and relation (4.2.57) taken at $\beta = \varepsilon + is\eta$, $\alpha = 0$. We obtain

$$|\Phi(0+i\eta)| = \lambda^{-(N-1)/4}\left|\lambda^{i(s/2+\delta)\eta}\right| |\sigma_\lambda^{\varepsilon+is\eta}(h)|$$

$$\le C_4\lambda^{-(N-1)/4-\varepsilon}|\lambda^{-is\eta}|\int_0^\lambda (\lambda-t)^{\varepsilon-1}|(\lambda-t)^{is\eta}|\,|\sigma_t^0(h)|dt$$

$$\le C_5\lambda^{-(N-1)/4-\varepsilon}\int_0^\lambda (\lambda-t)^{\varepsilon-1}t^{(N-1)/4}dt \le C_6,$$

since

$$\int_0^\lambda (\lambda-t)^{\varepsilon-1}t^{(N-1)/4}dt = \lambda^{\varepsilon+(N-1)/4}\int_0^1 (1-\tau)^{\varepsilon-1}\tau^{(N-1)/4}d\tau \le \frac{1}{\varepsilon}\lambda^{\varepsilon+(N-1)/4}.$$

Thus, inequality (4.2.55) has been proved.

To establish inequality (4.2.56), we make use of definition (4.2.54) for the function $\Phi(z)$, relation (4.2.57) taken at $\beta = s + \varepsilon + is\eta$, $\alpha = s$, and inequality (4.2.52), which we have assumed to hold true. In this particular case we obtain

$$|\Phi(1+i\eta)| = \lambda^{-(N-1)/4+s/2+\delta}\left|\lambda^{i(s/2+\delta)\eta}\right| |\sigma_\lambda^{s+\varepsilon+is\eta}(h)|$$

$$\leq C_7\lambda^{-(N-1)/4+s/2+\delta}\lambda^{-s-\varepsilon}|\lambda^{-is\eta}|\int_0^\lambda (\lambda-t)^{(s+\varepsilon)-s-1}|(\lambda-t)^{is\eta}|t^s|\sigma_t^s(h)|dt$$

$$\leq C_8\lambda^{-(N-1)/4-s/2+\delta-\varepsilon}\int_0^\lambda (\lambda-t)^{\varepsilon-1}t^{s+(N-1)/4-s/2-\delta}dt \leq C_9$$

since for $0<\delta<1$ the inequality $(N-1)/4+s/2-\delta > -1$ holds, implying

$$\int_0^\lambda (\lambda-t)^{\varepsilon-1}t^{(N-1)/4+s/2-\delta}dt$$

$$= \lambda^{(N-1)/4+s/2-\delta+\varepsilon}\int_0^1 (1-\tau)^{\varepsilon-1}\tau^{(N-1)/4+s/2-\delta}d\tau \leq C_{10}\lambda^{(N-1)/4+s/2-\delta+\varepsilon}.$$

Thus, inequality (4.2.56) has also been proved.

From inequalities (4.2.55), (4.2.56) and from the known Stein's interpolation lemma[11)] one infers that for any t in the interval $0<t<1$ there exists a constant C_{11} such that

$$|\Phi(t)| \leq C_{11}. \tag{4.2.58}$$

We consider now the number t defined as

$$t = \frac{s_0-\varepsilon}{s}. \tag{4.2.59}$$

As long as $s_0 < s$ and taking into account the right-hand side of inequality (4.2.53), the estimate $\varepsilon < s_0$ is valid; consequently, the number t as defined by (4.2.59) satisfies the condition $0<t<1$.

Therefore, having set $t = (s-\varepsilon)/s$ in (4.2.58) and allowing for $ts+\varepsilon = s_0$ by virtue of (4.2.59), we obtain

$$\left|\lambda^{-(N-1)/4+(s/2+\delta)((s_0-\varepsilon)/s)}\sigma_\lambda^{s_0}(h)\right| \leq C_{11}. \tag{4.2.60}$$

11) See, for example, A. Zygmund [79, Vol. 2, p. 149].

Now, to make certain that inequality (4.2.60) is contradictory to (4.2.50), we must show that the number $(s/2+\delta)\frac{s_0-\varepsilon}{s}$ admits the representation

$$\left(\frac{s}{2}+\delta\right)\frac{s_0-\varepsilon}{s}=\frac{s_0}{2}+\delta_1, \tag{4.2.61}$$

where δ_1 is a number in the interval $0<\delta_1<1$.

We put $\delta_1=\left(\frac{s}{2}+\delta\right)\frac{s_0-\varepsilon}{s}-\frac{s_0}{2}$.

By virtue of the right-hand inequality (4.2.53) we have

$$\delta_1=\frac{(s+2\delta)(s_0-\varepsilon)}{2s}-\frac{s_0}{2}>\frac{(s+2\delta)\left[s_0-s_0\frac{2\delta}{s+2\delta}\right]}{2s}-\frac{s_0}{2}=0;$$

therefore $\delta_1>0$.

Further, in consideration of $0<\delta<1$, we have

$$\delta_1=\frac{s_0}{2}+\delta\frac{s_0-\varepsilon}{s}-\left(\frac{s}{2}+\delta\right)\frac{\varepsilon}{s}-\frac{s_0}{2}<\delta\frac{s_0-\varepsilon}{s}=\delta t<\delta<1.$$

One infers that representation (4.2.61) is justified.

By virtue of this representation, inequality (4.2.60) is in contradiction with relation (4.2.50).

The contradiction we have arrived at completes the proof of the assertion formulated above and of Lemma 4.4.

4.2.5. Direct Proof of Theorem 4.1

Using Lemma 4.4 and invoking the arguments used in Section 2.4.4 irrespective of the form of an elliptic operator, we come to an assertion which we formulate, with no loss of generality, in the following manner: We assume that the extension $\widehat{A}$ of the elliptic operator (4.1.1) is a strictly positive one, that is, given a $\mu_0>0$, the condition $(\widehat{A}u,u)\geq\mu_0(u,u)$ is satisfied for any function $u(x)$ in the class $C_0^\infty(G)$.

LEMMA 4.5. *Assume that all the conditions of Lemma 4.2 are satisfied. If so, for any s in the half-segment $0\leq s<(N-1)/2$ and for any δ in the interval $0<\delta<(N-1)/4-s/2$, a positive number R can be fixed so small that the expression*

$$\int_E\left|\sum_{i=1}^{\widehat{m}}\int_{\mu_0}^{\lambda}u_i(x_0,t)u_i(y,t)t^{-((N-1)/4-s/2-\delta)}\left(1-\frac{t}{\lambda}\right)^s d\rho(t)\right|\rho_0(y)dy \tag{4.2.62}$$

is unbounded in value on the half-line $\lambda\geq\mu_0$.

The unboundedness of (4.2.62) and the resonance-type theorem, invoked in the foregoing section, imply that there exists a function $h(y)$ in the class $L_\infty(E)$ such that the expression

$$\sigma_\lambda^s(x_0) = \int_E h(y) \left[\sum_{i=1}^{\widehat{m}} \int_5^\lambda u_i(x_0,t) u_i(y,t) t^{-((N-1)/4-s/2-\delta)} \left(1 - \frac{t}{\lambda}\right)^s d\rho(t) \right] \rho_0(y) dy \tag{4.2.63}$$

is also unbounded in value on the half-line $\lambda \geq \mu_0$.

We put

$$\bar{h}(y) = \begin{cases} h(y) & \text{for} \quad y \in E, \\ 0 & \text{for} \quad y \in G \backslash E \end{cases}$$

and consider the function

$$F(x) = \widehat{A}^{-\alpha_0} \bar{h}(y), \tag{4.2.64}$$

subject to $\alpha_0 = (N-1)/4 - s/2 - \delta$. The unboundedness of (4.2.63) on the half-line $\lambda \geq \mu_0$ signifies that the Riesz means of order s for function (4.2.64) diverge at point x_0. In other words,

$$\overline{\lim_{\lambda \to +\infty}} \sigma_\lambda^s(F, x_0) = +\infty. \tag{4.2.65}$$

By virtue of Assertion 1 at $p = \infty$ (see Section 4.1.4), there is a function $f(x)$ belonging to the Zygmund–Hölder class $C^{(N-1)/2-s-2\delta}(G)$, finite in the layer $\widehat{E} = \{R^4/8 \leq r_{x_0 y} \leq 2R^4\}$, and such that the difference $f(x) - F(x)$ belongs to the class $\widetilde{\Delta}(\widehat{A})$ (defined in Section 4.1.3) which is an intersection of the domains of definition for all positive integer powers of the operator $\widehat{A}$.

Inasmuch as δ is an arbitrary positive number, one can fix, for any α in the interval $0 < \alpha < (N-1)/2 - s$, a $\delta > 0$ such that the function $f(x)$ belongs to the class $C^\alpha(G)$.

Furthermore, since the function $f(x)$ is finite in the layer $\widehat{E} = \{R^4/8 \leq r_{x_0 y} \leq 2R^4\}$, this function, given a sufficiently small R, is finite in domain G and becomes zero in the vicinity of the point x_0.

What we have left to do is to show that the Riesz means of order s are divergent at point x_0, that is, to prove that

$$\overline{\lim_{\lambda \to +\infty}} \sigma_\lambda^s(x_0, f) = +\infty. \tag{4.2.66}$$

Since $f(x) - F(x) \in \widetilde{\Delta}(\widehat{A})$, the spectral decomposition of the function $g(x) = f(x) - F(x)$ and, the more so, its Riesz means of any positive order are convergent at the point x_0.[12)] If so, (4.2.66) follows from (4.2.65).

This completes the proof of Theorem 4.1.

---

12) This follows from the fact that for any m the Fourier image $\widehat{g}_i(t)$ of the function $g(x) =$

4.3. THEOREMS OF POSITIVE TYPE

In this section we are concerned with theorems of positive type for an arbitrary self-adjoint nonnegative extension $\widehat{A}$ of the elliptic operator of second order (4.1.1) in an arbitrary N-dimensional domain G with weight $\rho_0(x)$. These theorems are analogous to Theorems 2.3, 2.4, and 2.5, earlier established in Chapter 2 for the Laplace operator.

For simplicity, we shall focus on the spectral decompositions rather than on their Riesz means. We assume also, with no loss of generality, that the self-adjoint extension under study is strictly positive, that is, subject to the condition $(\widehat{A}u, u) \geq \mu_0(u, u)$ for a certain $\mu_0 > 0$ and for all $u(x)$ in the class $C_0^\infty(G)$.

Furthermore, we assume that the coefficients of operator (4.1.1) and the weight function $\rho_0(x)$ satisfy the requirements as formulated in Section 4.1.1.

THEOREM 4.3. *(On conditions ensuring localization and on conditions ensuring a uniform convergence of spectral decompositions in the Nikol'skii classes). Assume that $N \geq 2$, G is an arbitrary domain in the space E^N, $\widehat{A}$ is an arbitrary self-adjoint nonnegative extension of operator (4.1.1) in domain G with weight $\rho_0(x)$, and $f(x)$ is an arbitrary function satisfying the following requirements:*

(1) $f(x)$ becomes zero outside the set G_{h_0} at a certain $h_0 > 0$[13]*;*

(2) the function $f(x)$ in the entire domain G belongs to the Nikol'skii class H_2^α at $\alpha \geq (N-1)/2$;

(3) in a certain domain D interior to, or occasionally coincident with, domain G, the function $f(x)$ belongs to the Nikol'skii class H_p^α for certain α and p subject to the conditions

$$\alpha \geq (N-1)/2, \qquad \alpha p > N, \quad p \geq 1. \tag{4.3.1}$$

If so, the spectral decomposition of the function $f(x)$ converges to this function as $\lambda \to \infty$ uniformly on any compact set K of domain D.

A comparison of Theorem 4.3 with Theorem 4.2 for the case $s = 0$ enables us to draw the following conclusions:

I. An ultimate condition for localization of the spectral decomposition of a (finite in domain G) function in the Nikol'skii class $H_2^\alpha(G)$ is the requirement that $\alpha \geq (N-1)/2$ (for $\alpha \geq (N-1)/2$, the localization of spectral decompositions takes place in accordance with Theorem 4.3; for $\alpha < (N-1)/2$, no localization of spectral decompositions occurs, according to Theorem 4.2).

$f(x) - F(x)$ is $(-1)^m(\widehat{L^m g})_i(t)t^{-m}$, and, therefore, the spectral decomposition of the function $f(x)$ converges uniformly on any compact set of domain G by virtue of the Cauchy–Buniakowski inequality, the Parseval equality for $L^m g(x)$, and by virtue of the convergence of series (4.1.51) uniform on any compact set of domain G for any $\delta > 0$.

[13] Recall that for any $h > 0$ and for any domain D we make use of the symbol D_h to denote a set of points in domain D removed from the boundary ∂D of domain D to a distance greater than h.

II. The definitive condition for uniform convergence of the spectral decompositions of the function in the Nikol'skii class $H_p^\alpha(G)$, finite in domain G, is the fulfillment of the three inequalities (4.3.1).

Indeed, the definiteness of the first inequality (4.3.1) follows from comparing Theorem 4.3 to Theorem 4.2, and the definiteness of the other two inequalities (4.3.1) is borne out by the same argumentation as in the case of the Laplace operator (see Section 2.2.1).

The following two theorems immediately follow from Theorem 4.3.

THEOREM 4.4. *(On conditions ensuring localization, and on conditions ensuring a uniform convergence of spectral decompositions in the Sobolev–Liouville classes and the Besov classes). Assume that $N \geq 2$, G is an arbitrary domain in the space E^N, $\hat{A}$ is an arbitrary self-adjoint nonnegative extension of operator (4.1.1) in domain G with weight $\rho_0(x)$, $f(x)$ is an arbitrary function satisfying the following requirements:*

(1) $f(x)$ becomes zero outside the set G_{h_0} at a certain $h_0 > 0$;

(2) in the entire domain G, the function $f(x)$ belongs to the Sobolev–Liouville class L_2^α (respectively, to the Besov class $B_{2,\theta}^\alpha$) at $\alpha \geq (N-1)/2$ (and at any $\theta \geq 1$);

(3) in a certain domain D interior to, or occasionally coincident with, G the function $f(x)$ belongs to the Sobolev–Liouville class L_p^α (respectively, to the Besov class $B_{p,\theta}^\alpha$) for certain α and p satisfying the three inequalities (4.3.1) (for any $\theta \geq 1$ in the case of the Besov class).

If so, the spectral decomposition of the function $f(x)$ converges to this function as $\lambda \to \infty$ uniformly on any compact set K of domain G.

A comparison of Theorem 4.4 with Theorem 4.2 taken at $s = 0$ allows one to draw two conclusions similar to those for the Nikol'skii class, namely:

I. The condition $\alpha \geq (N-1)/2$ is an ultimate condition for localization of the spectral decomposition of a function finite in domain G in the Sobolev–Liouville class $L_2^\alpha(G)$ and the Besov class $B_{2,\theta}^\alpha(G)$ (for any $\theta \geq 1$);

II. The three inequalities (4.1.1) are definitive conditions for uniform convergence of the spectral decomposition of a function, finite in domain G, in the Sobolev–Liouville class $L_p^\alpha(G)$ and in the Besov class $B_{p,\theta}^\alpha(G)$ (for any $\theta \geq 1$).

THEOREM 4.5. *(On conditions ensuring localization and uniform convergence of spectral decompositions in the Zygmund–Hölder classes): Assume that $N \geq 2$, G is an arbitrary domain in the space E^N, $\hat{A}$ is an arbitrary self-adjoint nonnegative extension of operator (4.1.1) in domain G with weight $\rho_0(x)$, $f(x)$ is an arbitrary function satisfying the following two requirements:*

(1) $f(x)$ becomes zero outside the set G_{h_0} at a certain $h_0 > 0$;

(2) $f(x)$ belongs to the Zygmund–Hölder class $C^\alpha(G)$ at a certain $\alpha \geq (N-1)/2$.

If so, the spectral decomposition of the function $f(x)$ converges to this function as $\lambda \to \infty$ uniformly on any compact set K of domain G.

Comparing Theorem 4.5 to Theorem 4.1 taken at $s = 0$ leads, similar to the case of the Laplace operator, to the conclusion that the requirement that $\alpha \geq (N-1)/2$

is a definitive condition both for localization and uniform convergence of the spectral decomposition of a function, finite in domain G, in the Zygmund–Hölder class $C^\alpha(G)$. Note that the Zygmund–Hölder classes fail to resolve a "gap" between the definitive conditions for localization and uniform convergence.

As has already been mentioned above, Theorems 4.4 and 4.5 are corollaries to Theorem 4.3, since, given any fixed $\alpha > 0$ and $p \geq 1$ and any $\theta \geq 1$, each of the classes $L_2^\alpha(G)$ and $B_{p,\theta}^\alpha(G)$ is contained, by virtue of embeddings (2.1.18) (see Section 2.1.3), in $H_p^\alpha(G)$ and, moreover, the class $C^\alpha(G)$ is contained in the class $H_p^\alpha(G)$ at any $p \geq 1$.

Our major goal in this section is to prove Theorem 4.3. In Sections 4.1.1–4.1.4 we prove the part of the theorem concerned with localization, and in Sections 4.1.5–4.1.7, we prove the statement about the uniform convergence of spectral decompositions.

4.3.1. Lemma on the Fourier Images of a Finite Function in the Nikol'skii Class

In this section, we are primarily concerned with a proof of the ancillary assertion about the Fourier images of a finite function in the main domain G on the Nikol'skii class $H_2^\alpha(G)$. This assertion is an analog of Lemma 2.6 (see Section 2.5.1).

However, we precede the proof of this assertion with the demonstration of an important ancillary relation.

LEMMA 4.6. *Let the function $F(x, y)$ be continuous and finite in domain $G \times G$ and the integral*

$$\int_\omega f(x + r\omega)d\omega$$

be defined by equality (4.1.8) (see Section 4.1.2). Then for every sufficiently small $R > 0$ the following relation holds:

$$\int_G \left[\int_\omega F(x, x + R\omega)d\omega \right] dx = \int_G \left[\int_\omega F(x + R\omega, x)d\omega \right] dx. \qquad (4.3.2)$$

PROOF: With reference to inequality (4.1.9) and denoting by r_{xy} the geodetic distance between x and y, we can write the left-hand side of (4.3.2) in the following form:

$$\int_G \left[\int_\omega F(x, x + R\omega) d\omega \right] dx$$

$$= \frac{1}{R^{N-1}} \frac{d}{dR} \left\{ \int_G \left[\int_0^R r^{N-1} \left(\int_\omega F(x, x + r\omega) d\omega \right) dr \right] dx \right\} \tag{4.3.3}$$

$$= \frac{1}{R^{N-1}} \frac{d}{dR} \left\{ \int_G \left[\int_{r_{xy} \leq R} F(x, y) dy \right] dx \right\}.$$

If R does not exceed the distance between the subset $G \times G$ (on which $F(x, y)$ is distinct from zero) and the boundary of $G \times G$, then, continuing $F(x, y)$ with zero beyond $G \times G$, we may assume that the outer integral with respect to x in the right-hand side of (4.3.3) is taken over the entire space E^N.

With this in mind and having defined a function $\eta_R(t)$ of the form

$$\eta_R(t) = \begin{cases} 1 & \text{for} \quad 0 \leq t \leq R, \\ 0 & \text{for} \quad t > R \end{cases}$$

we can rewrite the right-hand side of (4.3.3) in the following manner:

$$\frac{1}{R^{N-1}} \frac{d}{dR} \left\{ \int_G \left[\int_{r_{xy} \leq R} F(x, y) dy \right] dx \right\}$$
$$= \frac{1}{R^{N-1}} \frac{d}{dR} \left\{ \int_{E^N} \left[\int_{E^N} F(x, y) \eta_R(r_{xy}) dy \right] dx \right\}. \tag{4.3.4}$$

Applying the Fubini theorem to the right-hand side of (4.3.4) and taking into account the symmetry invariance of the geodetic distance r_{xy}, we obtain

$$\frac{1}{R^{N-1}}\frac{d}{dR}\left\{\int\limits_{E^N}\left[\int\limits_{E^N} F(x,y)\eta_R(r_{xy})dy\right]dx\right\}$$

$$= \frac{1}{R^{N-1}}\frac{d}{dR}\left\{\int\limits_{E^N}\left[\int\limits_{E^N} F(x,y)\eta_R(r_{xy})dx\right]dy\right\} \tag{4.3.5}$$

$$= \frac{1}{R^{N-1}}\frac{d}{dR}\left\{\int\limits_{E^N}\left[\int\limits_{r_{xy}\leq R} F(x,y)dx\right]dy\right\}.$$

Making use once again of equality (4.1.9) and keeping in mind that the function $F(x,y)$ equals zero outside $G \times G$, we can see that the right-hand side of (4.3.5) is

$$\frac{1}{R^{N-1}}\frac{d}{dR}\left\{\int\limits_{E^N}\left[\int\limits_{r_{xy}\leq R} F(x,y)dx\right]dy\right\}$$

$$= \frac{1}{R^{N-1}}\frac{d}{dR}\left\{\int\limits_{G}\left[\int\limits_{0}^{R} r^{N-1}\left(\int\limits_{\omega} F(y+t\omega,y)d\omega\right)dr\right]dy\right\} \tag{4.3.6}$$

$$= \int\limits_{G}\left[\int\limits_{\omega} F(y+r\omega,y)d\omega\right]dy.$$

Comparing equalities (4.3.3)–(4.3.6), we come to relation (4.3.2). This completes the proof of Lemma 4.6.

By convention, let the symbol E_λ denote an operator that brings every function $f(x)$ into correspondence with its spectral decomposition $E_\lambda f$, and the symbol I denote a unit operator. Also, we assume that $N \geq 2$.

LEMMA 4.7. *(On the Fourier images of a finite function in the Nikol'skii class). For an arbitrary $\alpha > 0$ and an arbitrary function $f(x)$, finite in domain G, in the Nikol'skii class $H_2^\alpha(G)$, there exists a constant C such that the inequality*

$$\|(I - E_\lambda)f\|_{L_2(G)} \leq C\lambda^{-\alpha/2}\|f\|_{H_2^\alpha(G)} \tag{4.3.7}$$

holds.

PROOF: 1^0. First, we consider the case $0 < \alpha < 1$.

We can focus our reasoning on a function $f(x)$ in the class $C_0^\infty(G)$ and provide, for this function, the proof of estimate (4.3.7) with a constant C independent of the function $f(x)$.

For an arbitrary $h > 0$ smaller than the geodetic distance between the support of the function $f(x)$ and the boundary of domain G, we consider the function

$$\varepsilon(x, h, f) = \varepsilon(x, h) = \frac{1}{\omega_N} \int_\omega f(x + h\omega) p(x + h\omega, x) d\omega, \tag{4.3.8}$$

where $p(x, y)$ is the function involved in the Moiseev mean-value formula (4.1.10) (see Section 4.1.2) and ω_N is the surface area of a Euclidean sphere in E^N of unit radius, that is,

$$\omega_N = 2(\pi)^{N/2}[\Gamma(N/2)]^{-1}.$$

Let us determine the Fourier image $\widehat{\varepsilon}_i(\lambda, h)$ of function (4.3.8) with respect to the fundamental function $u_i(x, \lambda)$.

Recalling the definition of a Fourier image and making use of relation (4.3.2), we obtain

$$\begin{aligned}
\widehat{\varepsilon}_i(\lambda, h) &= \int_G \varepsilon(x, h) u_i(x, \lambda) \rho_0(x) dx \\
&= \frac{1}{\omega_N} \int_G u_i(x, \lambda) \rho_0(x) \left[\int_\omega f(x + h\omega) p(x + h\omega, x) d\omega \right] dx \\
&= \frac{1}{\omega_N} \int_G f(x) \left[\int_\omega u_i(x + h\omega, \lambda) \rho_0(x + h\omega) p(x, x + h\omega) d\omega \right] dx.
\end{aligned} \tag{4.3.9}$$

Applying the Moiseev mean-value formula (4.1.10) to the bracketed term in the right-hand side of (4.3.9) and making use of ν to denote $\nu = (N-2)/2$ and of $\varphi(t)$ to denote the function

$$\varphi(t) = 2^\nu \Gamma(\nu + 1) t^{-\nu} J_\nu(t), \tag{4.3.10}$$

we arrive at the following expression for the Fourier image:

$$\widehat{\varepsilon}_i(\lambda, h) = \widehat{f}_i(\lambda) \varphi(h\sqrt{\lambda}) + P_i(\lambda, h), \tag{4.3.11}$$

where $\widehat{f}_i(\lambda) = \int_G f(x) u_i(x, \lambda) \rho_0(x) dx$ is the Fourier image for the function $f(x)$, and $P_i(\lambda, h)$ is generated by the additional term of the mean-value formula (4.1.10) and takes the form

$$P_i(\lambda,h) = \frac{1}{\omega_N}\int\limits_G f(x)\rho_0(x)\Big\{h^{-\nu}\int\limits_0^h t^\nu w_\nu(t\sqrt{\lambda},h\sqrt{\lambda}) \times\Big[\int\limits_\omega u_i(x+t\omega,\lambda)q(x,x+t\omega)d\omega\Big]dt\Big\}dx. \tag{4.3.12}$$

Our aim is to prove the estimate

$$\sum_{i=1}^{\widehat{m}}\int\limits_0^\infty[\varphi(h\sqrt{\lambda})-1]^2\left|\widehat{f}_i(\lambda)\right|^2 d\rho(\lambda) \le C_1h^{2\alpha}\|f\|^2_{H_2^\alpha(G)}. \tag{4.3.13}$$

Having established estimate (4.3.13), we see that the function $\varphi(t)$ in the form (4.3.10) implies the existence of a constant $C_2 > 0$ such that for all $t \ge 1$ the inequality $|\varphi(t)-1| \ge C_2$ holds; further, (4.3.13), in turn, implies

$$\sum_{i=1}^{\widehat{m}}\int\limits_{h\sqrt{t}\ge 1}\left|\widehat{f}_i(t)\right|^2 d\rho(t) \le C_3h^{2\alpha}\|f\|^2_{H_2^\alpha(G)};$$

from this inequality at $h = 1/\sqrt{\lambda}$, by virtue of the Parseval equality

$$\|(I-E_\lambda f\|^2_{L_2(G)} = \sum_{i=1}^{\widehat{m}}\int\limits_\lambda^\infty\left|\widehat{f}_i(t)\right|^2 d\rho(t)$$

we obtain the desired estimate (4.3.7).

Thus, to prove Lemma 4.7 for the particular case $0 < \alpha < 1$, it suffices to establish estimate (4.3.13).

We now proceed to establish this estimate.

From equality (4.3.11) the relation

$$[\varphi(h\sqrt{\lambda})-1]\widehat{f}_i(\lambda) = [\widehat{\varepsilon}_i(\lambda,h)-\widehat{f}_i(\lambda)] - P_i(\lambda,h)$$

follows, which in turn implies the inequality

$$\sum_{i=1}^{\widehat{m}}\int\limits_0^\infty[\varphi(h\sqrt{\lambda})-1]^2\widehat{f}_i^2(\lambda)d\rho(\lambda) \le 2\sum_{i=1}^{\widehat{m}}\int\limits_0^\infty\left[\widehat{\varepsilon}_i(\lambda,h)-\widehat{f}_i(\lambda)\right]^2 d\rho(\lambda) + 2\sum_{i=1}^{\widehat{m}}\int\limits_0^\infty P_i^2(\lambda,h)d\rho(\lambda). \tag{4.3.14}$$

A moment's consideration of (4.3.14) shows that in order to prove estimate (4.3.13), it suffices to establish the following two estimates:

$$\sum_{i=1}^{\widehat{m}} \int_0^\infty \left[\widehat{\varepsilon}_i(\lambda, h) - \widehat{f}_i(\lambda)\right]^2 d\rho(\lambda) \leq C_4 h^{2\alpha} \|f\|^2_{H_2^\alpha(G)}, \tag{4.3.15}$$

$$\sum_{i=1}^{\widehat{m}} \int_0^\infty P_i^2(\lambda, h) d\rho(\lambda) \leq C_5 h^{2\alpha} \|f\|^2_{H_2^\alpha(G)}. \tag{4.3.16}$$

To prove estimate (4.3.15), we make use of the Parseval equality

$$\sum_{i=1}^{\widehat{m}} \int_0^\infty \left[\widehat{\varepsilon}_i(\lambda, h) - \widehat{f}_i(\lambda)\right]^2 d\rho(\lambda) = \int_G \rho_0(x)[\varepsilon(x, h) - f(x)]^2 dx \tag{4.3.17}$$

and the relation

$$\begin{aligned} \varepsilon(x, h) - f(x) = \tfrac{1}{\omega_N} \int_\omega [f(x + h\omega) - f(x)] p(x + h\omega, x) d\omega \\ + f(x) \left[\frac{1}{\omega_N} \int_\omega p(x + h\omega, x) d\omega - 1\right]. \end{aligned} \tag{4.3.18}$$

Previously, we have established in Section 4.1.2 that the relation

$$\frac{1}{\omega_N} \int_\omega p(x + h\omega, x) d\omega = 1 + O(h) \tag{4.3.19}$$

holds uniformly with respect to x on an arbitrary compact set of domain G for all sufficiently small $h > 0$.

From relations (4.3.17), (4.3.18), and (4.3.19), from the fact that $0 < \alpha < 1$, from the definition of the norm in the space H_2^α, and from the inequality $(A + B)^2 \leq 2A^2 + 2B^2$, we obtain

$$\begin{aligned} &\sum_{i=1}^{\widehat{m}} \int_0^\infty [\widehat{\varepsilon}_i(\lambda, h) - \widehat{f}_i(\lambda)]^2 d\rho(\lambda) \\ &\leq C_6 \int_G \left[\int_\omega |f(x + h\omega) - f(x)|^2 d\omega\right] dx + C_7 h^2 \int_G |f(x)|^2 dx \\ &\leq C_8 h^{2\alpha} \|f\|^2_{H_2^\alpha(G)} + C_7 h^2 \|f\|^2_{L_2(G)} \leq C_9 h^{2\alpha} \|f\|^2_{H_2^\alpha(G)}. \end{aligned}$$

Thus, estimate (4.3.15) has been proved.

To complete the proof of Lemma 4.7 in the case $0 < \alpha < 1$, all we have to do is to prove estimate (4.3.16).

For every t in the half-segment $0 < t \leq h$ we introduce an ancillary function

$$Q(x,t) = \frac{1}{\rho_0(x)} \int\limits_{\omega} f(x+t\omega) q(x+t\omega, x) \rho_0(x+t\omega) d\omega \tag{4.3.20}$$

with its Fourier image defined as

$$\widetilde{Q}_i(\lambda,t) = \int\limits_{G} Q(x,t) u_i(x,\lambda) \rho_0(x) dx, \qquad i = 1,2,\dots,\widehat{m}.$$

Making use of relation (4.3.2), we obtain

$$\widehat{Q}_i(\lambda,t) = \int\limits_{G} f(x)\rho_0(x) \left[\int\limits_{\omega} u_i(x+t\omega,\lambda) q(x, x+t\omega) d\omega \right] dx. \tag{4.3.21}$$

Comparing (4.3.21) and (4.3.12) and keeping in mind that $\nu = N/2 - 1$, we arrive at the following relation:

$$P_i(\lambda,h) = \frac{1}{\omega_N} h^{1-N/2} \int\limits_0^h w_{(N-2)/2}(t\sqrt{\lambda}, h\sqrt{\lambda}) t^{N/2-1} \widehat{Q}_i(\lambda,t) dt. \tag{4.3.22}$$

We apply the Cauchy–Buniakowski inequality to the integral on the right-hand side of (4.3.22) to obtain

$$P_i^2(\lambda,h) \leq \omega_N^{-2} h^{2-N} \int\limits_0^h w^2_{(N-2)/2}(t\sqrt{\lambda}, h\sqrt{\lambda}) t^{N-2} dt \int\limits_0^h \widehat{Q}_i^2(\lambda,t) dt. \tag{4.3.23}$$

Recalling the well-known properties of cylindrical functions[14])

(a) for $\nu > 0$ and for all $\tau > 0$

$$J_\nu^2(\tau) = O(1), \qquad J_\nu^2(\tau) = O(\tau^{2\nu}),$$

$$Y_\nu^2(\tau) = \begin{cases} O(\tau^{-2\nu}) & \text{for} \quad 0 < \tau < 1, \\ O(1) & \text{for} \quad \tau \geq 1; \end{cases}$$

(b) for $\nu = 0$ and for all $\tau > 0$

$$Y_0(\tau) = \frac{2}{\pi} J_0(\tau) \ln\left(\frac{\tau}{2}\right) + O(1), \qquad J_0(\tau) = O(1),$$

---

[14]) See, for example, Bateman and Erdélyi [9, pp. 16–17].

we arrive at the following estimate, valid for all $0 < t < h$ and for all $\lambda > 0$:

$$w^2_{(N-2)/2}(t\sqrt{\lambda}, h\sqrt{\lambda}) = \begin{cases} O\left[\left(\frac{h}{t}\right)^{N-2}\right] & \text{for} \quad N \geq 3, \\ O\left[1 + \ln^2\left(\frac{h}{t}\right)\right] = O\left[\left(\frac{h}{t}\right)^{\delta}\right] & \text{for} \quad N = 2. \end{cases} \tag{4.3.24}$$

Here the δ on the right-hand side denotes an arbitrary number fixed in the interval $0 < \delta < 1$.

It follows from estimate (4.3.24) that, given any $N \geq 2$, we have

$$h^{2-N} \int_0^h w^2_{(N-2)/2}(t\sqrt{\lambda}, h\sqrt{\lambda}) t^{N-2} dt = O(h). \tag{4.3.25}$$

Substituting estimate (4.3.25) into the right-hand side of (4.3.23), we find that there is a constant C_{10} such that

$$P_i^2(\lambda, h) \leq C_{10} h \int_0^h \widehat{Q}_i^2(\lambda, t) dt. \tag{4.3.26}$$

In turn, it follows from (4.3.26) that

$$\sum_{i=1}^{\widehat{m}} \int_0^\infty P_i^2(\lambda, h) d\rho(\lambda) \leq C_{10} h \int_0^h \left[\sum_{i=1}^{\widehat{m}} \int_0^\infty \widehat{Q}_i^2(\lambda, t) d\rho(\lambda) \right] dt. \tag{4.3.27}$$

Now, to derive estimate (4.3.16) from inequality (4.3.27), it suffices to write the Parseval equality for function (4.3.20)

$$\sum_{i=1}^{\widehat{m}} \int_0^\infty \widehat{Q}_i^2(\lambda, t) d\rho(\lambda) = \int_G \rho_0(x) Q^2(x, t) dx \tag{4.3.28}$$

and to take into account that

$$\int_G \rho_0(x) Q^2(x, t) dx \leq C_{11} \|f\|^2_{L_2(G)}. \tag{4.3.29}$$

Comparing (4.3.27), (4.3.28), and (4.3.29) and keeping in mind that $0 < \alpha < 1$, we obtain

$$\sum_{i=1}^{\widehat{m}} \int_0^\infty P_i^2(\lambda, h) d\rho(\lambda) \leq C_{12} h^2 \|f\|^2_{L_2(G)} \leq C_{12} h^2 \|f\|^2_{H_2^\alpha(G)} \leq C_{12} h^{2\alpha} \|f\|^2_{H_\alpha^2(G)}.$$

Thus, estimate (4.3.16) has been proved, which completes the proof of Lemma 4.7 for the case $0 < \alpha < 1$.

2^0. Now we prove Lemma 4.7 for the case $\alpha \geq 1$.

We define, by analogy with Section 4.1.4, an operator

$$\widehat{A}^\tau f = \int_0^\infty \lambda^\tau dE_\lambda f \tag{4.3.30}$$

for any real $\tau > 0$ and recall the assertion formulated in that section: if $0 < 2\tau < \alpha$, then for any function $f(x)$ in the class [15] $\overset{0}{H}{}_2^\alpha(G)$ there exist functions $g(x)$ and $h(x)$ such that $\widehat{A}^\tau f = g + h$; here the function $g(x)$ is finite in domain G, belongs to the class $H_2^{\alpha-2\tau}(G)$, and satisfies the relation $||g||_{H_2^{\alpha-2\tau}(G)} \leq C||f||_{H_2^\alpha(G)}$, and the function $h(x)$ for any number n satisfies the relation $||\widehat{A}^n h||_{L_2(G)} \leq C_n||f||_{H_2^\alpha(G)}$.

For an arbitrary $\alpha \geq 1$ we fix τ subject to the condition $0 < \alpha - 2\tau < 1$. Then, with reference to the above case 1^0 and by virtue of the assertion we have just formulated, the following estimates hold:

$$||(I - E_\lambda)g||_{L_2(G)} \leq C\lambda^{-(\alpha/2-\tau)}||f||_{H_2^\alpha(G)},$$

$$||(I - E_\lambda)h||_{L_2(G)} \leq C_n\lambda^{-n}||f||_{H_2^\alpha(G)}.$$

We infer from these estimates that

$$||(I - E_\lambda)f||_{L_2(G)} = \left\|\widehat{A}^{-\tau}(I - E_\lambda)\widehat{A}^\tau f\right\|_{L_2(G)}$$

$$\leq \lambda^{-\tau}\left\|(I - E_\lambda)\widehat{A}^\tau f\right\|_{L_2(G)} \leq C\lambda^{-\alpha/2}||f||_{H_2^\alpha(G)}.$$

This completes the proof of Lemma 4.7.

4.3.2. Proof of the Main Local Asymptotic Estimate

LEMMA 4.8. *If a function $f(x)$ is compactly supported in domain G, belongs to the class $H_2^\alpha(G)$ at a certain $\alpha \geq (N-1)/2$, and becomes zero in a domain D interior to G, then, given sufficiently small $\delta > 0$, for any τ in the interval $(N-1)/4 - \delta/2 < \tau < (N-1)/4$ the asymptotic formula*

$$\widehat{A}^\tau E_\lambda f(x_0) = O(\lambda^\tau)||f||_{H_2^\alpha(G)} \tag{4.3.31}$$

[15] We recall that the symbol $\overset{0}{H}{}_p^\alpha(G)$ denotes a class of functions obtained by closing a functional set from $C_0^\infty(G)$ in the norm of the Nikol'skii class $H_p^\alpha(G)$.

holds uniformly with respect to x_0 on every compact set K of domain D, the symbol $\widehat{A}^\tau$ denoting an operator of the form (4.3.30).

PROOF: We fix an arbitrary compact set K of domain D, an arbitrary number R (which is less than half the distance between the compact set K and the boundary of D), and an arbitrary point x_0 of the compact set K; we consider the function (4.2.2) taken at $s = 0$.

It has been established in Section 4.2.1 that the Fourier image of this function, multiplied by $p(x_0, y)$, is determined by relation (4.2.7), which at $s = 0$ takes the form

$$\widehat{V}_i^\lambda(x_0, t) = u_i(x_0, t)\left[\delta_t^\lambda - \lambda^{N/4} t^{(2-N)/4} I_t^\lambda(R)\right] + \widetilde{V}_i^\lambda(x_0, t) \tag{4.3.32}$$

$$(i = 1, 2, \dots, \widehat{m}),$$

where

$$\delta_t^\lambda = \begin{cases} 1 & \text{for} \quad t < \lambda, \\ 1/2 & \text{for} \quad t = \lambda, \\ 0 & \text{for} \quad t > \lambda, \end{cases} \tag{4.3.33}$$

$$I_t^\lambda(R) = \int_R^\infty J_{N/2}(r\sqrt{\lambda}) J_{N/2-1}(r\sqrt{t})\,dr, \tag{4.3.34}$$

and the additional term $\widetilde{V}_i^\lambda(x_0, t)$ is

$$\begin{aligned} \widetilde{V}_i^\lambda(x_0, t) = (2\pi)^{-N/2}\lambda^{N/4} \int_0^R J_{N/2}(r\sqrt{\lambda}) \Bigg[\int_0^r \tau^\nu w_\nu(\tau\sqrt{t}, r\sqrt{t}) \\ \times \left(\int_\omega u_i(x_0 + \tau\omega, \lambda) q(x_0, x_0 + \tau\omega)\,d\omega\right) d\tau\Bigg] dr \end{aligned} \tag{4.3.35}$$

$$(i = 1, 2, \dots, \widehat{m}).$$

We multiply (4.3.32) by the Fourier image $\widehat{F}_i(\lambda)$ of an arbitrary function $F(x)$ in the class $L_2(G)$ and sum up the equality obtained over all i from 1 to $\widehat{m}$ and then integrate with respect to the spectral measure $\rho(t)$ between the limits for t from 0 to

$+\infty$. Making use of the generalized Parseval equality (4.1.5), we arrive at

$$E_\lambda F(x_0) = -\frac{1}{2}\sum_{i=1}^{\widehat{m}} F_i(\lambda)u_i(x_0,\lambda)[\rho(\lambda+0)-\rho(\lambda-0)]$$

$$+\int_G V^\lambda(x_0,x)F(x)\rho_0(x)dx + \lambda^{N/4}\sum_{i=1}^{\widehat{m}}\int_0^\infty t^{(2-N)/4}I_t^\lambda(R)\widehat{F}_i(t)u_i(x_0,t)d\rho(t)$$

$$-\sum_{i=1}^{\widehat{m}}\int_0^\infty \widetilde{V}_i^\lambda(x_0,t)\widehat{F}_i(t)d\rho(t). \tag{4.3.36}$$

In this equality the symbol $V^\lambda(x_0,y)$ signifies function (4.2.2) multiplied by $p(x_0,y)$ and taken at $s=0$.

Assume now that the function $f(x)$ is compactly supported in domain G and belongs to the class $H_2^\alpha(G)$ at $\alpha \ge (N-1)/2$. Further, we fix an arbitrary positive $\tau < (N-1)/4$ and put $F=\widehat{A}^\tau f$, where $\widehat{A}^\tau f$ is an operator of the form (4.3.30). If so, equality (4.3.36) takes the following form:

$$\widehat{A}^\tau E_\lambda f(x_0) = \int_G V^\lambda(x_0,x)\widehat{A}^\tau f(x)\rho_0(x)dx$$

$$-\frac{1}{2}\sum_{i=1}^{\widehat{m}} u_i(x_0,\lambda)\widehat{f}_i(\lambda)\lambda^\tau[\rho(\lambda+0)-\rho(\lambda-))]$$

$$+\lambda^{N/4}\sum_{i=1}^{\widehat{m}}\int_0^\infty t^{(2-N)/4}I_t^\lambda(R)u_i(x_0,t)\widetilde{f}_i(t)t^\tau d\rho(t) \tag{4.3.37}$$

$$-\sum_{i=1}^{\widehat{m}}\int_0^\infty \widetilde{V}_i^\lambda(x_0,t)\widehat{f}_i(t)t^\tau d\rho(t).$$

It suffices to show that if τ, given a sufficiently small $\delta>0$, satisfies the condition $(N-1)/4-\delta/2<\tau<(N-1)/4$, then each of the four summands on the right-hand side of (4.3.37), uniformly with respect to x_0 on the fixed compact set K, has the order of

$$O(\lambda^\tau)\|f\|_{H_2^\alpha(G)}. \tag{4.3.38}$$

Let us ensure, for one thing, that the first summand on the right-hand side of (4.3.37), that is, the integral

$$\int_G V^\lambda(x_0, x)\widehat{A}^\tau f(x)\rho_0(x)dx, \tag{4.3.39}$$

has the same order of magnitude.

We denote by K' a compact set of domain D such that the K' is contained in the previously fixed compact set K, where the distance between K and the boundary of K' is not smaller than R.[16)]

By virtue of the assertion formulated in Section 4.1.4, the function $\widehat{A}^\tau f(x)$ can be represented as the sum of two functions, $g(x) + h(x)$, of which the former becomes zero on the compact set K' (for such a function, considering that R is smaller than the distance from K to the boundary of K', expression (4.3.39) becomes zero for all $x_0 \in K$).

As regards the function $h(x)$, the estimate

$$\int_G V^\lambda(x_0, x)h(x)\rho_0(x)dx = O(\lambda^\tau)\|f\|_{H_2^\alpha(G)}$$

for it a priori holds uniformly with respect to x_0 on the compact set K by virtue of the assertion in Section 4.1.3.

It has been proved thereby that integral (4.3.39) has the order of (4.3.38) uniformly with respect to x_0 on the compact set K.

We now prove that, without allowing necessarily for $f(x)$ becoming zero in a subdomain D of domain G, *the remaining three summands on the right-hand side of (4.3.37) have the order of (4.3.38) uniformly with respect* x_0 *on any compact set* K *of the entire domain* G (not confined solely to any compact set K of subdomain D).

In order to prove that the second summand on the right-hand side of (4.3.37),

$$\frac{1}{2}\sum_{i=1}^{\widehat{m}} u_i(x_0, \lambda)\widehat{f}_i(\lambda)\lambda^\tau[\rho(\lambda+0) - \rho(\lambda-0)], \tag{4.3.40}$$

has the order of (4.3.38) uniformly with respect to x_0 on any compact set K of domain G, it suffices to show that the expression

$$\sum_{i=1}^{\widehat{m}} \int_{\lambda-\Delta\lambda}^{\lambda+\Delta\lambda} u_i(x_0, t)\widehat{f}_i(t)t^\tau d\rho(t)$$

is of precisely the same order on any compact set of domain G for all sufficiently small positive $\Delta\lambda$. This follows immediately from the Cauchy–Buniakowski inequality and

16) Such a compact set K' does indeed exist, inasmuch as the number R has been fixed smaller than half the distance between the compact set K and the boundary of D.

from the estimates

$$\sum_{i=1}^{\widehat{m}} \int_{\lambda-\Delta\lambda}^{\lambda+\Delta\lambda} \widehat{f}_i^2(t)d\rho(t) = O(\lambda^{-\alpha})\|f\|^2_{H_2^\alpha(G)},$$

$$\sum_{i=1}^{\widehat{m}} \int_{\lambda-\Delta\lambda}^{\lambda+\Delta\lambda} u_i^2(x_0,t)d\rho(t) = O(\lambda^{(N-1)/2}).$$

Of these, the former holds by virtue of Lemma 4.7 for any $0 < \Delta\lambda < \lambda/2$, and the latter holds (uniformly with respect to x_0 on any compact set of domain G) for any $0 < \Delta\lambda < \sqrt{\lambda}$ by virtue of the main estimate (4.1.14) (see Lemma 4.1, Section 4.1.3).

Thus, we have shown that expression (4.3.40) has the order of (4.3.38) uniformly with respect to x_0 on any compact set K of the main domain G.

Now we proceed to estimate the third summand on the right-hand side of (4.3.37). With no loss of generality, we may assume the self-adjoint extension $\widehat{A}$ in question to be strictly positive, that is, satisfying the condition $(\widehat{A}u, u) \geq \mu_0(u,u)$ for any function $u(x)$ from $C_0^\infty(G)$ and for a certain $\mu_0 \geq 0$; consequently, we can write the third summand of interest as

$$\lambda^{N/4}\sum_{i=1}^{\widehat{m}} \int_{\mu_0}^{\infty} t^{(2-N)/4} I_t^\lambda(R) u_i(x_0,t)\widehat{f}_i(t)t^\tau d\rho(t). \tag{4.3.41}$$

Making use of the technique we have worked out in Chapters 1 and 2 for the Laplace operator, we represent expression (4.3.41) as the sum of four terms $\mathcal{K}_1 + \mathcal{K}_2 + \mathcal{K}_3 + \mathcal{K}_4$, where

$$\mathcal{K}_1 = \lambda^{N/4}\sum_{i=1}^{\widehat{m}} \int_{\sqrt{\mu_0}\leq\sqrt{t}\leq\sqrt{\lambda}/2} t^{(2-N)/4} I_t^\lambda(R) u_i(x_0,t)\widehat{f}_i(t)t^\tau d\rho(t), \tag{4.3.42}$$

$$\mathcal{K}_2 = \lambda^{N/4}\sum_{i=1}^{\widehat{m}} \int_{1<|\sqrt{t}-\sqrt{\lambda}|<\sqrt{\lambda}/2} t^{(2-N)/4} I_t^\lambda(R) u_i(x_0,t)\widehat{f}_i(t)t^\tau d\rho(t), \tag{4.3.43}$$

$$\mathcal{K}_3 = \lambda^{N/4}\sum_{i=1}^{\widehat{m}} \int_{|\sqrt{t}-\sqrt{\lambda}|\leq 1} t^{(2-N)/4} I_t^\lambda(R) u_i(x_0,t)\widehat{f}_i(t)t^\tau d\rho(t), \tag{4.3.44}$$

$$\mathcal{K}_4 = \lambda^{N/4}\sum_{i=1}^{\widehat{m}} \int_{\sqrt{t}\geq 3\sqrt{\lambda}/2} t^{(2-N)/4} I_t^\lambda(R) u_i(x_0,t)\widehat{f}_i(t)t^\tau d\rho(t). \tag{4.3.45}$$

Invoking Lemma 4.7, estimates (4.1.14), (4.1.18), (4.1.49), (4.1.50) and the estimates

$$|I_t^\lambda(R)| = O(\lambda^{-1/4}t^{-1/4}), \quad |I_t^\lambda(R)| = O(\lambda^{-1/4}t^{-1/4}|\sqrt{t}-\sqrt{\lambda}|^{-1}), \tag{4.3.46}$$

valid for all $\lambda \geq 1, t \geq \mu_0$ (see estimates (2.4.10) and (2.4.11), Section 2.4.2), we shall prove that each of (4.3.42)–(4.3.45) has the order of (4.3.38).

To estimate (4.3.42) we make use of the latter of estimates (4.3.46) which, given $\sqrt{\mu_0} \leq \sqrt{t} \leq \sqrt{\lambda}/2$, can be rewritten as

$$|I_t^\lambda(R)| = O(\lambda^{-3/4}t^{-1/4}). \tag{4.3.47}$$

One observes that

$$\sum_{i=1}^{\widehat{m}} \int_{\mu_0}^{\infty} \widehat{f}_i^2(t) t^{\alpha-\gamma} d\rho(t) \leq C\|f\|^2_{H_2^\alpha(G)} \tag{4.3.48}$$

stems from estimate (4.37) for an arbitrary sufficiently small fixed $\gamma > 0$.[17]

Applying the Cauchy–Buniakowski inequality in (4.3.42) first to the integral and then to the sum and making use of estimates (4.3.47), (4.3.48), and (4.1.49), and keeping in mind that $\alpha \geq (N-1)/2$, we obtain, given $\tau + \gamma/2 > N/4 - 1/2$, $\gamma \leq 1/2$,

[17] To make certain of this, it is sufficient to apply the formula for integration by parts

$$\sum_{i=1}^{\widehat{m}} \int_{\mu_0}^{\infty} \widehat{f}_i^2(t) t^{\alpha-\gamma} d\rho(t) = \left[t^{\alpha-\gamma} \sum_{i=1}^{\widehat{m}} \int_t^{\infty} \widehat{f}_i^2(\tau) d\rho(\tau) \right] \Bigg|_{t=\mu_0}^{t=\infty}$$

$$-(\alpha-\gamma) \int_{\mu_0}^{\infty} t^{\alpha-\gamma-1} \left[\sum_{i=1}^{\widehat{m}} \int_t^{\infty} \widehat{f}_i^2(\tau) d\rho(\tau) \right] dt$$

and to use estimate (4.3.7) and the Parseval equality.

$$|\mathcal{K}_1| \leq \lambda^{(N-3)/4} \left\{ \sum_{i=1}^{\widehat{m}} \int\limits_{\sqrt{\mu_0}\leq\sqrt{t}\leq\sqrt{\lambda}/2} \widehat{f}_i^2(t) t^{\alpha-\gamma} d\rho(t) \right\}^{1/2}$$

$$\times \left\{ \sum_{i=1}^{\widehat{m}} \int\limits_{\sqrt{\mu_0}\leq\sqrt{t}\leq\sqrt{\lambda}/2} t^{(1-N)/2+2\tau+\delta-\alpha} u_i^2(x_0,t) d\rho(t) \right\}^{1/2}$$

$$\leq C\lambda^{(N-3)/4} \|f\|_{H_2^\alpha(G)} \left\{ \sum_{i=1}^{\widehat{m}} \int\limits_{\sqrt{\mu_0}\leq\sqrt{t}\leq\sqrt{\lambda}/2} t^{1-N+2\tau+\delta} u_i^2(x_0,t) d\rho(t) \right\}^{1/2}$$

$$\leq C_1 \lambda^{\tau-1/4+\gamma/2} \|f\|_{H_2^\alpha(G)} \leq C_1 \lambda^\tau \|f\|_{H_2^\alpha(G)}.$$

It has been proved thereby that quantity (4.3.42) has the order of (4.3.38).

To estimate quantity (4.3.43), we make use of the latter estimate (4.3.46), keeping in mind that $\sqrt{\lambda}/2 \leq \sqrt{t} \leq (3/2)\sqrt{\lambda}$.

Once again, we apply the Cauchy–Buniakowski inequality first to the integral and then to the sum, and, making use of estimate (4.3.7), we arrive at

$$|\mathcal{K}_2| \leq C_2 \lambda^\tau \left\{ \sum_{i=1}^{\widehat{m}} \int\limits_{\sqrt{t}\geq\sqrt{\lambda}/2} \widehat{f}_i^2(t) d\rho(t) \right\}^{1/2}$$

$$\times \left\{ \sum_{i=1}^{\widehat{m}} \int\limits_{1\leq|\sqrt{t}-\sqrt{\lambda}|\leq\sqrt{\lambda}/2} |\sqrt{t}-\sqrt{\lambda}|^{-2} u_i^2(x_0,t) d\rho(t) \right\}^{1/2}$$

$$\leq C_3 \lambda^{\tau-\alpha/2} \|f\|_{H_2^\alpha(G)} \left\{ \sum_{i=1}^{\widehat{m}} \int\limits_{1\leq|\sqrt{t}-\sqrt{\lambda}|\leq\sqrt{\lambda}/2} |\sqrt{t}-\sqrt{\lambda}|^{-2} u_i(x_0,t) d\rho(t) \right\}^{1/2} \tag{4.3.49}$$

To estimate the term within the braces on the right-hand side of (4.3.49), we denote by M the integer part of the number $\sqrt{\lambda}/2$ and, with reference to inequality

(4.1.14), majorize this term with the sum

$$\left\{\sum_{i=1}^{\widehat{m}} \int\limits_{1\le|\sqrt{t}-\sqrt{\lambda}|\le\sqrt{\lambda}/2} |\sqrt{t}-\sqrt{\lambda}|^{-2} u_i^2(x_0,t)\,d\rho(t)\right\}$$

$$\le \sum_{\ell=1}^{M} \left\{\sum_{i=1}^{\widehat{m}} \int\limits_{\ell\le|\sqrt{t}-\sqrt{\lambda}|\le\ell+1} |\sqrt{t}-\sqrt{\lambda}|^{-2} u_i^2(x_0,t)\,d\rho(t)\right\}$$

$$\le \sum_{\ell=1}^{M} \ell^{-2} \left[\sum_{i=1}^{\widehat{m}} \int\limits_{\ell\le|\sqrt{t}-\sqrt{\lambda}|\le\ell+1} u_i^2(x_0,t)\,d\rho(t)\right] \tag{4.3.50}$$

$$\le C_4\lambda^{(N-1)/2} \sum_{\ell=1}^{\infty} \ell^{-2} \le C_5\lambda^{(N-1)/2}.$$

Comparing (4.3.50) to (4.3.49) and keeping in mind that $\alpha \ge (N-1)/2$, we obtained the desired order of (4.3.38) for quantity (4.3.43).

To estimate quantity (4.3.44), we apply the Cauchy–Buniakowski inequality to it and make use of the former estimate (4.3.46), the relation $t = O(\lambda)$, estimate (4.3.7), and estimate (4.1.14).

Recalling that $\alpha \ge (N-1)/2$, we obtain

$$|\mathcal{K}_3| \le C_6\lambda^{\tau} \left\{\sum_{i=1}^{\widehat{m}} \int\limits_{\sqrt{t}>\sqrt{\lambda}-1} f_i^2(t)\,d\rho(t)\right\}^{1/2} \left\{\sum_{i=1}^{\widehat{m}} \int\limits_{|\sqrt{t}-\sqrt{\lambda}|\le 1} u_i^2(x_0,t)\,d\rho(t)\right\}^{1/2}$$

$$\le C_7\lambda^{(N-1)/4-\alpha/2+\tau}\|f\|_{H_2^{\alpha}(G)} \le C_7\lambda^{\tau}\|f\|_{H_2^{\alpha}(G)}.$$

This proves that quantity (4.3.44) has the order of (4.3.38).

Now it remains to prove that quantity (4.3.45) likewise possesses the order of (4.3.38).

We recall that at $\sqrt{t} \ge (3/2)\sqrt{\lambda}$ the latter estimate (4.3.46) can be rewritten as

$$|I_t^{\lambda}(R)| = O(\lambda^{-1/4}t^{-3/4}).$$

Once again, with reference to estimate (4.3.7) and estimate (4.1.50), we obtain

$$|\mathcal{K}_4| \le C_8\lambda^{(N-1)/4-\alpha/2+\tau}\|f\|_{H_2^\alpha(G)}\left\{\sum_{i=1}^{\widehat{m}}\int_{\sqrt{t}\ge\lambda\sqrt{\lambda}/2} t^{(N+1)/2}u_i(x_0,t)d\rho(t)\right\}^{1/2}$$

$$\le C_9\lambda^{(N-1)/4-\alpha/2+\tau}\|f\|_{H_2^\alpha(G)} \le C_9\lambda^{\tau}\|f\|_{H_2^\alpha(G)}$$

applying the Cauchy–Buniakowski inequality to (4.3.45) and keeping in mind that $\alpha \ge (N-1)/2$.

This has provided a final proof that the four quantities (4.3.42)–(4.3.45) are of the order of (4.3.38); consequently, the same is true of quantity (4.3.41).

Now, to have a complete proof of Lemma 4.8, all we have left to do is to show that the last term on the right-hand side of (4.3.37), generated by the remainder term in the mean-value formula, has the order of (4.3.38), that is, to prove the estimate

$$\sum_{i=1}^{\widehat{m}}\int_0^\infty \widetilde{V}_i^\lambda(x_0,t)\widehat{f}_i(t)t^\tau d\rho(t) = O(\lambda^\tau)\|f\|_{H_2^\alpha(G)} \tag{4.3.51}$$

uniform with respect to x_0 on a compact set K of domain G.

To obtain this estimate, we shall want the following two estimates, uniform with respect to x_0 on an arbitrary compact set K of domain G, for the quantity $\widetilde{V}_i^\lambda(x_0,t)$ as defined by relation (4.3.35):

$$\sum_{i=1}^{\widehat{m}}\int_0^\infty \left[\widetilde{V}_i^\lambda(x_0,t)\right]^2 d\rho(t) \le C_{10}\lambda^{(N-1)/2}, \tag{4.3.52}$$

$$\sum_{i=1}^{\widehat{m}}\int_0^{\lambda/2} \left[\widetilde{V}_i^\lambda(x_0,t)\right]^2 d\rho(t) \le C_{11}\lambda^{(N-1)/2-\delta} \tag{4.3.53}$$

(in the latter of these, the δ denotes a sufficiently small positive number).

Estimate (4.3.52) is identical with estimate (4.2.14) taken at $s = 0$; it has been shown in Section 4.2.2 that the latter estimate is uniform with respect to x_0 on every fixed compact set K of domain G.

The proof of estimate (4.3.53), uniform with respect to x_0 on every compact set K of domain G, will be given below as Lemma 4.9 in a separate section.

Meanwhile, we proceed with the proof of Lemma 4.8; based on estimates (4.3.52) and (4.3.53), we show that estimate (4.3.51) is uniform with respect to x_0 on an arbitrary compact set K of domain G.

We partition the integral on the left-hand side of (4.3.51) into the sum $I_1 + I_2$ of two integrals,

$$I_1 = \sum_{i=1}^{\widehat{m}} \int_0^{\lambda/4} \widetilde{V}_i^{\lambda}(x_0, t) \widehat{f}_i(t) t^{\tau} \, d\rho(t), \tag{4.3.54}$$

$$I_2 = \sum_{i=1}^{\widehat{m}} \int_{\lambda/4}^{\infty} \widetilde{V}_i^{\lambda}(x_0, t) \widehat{f}_i(t) t^{\tau} \, d\rho(t), \tag{4.3.55}$$

to ensure that both (4.3.54) and (4.3.55) are uniform with respect to x_0 on an arbitrary compact set of domain G and have the order of (4.3.38).

We denote by n_0 the least integer satisfying the condition $2^{n_0+1} \geq \lambda/4$; one will observe that the inequality

$$|I_1| \leq \sum_{i=1}^{\widehat{m}} \int_0^{1} \left| \widetilde{V}_i^{\lambda}(x_0, t) \widehat{f}_i(t) t^{\tau} \right| d\rho(t) + \sum_{\ell=0}^{n_0} \sum_{i=1}^{\widehat{m}} \int_{2^{\ell}}^{2^{\ell+1}} \left| \widetilde{V}_i^{\lambda}(x_0, t) \widehat{f}_i(t) t^{\tau} \right| d\rho(t) \tag{4.3.56}$$

stems from expression (4.3.54).

In estimating the integrals on the right-hand side of (4.3.56), we shall make use of the following two inequalities:

$$\sum_{i=1}^{\widehat{m}} \int_0^{1} \widehat{f}_i^2(t) t^{2\tau} \, d\rho(t) \leq C_{12} \|f\|^2_{L_2(G)} \leq C_{12} \|f\|^2_{H_2^{\alpha}(G)},$$

$$\sum_{i=1}^{\widehat{m}} \int_{2^{\ell}}^{2^{\ell+1}} \widehat{f}_i^2(t) t^{2\tau} \, d\rho(t) \leq 2^{(\ell+1)2\tau} \sum_{i=1}^{\widehat{m}} \int_{2^{\ell}}^{2^{\ell+1}} \widehat{f}_i^2(t) t^{\tau} \, d\rho(t) = O(2^{\ell(2\tau-\alpha)}) \|f\|_{H_2^{\alpha}(G)}.$$

The former of these stems from the Parseval equality for the function $f(x)$, and the latter, from the estimate (4.3.7) as established in Section 4.3.1.

Applying the Cauchy–Buniakowski inequality first to the integral and then to the i-indexed sum on the right-hand side of (4.3.56) and making use of estimate (4.3.53) and of the two inequalities above, we obtain

$$|I_1| \le \left\{ \sum_{i=1}^{\widehat{m}} \int_0^1 \left[\widetilde{V}_i^\lambda(x_0,t) \right]^2 d\rho(t) \right\}^{1/2} \left\{ \sum_{i=1}^{\widehat{m}} \int_0^1 \widehat{f}_i^2(t) t^{2\tau} d\rho(t) \right\}^{1/2}$$

$$+ \sum_{\ell=0}^{n_0} \left\{ \sum_{i=1}^{\widehat{m}} \int_{2^\ell}^{2^{\ell+1}} \left[\widetilde{V}_i^\lambda(x_0,t) \right]^2 d\rho(t) \right\}^{1/2} \left\{ \sum_{i=1}^{\widehat{m}} \int_{2^\ell}^{2^{\ell+1}} \widehat{f}_i^2(t) t^{2\tau} d\rho(t) \right\}^{1/2}$$

$$= O(\lambda^{(N-1)/4-\delta/2}) \|f\|_{H_2^\alpha(G)} \left[1 + \sum_{\ell=0}^{n_0} 2^{\ell(\tau-\alpha/2)} \right].$$

From this inequality, taking into account that $(N-1)/4 - \delta/2 < \tau$ and $\tau < (N-1)/4 \le \alpha/2$, we obtain for I_1 the order of (4.3.38) uniform with respect to x_0 on an arbitrary compact set of domain G.

It remains now to ensure that the same order of (4.3.38) applies to quantity (4.3.55) uniformly with respect to x_0 on any compact set of domain G.

We represent (4.3.55) as the sum

$$I_2 = \sum_{i=1}^{\widehat{m}} \sum_{\ell=-2}^{\infty} \int_{2^\ell \lambda}^{2^{\ell+1}\lambda} \widetilde{V}_i^\lambda(x_0,t) \widehat{f}_i^2(t) t^\tau d\rho(t). \tag{4.3.57}$$

Factoring the greatest value of t^τ outside the integral sign and applying the Cauchy–Buniakowski inequality first to the integral and then to both the i- and l-indexed sums on the right-hand side of (4.3.57), we obtain

$$|I_2| \le \sum_{i=1}^{\widehat{m}} \sum_{\ell=-2}^{\infty} 2^{\tau(\ell+1)} \lambda^\tau \left\{ \int_{2^\ell\lambda}^{2^{\ell+1}\lambda} \left[\widetilde{V}_i^\lambda(x_0,t) \right]^2 d\rho(t) \right\}^{1/2} \left\{ \int_{2^\ell\lambda}^{2^{\ell+1}\lambda} \widehat{f}_i^2(t) t^{2\tau} d\rho(t) \right\}^{1/2}$$

$$\le (2\lambda)^\tau \left\{ \sum_{i=1}^{\widehat{m}} \sum_{\ell=-2}^{\infty} \int_{2^\ell\lambda}^{2^{\ell+1}\lambda} \left[\widetilde{V}_i^\lambda(x_0,t) \right]^2 d\rho(t) \right\}^{1/2}$$

$$\times \left\{ \sum_{i=1}^{\widehat{m}} \sum_{\ell=-2}^{\infty} 2^{2\tau\ell} \int_{2^\ell\lambda}^{2^{\ell+1}\lambda} \widehat{f}_i^2(t) d\rho(t) \right\}^{1/2} \tag{4.3.58}$$

One observes that the inequality

$$\sum_{i=1}^{\widehat{m}}\sum_{\ell=-2}^{\infty}\int_{2^{\ell}\lambda}^{2^{\ell+1}\lambda}\left[\widetilde{V}_i^{\lambda}(x_0,t)\right]^2 d\rho(t)$$
$$=\sum_{i=1}^{\widehat{m}}\int_{\lambda/4}^{\infty}\left[\widetilde{V}_i^{\lambda}(x_0,t)\right]^2 d\rho(t)\le C_{10}\lambda^{(N-1)/2} \tag{4.3.59}$$

holds by virtue of estimate (4.3.52) uniformly with respect to x_0 on an arbitrary compact set of domain G. Also,

$$\sum_{i=1}^{\widehat{m}}\int_{2^{\ell}\lambda}^{2^{\ell+1}\lambda}\widehat{f}_i^2(t)d\rho(t)\le\sum_{i=1}^{\widehat{m}}\int_{2^{\ell}\lambda}^{\infty}\widehat{f}_i^2(t)d\rho(t)=O[(2^{\ell}\lambda)^{-\alpha}]\|f\|_{H_2^{\alpha}(G)}$$

by virtue of estimate (4.3.7).

Applying the above estimate and estimate (4.3.59) to the right-hand side of (4.3.58), we obtain finally

$$|I_2|\le C_{13}\lambda^{\tau+(N-1)/4-\alpha/2}\left\{\sum_{\ell=-2}^{\infty}2^{(2\tau-\alpha)\ell}\right\}^{1/2}\|f\|_{H_2^{\alpha}(G)}. \tag{4.3.60}$$

Recall that $\alpha\ge(N-1)/2$ and $\tau<(N-1)/4$, that is, $2\tau-\alpha<0$, and the expression within the braces in (4.3.60) equals $O(1)$.

From (4.3.60) we obtain

$$|I_2|\le C_{14}\lambda^{\lambda}\|f\|_{H_2^{\alpha}(G)}$$

for any $\tau<(N-1)/4$ uniformly with respect to x_0 on an arbitrary compact set of domain G.

Estimate (4.3.51) has thus been proved, which completes the proof of Lemma 4.8 provided that estimate (4.3.53) remains valid with respect to x_0 on any compact set of domain G.

Before proving estimate (4.3.53), we wish to make a remark.

REMARK. In verifying estimate (4.3.38) for the second, third, and fourth terms in the right-hand side of (4.3.37), uniform with respect to x_0 on an arbitrary compact set of the entire domain G, we have made use of inequality (4.3.7) only.

Thus, if we neglect the function $f(x)$ being compactly supported in domain G and merely require that the function $f(x)$ belong to the class $H_2^{\alpha}(G)$ at $\alpha\ge(N-1)/2$ and inequality (4.3.7) apply to this function, then the relation

$$\widehat{A}^{\tau}E_{\lambda}f(x_0)=\int_G V^{\lambda}(x_0,x)\widehat{A}^{\tau}f(x)\rho_0(x)dx+O(\lambda^{\tau})\|f\|_{H_2^{\alpha}(G)} \tag{4.3.61}$$

holds for this function uniformly with respect to x_0 on an arbitrary compact set of domain G.

4.3.3. Proof of Estimate (4.3.53)

LEMMA 4.9. *Estimate (4.3.53) holds for a certain sufficiently small positive number δ uniformly with respect to x_0 on an arbitrarily fixed compact set of domain G.*

PROOF: We will follow a line of reasoning similar to that we have adopted in Section 4.2.2 for proving inequality (4.2.14). Accordingly, it will suffice to show that each of the five expressions below

$$I_1(R,\lambda)=\sum_{i=1}^{\widehat{m}}\int_{1/R^2}^{\lambda/2}\left\{\int_{1/\sqrt{t}}^{R}\left[\int_{r_{x_0y}\le\tau}\frac{u_i(y,t)q(x_0,y)}{r_{x_0y}^{N/2-\delta}}dy\right]\right.$$

$$\left.\times\left[\int_{\tau}^{R}J_{N/2}(r\sqrt{\lambda})\frac{d}{d\tau}(\tau^{-\delta}w_\nu(\tau\sqrt{t},r\sqrt{t}))dr\right]d\tau\right\}^2 d\rho(t), \tag{4.3.62}$$

$$I_2(R,\lambda)=\sum_{i=1}^{\widehat{m}}\int_{1/R^2}^{\lambda/2}\left\{\int_{0}^{1/\sqrt{t}}\left[\int_{r_{x_0y}\le\tau}\frac{u_i(y,t)q(x_0,y)}{r_{x_0y}^{N/2-\delta}}dy\right]\right.$$

$$\times\left[\sqrt{t}\tau^{-\delta}Y_{\nu+1}(\tau\sqrt{t})-(\nu-\delta)\tau^{-\delta-1}Y_\nu(\tau\sqrt{t})\right] \tag{4.3.63}$$

$$\left.\times\left[\int_{\tau}^{R}J_{N/2}(r\sqrt{\lambda})J_\nu(r\sqrt{t})dr\right]d\tau\right\}^2 d\rho(t),$$

$$I_3(R,\lambda) = \sum_{i=1}^{\widehat{m}} \int_{1/R^2}^{\lambda/2} \left\{ \int_0^{1/\sqrt{t}} \left[\int_{r_{x_0 y} \le \tau} \frac{u_i(y,t)q(x_0,y)}{r_{x_0 y}^{N/2-\delta}} dy \right] \right.$$

$$\times \left[\sqrt{t}\tau^{-\delta} J_{\nu+1}(\tau\sqrt{t}) - (\nu-\delta)\tau^{-\delta-1} J_\nu(\tau\sqrt{t}) \right] \tag{4.3.64}$$

$$\left. \times \left[\int_\tau^{1/\sqrt{t}} J_{N/2}(r\sqrt{\lambda}) Y_\nu(r\sqrt{t}))dr \right] d\tau \right\}^2 d\rho(t),$$

$$I_4(R,\lambda) = \sum_{i=1}^{\widehat{m}} \int_{1/R^2}^{\lambda/2} \left\{ \int_0^{1/\sqrt{t}} \left[\int_{r_{x_0 y} \le \tau} \frac{u_i(y,t)q(x_0,y)}{r_{x_0 y}^{N/2-\delta}} dy \right] \right.$$

$$\times \left[\sqrt{t}\tau^{-\delta} J_{\nu+1}(\tau\sqrt{t}) - (\nu-\delta)\tau^{-\delta-1} J_\nu(\tau\sqrt{t}) \right] \tag{4.3.65}$$

$$\left. \times \left[\int_{1/\sqrt{t}}^{R} J_{N/2}(r\sqrt{\lambda}) Y_\nu(r\sqrt{t}))dr \right] d\tau \right\}^2 d\rho(t),$$

$$I_5(R,\lambda) = \sum_{i=1}^{\widehat{m}} \int_0^{1/R^2} \left\{ \int_0^{R} \left[\int_{r_{x_0 y} \le \tau} \frac{u_i(y,t)q(x_0,y)}{r_{x_0 y}^{N/2-\delta}} dy \right] \right.$$

$$\left. \times \left[\int_\tau^{R} J_{N/2}(r\sqrt{\lambda}) \frac{d}{d\tau}(\tau^{-\delta} w_\nu(\tau\sqrt{t}, r\sqrt{t}))dr \right] d\tau \right\}^2 d\rho(t), \tag{4.3.66}$$

given a certain $\delta > 0$, has the order of $O(\lambda^{-\delta-1/2})$ uniformly with respect to x_0 on an arbitrary compact set of domain G.

By virtue of the scheme in Section 4.2.2, it will suffice, in place of estimates (4.2.28), (4.2.31), (4.2.34), (4.2.35), and (4.2.37), to establish the following five estimates:

$$\left| \int_\tau^R J_{N/2}(r\sqrt{\lambda}) \frac{d}{d\tau} \left[\tau^{-\delta} w_\nu(\tau\sqrt{t}, r\sqrt{t}) \right] dr \right| = O(\lambda^{-1/4-\delta/2} \tau^{-1/2-\delta}), \tag{4.3.67}$$

$$\left|\left[\sqrt{t}\tau^{-\delta}Y_{\nu+1}(\tau\sqrt{t})-(\nu-\delta)\tau^{-\delta-1}Y_\nu(\tau\sqrt{t})\right]\right.$$
$$\left.\times\int_\tau^R J_{N/2}(r\sqrt{\lambda})J_\nu(r\sqrt{t})dr\right|=O(\lambda^{-1/4-\delta/2}\sqrt{t}^{-\nu-\delta-1/2}\tau^{-\nu-1-4\delta}),\tag{4.3.68}$$

$$\left|\left[\sqrt{t}\tau^{-\delta}J_{\nu+1}(\tau\sqrt{t})-(\nu-\delta)\tau^{-\delta-1}J_\nu(\tau\sqrt{t})\right]\right.$$
$$\left.\times\int_\tau^{1/\sqrt{t}} J_{N/2}(r\sqrt{\lambda})Y_\nu(r\sqrt{t})dr\right|=O(\lambda^{-1/4-\delta/2}\sqrt{t}^{-\nu-\delta-1/2}\tau^{-\nu-1-4\delta}),\tag{4.3.69}$$

$$\left|\left[\sqrt{t}\tau^{-\delta}J_{\nu+1}(\tau\sqrt{t})-(\nu-\delta)\tau^{-\delta-1}J_\nu(\tau\sqrt{t})\right]\right.$$
$$\left.\times\int_{1/\sqrt{t}}^R J_{N/2}(r\sqrt{\lambda})Y_\nu(r\sqrt{t})dr\right|=O(\lambda^{-1/4-\delta/2}\sqrt{t}^{-\nu-\delta-1/2}\tau^{-\nu-1-4\delta}),\tag{4.3.70}$$

$$\left|\int_\tau^R J_{N/2}(r\sqrt{\lambda})\frac{d}{d\tau}[\tau^{-\delta}w_\nu(\tau\sqrt{t},r\sqrt{t})]dr\right|=O(\lambda^{-1/4-\delta}\tau^{-3\delta-\nu-1}R^{\nu+1/2}).\tag{4.3.71}$$

It is assumed that $r\geq\tau\geq 1/\sqrt{t}$ in (4.3.67), $\tau\leq 1/\sqrt{t}$ in (4.3.68), $\tau\leq r\leq 1/\sqrt{t}$ in (4.3.69), $r\geq 1/\sqrt{t}\geq\tau$ in (4.3.70), and $\tau\leq r\leq R\leq 1/\sqrt{t}$ in (4.3.71).

Let us convince ourselves of the validity of estimates (4.3.67)–(4.3.71).

To proceed with the proof of estimate (4.3.67), we note, first of all, that

$$\begin{aligned}&\frac{d}{d\tau}[\tau^{-\delta}w_\nu(\tau\sqrt{t},r\sqrt{t})]\\&\quad=J_\nu(r\sqrt{t})[\sqrt{t}\tau^{-\delta}Y_{\nu+1}(\tau\sqrt{t})-(\nu-\delta)\tau^{-\delta-1}Y_\nu(\tau\sqrt{t})]\\&\quad-Y_\nu(r\sqrt{t})[\sqrt{t}\tau^{-\delta}J_{\nu+1}(\tau\sqrt{t})-(\nu-\delta)\tau^{-\delta-1}J_\nu(\tau\sqrt{t})].\end{aligned}\tag{4.3.72}$$

Further, we take into account that the two estimates

$$\begin{cases}|\sqrt{t}\tau^{-\delta}Y_{\nu+1}(\tau\sqrt{t})-(\nu-\delta)\tau^{-\delta-1}Y_\nu(\tau\sqrt{t})|=O(t^{1/4}\tau^{-\delta-1/2}),\\|\sqrt{t}\tau^{-\delta}J_{\nu+1}(\tau\sqrt{t})-(\nu-\delta)\tau^{-\delta-1}J_\nu(\tau\sqrt{t})|=O(t^{1/4}\tau^{-\delta-1/2}),\end{cases}\tag{4.3.73}$$

a priori hold at $\tau \geq 1/\sqrt{t}$.[18]

With the aid of relations (4.3.72) and (4.3.73), we shall reduce the proof of estimate (4.3.67) to establishing the following two estimates at $r \geq \tau \geq 1/\sqrt{t}$, $t \leq \lambda/2$:

$$|I_t^\lambda(\tau, R)| = \left|\int_\tau^R J_{N/2}(r\sqrt{\lambda})J_\nu(r\sqrt{t})dr\right| = O(\lambda^{-1/4-\delta/2}t^{-1/4}\tau^{-\delta}), \tag{4.3.74}$$

$$|K_t^\lambda(\tau, R)| = \left|\int_\tau^R J_{N/2}(r\sqrt{\lambda})Y_\nu(r\sqrt{t})dr\right| = O(\lambda^{-1/4-\delta/2}t^{-1/4}\tau^{-\delta}). \tag{4.3.75}$$

Estimates (4.3.74) and (4.3.75) are established in a similar manner, and we shall focus on the former of these only.

Integrating (4.3.74) two times by parts and making use of the recurrence relations for cylindrical functions, we obtain, in much the same manner as in Section 2.4.2 (see the derivation of (2.4.44)),

$$\begin{aligned} I_t^\lambda(\tau, R)\left(1 - \frac{t}{\lambda}\right) &= -\frac{1}{\sqrt{\lambda}}[J_\nu(r\sqrt{\lambda})J_\nu(r\sqrt{t})]\Big|_{r=\tau}^{r=R} \\ &- \frac{\sqrt{t}}{\lambda}[J_{\nu-1}(r\sqrt{\lambda})J_{\nu-1}(r\sqrt{t})]\Big|_{r=\tau}^{r=R} \\ &+(N-2)\frac{t}{\lambda^{3/2}}\int_\tau^R J_\nu(r\sqrt{\lambda})J_{\nu-2}(r\sqrt{t})\frac{dr}{r} \\ &-(N-4)\frac{\sqrt{t}}{\lambda}\int_\tau^R J_{N/2}(r\sqrt{\lambda})J_{\nu-1}(r\sqrt{t})\frac{dr}{r}. \end{aligned} \tag{4.3.76}$$

Since, in this particular case, $r\sqrt{t} \geq \tau\sqrt{t} \geq 1$, all the cylindrical functions on the right-hand side of (4.3.76) can be majorized using the estimate $|J_\nu(\rho)| = O(\rho^{-1/2})$.

Allowing for $t \leq \lambda/2$ and $1 - t/\lambda \geq 1/2$, we obtain from (4.3.76)

$$|I_t^\lambda(\tau, R)| = O(\lambda^{-3/4}t^{-1/4}\tau^{-1}). \tag{4.3.77}$$

Now, keeping in mind that $\tau^{-1} \leq \sqrt{t} \leq \sqrt{\lambda/2}$ and with reference to (4.3.77), we can ensure the validity of the relation $\tau^{-1} = \tau^{-\delta}\tau^{-1+\delta} = O(\tau^{-\delta}\lambda^{1/2-\delta/2})$.

This completes the derivation of estimate (4.3.74). Inasmuch as estimate (4.3.75) is derived in much the same manner, estimate (4.3.67) stems from relations (4.3.73), (4.3.74), and (4.3.75).

---

[18] These estimates stem from the asymptotic estimates $J_\nu(\rho) = O(\rho^{-1/2})$ and $Y_\nu(\rho) = O(\rho^{-1/2})$, which are valid for all $\rho \geq 1$. It stands to reason that the estimates of O-terms in (4.3.73) are dependent on both ν and δ.

Next, we are concerned with verifying estimate (4.3.68). Allowing for $\tau \le 1/\sqrt{t}$ and majorizing the Neyman functions through $|Y_\nu(\tau\sqrt{t})| \le C(\nu\sqrt{t})^{-\nu-\delta}$ with a certain $\delta > 0$, we arrive at the relation

$$|\sqrt{t}\,\tau^{-\delta}Y_{\nu+1}(\tau\sqrt{t}) - (\nu-\delta)\tau^{-\delta-1}Y_\nu(\tau\sqrt{t})| = O(\tau^{-\nu-1-2\delta}\sqrt{t}^{-\nu-\delta}). \qquad (4.3.78)$$

It follows from (4.3.78) that, in order to prove estimate (4.3.68), it suffices to establish the estimate

$$|I_t^\lambda(\tau,R)| = \left|\int_\tau^R J_{N/2}(r\sqrt{\lambda})J_\nu(r\sqrt{t})dr\right| = O(\lambda^{-1/4-\delta/2}\sqrt{t}^{-1/2}\tau^{-2\delta}, \qquad (4.3.79)$$

for $\tau \le 1/\sqrt{t}$ and for all sufficiently large δ. (Here one may assume that $\sqrt{\lambda} \ge R$, so that $\tau \le \sqrt{\lambda}$.)

To establish estimate (4.3.79), we take the integral on the left-hand side of (4.3.79) one time by parts. We obtain thus the equality

$$\begin{aligned} I_t^\lambda(\tau,R) &= -\frac{1}{\sqrt{\lambda}}[J_\nu(r\sqrt{\lambda})J_\nu(r\sqrt{t})]\Big|_{r=\tau}^{r=R} \\ &\quad + \frac{\sqrt{t}}{\sqrt{\lambda}}\int_\tau^R J_\nu(r\sqrt{\lambda})J_{\nu-1}(r\sqrt{t})dr. \end{aligned} \qquad (4.3.80)$$

It remains to prove that the right-hand side of (4.3.80) has the order as shown in the right-hand side of (4.3.79). We majorize all the Bessel functions substituted in the right-hand side of (4.3.80) with unity and obtain, with allowance, in this particular case, for

$$\begin{aligned} \frac{1}{\sqrt{\lambda}} &\le \lambda^{-1/4-\delta/2}t^{-1/4+\delta/2} \le \lambda^{-1/4-\delta/2}t^{-1/4+\delta/2}(\tau\sqrt{t})^{-\delta} \\ &= \lambda^{-1/4-\delta/2}t^{-1/4}\tau^{-\delta} = O(\lambda^{-1/4-\delta/2}\sqrt{t}^{-1/2}\tau^{-2\delta}), \end{aligned} \qquad (4.3.81)$$

that the substitutions in (4.3.80) have the order shown on the right-hand side of (4.3.79).

The latter term on the right-hand side of (4.3.80) is estimated thus as

$$\begin{aligned} \left|\frac{\sqrt{t}}{\sqrt{\lambda}}\int_\tau^R J_\nu(r\sqrt{\lambda})J_{\nu-1}(r\sqrt{t})dr\right| &\le C_1\frac{\sqrt{t}}{\sqrt{\lambda}}\int_\tau^R (r\sqrt{\lambda})^{-1/2}(r\sqrt{t})^{-1/2}dr \\ &\le C_2\frac{R^\delta}{\sqrt{\lambda}}\int_\tau^R \frac{dr}{r^{1+\delta}} \le C_3\tau^{-\delta}\sqrt{\lambda}^{-1} = O(\lambda^{-1/4-\delta/2}\sqrt{t}^{-1/2}\tau^{-2\delta}). \end{aligned} \qquad (4.3.82)$$

This completes the proof of estimate (4.3.68).

We now proceed to estimate (4.3.69), this time allowing for $\tau \le r \le 1/\sqrt{t}$. It is obvious that

$$|\sqrt{t}\,\tau^{-\delta}J_{\nu+1}(\tau\sqrt{t}) - (\nu-\delta)\tau^{-\delta-1}J_\nu(\tau\sqrt{t})| = O(\tau^{-\delta-1}). \tag{4.3.83}$$

Therefore, to establish estimate (4.3.69), it suffices to show that

$$\left|\int_\tau^{1/\sqrt{t}} J_{N/2}(t\sqrt{\lambda})Y_\nu(r\sqrt{t})dr\right| = O(\lambda^{-1/4-\delta/2}\sqrt{t}^{-\nu-\delta-1/2}\tau^{-\nu-3\delta}). \tag{4.3.84}$$

To verify (4.3.84), we make use of the formula below, derived through integration by parts:

$$\int_\tau^{1/\sqrt{t}} J_{N/2}(r\sqrt{\lambda})Y_\nu(r\sqrt{t})dr = -\frac{1}{\sqrt{\lambda}}[J_\nu(r\sqrt{\lambda})Y_\nu(r\sqrt{t})]\Big|_{r=\tau}^{r=1/\sqrt{t}} + \frac{\sqrt{t}}{\sqrt{\lambda}}\int_\tau^{1/\sqrt{t}} J_\nu(r\sqrt{\lambda})Y_{\nu-1}(r\sqrt{t})dr. \tag{4.3.85}$$

To estimate the substitution on the right-hand side of (4.3.85), we majorize the Bessel function by unity and the Neyman function by a quantity of the order of its argument raised to the power of $-\nu-2\delta$, where $\delta > 0$. We obtain

$$\begin{aligned}\left|\frac{1}{\sqrt{\lambda}}[J_\nu(r\sqrt{\lambda})Y_\nu(r\sqrt{t})]\Big|_{r=\tau}^{r=1/\sqrt{t}}\right| &= O(\lambda^{-1/2}(\tau\sqrt{t})^{-\nu-2\delta}) \\ = O(\lambda^{-1/4-\delta/2}\sqrt{t}^{-\nu-\delta-1/2}\tau^{-\nu-2\delta}) &= O(\lambda^{-1/4-\delta/2}\sqrt{t}^{-\nu-\delta-1/2}\tau^{-\nu-3\delta}).\end{aligned} \tag{4.3.86}$$

To estimate the integral on the right-hand side of (4.3.85), we make use of the estimates

$$|J_\nu(r\sqrt{\lambda})| = O(r^{-1/2}\lambda^{-1/4}), \qquad |Y_{\nu-1}(r\sqrt{t})| = O(r^{-\nu-1/2-\delta}\sqrt{t}^{(-\nu-1/2-\delta)})$$

under the sign of this integral.

The result will be

$$\begin{aligned}\left|\frac{\sqrt{t}}{\sqrt{\lambda}}\int_\tau^{1/\sqrt{t}} J_\nu(r\sqrt{\lambda})Y_{\nu-1}(r\sqrt{t})dr\right| &= O(\lambda^{-1/4}\sqrt{t}^{-\nu-1/2-\delta}\tau^{-\nu-\delta}(t/\lambda)^{\delta/2}) \\ O(\lambda^{-1/4-\delta/2}\sqrt{t}^{-\nu-\delta-1/2}\tau^{-\nu-2\delta}) &= O(\lambda^{-1/4-\delta/2}\sqrt{t}^{-\nu-\delta-1/2}\tau^{-\nu-3\delta}).\end{aligned} \tag{4.3.87}$$

Relations (4.3.86) and (4.3.87) thus complete the proof of estimate (4.3.69).

Let us now turn to the verification of estimate (4.3.70), taking into account this time that $r \geq 1/\sqrt{t} \geq \tau$. Considering that estimate (4.3.83) remains valid, to establish estimate (4.30.70), it will suffice to show that the following relation holds:

$$\left| \int_{1/\sqrt{t}}^{R} J_{N/2}(r\sqrt{\lambda})Y_\nu(r\sqrt{t})dr \right| = O(\lambda^{-1/4-\delta/2}\sqrt{t}^{-\nu-1/2-\delta}\tau^{-\nu-3\delta}). \tag{4.3.88}$$

To estimate the integral in (4.3.88), we make use of the formula for integration by parts,

$$\int_{1/\sqrt{t}}^{R} J_{N/2}(r\sqrt{\lambda})Y_\nu(r\sqrt{t})dr = -\frac{1}{\sqrt{\lambda}}[J_\nu(r\sqrt{\lambda})Y_\nu(r\sqrt{t})]\Big|_{r=1/\sqrt{t}}^{r=R} + \frac{\sqrt{t}}{\sqrt{\lambda}} \int_{1/\sqrt{t}}^{R} J_\nu(r\sqrt{\lambda})Y_{\nu-1}(r\sqrt{t})dr. \tag{4.3.89}$$

To estimate the substitution on the right-hand side of (4.3.89) subject to $r\sqrt{t} \geq 1$, we majorize each cylindrical function with unity, taking into account that the quantity $1/\sqrt{\lambda}$ has, by virtue of relations (4.3.81), the order specified on the right-hand side of (4.3.81) and, consequently, the order shown on the right-hand side of (4.3.88).

To estimate the integral on the right-hand side of (4.3.89), we make use of the estimates

$$|J_\nu(r\sqrt{\lambda})| = O(r^{-1/2}\lambda^{-1/4}), \qquad |Y_\nu(r\sqrt{t})| = O(r^{-1/2}t^{-1/4})$$

under the sign of this integral, following the line of reasoning previously applied to (4.3.82). We find that this integral, taking into account its cofactor, has the order shown on the right-hand side of (4.3.82) and, consequently, the order shown on the right-hand side of (4.3.88).

This completes the proof of estimate (4.3.70).

Now, to complete the proof of Lemma 4.9, all we have to do is to establish estimate (4.3.71), taking into account that for this time $\tau \leq r \leq R \leq 1/\sqrt{t}$, assuming that $t \geq 1$.

By virtue of relation (4.3.72) and with regard for the simple estimates[19)]

$$|\sqrt{t}\tau^{-\delta}Y_{\nu+1}(\tau\sqrt{t}) - (\nu - \delta)\tau^{-\delta-1}Y_\nu(\tau\sqrt{t})| = O(\tau^{-\nu-1-2\delta}\sqrt{t}^{-\nu-\delta}),$$

---

19) With a view to encompassing the case $\nu = 0$, we make use of the estimate $|Y_\nu(\tau\sqrt{t})| = O[(\tau\sqrt{t})^{-\nu-\delta}]$ at $\delta > 0$.

$$|\sqrt{t}\tau^{-\delta}J_{\nu+1}(\tau\sqrt{t}) - (\nu-\delta)\tau^{-\delta-1}J_\nu(\tau\sqrt{t})| = O(\tau^{\nu-1-\delta}\sqrt{t}^{\nu}),$$

it suffices to establish the two relations

$$\left|\int_\tau^R J_{N/2}(r\sqrt{\lambda})J_\nu(r\sqrt{t})dr\right| = O(\lambda^{-1/4-\delta/2}\tau^{-\delta}\sqrt{t}^{\nu+\delta}R^{\nu+1/2}), \tag{4.3.90}$$

$$\left|\int_\tau^R J_{N/2}(r\sqrt{\lambda})Y_\nu(r\sqrt{t})dr\right| = O(\lambda^{-1/4-\delta/2}\tau^{-2\nu-\delta}\sqrt{t}^{-\nu}R^{\nu+1/2}) \tag{4.3.91}$$

in order to verify estimate (4.3.71).

First, we will establish estimate (4.3.90). In doing so, we will start from relation (4.3.80), estimating separately the substitution and the integral on the right-hand side.

In estimating the substitution, we majorize the first Bessel function with unity and the second Bessel function with a quantity of the order of its argument raised to the power ν. We obtain that the substitution in question has the order $O(\lambda^{-1/2}R^\nu\sqrt{t}^{\nu})$ and, consequently, allowing for $t \geq 1$, the order shown on the right-hand side of (4.3.90).

To estimate the integral on the right-hand side of (4.3.80), we make use of the estimates

$$|J_\nu(r\sqrt{\lambda})| \leq 1, \qquad |J_{\nu-1}(r\sqrt{t})| = O(r^{\nu-1}\sqrt{t}^{\nu-1})$$

under the sign of this integral.

We obtain, with the aid of these estimates,

$$\left|\frac{\sqrt{t}}{\sqrt{\lambda}}\int_\tau^R J_\nu(r\sqrt{\lambda})J_{\nu-1}(r\sqrt{t})dr\right| = \begin{cases} O(\lambda^{-1/2}\sqrt{t}^{\nu}R^\nu) & \text{for } \nu > 0, \\ O(\lambda^{-1/2}(\tau R)^{-\delta}) & \text{for } \nu = 0. \end{cases}$$

In both cases, we arrive at the order shown on the right-hand side of (4.3.90). This completes the proof of estimate (4.3.90).

Let us now turn to estimate (4.3.91).

We start from the formula for integration by parts

$$\int_\tau^R J_{N/2}(r\sqrt{\lambda})Y_\nu(r\sqrt{t})dr = -\frac{1}{\sqrt{\lambda}}[J_\nu(r\sqrt{\lambda})Y_\nu(r\sqrt{t})]\Big|_{r=1/\sqrt{t}}^{r=R} + \frac{\sqrt{t}}{\sqrt{\lambda}}\int_\tau^R J_\nu(r\sqrt{\lambda})Y_{\nu-1}(r\sqrt{t})dr. \tag{4.3.92}$$

In estimating the substitution on the right-hand side of (4.3.92), we majorize the Bessel function with unity and the Neyman function with a quantity of the order of its argument raised to the power of $-\nu-\delta$. We find that the substitution of interest has the order $O(\lambda^{-1/2}\tau^{-\nu-\delta}\sqrt{t}^{-\nu-\delta})$ and, consequently, with consideration for $t \geq 1$, $(R/\tau)^{\nu} \leq 1$, the order shown on the right-hand side of (4.3.91).

To estimate the integral on the right-hand side of (4.3.92), we make use of the estimates

$$|J_\nu(r\sqrt{\lambda})| = O(r^{-1/2}\lambda^{-1/4}), \qquad |Y_\nu(r\sqrt{t})| = O[(r\sqrt{t})^{-\nu-1/2-\delta}], \qquad \delta > 0,$$

under the sign of this integral.

With the aid of these estimates and the inequality $t^{1/4} \leq \lambda^{1/4}$ we obtain that this integral has the order $O(\lambda^{-1/2}\tau^{-\nu-\delta}\sqrt{t}^{-\nu-\delta})$ and, consequently, the order shown on the right-hand side of (4.3.91).

This provides a complete proof of Lemma 4.9.

4.3.4. Direct Proof of the Localization Theorem

To prove the spectral decomposition localization theorem, we shall want a simple lemma.

LEMMA 4.10. *If the estimate*

$$\widehat{A}^{\tau} E_\lambda f(x_0) = O(\lambda^{\tau})\|f\|_{H_p^{\alpha}(G)}, \tag{4.3.93}$$

uniform with respect to x_0 on an arbitrary compact set of a certain domain D, is valid for a certain $\tau_0 = \sigma > 0$, then this estimate, uniform on an arbitrary compact set of domain D, remains valid for any $\tau > 0$.

PROOF: Assume that the estimate (4.3.93), uniform with respect to x_0 on an arbitrary compact set of a certain domain D, holds at a certain $\tau_0 = \sigma > 0$. We fix an arbitrary $\tau > 0$.

In the equality

$$\widehat{A}^{\tau} E_\lambda f(x_0) = \int_0^\lambda t^{\tau} dE_t f(x_0) = \int_0^\lambda t^{\tau-\sigma} t^{\sigma} dE_t f(x_0)$$

we integrate the last integral by parts to obtain the relation

$$\widehat{A}^{\tau} E_\lambda f(x_0) = \lambda^{\tau-\sigma}[\widehat{A}^{\sigma} E_\lambda f(x_0)] + (\sigma-\tau)\int_0^\lambda t^{\tau-\sigma-1}[\widehat{A}^{\sigma} E_t f(x_0)]dt. \tag{4.3.94}$$

Using estimate (4.3.93) at $\tau = \tau_0 = \sigma$ for the bracketed expressions on the right-hand side of (4.3.94), we can show the validity of estimate (4.3.93) uniform with respect to x_0 on an arbitrary compact set of domain D at an arbitrarily fixed $\tau > 0$.

Lemma 4.10 has thus been proved.

Let us now proceed to a direct proof of the spectral decomposition localization theorem.

Assume that a function $f(x)$, compactly supported in domain G, belongs to the class $H_2^\alpha(G)$ and becomes zero in a domain D interior to G.

Then, by virtue of Lemmas 4.8 and 4.10 for any $\tau > 0$ uniformly with respect to x_0 on an arbitrary compact set of domain D, the estimate (4.3.93) taken at $p = 2$ holds.

On the other hand, by virtue of the known theorem first proved by Gårding [19] and Levitan [59], the Riesz means $E_\lambda^s f(x_0)$ of order $s \geq (N-1)/2$ of any function $f(x)$ satisfying the specified conditions converge to zero uniformly with respect to x_0 on an arbitrary compact of domain D. We denote by n the least integer greater than or equal to $(N-1)/2$. Then, making use of the Riesz means of order n, represented by the spectral decomposition $E_\lambda f(x_0)$, and of the binomial formula, we can write the following chain of equalities:

$$\begin{aligned} E_\lambda^n f(x_0) &= \int_0^\lambda \left(1 - \frac{t}{\lambda}\right)^n dE_t f(x_0) \\ &= E_\lambda f(x_0) + \sum_{k=1}^n (-1)^k C_n^k \frac{1}{\lambda^k} \int_0^\lambda t^k dE_t f(x_0) \\ &= E_\lambda f(x_0) + \sum_{k=1}^n (-1)^k C_n^k \frac{\widehat{A}^k E_\lambda f(x_0)}{\lambda^k}. \end{aligned} \tag{4.3.95}$$

From (4.3.95), making use of estimate (4.3.93) at $p = 2$, valid for any $\tau > 0$, we infer that there is a constant C_0 such that

$$|E_\lambda^n f(x_0) - E_\lambda f(x_0)| \leq C_0 ||f||_{H_2^\alpha(G)} \tag{4.3.96}$$

uniformly with respect to x_0 on an arbitrary compact of domain D.

We fix an arbitrary $\varepsilon > 0$ and a function $f_1(x)$ in the class $C_0^\infty(G)$, this function becoming zero in domain D and being such that

$$||f - f_1||_{H_2^\alpha(G)} < \frac{\varepsilon}{4C_0}, \tag{4.3.97}$$

where C_0 is the constant in estimate (4.3.96). Then, with $f(x)$ replaced by the difference $f(x) - f_1(x)$ in estimate (4.3.96) and making use of (4.3.97), we obtain the

inequality

$$|[E_\lambda^n f(x_0) - E_\lambda f(x_0)] - [E_\lambda^n f_1(x_0) - E_\lambda f_1(x_0)]| < \frac{\varepsilon}{4}, \tag{4.3.98}$$

uniform with respect to x_0 on an arbitrary compact set of domain D.

We recall now that for the function $f_1(x)$ in the class $C_0^\infty(G)$ becoming zero in domain D, there is a number $\Lambda_1 > 0$ such that the inequalities

$$|E_\lambda f_1(x_0)| < \frac{\varepsilon}{4}, \qquad |E_\lambda^n f_1(x_0)| < \frac{\varepsilon}{4} \tag{4.3.99}$$

hold at $\lambda \geq \Lambda_1$ uniformly with respect to x_0 on an arbitrary compact set of domain D.

Further, we recall, by virtue of the above-mentioned Gårding–Levitan theorem, that there is a $\Lambda_2 > 0$ such that, given $\lambda \geq \Lambda_2$, we have

$$|E_\lambda^n f(x_0)| < \frac{\varepsilon}{4} \tag{4.3.100}$$

uniformly with respect to x_0 on an arbitrary compact set of domain D.

A comparison of inequalities (4.3.98)–(4.3.100) shows that for $\lambda \geq \Lambda$, where $\Lambda = \max\{\Lambda_1, \Lambda_2\}$, the inequality

$$|E_\lambda f(x_0)| < \varepsilon$$

holds uniformly with respect to x_0 on an arbitrary compact set of domain D. This completes the proof of the spectral decomposition localization theorem.

4.3.5. The Mean Geodetic-Sphere Estimate for a Function of the Nikol'skii Class

In this section we are concerned with an ancillary theorem which we shall need in establishing conditions for the uniform convergence of spectral decompositions.

We fix an arbitrary compact set K on an arbitrary subdomain D of domain G, an arbitrary point x_0 of the compact set K, and an arbitrary, sufficiently small number R which is smaller than half the distance between the compact set K and the boundary of D.

Within a small vicinity of each point x_0 we introduce the geodetic coordinates as specified in Section 4.1.2.

LEMMA 4.11. *Assume that $N \geq 2$, a function $F(x)$ is compactly supported in domain D, continued with zero to $G \backslash D$, and belongs to the Nikol'skii class $H_p^\alpha(G)$ with order of differentiability α and degree of summability p, subject to the conditions*

$$0 < \alpha < 1, \quad \alpha < (N-1)/2, \quad p > 2N/(N-1).$$

Then for the function of the geodetic distance $r = r_{x_0 y}$

$$\varphi(r) = r^{(N-3)/2} \int_\omega F(x_0 + r\omega)\, d\omega \tag{4.3.101}$$

for all sufficiently small $h > 0$ the estimate below holds

$$\int_0^R |\varphi(r+h) - \varphi(r)|dr \leq Ch^\alpha \|F\|_{H_p^\alpha(G)} \tag{4.3.102}$$

uniformly with respect to x_0 on the compact set K of domain D.

PROOF: Relation (4.3.101) can be rewritten, by virtue of relation (4.1.8), in the form

$$\varphi(r) = r^{-(N+1)/2} \int_\omega F(x_0, r, \omega) J(x_0, r, \omega) d\omega, \tag{4.3.103}$$

where $J(x_0, r, \omega)$ refers to the Jacobian of the system of coordinates in question.

From relation (4.3.103), the increment of the function $\varphi(r)$ is represented as

$$\varphi(r+h) - \varphi(r) = A(r,h) + B(r,h), \tag{4.3.104}$$

where

$$A(r,h) = r^{-(N+1)/2} \int_\omega [F(x_0, r+h, \omega) - F(x_0, r, \omega)] J(x_0, r, \omega) d\omega, \tag{4.3.105}$$

$$B(r,h) = \int_\omega F(x_0, r, \omega) \Big[(r+h)^{-(N+1)/2} J(x_0, r+h, \omega) - r^{-(N+1)/2} J(x_0, r, \omega) \Big] d\omega. \tag{4.3.106}$$

We show that the integral of the modulus of either (4.3.105) or (4.3.106) with respect to the variable r taken between the limits from 0 to R has the order specified on the right-hand side of (4.3.102).

To estimate the integral of the modulus of (4.3.105), we invoke the estimate

$$\|F(x_0, r+h, \omega) - F(x_0, r, \omega)\|_{L_p(G)} \leq C_1 h^\alpha \|F\|_{H_p^\alpha(D)}, \tag{4.3.107}$$

valid for all x_0 in the subset D_{2h} (the norm in this estimate is taken with respect to coordinates r, ω).[20]

---

[20] It is to be kept in mind that the symbol D_{2h} denotes a subset of points removed from the boundary of D to a distance larger than $2h > 0$.

This estimate is proved in much the same manner as estimate (2.5.53) has been proved (see Lemma 2.8, Section 2.5.2); thus we may omit its proof from consideration.

It follows from relation (4.3.105) that

$$\int_0^R |A(r,h)|dr \le \int_{r_{x_0x}\le R} |F(x_0, r+h, \omega) - F(x_0, r, \omega)| r_{x_0x}^{-(N+1)/2} dx.$$

Applying the Hölder inequality to the right-hand side of the last inequality at $p > 2N/(N-1)$, $q = p/(p-1) \le 2N/(N+1)$ and making use of estimate (4.3.107), we obtain

$$\int_0^R |A(r,h)|dr \le C_2 \|F(x_0, r+h, \omega) - F(x_0, r, \omega)\|_{L_p(D)} \le C_3 h^\alpha \|F\|_{H_p^\alpha(D)}. \tag{4.3.108}$$

We have proved thereby that the integral of the modulus of (4.3.105) has the required order.

It remains to prove that the integral of the modulus of (4.3.106) has the same order.

We ensure first that to this end it suffices to establish that the estimate

$$\int_0^R |B(r,h)|dr \le C_4 h^\alpha \Big[\int_{r_{x_0x}\le R} |F(x)| r_{x_0x}^{-(N+1)/2-\alpha} dx + \int_{h\le r_{x_0x}\le R+h} |F(x_0, r_{x_0x} - h, \omega)| r_{x_0x}^{-(N+1)/2-\alpha} dx \Big] \tag{4.3.109}$$

holds for all sufficiently small $h > 0$ uniformly with respect to x_0 on a compact set K of domain D.

Indeed, having established estimate (4.3.109), we obtain, choosing $p_1 > p$ subject to the condition[21)]

$$\frac{N}{p} - \alpha < \frac{N}{p_1} < \frac{N-1}{2} - \alpha, \tag{4.3.110}$$

the relation

$$||F||_{L_1(D)} \leq C_5 ||F||_{H^{\alpha_1}_{p_1}(D)} \leq C_6 ||F||_{H^{\alpha}_{p}(D)}. \tag{4.3.111}$$

Here we make use of the embedding theorem for the Nikol'skii class,[22)] $H^{\alpha}_{p} \to H^{\alpha_1}_{p_1}$, at $\alpha_1 - \alpha - N(p^{-1} - p_1^{-1}) > 0$ and take into account that $H^{\alpha_1}_{p_1} \to L_{p_1}$.

Having p_1 and $q_1 = p_1/(p_1 - 1)$, we apply the Hölder inequality to each integral on the right-hand side, taking into account that $q_1[(N+1)/2 + \alpha] > N$ by virtue of the right-hand inequality (4.3.110) and making use of estimate (4.3.111), to obtain

$$\int_0^R |B(r,h)|dr \leq C_7 h^{\alpha} ||F||_{L_{p_1}(D)} \leq C_8 h^{\alpha} ||F||_{H^{\alpha}_{p}(D)}. \tag{4.3.112}$$

It is seen that this estimate holds for the integral of the modulus of (4.3.106).

Thus, to complete the proof of Lemma 4.11 all we have to do is to establish estimate (4.3.109) uniform with respect to x_0 on the compact set K of domain D.

We represent the integral on the left-hand side of (4.3.109) as the sum of two integrals:

$$\int_0^R |B(r,h)|dr = \int_0^{2h} |B(r,h)|dr + \int_{2h}^R |B(r,h)|dr. \tag{4.3.113}$$

The following chain of inequalities holds for the former integral on the right-hand

---

21) Such a choice of p_1 is possible considering the conditions of $p > \frac{2N}{(N+1)}$, $0 < \alpha < \frac{(N-1)}{2}$.

22) Embedding theorem (2.1.24), see Section 2.1.3. The fact that $\alpha_1 > 0$ stems from the left-hand inequality (4.3.110).

side of (4.3.113):

$$\int_0^{2h} |B(r,h)|dr \le \int_0^{2h} dr \int_\omega |F(x_0,r,\omega)|(r+h)^{-(N+1)/2} J(x_0,r+h,\omega)d\omega$$

$$+\int_0^{2h} dr \int_\omega |F(x_0,r,\omega)|r^{-(N+1)/2} J(x_0,r,\omega)d\omega$$

$$\le 3^\alpha \int_0^{2h} dr \int_\omega |F(x_0,r,\omega)|(r+h)^{-(N+1)/2-\alpha} J(x_0,r+h,\omega)d\omega$$

$$+2^\alpha \int_0^{2h} dr \int_\omega |F(x_0,r,\omega)|r^{-(N+1)/2-\alpha} J(x_0,r,\omega)d\omega$$

$$\le 3h^\alpha \int\limits_{h\le r_{x_0x}\le 3h} |F(x_0,r_{x_0x}-h,\omega)|r_{x_0x}^{-(N+1)/2-\alpha}dx$$

$$+2h^\alpha \int\limits_{0\le r_{x_0x}\le 2h} |F(x)|r_{x_0x}^{-(N+1)/2-\alpha}dx \le C_9 h^\alpha\Bigg[\int\limits_{r_{x_0x}\le R} |F(x)|r_{x_0x}^{-(N+1)/2-\alpha}dx$$

$$+\int\limits_{h\le r_{x_0x}\le R+h} |F(x_0,r_{x_0x}-h,\omega)|r_{x_0x}^{-(N+1)/2-\alpha}dx\Bigg].$$

We have thus obtained an estimate of the form (4.3.109) for the former integral on the right-hand side of (4.3.113).

We now proceed to an estimate of the form (4.3.109) for the latter integral on the right-hand side of (4.3.113).

Substituting $B(r,h)$ as defined by (4.3.106) into this integral, we arrive at the

inequality

$$\int_{2h}^{R} |B(r,h)|dr \le \int_{2h}^{R} dr \int_{\omega} |F(x_0, r, \omega)|$$

$$\times \left|(r+h)^{-(N+1)/2} J(x_0, r+h, \omega) - r^{-(N+1)/2} J(x_0, r, \omega)\right| d\omega$$

$$\le \int_{2h}^{R} dr \int_{\omega} |F(x_0, r, \omega)| J(x_0, r+h, \omega) \left|(r+h)^{-(N+1)/2} - r^{-(N+1)/2}\right| d\omega$$

$$+ \int_{2h}^{R} dr \int_{\omega} |F(x_0, r, \omega)| r^{-(N+1)/2} |J(x_0, r+h, \omega) - J(x_0, r, \omega)| d\omega. \tag{4.3.114}$$

Considering that

$$\left|(r+h)^{-(N+1)/2} - r^{-(N+1)/2}\right| = (r+h)^{-(N+1)/2} \left|1 - \left(\frac{r+h}{r}\right)^{(N+1)/2}\right|$$

$$\le C_{10}(r+h)^{-(N+1)/2} \left(\frac{h}{r}\right) \le C_{10}(r+h)^{-(N+1)/2} \left(\frac{h}{r}\right)^{\alpha}$$

$$\le C_{10}(r+h)^{-(N+1)/2-\alpha} h^{\alpha}$$

we obtain, for the former integral on the right-hand side of (4.3.114), the estimate

$$\int_{2h}^{R} dr \int_{\omega} |F(x_0, r, \omega)| J(x_0, r+h, \omega) \left|(r+h)^{-(N+1)/2} - r^{-(N+1)/2}\right| d\omega$$

$$\le C_{10} h^{\alpha} \int_{3h \le r_{x_0 x} \le R+h} |F(x_0, r_{x_0 x} - h, \omega)| r^{-(N+1)/2-\alpha} dx,$$

that is, an estimate of the form (4.3.109) remains valid for this integral also.

Now we have to obtain an estimate of the form (4.3.109) for the latter integral on the right-hand side of (4.3.114).

To this end, let us estimate the difference

$$J(x_0, r+h, \omega) - J(x_0, r, \omega).$$

We make use of the familiar representation of a Jacobian in the Riemannian coordinates[23])

$$J(x_0, r, \omega) = \widetilde{J}(x_0, r, \omega)\widehat{J}(\omega), \tag{4.3.115}$$

where

$$\widehat{J}(\omega) = \sin^{N-2}\omega_1 \sin^{N-3}\omega_2 \ldots \sin\omega_{N-1},$$

and $\widetilde{J}(x_0, r, \omega) > 0$ for all $r > 0$; recall that there exist strictly positive constants a_1, a_2, b_1, and b_2 such that the inequalities

$$a_1 r^{N-1} \le \widetilde{J}(x_0, r, \omega) \le a_2 r^{N-1},$$

$$b_1 r^{N-2} \le \left| \frac{d\widetilde{J}}{dr}(x_0, r, \omega) \right| \le b_2 r^{N-2} \tag{4.3.116}$$

hold for all sufficiently small positive r and for all points x_0 of the compact set K.

Applying the Lagrange theorem to the difference $\widetilde{J}(x_0, r+h, \omega) - \widetilde{J}(x_0, r, \omega)$, we find that there is a number θ in the interval of $0 < \theta < 1$ such that

$$\widetilde{J}(x_0, r+h, \omega) - \widetilde{J}(x_0, r, \omega) = h\frac{\partial \widetilde{J}}{\partial r}(x_0, r+\theta h, \omega).$$

From this equality, allowing for $r \ge 2h$ and with reference to the latter estimate (4.3.116), we obtain the inequality

$$\left| \widetilde{J}(x_0, r+h, \omega) - \widetilde{J}(x_0, r, \omega) \right| \le C_{11} h\, r^{N-2}. \tag{4.3.117}$$

From inequality (4.3.117) and from the former estimate (4.3.116), we get the following inequality:

$$\frac{|\widetilde{J}(x_0, r+h, \omega) - \widetilde{J}(x_0, r, \omega)|}{\widetilde{J}(x_0, r, \omega)} \le C_{12}\frac{h}{r},$$

which can be rewritten, by virtue of $0 < \alpha < 1$ and $r \ge 2h$, as

$$\frac{|\widetilde{J}(x_0, r+h, \omega) - \widetilde{J}(x_0, r, \omega)|}{\widetilde{J}(x_0, r, \omega)} \le C_{12}\left(\frac{h}{r}\right)^{\alpha}. \tag{4.3.118}$$

In turn, inequality (4.3.118) and representation (4.3.115) lead to the inequality

$$\frac{|J(x_0, r+h, \omega) - J(x_0, r, \omega)|}{\widetilde{J}(x_0, r, \omega)} \le C_{12}\left(\frac{h}{r}\right)^{\alpha}. \tag{4.3.119}$$

---

23) See, for example, Rashevsky [68, p. 560].

With the aid of (4.3.119), the required estimate of the form (4.3.109) follows immediately for the latter integral on the right-hand side of (4.3.114). Indeed, by virtue of (4.3.119), we have

$$\int_{2h}^{R} dr \int_{\omega} |F(x_0, r, \omega)| r^{-(N+1)/2} |J(x_0, r+2h, \omega) - J(x_0, r, \omega)| d\omega$$

$$\leq C_{12} h^{\alpha} \int_{2h \leq r_{x_0 x} \leq R} |F(x)| r_{x_0 x}^{-(N+1)/2-\alpha} dx.$$

This completes the proof of Lemma 4.11.

4.3.6. Proof of the Main Asymptotic Estimate

In this section we shall establish the main asymptotic formula which leads to a proof of the positive Theorem 4.3 within the standard framework as outlined in Section 2.5.4.

LEMMA 4.12. *Assume that $N \geq 2$, and a function $f(x)$ satisfies the three requirements that have been formulated for Theorem 4.3, namely: (1) $f(x)$ becomes zero outside the set G_{h_0}, given a certain $h_0 > 0$; (2) $f(x)$ belongs to the Nikol'skii class H_2^{α} in the entire domain G at $\alpha \geq (N-1)/2$; (3) $f(x)$ belongs to the Nikol'skii class H_p^{α} within a certain domain D interior to G at certain α and p subject to*

$$\alpha \geq (N-1)/2, \qquad \alpha p > N, \qquad p \geq 1. \tag{4.3.120}$$

If so, the asymptotic formula

$$\widehat{A}^{\tau} E_{\lambda} f(x_0) = O(\lambda^{\tau}) \left[||f||_{H_2^{\alpha}(G)} + ||f||_{H_p^{\alpha}(D)} \right] \tag{4.3.121}$$

holds for any $\tau > 0$ uniformly with respect to x_0 on an arbitrary compact set K of domain D.

PROOF: We note, for one thing, that the asymptotic formula (4.3.121) is sufficient for the proof at $\alpha = (N-1)/2$, $\alpha p > N$, that is, at $p > 2N/(N-1)$. Then, by virtue of the embedding theorem for the Nikol'skii classes (Theorem (2.1.24), see Section 2.1.3), formula (4.3.121) will hold for any α and p satisfying the three requirements (4.3.120).

We note, for another thing, that it suffices, by virtue of Lemma 4.10, to prove estimate (4.3.121) uniform with respect to x_0 on an arbitrary compact set K of domain D for at least one $\tau_0 > 0$. We wish to prove this estimate for values of τ lying within the interval $(N-1)/4 - \delta < \tau < (N-1)/4$, where δ is a sufficiently small positive number.

In domain D, we fix an arbitrary compact set K and two other compact sets K' and K'' such that K' is strictly interior to K, and K'' is strictly interior to K'.

We fix a function $\eta(x)$ of the class $C_0^\infty(G)$ such that $\eta(x) = 1$ in the compact set K' and $\eta(x) = 0$ outside the compact set K''; further, we represent the function $f(x)$ as the sum of two functions, $f_1(x) + f_2(x)$, such that $f_1(x) = \eta(x)f(x)$, $f_2(x) = [1 - \eta(x)]f(x)$.

Now, taking into account that there exist constants C_1 and C_2 such that

$$\begin{aligned} ||f_1||_{H_p^\alpha(D)} &\leq C_1||f||_{H_p^\alpha(D)}, \\ ||f_2||_{H_2^\alpha(D)} &\leq C_2||f||_{H_2^\alpha(D)}, \end{aligned} \tag{4.3.122}$$

and that the function $f_2(x)$ is compactly supported in G and becomes zero on the compact set K', we obtain, by virtue of Lemma 4.8, the following estimate:

$$\widehat{A}^\tau E_\lambda f_2(x_0) = O(\lambda^\tau)||f_2||_{H_2^\alpha(G)} = O(\lambda^\tau)||f||_{H_2^\alpha(G)}. \tag{4.3.123}$$

For the function $f_1(x)$, by virtue of the same Lemma 4.8, the representation of the form (4.3.37)

$$\begin{aligned} \widehat{A}^\tau E_\lambda f_1(x_0) = &\int_G V^\lambda(x_0, x)\widehat{A}^\tau f_1(x)\rho_0(x)dx \\ &- \frac{1}{2}\sum_{i=1}^{\widehat{m}} \int_0^\infty u_i(x_0, t)\widehat{f}_i^{(1)}(\lambda)\lambda^\tau[\rho(\lambda+0) - \rho(\lambda-0)] \\ &+ \lambda^{N/4}\sum_{i=1}^{\widehat{m}} \int_0^\infty t^{(2-N)/4} I_t^\lambda(R)u_i(x_0, t)\widehat{f}_i^{(1)}(t)t^\tau\, d\rho(t) \\ &- \sum_{i=1}^{\widehat{m}} \int_0^\infty \widetilde{V}_i^\lambda(x_0, t)\widehat{f}_i^{(1)}(t)t^\tau\, d\rho(t) \end{aligned} \tag{4.3.124}$$

holds. (For notations, see Section 4.3.2; the symbol $\widehat{f}_i^{(1)}(t)$ stands for the Fourier image of the function $f_1(x)$.)

It has been shown in Lemma 4.8 that if the function $f_1(x)$ belongs to $H_2^\alpha(G)$ and is compactly supported in G, then each of the last three expressions on the right-hand side of (4.3.124) has the same order[24])

$$O(\lambda^\tau)||f_1||_{H_2^\alpha(G)} = O(\lambda^\tau)||f||_{H_2^\alpha(G)}.$$

[24]) Here we make use of the former estimate in (4.3.122).

Relation (4.3.124) can be rewritten as

$$\widehat{A}^{\tau} E_{\lambda} f_1(x_0) = \int\limits_G V^{\lambda}(x_0, x)\widehat{A}^{\tau} f_1(x)\rho_0(x)dx + O(\lambda^{\tau})||f||_{H_2^{\alpha}(G)}. \tag{4.3.125}$$

Now, let us consider the function

$$F(x) = \eta(x)p(x_0, x)\widehat{A}^{\tau} f_1(x), \tag{4.3.126}$$

where $p(x_0, x)$ is the function defined for the mean-value formula (4.1.10) (see Section 4.1.2); $\eta(x)$ is the same function as above.

By virtue of Assertion 2 (see Section 4.1.4), the function $F(x)$ belongs to the Nikol'skii class $H_p^{\alpha-2\tau}(D)$ subject to the relation

$$||F||_{H_p^{\alpha-2\tau}(D)} \leq C_3||f_1||_{H_p^{\alpha}(D)} \leq C_4||f||_{H_p^{\alpha}(D)}. \tag{4.3.127}$$

If, in the definition of the function $V^{\lambda}(x_0, x)$, the number R has been chosen smaller than the geodetic distance between the compact set K and the boundary of the compact set K', then relation (4.3.125) can be rewritten with reference to equality (4.3.126) as

$$\widehat{A}^{\tau} E_{\lambda} f_1(x_0) = \left(\frac{\lambda}{2\pi}\right)^{N/2} \int\limits_{r_{x_0x}\leq R} \left(r_{x_0x}\sqrt{\lambda}\right)^{-N/2} J_{N/2}(r_{x_0x}\sqrt{\lambda})F(x)dx \tag{4.3.128}$$
$$+O(\lambda^{\tau})||f||_{H_2^{\alpha}(G)}.$$

We consider now the function $\varphi(r)$ defined by relation (4.3.101), assuming that $F(x)$ is a function of the form (4.3.126). Given such a function $\varphi(r)$, all conditions of Lemma 4.11 will be fulfilled for $\kappa = \alpha - 2\tau$, rather than for α.

By virtue of Lemma 4.11, for $\varphi(r)$, given a sufficiently small $h > 0$, the estimate

$$\int\limits_0^R |\varphi(r+h) - \varphi(r)| \leq Ch^{\kappa}||F||_{H_p^{\kappa}(D)} \tag{4.3.129}$$

holds uniformly with respect to x_0 on the compact set K of domain D; here $\kappa = \alpha - 2\tau = (N-1)/2 - 2\tau$.

In terms of the function $\varphi(r)$, relation (4.3.128) takes the form

$$\widehat{A}^{\tau} E_{\lambda} f_1(x_0) = (2\pi)^{-N/2}\lambda^{(N-1)/4} \int\limits_0^R (r\sqrt{\lambda})^{1/2} J_{N/2}(r\sqrt{\lambda})\varphi(r)dr \tag{4.3.130}$$
$$+O(\lambda^{\tau})||f||_{H_2^{\alpha}(G)}.$$

One infers from relations (4.3.123) and (4.3.130) and from the fact that $f(x) = f_1(x) + f_2(x)$ that in order to establish estimate (4.3.121), one must prove that the equality

$$\int_0^R (r\sqrt{\lambda})^{1/2} J_{N/2}(r\sqrt{\lambda})\varphi(r)dr = O(\lambda^{-\kappa/2})\|F\|_{H_p^\kappa(D)} \tag{4.3.131}$$

holds uniformly with respect to x_0 on the compact set K, where $\kappa = \alpha - 2\tau$.

A proof of (4.3.131) has been given earlier (see Lemma 2.10, equality (2.5.92), Section 2.5.2), based on the fact that, for the function $\varphi(r)$, relation (4.3.129) and the three relations below[25]

$$\left|\int_0^R r^{-\kappa}\varphi(r)dr\right| \le C_5\|F\|_{H_p^\kappa(D)}, \tag{4.3.132}$$

$$\int_0^h |\varphi(r)|dr \le C_6 h^\kappa \|F\|_{H_p^\kappa(D)}, \tag{4.3.133}$$

$$\int_R^{R+h} |\varphi(r)| \le C_7 h^\kappa \|F\|_{H_p^\kappa(D)}, \tag{4.3.134}$$

are valid, uniformly with respect to x_0 on the compact set K for all sufficiently small $h > 0$.

Thus, to prove Lemma 4.12, we must ensure that estimates (4.3.132)–(4.3.134) do hold indeed.

We first consider estimate (4.3.132). Since, by definition, $F(x) \in H_p^\kappa(D)$ for $p > 2N/(N-1)$, there is a sufficiently small $\delta > 0$ such that $F(x) \in H_p^\kappa(D)$ at $p = 2N/(N-1-2\delta)$.

The number q conjugate to p is

$$q = \frac{p}{p-1} = \frac{2N}{N+1+2\delta}. \tag{4.3.135}$$

Invoking the embedding theorem (2.1.24) in Section 2.1.3, we have

$$H_p^\kappa \to H_{p'}^{\kappa - N(1/p-1/p')},$$

where p' is any number such that $\kappa - N(1/p - 1/p') > 0$, or, what is the same,

$$\frac{1}{p'} > \frac{1}{p} - \frac{\kappa}{N} = \frac{N-1-2\delta}{2N} - \frac{\kappa}{N} = \frac{N-1-2\delta-2\kappa}{2N}. \tag{4.3.136}$$

---

[25] See in Section 2.5.2, respectively, relations (2.5.102), (2.5.114), and (2.5.115).

To fulfill (4.3.136), it suffices for the $\delta > 0$ chosen above to set

$$\frac{1}{p'} = \frac{N-1-\delta-2\kappa}{2N}. \qquad (4.3.137)$$

Then for q', conjugate to p', we obtain

$$\frac{1}{q'} = 1 - \frac{1}{p'} = \frac{N+1+\delta+2\kappa}{2N},$$

so that

$$q' = \frac{2N}{N+1+\delta+2\kappa}. \qquad (4.3.138)$$

Keeping in mind that $\varphi(r)$ is defined by relation (4.3.103), we shall write the left-hand side of (4.3.132) in the form

$$\left|\int_0^R r^{-\kappa}\varphi(r)dr\right| = \left|\int_{r_{x_0 x}\le R} r^{-(N+1)/2-\kappa}F(x)dx\right|. \qquad (4.3.139)$$

To obtain estimate (4.3.132) uniform with respect to x_0 on the compact set K, we apply the Hölder inequality to the integral on the right-hand of (4.3.139) at p' and q', as defined by relations (4.3.137) and (4.3.138), taking into account that $F(x) \in L_{p'}(D)$ subject to

$$\|F\|_{L_{p'}(D)} = O\left(\|F\|_{H_p^\kappa(D)}\right),$$

and keeping in mind that

$$\left(\frac{N+1}{2}+\kappa\right)q' = \frac{\frac{N+1}{2}+\kappa}{\frac{N+1}{2}+\kappa+\frac{\delta}{2}}N < N.$$

To obtain estimate (4.3.133) from estimate (4.3.132), we recall that

$$\int_0^h |\varphi(r)|dr = \int_0^h r^\kappa[r^{-\kappa}|\varphi(r)|]dr$$

$$\le h^\kappa \int_0^R r^{-\kappa}|\varphi(r)|dr = O(h^\kappa)\|F\|_{H_p^\kappa(D)}.$$

Estimate (4.3.134) is easily established using (4.3.129) and (4.3.133). Indeed,

$$\int_R^{R+h} |\varphi(r)|dr - \int_0^h |\varphi(r)|dr = \int_R^{R+h} |\varphi(r)|dr - \int_0^R |\varphi(r)|dr$$

$$= \int_0^R |\varphi(r+h)|dr - \int_0^R |\varphi(r)|dr \leq \int_0^R |\varphi(r+h) - \varphi(r)|dr,$$

so that

$$\int_R^{R+h} |\varphi(r)|dr \leq \int_0^h |\varphi(r)|dr + \int_0^R |\varphi(r+h) - \varphi(r)|dr.$$

One infers from this inequality and from (4.3.133) and (4.3.129) that estimate (4.3.134) holds uniformly with respect to x_0 on the compact set K.

This completes the proof of Lemma 4.12.

4.3.7. Direct Proof of the Uniform Convergence Theorem

Assume that all the conditions of Theorem 4.3 are fulfilled. We denote by n the least integer satisfying $n \geq (N-1)/2$. In Section 4.3.4, the relation

$$E_\lambda^n f(x_0) - E_\lambda f(x_0) = \sum_{k=1}^{n} (-1)^k C_n^k \frac{\widehat{A}^k E_\lambda f(x_0)}{\lambda^k} \tag{4.3.140}$$

has been derived for the Riesz means difference of order n of a spectral decomposition and for the spectral decomposition itself (see relation (4.3.95), Section 4.3.4).

Applying, to the right-hand side of (4.3.14)), the asymptotic estimate (4.3.121), which is valid for any $\tau > 0$ and is uniform with respect to x_0 on any fixed compact set K of domain D, we find that there is a constant C_0 such that

$$|E_\lambda^n f(x_0) - E_\lambda f(x_0)| \leq C_0 \left[\|f\|_{H_2^\alpha(G)} + \|f\|_{H_p^\alpha(D)} \right] \tag{4.3.141}$$

uniformly with respect x_0 on the compact set K.

We now fix an arbitrary $\varepsilon > 0$ and define a function $f_1(x)$ in the class $C_0^\infty(G)$ such that

$$\|f - f_1\|_{H_2^\alpha(G)} + \|f - f_1\|_{H_p^\alpha(D)} < \frac{\varepsilon}{4C_0}, \tag{4.3.142}$$

where C_0 is the constant in (4.3.141).

Having replaced the $f(x)$ in (4.3.141) by $f(x) - f_1(x)$, we obtain, with reference to estimate (4.3.142), the inequality

$$|[E_\lambda^n f(x_0) - E_\lambda f(x_0)] - [E_\lambda^n f_1(x_0) - E_\lambda f_1(x_0)]| < \frac{\varepsilon}{4},$$

or, what is the same,

$$\begin{aligned}&\big|\{[E_\lambda^n f(x_0) - f(x_0)] - [E_\lambda f(x_0) - f(x_0)]\}\\ &\qquad - \{[E_\lambda^n f_1(x_0) - f_1(x_0)] - [E_\lambda f_1(x_0) - f_1(x_0)]\}\big| < \frac{\varepsilon}{4}.\end{aligned} \tag{4.3.143}$$

We recall, for one thing, that for the function $f_1(x)$ in the class $C_0^\infty(G)$ there is a number $\Lambda_1 > 0$ such that for $\lambda > \Lambda_1$ the inequalities

$$|E_\lambda^n f_1(x_0) - f_1(x_0)| < \frac{\varepsilon}{4}, \quad |E_\lambda f_1(x_0) - f_1(x_0)| < \frac{\varepsilon}{4} \tag{4.3.144}$$

hold uniformly with respect to x_0 on any compact set K of domain D.

We recall, for another thing, that, invoking the theorems of Gårding and Levitan (referred to in Section 4.3.4), there is a number $\Lambda_2 > 0$ such that at $\lambda \geq \Lambda_2$ the relation

$$|E_\lambda^n f(x_0) - f(x_0)| < \frac{\varepsilon}{4} \tag{4.3.145}$$

holds uniformly with respect to x_0 on any compact set K of domain D.[26)]

Having set $\Lambda = \max\{\Lambda_1, \Lambda_2\}$, a comparison of inequalities (4.3.143)–(4.3.145) implies that, given $\lambda \geq \Lambda$, the inequality

$$|E_\lambda f(x_0) - f(x_0)| < \varepsilon$$

holds uniformly with respect to x_0 on an arbitrary compact set K of domain D, which establishes the theorem on the uniform convergence of spectral decompositions.

This completes the proof of Theorem 4.3.

COMMENTS ON CHAPTER 4

The mean-value formula of Moiseev [62] and the properties of fractional powers of the basic integral operator (Alimov [3]) have been major auxiliary tools used in this chapter.

---

26) Also, we take into account that, according to the embedding theorem, the function $f(x)$ in the class $H_p^\alpha(D)$ at $\alpha p > N$ is continuous on any compact set of domain D.

The material of Section 4.1.3 has been reported for the first time.

Section 4.2 is a revised presentation of the results first reported in Il'in [36].

Section 4.3 is based on the results reported by Il'in and Alimov [42] and, in part, on the results of an earlier paper of Il'in and Moiseev [45].

Conditions for the spectral decomposition convergence, less general than those dealt with in Section 4.3, were reported by Bergendal [10].

The results in Section 4.3 were generalized by Kostyuchenko and Mityagin [54] to elliptic operators of order $2m$ with constant coefficients, and by Alimov [4] to general elliptic operators of order $2m$ with smooth coefficients.

The conditions that guarantee the convergence of spectral decompositions uniform entirely within a closed domain were established by Moiseev [63] for eigenfunctions of the first boundary-value problem for an elliptic operator of second order and for a compactly supported function.

Appendix 1

Conditions for the Uniform Convergence of Multiple Trigonometric Fourier Series with Spherical Partial Sums

Let domain $G = T^N$ be an N-dimensional cube $-\pi \leq x_k \leq \pi$, $k = 1, 2, \ldots, N$.

Further, let $\bar{n} = (n_1, n_2, \ldots, n_N)$ be a vector with integer-valued coordinates n_1, n_2, ..., n_N, and $\bar{x} = (x_1, x_2, \ldots, x_N)$ an arbitrary vector; $\bar{n} = \sqrt{n_1^2 + n_2^2 + \ldots + n_N^2}$; the symbol $(\bar{n}, \bar{x})$ is used to denote the scalar product

$$(\bar{n}, \bar{x}) = n_1 x_1 + n_2 x_2 + \ldots + n_N x_N. \tag{A.1.1}$$

Each function $f(x)$ in the class $L_1(T^N)$ is associated with a multiple trigonometric Fourier series of the form

$$\sum f_{\bar{n}} e^{i(\bar{n},\bar{x})},$$

where the Fourier coefficient $f_{\bar{n}}$ is

$$f_{\bar{n}} = (2\pi)^{-N} \int\limits_{T^N} f(y) e^{-i(\bar{n},\bar{y})} dy. \tag{A.1.2}$$

Inasmuch as the multiple trigonometric system represents (to within a normalization factor) a special case of the Laplace operator FSF in the cube T^N (see Chapter 1) and inasmuch as the convergence of a spectral decomposition in such a FSF corresponds to the convergence (as $\lambda \to \infty$) of the so-called spherical partial sums of a multiple trigonometric Fourier series

$$S_\lambda(x, f) = \sum_{|\bar{n}|<\lambda} f_{\bar{n}} e^{i(\bar{n},\bar{x})}, \tag{A.1.3}$$

it is to be inferred from the results of Chapters 1 and 2 that *if a function $f(x)$ is compactly supported in an N-dimensional cube T^N, then for any $N > 1$ in the Nikol'skii $H_p^\alpha(T^N)$, Sobolev–Liouville $L_p^\alpha(T^N)$, Besov $B_{p,\theta}^\alpha(T^N)$ classes (for any $\theta \geq 1$) and in the Zygmund–Hölder class $C^\alpha(T^N)$, the three inequalities*

$$\alpha \geq (N-1)/2, \qquad \alpha p > N, \qquad p \geq 1 \tag{A.1.4}$$

are the ultimate conditions for a uniform (on any compact set of the cube T^N as $\lambda \to \infty$) convergence of the spherical partial sums (A.1.3).[1)]

It is interest to establish the ultimate conditions for a uniform (within the closed cube T^N) convergence of the spherical partial sums (A.1.3) without the necessary assumption that the function be compactly supported in this cube.

We shall establish such conditions, restricting ourselves to the case of an odd number $N > 1$ of dimensions in terms of the periodic Sobolev–Liouville classes.

THEOREM A.1. *If a function $f(x)$ of an odd number $N > 1$ of variables belongs to a periodic (with a period of 2π in each variable) Sobolev–Liouville class L_p^α with arbitrary α and p satisfying the three inequalities (A.1.4), then the spherical partial sums (A.1.3) of a multiple trigonometric Fourier series of this function converge, as $\lambda \to \infty$, to the same function uniformly within the closed cube T^N.*

We note, for one thing, that, by virtue of the embedding theorem for the Sobolev–Liouville classes, it suffices to have the proof of Theorem A.1 carried out at $\alpha = (N-1)/2$, $p > 2N/(N-1)$, that is, for a function $f(x)$ which belongs to a periodic Sobolev–Liouville class $L_p^{(N-1)/2}$ at $p > 2N/(N-1)$.

We note, for another thing, that instead of the N-dimensional cube T^N with arbitrary real $\delta_1, \delta_2, \ldots, \delta_N$ we can consider another N-dimensional cube $T^N_{\delta_1\delta_2\ldots\delta_N}$, defined by the relations $-\pi + \delta_k \leq x_k \leq \pi + \delta_k$, $k = 1, 2, \ldots, N$.

Since the multiple trigonometric system forms, to within the same normalization factors, FSF of the Laplace operator in any such cube $T^N_{\delta_1\delta_2\ldots\delta_N}$ and since, by virtue of the periodicity of functions $f(x)$ and $e^{-i(\bar{n},\bar{x})}$, we have

$$f_{\bar{n}} = (2\pi)^{-N} \int_{T^N} f(y) e^{-i(\bar{n},\bar{y})} dy = (2\pi)^{-N} \int_{T^N_{\delta_1\delta_2\ldots\delta_N}} f(y) e^{-i(\bar{n},\bar{y})} dy,$$

in order to prove that the spherical partial sums (A.1.3) converge as $\lambda \to \infty$ uniformly within the entire closed cube T^N, it will suffice to show that for the function of the

[1)] For the Zygmund–Hölder classes $C^\alpha(T^N)$, three inequalities (A.1.4) degenerate into the inequality $\alpha \geq (N-1)/2$.

aforesaid Sobolev–Liouville class, these spherical partial sums converge as $\lambda \to \infty$ uniformly on any compact set of an N-dimensional cube $T^N_{\delta_1\delta_2\ldots\delta_N}$.

To this end (see Sections 4.3.2, 4.3.6, 4.3.7) it suffices to show that the inequality[2]

$$\sum |f_{\bar{n}}|^2 \, |\bar{n}|^{N-1} \leq C\|f\|_{L_2^{(N-1)/2}(T^N)} \tag{A.1.5}$$

holds for the function $f(x)$ of the Sobolev–Liouville class in question.

Let us now turn to inequality (A.1.5).

We note that, for any function $f(x)$ in the periodic Sobolev–Liouville class L_2^1 and for any element $u_{\bar{n}}(x) = e^{i(\bar{n},\bar{x})}$ of a multiple trigonometric system, the equality

$$\int_{T^N} \left[\sum_{k=1}^{N} \frac{\partial f(x)}{\partial x_k} \frac{\partial u_{\bar{n}}(x)}{\partial x_k}\right] dx = |\bar{n}|^2 \int_{T^N} f(x) u_{\bar{n}}(x) dx \tag{A.1.6}$$

holds by virtue of the first Green's formula and the relation $\Delta u_{\bar{n}}(x) + |\bar{n}|^2 u_{\bar{n}}(x) = 0$.

In particular, given $f(x) = u_{\bar{m}}(x)$, we obtain from (A.1.6)

$$\int_{T^N} \left[\sum_{k=1}^{N} \frac{\partial u_{\bar{m}}(x)}{\partial x_k} \frac{\partial u_{\bar{n}}(x)}{\partial x_k}\right] dx = \begin{cases} |\bar{n}|^2, & \text{when vectors } n \text{ and } m \text{ are coincident;} \\ 0, & \text{otherwise.} \end{cases} \tag{A.1.7}$$

We wish to prove the following ancillary lemma.

LEMMA A.1. *If a function $\Phi(x)$ belongs to the periodic Sobolev–Liouville class L_2^1 [respectively, L_2^2], then the inequality*

$$\sum |\Phi_{\bar{n}}|^2 |\bar{n}|^2 \leq \int_{T^N} \sum_{k=1}^{N} \left|\frac{\partial \Phi}{\partial k_k}\right|^2 dx \tag{A.1.8}$$

$$\left[\text{respectively,} \quad \sum |\Phi_{\bar{n}}|^2 |\bar{n}|^4 \leq \int_{T^N} |\Delta\Phi|^2 dx\right] \tag{A.1.9}$$

holds for its Fourier coefficients in the multiple trigonometric system $\Phi_{\bar{n}}$.

PROOF: In the case $\Phi \in L_2^1$, for any $\lambda > 0$ we consider the nonnegative quantity

$$I = \int_{T^N} \left\{\sum_{k=1}^{N} \frac{\partial}{\partial x_k} \left[\Phi - \sum_{|\bar{n}|<\lambda} \Phi_{\bar{n}} \, u_{\bar{n}}(x)\right]\right\}^2 dx \geq 0$$

---

[2] The validity of this inequality implies the validity of inequality (4.3.61) with the Nikol'skii-class norm on its right-hand side replaced by a Sobolev–Liouville-class norm.

and, making use of relations (A.1.6) and (A.1.7), rewrite it as

$$I = \int\limits_{T^N} \sum_{k=1}^{N} \left| \frac{\partial \Phi(x)}{\partial x_k} \right|^2 dx - \sum_{|\bar{n}|<\lambda} \Phi_{\bar{n}}^2 \, |u_{\bar{n}}(x)|^2 \geq 0.$$

By virtue of the arbitrary choice of $\lambda > 0$, the last inequality implies inequality (A.1.8).

In the case $\Phi \in L_2^2$, inequality (A.1.9) is derived from the Bessel inequality for the function $\Delta\Phi \in L_2$, taking into account that, by virtue of the second Green's formula,

$$(\Delta\Phi)_{\bar{n}} = \int\limits_{T^N} u_{\bar{n}}(x)\Delta\Phi dx = -\int\limits_{T^N} \Phi\Delta u_{\bar{n}} dx = -|\bar{n}|^2 \sum_{T^N} \Phi u_{\bar{n}} dx = -|\bar{n}|^2 \Phi_{\bar{n}}. \quad \text{(A.1.10)}$$

Lemma A.1 is thus proved.

Let us now return to the proof of inequality (A.1.5).

Assume that the function $f(x)$, given an odd $N > 1$, belongs to the periodic Sobolev–Liouville class $L_2^{(N-1)/2}$. We prove that for the Fourier coefficient $f_{\bar{n}}$ of the function $f(x)$ the following inequality holds:

$$\sum |f_{\bar{n}}|^2 |\bar{n}|^{N-1} \leq \begin{cases} \displaystyle\int\limits_{T^N} \left|\Delta^{(N-1)/4} f\right|^2 dx & \text{for even} \quad (N-1)/2, \\ \displaystyle\int\limits_{T^N} \sum_{k=1}^{N} \left| \frac{\partial}{\partial x_k} \left(\Delta^{(N-3)/4} f\right) \right|^2 dx & \text{for odd} \quad (N-1)/2. \end{cases}$$

(A.1.11)

In the case of an even $(N-1)/2$, that is, with $N = 5, 9, 13, \ldots$, inequality (A.1.11) is obtained by applying successively relation (A.1.10) to the functions f, Δf, $\ldots$, $\Delta^{(N-5)/4} f$, and the Bessel inequality to the function $\Delta^{(N-1)/4} f$.

In the case of an odd $(N-1)/2$, that is, with $N = 3, 7, 11, \ldots$, inequality (A.1.11) with $N = 3$ is equivalent to (A.1.8), whereas with $N = 7, 11, \ldots$, it is obtained by applying succesively relation (A.1.10) to the functions f, Δf, $\ldots$, $\Delta^{(N-7)/4} f$, and inequality (A.1.8) to the function $\Phi = \Delta^{(N-3)/4} f$.

Inequality (A.1.11) is proved.

Since the quantities on the right-hand side of (A.1.11) are majorized by the right-hand side of (A.1.5), inequality (A.1.5) is thus proved, which completes the proof of Theorem A.1.

REMARK. The fact that the condition for the uniform convergence (see Theorem A.1) is ultimate in terms of the periodic Sobolev–Liouville classes immediately follows from the negative Theorem 2.2 (see Chapter 2).

Appendix 2

Conditions for the Uniform Convergence of Decompositions in Eigenfunctions of the First, Second, and Third Boundary-Value Problems for an Elliptic Operator of Second Order

We shall consider here an arbitrary bounded domain G of any odd number N ($N > 1$) of dimensions and establish, for a function $f(x)$ arbitrary and compactly unsupported in this domain, the ultimate conditions in the Sobolev–Liouville class $L_p^\alpha(G)$ that guarantee the uniform, on any compact set of domain G, convergence of decompositions in the eigenfunctions of the first, second, and third boundary-value problems for an elliptic operator of second order specified in domain G

$$Lu = -\sum_{i,j=1}^{N} \frac{\partial}{\partial x_i}\left[a_{ij}(x)\frac{\partial u}{\partial x_j}\right] + c(x)u(x). \tag{A.2.1}$$

We assume that the coefficients of elliptic operator (A.2.1) satisfy the conditions formulated at the beginning of Chapter 4; further, we assume that they satisfy the requirement that $c(x) \geq 0$ in G, and that the boundary ∂G in domain G is piecewise smooth, and the eigenfunctions of elliptic operator (A.2.1) satisfy, at the boundary ∂G, either the boundary condition of the first kind

$$u|_{\partial G} = 0, \tag{A.2.2}$$

or the boundary condition of the second kind

$$\left.\frac{\partial u}{\partial \nu}\right|_{\partial G} = 0, \tag{A.2.3}$$

where ν is the conormal for operator (A.2.1) to the surface ∂G (see, for example,

Miranda [60], p. 15), or the boundary condition of the third kind

$$\left(\frac{\partial u}{\partial \nu} + h(s)u\right)\Big|_{\partial G} = 0, \tag{A.2.4}$$

where $h(s)$ is a continuous and nonnegative function defined along the ∂G.

Satisfying the specified boundary conditions may also be understood in a generalized sense. For the boundary condition of the first kind (A.2.2) this signifies that the function $u(x)$ belongs to the class $\overset{0}{L}{}_2^1(G)$ obtained by closing the set $C_0^\infty(G)$ in the norm of the Sobolev–Liouville space $L_2^1(G)$; satisfying the boundary condition of the second or third kind (A.2.4) is understood in a generalized sense (see, for example, the well-known monograph of Sobolev [74, pp. 110-111]).

THEOREM A.2. *Assume that G is a bounded domain of any odd number N ($N > 1$) of dimensions with a piecewise smooth boundary, $\{u_k(x)\}$ is a system of the eigenfunctions of elliptic operator (A.2.1) satisfying one of the three boundary conditions (A.2.2)–(A.2.4), and $f(x)$ is an arbitrary function satisfying the following two conditions:*

1^0. $f(x)$ belongs to the Sobolev–Liouville class $L_p^\alpha(G)$ with arbitrary α and p satisfying the three inequalities:

$$\alpha \geq (N-1)/2, \qquad \alpha p \geq N, \qquad p \geq 1; \tag{A.2.5}$$

2^0. Each of the functions f, Lf, $L^2 f$, ..., $L^s f$, where L is the elliptic operator (A.2.1), and the integer s equals[3] *$[(N-3)/4]$ for the first boundary-value problem and equals*[4] *$[(N-5)/4]$ for the second and third boundary-value problems, satisfies (at least in the generalized sense) the respective homogeneous boundary condition. Then the spectral decomposition of the function $f(x)$ in the system $\{u_k(x)\}$ converges to this function uniformly on any compact set K of domain G.*

Prior to proving Theorem A.2, let us discuss in some detail the definitiveness of requirements 1^0 and 2^0 imposed in this theorem on the function to be decomposed.

The definitiveness of requirement 1^0, that is, the definitiveness of the three inequalities (A.2.5), has been established by us in Chapter 4 for an arbitrary spectral decomposition.

The fact that requirement 2^0 is definitive is remarkable, since, as will be shown below in Theorem A.3, for the eigenfunctions of the first and the second boundary-value problems for the Laplace operator $L = -\Delta$ in an N-dimensional ball G, there exists a function $f(x)$ in the class $C^\infty(G)$ such that for this function all the functions f, $-\Delta f$, ..., $(-\Delta)^s f$, except for one (no matter which!), satisfy the respective homogeneous boundary condition, and the spectral decomposition of this function,

[3] The symbol $[\alpha]$ denotes the integral part of the number α.

[4] For $N = 3$, the requirement 2^0 for the second and third boundary-value problems may be omitted.

whatever may be the sequential order of its terms, does not converge at the center of the ball G (there exists an infinite subsequence of the spectral decomposition terms that do not tend to zero at the center of the ball G).

Thus, requirement 2^0 is a definitive one, even allowing for the existence of at least one sequential order for the terms of the spectral decomposition ensuring its convergence at the interior points of the domain.

We now proceed with the proof of Theorem A.2 and with a correct formulation and proof of Theorem A.3, establishing the definitiveness of requirement 2^0 to the function $f(x)$ in Theorem A.2.

We precede Theorem A.2 by two simple lemmas.

LEMMA A.2. *Let the function $\Phi(x)$ for the second and third boundary-value problems belong to the Sobolev–Liouville class $L_2^1(G)$ only, and let this function satisfy, for the first boundary-value problem, a boundary condition of the first kind in the generalized sense, that is, belong to the class $\overset{0}{L}{}_2^1(G)$.*[5] *If so, the following inequalities hold:*

(a) for the first boundary-value problem[6]

$$\sum_{k=1}^{\infty} \Phi_k^2 \lambda_k \leq \int_G \left[\sum_{i,j=1}^{N} a_{ij}(x) \frac{\partial \Phi}{\partial x_i} \frac{\partial \Phi}{\partial x_j} + c(x)\Phi^2 \right] dx; \tag{A.2.6}$$

(b) for the second and third boundary-value problems

$$\sum_{k=1}^{\infty} \Phi_k^2 \lambda_k \leq \int_G \left[\sum_{i,j=1}^{N} a_{ij}(x) \frac{\partial \Phi}{\partial x_i} \frac{\partial \Phi}{\partial x_j} + c(x)\Phi^2 \right] dx + \int_{\partial G} h(s)\Phi^2(s) ds. \tag{A.2.7}$$

PROOF OF THEOREM A.2: We note that in the case of the first boundary-value problem for any function $\Phi(x)$ in the class $\overset{0}{L}{}_2^1(G)$ and for any eigenfunction $u_k(x)$, the following identity holds:

$$\int_G \left[\sum_{i,j=1}^{N} a_{ij}(x) \frac{\partial u_k}{\partial x_i} \frac{\partial \Phi}{\partial x_j} + c(x) u_k \Phi \right] dx = \lambda_k \int_G \Phi u_k dx, \tag{A.2.8}$$

and, in particular, for $\Phi(x) = u_\ell(x)$,

$$\int_G \left[\sum_{i,j=1}^{N} a_{ij}(x) \frac{\partial u_k}{\partial x_i} \frac{\partial u_\ell}{\partial x_j} + c(x) u_k u_\ell \right] dx = \begin{cases} \lambda_k & \text{for } k = \ell, \\ 0 & \text{for } k \neq \ell. \end{cases} \tag{A.2.9}$$

---

[5] We recall that the symbol $\overset{0}{L}{}_2^1(G)$ denotes the closure of a set $C_0^\infty(G)$ in the norm of the space $L_2^1(G)$.

[6] Here the symbol Φ_k denotes the Fourier coefficient of the function $\Phi(x)$ with respect to the eigenfunction $u_k(x)$, and λ_k is the eigenvalue corresponding to this eigenfunction.

In the case of the second and third boundary-value problems, for any function $\Phi(x)$ belonging to the class $L_2^1(G)$ only and for any eigenfunction $u_k(x)$ the identity

$$\int_G \left[\sum_{i,j=1}^{N} a_{ij}(x)\frac{\partial u_k}{\partial x_i}\frac{\partial \Phi}{\partial x_j} + c(x)u_k\Phi\right]dx + \int_{\partial G} h(s)u_k(s)\Phi(s)ds = \lambda_k \int_G \Phi u_k dx \tag{A.2.10}$$

holds, and, in particular, with $\Phi(x) = u_\ell(x)$

$$\int_G \left[\sum_{i,j=1}^{N} a_{ij}(x)\frac{\partial u_k}{\partial x_i}\frac{\partial u_\ell}{\partial x_j} + c(x)u_k u_\ell\right]dx + \int_{\partial G} h(s)u_k(s)u_\ell ds = \begin{cases} \lambda_k & \text{for} \quad k = \ell, \\ 0 & \text{for} \quad k \neq \ell. \end{cases} \tag{A.2.11}$$

Given an arbitrary n, we consider the nonnegative quantity

$$I = \int_G \{\sum_{i,j=1}^{N} a_{ij}(x)\frac{\partial}{\partial x_i}\left[\Phi(x) - \sum_{k=1}^{n}\Phi_k u_k(x)\right]\frac{\partial}{\partial x_j}\left[\Phi(x) - \sum_{\ell=1}^{n}\Phi_\ell u_\ell(x)\right] + c(x)\left[\Phi(x) - \sum_{k=1}^{n}\Phi_k u_k(x)\right]^2\}dx \geq 0$$

for the first boundary-value problem, and the nonnegative quantity

$$I = \int_G \{\sum_{i,j=1}^{N} a_{ij}(x)\frac{\partial}{\partial x_i}\left[\Phi(x) - \sum_{k=1}^{n}\Phi_k u_k(x)\right]\frac{\partial}{\partial x_j}\left[\Phi(x) - \sum_{\ell=1}^{n}\Phi_\ell u_\ell(x)\right]$$

$$+ c(x)\left[\Phi(x) - \sum_{k=1}^{n}\Phi_k u_k(x)\right]^2\}dx + \int_{\partial G} h(s)\left[\Phi(s) - \sum_{k=1}^{n}\Phi_k u_k(s)\right]^2 ds \geq 0$$

for the second and third boundary-value problems. Making use of relations (A.2.8)–(A.2.11), we obtain for the first boundary-value problem

$$I = \int_G \left[\sum_{i,j=1}^{N} a_{ij}(x)\frac{\partial \Phi}{\partial x_i}\frac{\partial \Phi}{\partial x_j} + c(x)\Phi^2(x)\right]dx - \sum_{k=1}^{n}\Phi_k^2\lambda_k \geq 0,$$

and for the second and third boundary-value problems

$$I = \int_G \left[\sum_{i,j=1}^{N} a_{ij}(x) \frac{\partial \Phi}{\partial x_i} \frac{\partial \Phi}{\partial x_j} + c(x)\Phi^2(x) \right] dx + \int_{\partial G} h(s)\Phi^2(s)ds -$$

$$- \sum_{k=1}^{n} \Phi_k^2 \lambda_k \geq 0.$$

Since n has been chosen arbitrarily, Lemma A.2 is thus proved.

LEMMA A.3. *Assume that a function $\Phi(x)$ belongs to the Sobolev–Liouville class $L_2^2(G)$ and in a generalized sense satisfies the respective homogeneous boundary condition of one of the three kinds. Then the Fourier coefficient $(L\Phi)_k$ of the function $L\Phi(x)$ and the Fourier coefficient Φ_k of the function $\Phi(x)$ are related as*

$$(L\Phi)_k = \lambda_k \Phi_k. \tag{A.2.12}$$

PROOF: Applying the second Green's formula to the functions $\Phi(x)$ and $u_k(x)$, we obtain the relation

$$\int_G u_k(x) L\Phi(x) dx = \int_G \Phi(x) L u_k(x) dx = \lambda_k \int_G \Phi(x) u_k(x) dx,$$

which is equivalent to equality (A.2.12).

PROOF OF THEOREM A.2: By virtue of the embedding theorem for the Sobolev–Liouville classes, it suffices to show that the inequality[7)]

$$\sum_{k=1}^{\infty} f_k^2 \lambda_k^{(N-1)/2} \leq C \|f\|^2_{L_2^{(N-1)/2}(G)} \tag{A.2.13}$$

holds for a function $f(x)$ that belongs to the Sobolev–Liouville class $L_2^{(N-1)/2}(G)$ at $p > 2N/(N-1)$ and satisfies the requirement 2^0 of Theorem A.2.

In turn, it suffices to show that for a function $f(x)$ subject to the above condition, the inequality below holds

---

7) The validity of this inequality implies the validity of inequality (4.3.61) from Section 4.3.2 with the norm in the class $H_2^{(N-1)/2}(G)$ on the right-hand side of (4.3.61) replaced by the norm in the class $L_2^{(N-1)/2}(G)$. The proof of convergence given in Sections 4.3.6 and 4.3.7 is based on inequality (4.3.61) solely.

$$
\sum_{k=1}^{\infty} f_k^2 \lambda_k^{(N-1)/2}
$$

$$
\leq \begin{cases}
\displaystyle\int_G [L^{(N-1)/4} f]^2 dx \quad \text{for even} \quad (N-1)/2, \\[2ex]
\displaystyle\int_G \left\{ \sum_{i,j=1}^{N} a_{ij}(x) \frac{\partial}{\partial x_i} \left(L^{(N-3)/4} f \right) \frac{\partial}{\partial x_j} \left(L^{(N-3)/4} f \right) + c(x) \left(L^{(N-3)/4} f \right)^2 \right\} dx \\
\qquad \text{for odd} \quad (N-1)/2 \quad \text{and for the first} \\
\qquad \text{boundary-value problem} \\[2ex]
\displaystyle\int_G \left\{ \sum_{i,j=1}^{N} a_{ij}(x) \frac{\partial}{\partial x_i} \left(L^{(N-3)/4} f \right) \frac{\partial}{\partial x_j} \left(L^{(N-3)/4} f \right) + c(x) \left(L^{(N-3)/4} f \right)^2 \right\} dx \\
\displaystyle + \int_{\partial G} h(s) \left(L^{(N-3)/4} f \right)^2 ds \\
\qquad \text{for odd} \quad (N-1)/2 \quad \text{and for the} \\
\qquad \text{second and third boundary-value problems,}
\end{cases}
\tag{A.2.14}
$$

in view of the fact that the quantities on the right-hand side of (A.2.14) are majorized by the right-hand side of (A.2.13).

For an even $(N-1)/2$, that is, in the case $N = 5, 9, 13, \ldots$, inequality (A.2.14) is derived by successively applying relation (A.2.12) to the functions f, Lf, $\ldots$, $L^{(N-5)/4}f$, and the Bessel inequality to the function $L^{(N-1)/4}f$ belonging to the class $L_2(G)$.

For an odd $(N-1)/2$, that is, in the case $N = 3, 7, 11, \ldots$, inequality (A.2.14) for $N = 3$ is equivalent to (A.2.6) (respectively, to (A.2.7)); for $N = 7, 11, 15, \ldots$, it is derived by successively applying relation (A.2.12) to the functions f, Lf, $\ldots$, $L^{(N-7)/4}f$, and inequality (A.2.6) (respectively, (A.2.7)) to the function $\Phi = L^{(N-3)/4}f$.

Theorem A.2 is thus proved.

THEOREM A.3. *Assume that domain G is an N-dimensional ball of radius R ($R > 0$) of an odd number N ($N > 1$) of dimensions, and considered in this ball are the eigenfunctions of a Laplace operator $L = -\Delta$ with a homogeneous boundary condition of the first (or second) kind.*

Then there exists a function $f(x)$ satisfying the following three conditions:

(1) in the ball G, $f(x)$ belongs to the class C^∞;

(2) all the functions f, $(-\Delta)f$, ..., $(-\Delta)^s f$, except for one (no matter which!), involved in requirement 2^0, satisfy the respective homogeneous boundary condition;

(3) there exists an infinite sequence of the Fourier series terms of function $f(x)$ not tending to zero at the center x_0 of the ball G.

For such a function $f(x)$ its Fourier series does not converge at the center x_0 of the ball G no matter what the sequential order of its terms.

PROOF OF THEOREM A.3. We introduce spherical coordinates in the N-dimensional ball G of radius R ($R > 0$) in such a manner as to place the origin of the spherical frame of reference at the center x_0 of the ball G; we denote the spherical radius by r. We assume that the function f to be decomposed is dependent on the spherical radius r only. In this case, the expansion is carried out *only in the eigenfunctions of the N-dimensional ball of the spherical symmetry,* considering that the Fourier coefficients in the remaining eigenfunctions are equal to zero.

The normalized, spherically symmetric eigenfunctions of the N-dimensional ball of radius R are written as:

(a) for the first boundary-value problem $u(R) = 0$

$$u_n(r) = \frac{1}{\sqrt{\omega_N}} \frac{\sqrt{2}}{R} \frac{1}{J_{(N-4)/2}(\mu_n)} r^{(2-N)/2} J_{(N-2)/2}\left(\frac{r}{R}\mu_n\right) \quad (n = 1, 2, \ldots), \tag{A.2.15}$$

where μ_n are the positive zeros of the Bessel function $J_{(N-2)/2}(x)$ numbered in increasing order, $\omega_N = 2(\pi)^{N/2}[\Gamma(N/2)]^{-1}$;

(b) for the second boundary-value problem $\frac{\partial u}{\partial r}(R) = 0$

$$u_n(r) = \frac{1}{\sqrt{\omega_N}} \frac{\sqrt{2}}{R} \frac{1}{J_{(N-2)/2}(\mu_n)} r^{(2-N)/2} J_{(N-2)/2}\left(\frac{r}{R}\mu_n\right) \quad (n = 1, 2, \ldots), \tag{A.2.16}$$

where μ_n are the positive zeros of the Bessel function $J_{N/2}(x)$ numbered in increasing order.

It is seen, therefore, that the Fourier coefficient, dependent on the spherical radius of the function $f(r)$, is defined by the formula

$$f_n = \omega_N \int_0^R f(r) u_n(r) r^{N-1} dr = \frac{\sqrt{2\omega_N}}{R} \frac{1}{J_{(N-4)/2}(\mu_n)} \int_0^R r^{N/2} f(r) J_{(N-2)/2}\left(\frac{r}{R}\mu_n\right) dr \tag{A.2.17}$$

for the first boundary-value problem, and by the formula

$$f_n = \frac{\sqrt{2\omega_N}}{R} \frac{1}{J_{(N-2)/2}(\mu_n)} \int_0^R r^{N/2} f(r) J_{(N-2)/2}\left(\frac{r}{R}\mu_n\right) dr \tag{A.2.18}$$

for the second boundary-value problem.

In what follows we shall want values for eigenfunctions (A.2.15) and (A.2.16) at the center of the ball, that is, at $r = 0$.

We infer, from the asymptotic behavior of the Bessel functions and from the definition of their zeros, that the relation

$$J_{(N-4)/2}(\mu_n) = (-1)^n \sqrt{\frac{2}{\pi}} \frac{1}{\sqrt{\mu_n}} + o\left(\frac{1}{\sqrt{\mu_n}}\right) \tag{A.2.19}$$

holds for the first boundary-value problem, and

$$J_{(N-2)/2}(\mu_n) = (-1)^n \sqrt{\frac{2}{\pi}} \frac{1}{\sqrt{\mu_n}} + o\left(\frac{1}{\sqrt{\mu_n}}\right) \tag{A.2.20}$$

holds for the second boundary-value problem.

Further, following the same line of reasoning, we come to the conclusion that the eigenfunction $u_n(R)$ for the first and second boundary-value problems has its value at the center of the ball

$$u_n(0) = C_n \mu_n^{(N-1)/2}, \tag{A.2.21}$$

such that

$$\lim_{n\to\infty} |C_n| = \frac{1}{2}\sqrt{\frac{\pi}{\omega_N}}(2R)^{-N/2}\left[\Gamma\left(\frac{N}{2}\right)\right]^{-1}.$$

It is important to remember that *the values of C_n do not tend to zero as $n \to \infty$.*

Let us now construct the function f whose existence has been asserted by Theorem A.3.

We consider the first boundary-value problem.

Let k be an arbitrary integer subject to $0 \le k \le [(N-3)/2]$.

We define the function $f(r)$ as a solution of the equation $\Delta^k f = 1$ within the ball G satisfying the boundary conditions $f(R) = \Delta f(R) = \ldots = \Delta^{k-1} f(R) = 0$.

The function $f(r)$ will be sought in the form of a polynomial

$$f(r) = a_0 + a_1 r^2 + a_2 r^4 + \ldots + a_k r^{2k},$$

whose coefficients are readily determined from the conditions

$$f(R) = \Delta f(R) = \ldots = \Delta^{k-1} f(R) = 0, \quad \Delta^k f(R) = 1.$$

The function thus constructed belongs to the class $C^\infty(G)$ such that all the functions

$$f,\ (-\Delta)f,\ (-\Delta)^2 f,\ \ldots,\ (-\Delta)^{[(N-3)/4]} f$$

involved in requirement 2^0 of Theorem A.2 become zero at $r = R$, except for the function $(-\Delta)^k f$.[8] One should keep in mind that $(-\Delta)^k f(R) = (-1)^k \neq 0$.

[8] The fact that the functions $(-\Delta)^\ell f(r)$, $\ell > k$, become zero at $r = R$ follows from the fact that they are identically equal to zero.

Let us estimate the nth term of the Fourier series for the function $f(r)$ at the center of the ball G, that is, at $r = 0$.

Applying Lemma A.3 in succession to the functions f, $(-\Delta)f$, ..., $(-\Delta)^{k-1}f$, we find that the Fourier coefficients of the functions $f(r)$ and $(-\Delta)^k f(r)$ are related as

$$f_n = ((-\Delta)^k f)_n \left(\frac{R}{\mu_n}\right)^{2k}. \tag{A.2.22}$$

Inasmuch as $(-\Delta)^k f(r) = (-1)^k$, the Fourier coefficient for this function is trivially found using relation (A.2.17)[9)]

$$\begin{aligned}((-\Delta)^k f)_n &= \frac{\sqrt{2\omega_N}}{R}\frac{(-1)^k}{J_{(N-4)/2}(\mu_n)}\int_0^R r^{N/2}J_{(N-2)/2}\left(\frac{r}{R}\mu_n\right)dr \\ &= \frac{\sqrt{2\omega_N}}{\mu_N}\frac{(-1)^k}{J_{(N-4)/2}(\mu_n)}\left[r^{N/2}J_{N/2}\left(\frac{r}{R}\mu_n\right)\right]\Big|_{r=0}^{r=R} \\ &= \frac{\sqrt{2\omega_N}}{\mu_n}R^{N/2}(-1)^k\frac{J_{N/2}(\mu_n)}{J_{N/2-2}(\mu_n)} = -(-1)^k\frac{\sqrt{2\omega_N}}{\mu_n}R^{N/2}.\end{aligned} \tag{A.2.23}$$

We obtain, from relations (A.2.21), (A.2.22), and (A.2.23), that the nth term of the Fourier series at the center of the ball is written as

$$f_n u_n(0) = -\sqrt{2\omega_N}\,R^{N/2+2k}(-1)^k C_n \mu_n^{(N-3)/2-2k}. \tag{A.2.24}$$

Inasmuch as the value of C_n does not tend to zero as $n \to \infty$ and $k \le [(N-3)/4]$, the nth term of the Fourier series at the center of the ball (A.2.24) does not tend to zero as $n \to \infty$ either.

This completes the proof of Theorem A.3 for the first boundary-value problem.

Let us now turn to the case of the second boundary-value problem.

Let k be an arbitrary fixed number subject to $0 \le k \le [(N-5)/4]$.[10)]

This time we define the function $f(r)$ as a solution of the equation $\Delta^k f = r^2$ within the ball G satisfying the boundary conditions

$$\frac{d}{dr}f(R) = \frac{d}{dr}\Delta f(R) = \ldots = \frac{d}{dr}\Delta^{k-1}f(R) = 0.$$

---

9) Here we make use of the recurrence relations for Bessel functions

$$\int r^\nu J_{\nu-1}\left(\frac{r}{R}\mu_n\right)dr = \frac{R}{\mu_n}r^\nu J_\nu\left(\frac{r}{R}\mu_n\right), \quad J_{N/2}(\mu_n) + J_{N/2-2}(\mu_n) = \frac{N-2}{\mu_n}J_{N/2-1}(\mu_n) = 0.$$

10) Since at $N = 3$ the requirement 2^0 of Theorem A.2 for a boundary condition of the second kind is not necessary, we assume in the sequel that $N \ge 5$.

This function will be sought as a polynomial

$$f(r) = a_1 r^2 + a_2 r^4 + \ldots + a_{k+1} r^{2k+2},$$

with its coefficients $a_1, a_1, \ldots, a_{k+1}$ easily determined from the conditions

$$\frac{d}{dr} f(R) = \frac{d}{dr} \Delta f(R) = \ldots = \frac{d}{dr} \Delta^{k-1} f(R) = 0, \qquad \frac{d}{dr} \Delta^k f(R) = 2R.$$

The function thus constructed belongs to the class $C\infty(G)$; for this function, the functions

$$f,\ (-\Delta)f,\ (-\Delta)^2 f,\ \ldots,\ (-\Delta)^{[(N-5)/4]} f$$

involved in the requirement 2^0 of Theorem A.2, except for the function $(-\Delta)^k f$, satisfy the homogeneous boundary condition of the second kind.

As is seen,

$$\frac{d}{dr}(-\Delta)^k f(R) = (-1)^k 2R \neq 0.$$

Let us estimate the nth term of the Fourier series of the function $f(r)$ at the center of the ball, that is, at $r = 0$.

To determine the Fourier coefficient f_n, we make use of relation (A.2.22); one will observe that $\Delta^k f(r) = r^2$, and, therefore, the Fourier coefficient $((-\Delta)^k f)_n$ is easily determined by the use of relation (A.2.18). Indeed,

$$((-\Delta)^k f)_n = \frac{\sqrt{2\omega_N}}{R} \frac{(-1)^k}{J_{(N-2)/2}(\mu_n)} \int_0^R r^{N/2+2} J_{N/2-1}\left(\frac{r}{R}\mu_n\right) dr. \qquad \text{(A.2.25)}$$

The integral on the right-hand side of (A.2.25) is taken by parts using the recurrence relation

$$\int J_\nu \left(\frac{\mu_n}{R} r\right) r^\nu dr = \frac{R}{\mu_n} r^\nu J_\nu \left(\frac{\mu_n}{R} r\right).$$

Taking into account that $J_{N/2}(\mu_n) = 0$, we obtain the relation

$$\int_0^R r^{N/2+1} J_{N/2-1}\left(\frac{r}{R}\mu_n\right) dr = \left[r^2 \frac{R}{\mu_n} r^{N/2} J_{N/2}\left(\frac{r}{R}\mu_n\right)\right] \Big|_{r=0}^{r=R}$$

$$-2\frac{R}{\mu_n} \int_0^R r^{N/2+2} J_{N/2}\left(\frac{r}{R}\mu_n\right) dr = -2\frac{R^2}{\mu_n^2} \left[r^{N/2+1} J_{N/2+1}\left(\frac{r}{R}\mu_n\right)\right] \Big|_{r=0}^{r=R} \qquad \text{(A.2.26)}$$

$$= -2\frac{R^2}{\mu_n} R^{N/2+1} J_{N/2+1}(\mu_n) = 2\frac{R^{N/2+3}}{\mu_n^2} J_{N/2-1}(\mu_n).$$

We obtain from (A.2.25) and (A.2.26) that

$$((-\Delta)^k f)_n = (-1)^k \sqrt{2\omega_N}\frac{2R^{N/2+2}}{\mu_n^2}. \tag{A.2.27}$$

Finally, from relation (A.2.21), (A.2.22), and (A.2.27) we obtain the following expression for the nth term of the Fourier series at the center of the ball:

$$f_n u_n(0) = \sqrt{2\omega_N} 2R^{N/2+2k+2}(-1)^k C_n \mu_n^{(N-5)/2-2k}.$$

Inasmuch as $k \leq [(N-5)/4]$ and the value of C_n does not tend to zero as $n \to \infty$, the nth term of the Fourier series written above does not tend to zero as $n \to \infty$ either.

We have cleared up the second boundary-value problem.

This completes the proof of Theorem A.3.

Epilogue

The reader who has deemed it worthwhile to spend his time on this monograph may, naturally, ask himself the question: Is it possible, starting from the novel ideas and methods expounded, to trace a course to the spectral theory of *non-self-adjoint* differential operators?

The answer will be in the affirmative. We now cast a cursory glance at the major results in the spectral theory of non-self-adjoint differential operators that have been obtained in pursuance of certain concepts presented in this monograph.

First of all, one must mention the difficulties the mathematician will inevitably meet, even for distant approaches to such a challenging problem. As is known, the system of eigenfunctions of any self-adjoint extension of a formally self-adjoint differential operator with point spectrum always forms a basis (which may be even an orthonormal one) allowing the expansion of an arbitrary function from the class L_2. By contrast, with a non-self-adjoint differential operator, the system of eigenfunctions forms no basis in which an arbitrary function from the class L_2 might be expanded; the more so, it is not complete in L_2.[1] Therefore, if one wishes, in the case of a non-self-adjoint differential operator, to constract a system exhibiting the property of being complete (and, the more so, the property of being a basis), one must by necessity augment the system of all eigenfunctions with a system of judiciously selected *associated* functions. The union of the system of all eigenfunctions with the system of all associated functions of a particular differential operator is conventionally called the system of *root functions* of the given operator.

A great merit of Keldysh [80][2] has been his establishing the completeness in L_2 of a specially constructed system of root functions (coined by him a canonical system) for a vast class of boundary-value problems of non-self-adjoint differential operators, as well as for certain abstract non-self-adjoint operators.

However, the Keldysh theory fails to give an answer to the question (quite important from the standpoint of both theory and practical applications[3]) as to whether

[1] In other words, an arbitrary function in the class L_2 eludes approximation in the metric L_2 by a linear combination of eigenfunctions.

[2] See the list of references appended to the epilogue.

[3] For example, in specifying the conditions for turbulent plasma stability, in designing nuclear

the canonical system of root functions is capable of forming a basis in L_2 allowing for the expansion of an arbitrary function in the class L_2.

An answer to this question can be found in our papers [81]–[83]; the major relevant ideas have been outlined in Chapter 1 of the present monograph.

Let L be a non-self-adjoint ordinary differential operator defined on the interval $G = (0,1)$; for simplicity, we assume the order n of the operator to be even:

$$Lu = u^{(n)} + p_1(x)u^{(n)} + p_2(x)u^{(n-2)} + \ldots + p_n(x)u. \tag{1}$$

The major idea underlying our spectral theory is, in close analogy with Chapter 1 of the monograph, that we renounce specifying any boundary conditions and focus our attention on the generalized root functions of operator (1) which are regular solutions only of the corresponding differential equation with complex spectral parameter.

Otherwise stated, we take, for *the system of generalized root functions* of operator (1), an arbitrary system of complex-valued functions $\{u_k(x)\}$; of these, each belongs to the class $C^{(n)}(G)$ and, given a complex λ_k, satisfies the differential equation

$$Lu_k + \lambda_k u_k = \theta_k u_{k-1} \tag{2}$$

on the interval $G = (0,1)$; here θ_k is either zero or one (in the latter case, an additional requirement is $\lambda_k = \lambda_{k-1}$), and also $\theta_1 = 1$.

Such an approach implies the employment of (i) systems of root functions for all boundary-value problems with point spectrum, (ii) exponent-type systems, with no boundary conditions specified, (iii) also systems produced by the union of the subset of root functions of two different boundary-value problems.

We use the symbol μ_k to denote the spectral root of even power n for a complex number λ_k such that, given $-\pi \leq \varphi \leq \pi$, the number $[(-1)^{(n+2)/2}\lambda_k] = \rho e^{i\varphi}$, and, consequently,

$$\mu_k = \rho^{1/n} e^{(i\varphi)/n}.$$

A requirement in [81]–[85] has been that the system of generalized root functions of operator (1) must satisfy, at a fixed $p \geq 1$, two conditions A:

(1) the system should be closed and minimal in $L_p(G)$;

(2) the above-mentioned numbers μ_k should satisfy the following two inequalities[4]:

$$|\mathrm{Im}\,\mu_n| \leq C_1 \qquad \text{(for all } k), \tag{3}$$

$$\sum_{\mu \leq |\mu_k| \leq \mu+1} 1 \leq C_2 \qquad \text{(for all real } \quad \mu \geq 0). \tag{4}$$

Condition (4) implies that all elements $u_k(x)$ in the system of generalized root functions of operator (1) are numbered in the order of nondecreasing values of $|\mu_k|$.

---

reactors, and other applications.

[4] As was shown later, inequality (3) is a necessary condition for the system $\{u_k(x)\}$ to be a basis in $L_2(G)$, and inequality (4) is a necessary condition for the system $\{u_k(x)\}$ to be a Riesz basis in $L_2(G)$.

The former of the two conditions A enables us to assert that there exists one and only one system $\{v_k(x)\}$ biorthogonally adjoint to system $\{u_k(x)\}$, such that each element $v_k(x)$ of this system belongs to the class $L_q(G)$ at $q = p/(p-1)$ ($q = \infty$ at $p = 1$) and, given any k and ℓ, the relation below holds:

$$(u_k, v_\ell) = \int_G u_k(x)\overline{v_\ell(x)}dx = \begin{cases} 1 & \text{for} \quad k = \ell, \\ 0 & \text{for} \quad k \neq \ell. \end{cases}$$

We construct, for an arbitrary function $f(x)$ from the class $L_p(G)$ with the same fixed $p \geq 1$ as specified by conditions A above, a partial sum of order m of the spectral biorthogonal series expansion in the system $\{u_k(x)\}$,

$$\sigma_m(x, f) = \sum_{k=1}^{m} (f, v_k)u_k(x). \tag{5}$$

We say that a system of generalized root functions $\{u_k(x)\}$ possesses the basis property in L_p if for an arbitrary function $f(x)$ in the class $L_p(G)$ and for an arbitrary compact set K within the interval G the following condition holds:

$$\lim_{m\to\infty} \|f(x) - \sigma_m(x, f)\|_{L_p(K)} = 0.$$

To verify that the spectral decomposition (5) is equiconvergent with the expansion of the same function $f(x)$ in a conventional trigonometric Fourier series, we construct a modified partial sum of the Fourier-expanded function $f(x)$

$$S_\tau(x, f) = \frac{1}{\pi} \int_G \frac{\sin[\tau(x-y)]}{x-y} f(y)dy$$

of order $\tau = |\mu_m|$.

We say that two expansions of a function $f(x)$ — one in the system $\{u_k(x)\}$ and the other one in the trigonometric series — are uniformly equiconvergent on any compact set of interval G if the difference

$$R_m(x, f) = \sigma_m(x, f) - \exp\left(-\frac{1}{n}\int_0^x p_1(\xi)d\xi\right) S_{|\mu_m|}\left(x, f(x)\exp\left(\frac{1}{n}\int_0^x p_1(\xi)d\xi\right)\right) \tag{6}$$

tends to zero as $m \to \infty$ uniformly with respect to x on any compact set of interval G.[5)]

[5)] Here $p_1(x)$ is the coefficient of $u^{(n-1)}$ in the operator expression (1).

Note that if the coefficient $p_1(x)$ of operator (1) is identically equal to zero, the difference (6) reduces to

$$R_m(x,f) = \sigma_m(x,f) - S_{|\mu_m|}(x,f).$$

The following two theorems are major results reported in [81]–[83].[6)]

THEOREM 1. *In order for an arbitrary system $\{u_k(x)\}$ of generalized root functions of the operator (1) satisfying the two conditions A at a fixed $p > 1$ to possess the basis property in L_p, it is necessary and sufficient that there exist, for any compact set K_0 of interval G, a constant $C(K_0)$ such that the inequality*

$$\|u_k(x)\|_{L_p(K_0)} \|v_k(x)\|_{L_q(G)} \le C(K_0) \tag{7}$$

holds for all indices k, with $q = p/(p-1)$.

THEOREM 2. *Assume that an arbitrary function $f(x)$ of the class $L_p(G)$ at a fixed $p \ge 1$ can be expanded in an arbitrary system of the generalized root functions of operator (1) subject to the two conditions* A *at $p \ge 1$ and in a trigonometric series. In order for the two expansions to be uniformly equiconvergent on any compact set of interval G, it is necessary and sufficient that there exist, for any compact set K_0 of interval G, a constant $C(K_0)$ such that inequality (7) holds for all indices k with $q = p/(p-1)$ ($q = \infty$ at $p = 1$).*

We wish to make a few remarks concerning the constructibility of conditions and the range of applicability of Theorems 1 and 2.

First, we note that all the conditions of Theorems 1 and 2 are easily verified for a particular boundary-value problem. Indeed, the completeness of system $\{u_k(x)\}$ is ascertained using the Keldysh theorem [80]; the minimality of the system is seen in the root functions of the adjoint boundary-value problem forming a biorthogonally adjoint system, and the fulfillment of inequalities (3), (4), and (7) is tested with the aid of the leading terms of the conventional asymptotic expansions of the root functions and eigenvalues of concrete boundary-value problems in powers of $1/\mu_k$.

Second, a particular case of the system of generalized root functions of the simplest differential operator $Lu = u''$ or even $Lu = u'$ is the system of exponents

$$\overset{0}{u}_k(x) = e^{i\lambda_k x}$$

$$\overset{\ell}{u}_k(x) = \frac{x^\ell e^{i\lambda_k x}}{\ell!} \qquad (\ell = 1, 2, \ldots, m_k - 1),$$

which has been extensively studied by many mathematicians.[7)]

---

6) It is implied in these theorems that the coefficients $p_\ell(x)$ in the operator expression (1) belong to the classes $C^{(n-\ell+1)}(G)$.

7) One may mention N. Wiener and R. Paley, S. Werblunski, N. Levinson, V. E. Katsnelson, B. Ya. Levin, B. S. Pavlov, S. K. Khrushchev, N. K. Nikol'skii, and A. M. Sedletskii, to name but a few.

Nonetheless, both Theorems 1 and 2 are novel also for the system of exponents, since they appeal to statements that were not considered by the previous authors.

Further, a system of generalized root functions $\{u_k(x)\}$ for the simplest differential operator $Lu = u''$ has been dealt with in [84]. This system, demonstrative of a wider applicability of Theorems 1 and 2, is constructed through the union of the subset of root functions of this operator satisfying the periodicity conditions, and of the subset of root functions of the same operator satisfying a condition of antiperiodicity. An essential point is that for the system in question, all conditions of Theorem 1 for any $p > 1$ and those of Theorem 2 for any $p \geq 1$ are fulfilled.

Finally, we emphasize that Theorem 2 on the uniform equiconvergence of spectral expansion and trigonometric series is the strongest and simultaneously the most difficult to prove at $p = 1$, that is, in the class L_1. The proof of the necessity for this theorem as expounded in [83] has been a matter of spectral concern. The equiconvergence problem, starting from the famous paper by Steklov [86], was a subject of intense interest for many well-reputed mathematicians — the pioneer V. A. Steklov himself, Ya. D. Tamarkin, E. C. Titchmarsh, A. Haar, B. M. Levitan, and Ya. L. Geronimus, to name but a few. Theorem 2 as formulated above summarizes all the previous results (occasionally, in a stronger sense) and establishes an exact bound beyond which no equiconvergence is operable.

A question arises in a natural manner as to the basis property conditions (for example, in L_2) for the canonical system of root functions constructed by Keldysh in [80]. Using Theorem 1, we readily arrive at a somewhat unexpected result below.

THEOREM 3. *Suppose, a system of generalized root functions of operator (1) satisfies, at $p = 2$, the same conditions as those in Theorem 1, and the total number of associated functions in this system is not finite, If so, there always exists a closed system of generalized root functions of operator (1), minimal in $L_2(G)$ and canonical (in the Keldysh sense), which possesses no property of being a basis in L_2.*

Theorem 3 thus provides evidence that the concept of a canonical system of root functions as put forward by Keldysh is a good tool in an analysis for completeness, but fails in an analysis for the basis property of a system of root functions.

In view of this contingency, the concept of a *reduced* system for root functions of operator (1) was introduced in [85] and [87]. The system in question is assumed to exhibit the property of being a basis any time when this property is implemented at least in a single choice of root functions. An algorithm for constructing the reduced system of root functions was also proposed and a proof was given to the effect that such a system has the basis property if an inequality of the type (7) is fulfilled for it (at $p = 2$).

The equiconvergence Theorem 2 holds for operator (1) only if the coefficients of this operator are sufficiently smooth. It was recently shown [88] that, with the classical Schroedinger operator $Lu = u'' + q(x)u$ acting for operator (1), Theorem 2 holds for an arbitrary complex-valued potential $q(x)$ from the class $L_1(G)$ (it was assumed that the root functions $u_k(x)$ belonged to the class $W_1^2(G)$ only and that the respective equation (2) was satisfied almost everywhere on G). It has also been ascertained, as a particular case, that if the potential $q(x)$ from the $L_1(G)$ class is real-valued and if an arbitrary self-adjoint extension of the Schroedinger operator with point spectrum is dealt with, then the statement of Theorem 2 on uniform equiconvergence is always valid at $p = 1$, that is, for an arbitrary function in the class $L_1(G)$.

In a recent paper [89], the equiconvergence problem for the Schroedinger operator

$$LU = U'' + Q(x)U \tag{8}$$

has been studied; here $Q(x)$ is an $s \times s$ non-Hermitian matrix all of whose complex-valued elements $Q_{ij}(x)$ are summable only within the main interval $G = (0, 1)$.

In this case, it appears expedient to consider the class $L_p^s(G)$ with norm

$$\|f\|_{L_p^s(G)} = \left[\int_G \sum_{j=1}^{s} |f_j(x)|^p dx \right]^{1/p} \tag{9}$$

in the space of s-component vector functions $f(x) = \{f_1(x), f_2(x), \dots, f_s(x)\}$ for any $p \geq 1$ and to define the scalar product $\langle t, g \rangle$ of two s-component vector functions $f(x) = \{f_1(x), \dots, f_s(x)\}$ and $g(x) = \{g_1(x), \dots, g_s(x)\}$ as

$$\langle f, g \rangle = \int_G \sum_{j=1}^{s} f_j(x)\overline{g_j(x)}dx; \tag{10}$$

of these two vector functions, the former belongs to the class $L_p^s(G)$ at any $p \geq 1$, and the latter belongs to the class $L_q^s(G)$ at $q = p/(p-1)$ (with $q = \infty$ and $p = 1$).

In complete analogy with the above, we now consider for the system of generalized root vector functions of the Schroedinger operator (8), an arbitrary system of s-component vector functions $\{U^k(x)\} = \{U_1^k(x), U_2^k(x), \dots, U_s^k(x)\}$, where each component $U_j^k(x)$ belongs to the class $W_1^2(G)$ and, given a complex number λ_k, satifies the equation

$$LU^k + \lambda_k U^k = \theta_k U^{k-1} \tag{11}$$

almost everywhere in the interval G. Here L is the Schroedinger operator (8), and θ_k is the same as in Eq. (2).

For the two conditions A at a fixed $p \geq 1$ we take the same two conditions A that have been formulated earlier, the only distinction being that the completeness and minimality of the system $\{U^k(x)\}$ are presumed to be confined to the class $L_p(G)$, rather than to the class $L_p^s(G)$.

We construct a partial vector sum of order m of the expansion in system $\{U^k(x)\}$

$$\sigma^m(x, f) = \sum_{k=1}^{m} \langle f, V^k \rangle U^k(x) \tag{12}$$

for an arbitrary s-component vector function $f(x) = \{f_1(x), \dots, f_s(x)\}$ belonging to the class $L_p^s(G)$ for the same $p \geq 1$ as in conditions A. Here $\{V^k(x)\} = \{V_1^k(x), \dots, V_s^k(x)\}$ is a system biorthogonally adjoint to the system $\{U^k(x)\}$ in the sense of scalar product (10).

For each $j = 1, 2, \dots, s$ we consider the jth component of the partial vector sum (12)

$$\sigma_j^m(x, f) = \sum_{k=1}^{m} (f, V^k) U_j^k(x) \tag{13}$$

and compare it with the modified partial sum of the trigonometric series for the respective jth component $f_j(x)$ of the expanded vector function $f(x)$[8)]

$$S_{|\mu_m|}(x, f_j) = \frac{1}{\pi} \int_G \frac{\sin[|\mu_m|(x-y)]}{x-y} f_j(y)dy \tag{14}$$

of order $|\mu_m|$.

The following statement is a major result of [89].

THEOREM 4. *Let the potential $Q(x)$ of Schroedinger operator (8) represent an $s \times s$ arbitrary non-Hermitian matrix with complex-valued elements which are summable only on the interval G, and let $\{U^k(x)\}$ be an arbitrary system of generalized root vector functions of the operator (8) satisfying the above two conditions A at a fixed $p \geq 1$. Then in order for the difference between each jth component (13) of an arbitrary vector function $f_j(x)$ of the class $L_p^s(G)$ expanded in the system $\{U^k(x)\}$ and the respective jth component $f_j(x)$ of this function $f(x)$ expanded in the trigonometric series to tend to zero as $m \to \infty$ uniformly on any compact set of interval G, it is necessary and sufficient that there exist, for any compact set K_0 of interval G_0, a constant $C(K_0)$ such that the inequality*

$$\|U^k(x)\|_{L_p^s(K_0)} \|V^k(x)\|_{L_q^s(G)} \leq C(K_0) \tag{15}$$

holds for all indices k, where $q = p/(p-1)$ ($q = \infty$ at $p = 1$).

The property of equiconvergence as established by Theorem 4 lends itself, in a natural manner, to the definition of a *component-wise equiconvergence*, with the respective component of the vector function expanded in a trigonometric series.

The following remarkable statement is a corollary to Theorem 4.

THEOREM 5. *Assume that the potential $Q(x)$ of Schroedinger operator (8) and the system of its generalized root vector functions $\{U^k(x)\}$ satisfy, for a fixed $p \geq 1$, the same conditions as those in Theorem 4. Then, inequality (15) being valid for the expansion of an arbitrary function $f(x)$ of the class $L_p^s(G)$ in the system $\{U^k(x)\}$, the component-wise principle of localization holds true: the convergence or divergence of*

---

8) Here μ_m is the square root of λ_m for which $\operatorname{Re} \mu_m \geq 0$.

the jth expansion component (13) at a given point x_0 of interval G is dependent on the behavior, in a small vicinity of x_0, of the jth component $f_j(x)$ of the vector function $f(x)$ only and is independent of any other component of $f(x)$ — regardless of the fact that the coefficients $\langle f, V^k\rangle$ of expansion (13) may be dependent on all the components of the expandable vector function.

One will note a special important case where the matrix $Q(x)$ is symmetric and its elements are real-valued functions summable on G. If so, all the conditions of Theorems 4 and 5 are fulfilled at $p = 1$ for an arbitrary self-adjoint nonnegative extension of the Schroedinger operator (8) with point spectrum, that is, the component-wise equiconvergence with trigonometric series and the component-wise principle of localization apply to any vector function $f(x)$ in the class $L_1^s(G)$.

To conclude, we observe that the conditions of Theorems 4 and 5 are constructible and amenable to verification for a particular boundary-value problem.

Now we proceed to a brief overview of the works [90]–[93] focused on establishing the conditions that enable a system of the generalized root functions of the differential operator L to form the so-called *Riesz basis*[9)] within the entire closed main interval.

In specifying conditions for the Riesz basis property, we will henceforth require that the system of generalized root functions $\{u_k(x)\}$ (understood in the previous sense) of the differential operator L satisfies an additional condition, viz., the system in question should possess (in the former sense) a system of generalized root functions of the differential operator L^* for the biorthogonally adjoint system $\{v_k(x)\}$; here L^* is an operator formally adjoint to the differential operator L.

In [90], we have considered, on the interval $G = (0, 1)$, the differential operator of second order

$$Lu = u'' + p_1(x)u' + p_2(x)u \tag{16}$$

subject to minimal requirements for smoothness of its coefficients:

$$p_1(x) \in W_1^1(G), \qquad p_2(x) \in L_1(G).$$

The generalized root functions $u_k(x)$ belong to the class $W_1^2(G)$ only and satisfy the respective equation (2) almost everywhere on G.

The major result of [90] asserts that *in order for the system $\{u_k(x)\}$ of generalized root functions of operator (16), closed in $L_2(G)$ and obeying inequality (3), to be a Riesz basis in $L_2(G)$, it is necessary and sufficient that two inequalities hold, namely, inequality (4) and the inequality below*

$$\|u_k(x)\|_{L_2(G)} \|v_k(x)\|_{L_2(G)} \le C.$$

Another essential fact has been established in [98]: *The condition for the Riesz basis property or for the conventional basis property of a system of root functions,*

---

9) The Riesz basis is a basis equivalent to an orthonormal basis, that is, a basis which is converted to an orthonormal one by means of a linear restricted nonsingular transformation. For more details the reader is referred to the monograph [94].

as well as the condition for the equiconvergence of an expansion in this function and a trigonometric series, cannot be expressed in the traditional form of specifying the type of boundary condition. Thus, the root functions of operator (16) on the interval $G = ((0,1)$ under the same boundary conditions

$$u(0) = 0, \qquad u'(0) = u'(1)$$

at $p_1 \equiv 0$ and $p_2 \equiv 0$ and under a proper choice of associated functions form a Riesz basis in $L_2(G)$, and a basis in $L_p(G)$ at any $p > 1$; the expansion of any function from the $L_1(G)$ in these root functions is equiconvergent with the expansion of this function in a trigonometric series uniformly on any compact set of interval $(0,1)$. By contrast, at $p_1(x) = \varepsilon(x - 1/2)$, $p_2(x) = \frac{\varepsilon^2}{4}(x - 1/2)^2 + \frac{\varepsilon}{4}$, where ε is any number in the half-interval $0 < \varepsilon \leq 12$, the root functions (whatever they might be) form neither a Riesz basis in $L_2(G)$, nor a basis in $L_p(G)$ for any $p > 1$, and a function $f(x)$ from the $L_p(G)$ class, expanded in these functions at any $p > 1$, exhibits no equiconvergence with the expansion of this function in a trigonometric series.

Since $\varepsilon > 0$ can be taken arbitrarily small, this signifies that, given the same boundary conditions, the basis property and the trigonometric series equiconvergence are liable to variation under arbitrary small variations of the coefficients.

This example is illustrative of the fact that, given he same boundary conditions, the feasibility or infeasibility of the basis property is determined by the coefficients of the differential operator.

In [91], the major result obtained in [90] has been transferred to the case of the discontinuous operator (16), that is, when the root functions $u_k(x)$ satisfy the respective differential equation (2) not almost everywhere on the entire interval $G = (0,1)$, but only almost everywhere on each partial interval $(\xi_{\ell-1}, \xi)$ arising from partitioning the interval $G = (0,1)$ at points

$$0 = \xi_0 < \xi_1 < \xi_2 < \ldots < \xi_s < \xi_{s+1} = 1. \tag{17}$$

In doing so, no constraint is imposed on joining the root functions at the points ξ_1, $\xi_2, \ldots, \xi_s$.

The major objective in [91] was to define the conditions of the Riesz basis property for problems with the so-called *non-local boundary conditions*, for example, with conditions of the form

$$u(1) = \sum_{\ell=1}^{s} \alpha_\ell u(\xi_\ell) \qquad \text{or} \qquad u'(0) = \sum_{\ell=1}^{s} \beta_\ell u'(\xi_\ell).$$

Problems that are adjoint to problems thus defined are exactly those problems with a discontinuous operator.

Two issues have been dealt with in [92]: (i) conditions for the Riesz basis property in $L_2^s(G)$ of the systems of generalized root vector functions $\{U^k((x)\}$ of the Schroedinger operator (8) possessing a non-Hermitian matrix potential $Q(x)$ with its complex-valued elements $Q_{ij}(x)$ summable on the interval G only; (ii) the basis property of a system of the generalized root vector functions of a Schroedinger operator (8), dependent on the parameter t, and its relation to a system of integrals of motion in a nonlinear evolution equation which admits of the known Lax representation [95]

$$L_t = [A, L] = AL - LA. \tag{18}$$

The solution of the former problem (i) is quite analogous to the major result obtained in [90]: Assume given a system of generalized root vector functions of the Schroedinger operator (8) $\{U^k(x)\}$, closed in $L_2^s(G)$ and obeying inequality (3); then in order for this system to be a Riesz basis in $L_2^s(G)$, it is necessary and sufficient that two inequalities be satisfied, namely, inequality (4) and the inequality below

$$\|U^k(x)\|_{L_2^s(G)}\, \|V^k(x)\|_{L_2^s(G)} \leq G.$$

In [93], the formulated result has been applied to the case of a discontinuous Schroedinger operator (8), that is, when the root vector functions $U^k(x)$ do not necessarily satisfy Eq. (11) almost everywhere within the entire interval $G = (0, 1)$, but satisfy (11) almost everywhere on each interval $(\xi_{\ell-1}, \xi_\ell)$ generated by partitioning the interval $G = (0, 1)$ at points (17). Here arbitrary conditions for joining the root vector functions at points ξ_1, ξ_2, ..., ξ_s are permissible.

Let us now turn to the previous problem (ii) that has been dealt with in [92].

For one thing, an isospectral problem, equivalent to the Lax representation (18), has been formulated for a non-self-adjoint Schroedinger operator with matrix potential (8), dependent on the parameter t.

Assume that the problem for the t-parameter-dependent Schroedinger operator (8) has a point spectrum and is an isospectral one (that is, its spectrum $\{\lambda_k\}$ is independent of t). Denoting the generalized root vector functions in this problem by $U^k(x, t)$, we show further that each vector function $U^k(x, t)$ almost everywhere on the interval G is a solution of Eq. (11):

$$LU^k + \lambda_k U^k = \theta_k U^{k-1}, \tag{11}$$

where the numbers θ_k have the same meaning as defined in Eq. (2).

It has been proved, therefore, that *if a system of generalized root vector functions $\{U^k(x, t)\}$ is closed in $L_2^s(G)$ for any $t \geq 0$, then the Lax equation (18) is equivalent*[10] *to a system of two equations, one of which is Eq. (11) and the other of which takes the form*

$$U_t^k = AU^k + C_k U^k, \tag{19}$$

[10] The equivalence of the Lax equation (18) to the system (11), (19) is understood in the sense that each generalized root vector function $U^k(x, t)$ satisfies the system of equations (11), (19) if and only if any function f in the class $D(L) \cup D([A, L])$ obeys the relation $L_t f - [A, L]f$.

where C_k are constants subject to the condition

$$C_k = C_{k-1}$$

for all k with $\theta \neq 0$.

Assume now that in the Lax equation (18), the operator L is the familiar Schroedinger operator (8) and A is an operator defined as

$$A = C\frac{\partial^3}{\partial x^3} + D\frac{\partial}{\partial x} + \frac{\partial}{\partial x}D,$$

where C and D are operators satisfying the conditions

$$CQ = QC, \qquad CQ = -4D.$$

A moment's reflection will show that the Lax equation (18) converts into the Korteweg-de Vries matrix equation

$$Q_t = \frac{1}{4}CQ_{xxx} - \frac{3}{4}C(Q^2)_x. \tag{20}$$

The following result has been obtained in [92].

THEOREM 6. *If the system $\{U^k(x,t)\}$ of generalized root vector functions of the t-dependent Schroedinger operator (8) forms a basis in $L_2^s(G)$ at $t = 0$, then the integrals of motion of the Korteweg-de Vries matrix equation (20) are expressed as*

$$\int_G P_{r,\ell}(Q)dx,$$

where $P_{r,\ell}(Q)$ are polynomials of Q and its derivatives.

As concluding remarks of our epilogue, we note that the special estimates related to the norms of generalized root functions or vector functions of differential operators and first established by the author of this monograph have played a pivotal role in obtaining most of the above results.

For the generalized root functions of an ordinary differential equation (1) of order n satisfying Eq. (2), these estimates as established in [81] take the following form: *if K' and K'' are two compact sets of interval G such that K' is strictly interior to K'', then there exists a constant $C = C(K', K'')$ such that for all indices k one has*

$$\|\theta_k u_{k-1}\|_{L_2(K')} \leq C|\mu_k|^{n-1}\|u_k\|_{L_2K'')}. \tag{21}$$

One can easily ensure that the exponent $n-1$ in estimate (21) is an exact one. Estimate (21) has been coined by us an *anti-a priori-type estimate* as distinct from the a priori estimates since the latter estimate the solution of an equation through its right-hand side, whereas the inequality (21) estimates the right-hand side of differential equation (2) through a solution of the equation.

For the generalized root vector functions of the Schroedinger operator (8) with matrix potential, a stronger anti-a priori estimate

$$\|\theta_k U^{k-1}\|_{L_2^s(G)} \leq C|\mu_k|\,\|U^k\|_{L_2^s(G)} \tag{22}$$

has been established in [92]; as has been ascertained, the $L_2^s(G)$ norm in (22) can be replaced by the $L_p^s(G)$ norm for any $p \geq 1$.

In [96], an anti-a priori estimate has been established for the generalized root functions for an elliptic operator of second order in an arbitrary bounded N-dimensional domain G,

$$Lu = \sum_{i,j=1}^{N} \frac{\partial}{\partial x_i}\left[a_{ij}(x)\frac{\partial u}{\partial x_j}\right] + \sum_{i=1}^{N} b_i(x)\frac{\partial u}{\partial x_i} + c(x)u. \tag{23}$$

For the generalized root functions $u_k(x)$ of operator (23) which are solutions of the equation

$$Lu_k + \lambda_k u_k = \theta_k u_{k-1}$$

(with θ_k the same as in Eq. (2)), the anti-a priori estimate, exact to within an order of magnitude, is defined as follows: *For any two N-dimensional compact sets K' and K'' of domain G such that the compact set K' is strictly interior to the compact set K'', there exists a constant $C = C(K', K'')$ such that*

$$\|\theta_k u_{k-1}\|_{L_2(K')} \leq C|\sqrt{\lambda_k}|\,\|u_k\|_{L_2(K'')} \tag{24}$$

holds for all indices k.

Later, N. Yu. Kapustin and A. S. Makin, in pursuance of this idea, established anti-a priori estimates for the generalized root functions of a parabolic and even a hypoelliptic operator.

In conclusion, we remark that in [97] we have established, by making use of the anti-a priori estimate (24), conditions for the absolute and uniform convergence and for the convergence in $L_2(G)$ of the expansions in generalized root functions of the elliptic operator (23).

We hope that the results outlined in this epilogue will be represented *in extenso* in a second volume of this monograph.

References

1. S. Agmon. Asymptotic Formulas with Remainder Estimates for Eigenvalues of Elliptic Operators, *Arch. Rat. Mech. Anal.*, 1968, Vol. 28, pp. 165–183.

2. G. Alexits. Convergence Problems of Orthogonal Series, Akademiai Kiado, Budapest, 1961; Russian transl., IL, Moscow, 1963.

3. Sh. A. Alimov. The Fractional Powers of Elliptic Operators and the Isomorphism of Classes of Differentiable Functions, *Differents. Uravn.*, 1972, Vol. 8, No. 9, pp. 1609–1629.

4. Sh. A. Alimov. On the Spectral Decompositions of Functions in H_p^α. *Mat. Sb.*, 1976, Vol. 101 (143) No. 1, pp. 3–20.

5. Sh. A. Alimov, V. A. Il'in and E. M. Nikishin. On the Convergence of Multiple Trigonometric Series and Spectral Decompositions, *I. Usp. Mat. Nauk*, 1976, Vol. 31, No. 6, pp. 28–83.

6. Sh. A. Alimov, V. A. Il'in and E. M. Nikishin. On the Convergence of Multiple Trigonometric Series and Spectral Decompositions, II. *Usp. Mat. Nauk*, Vol. 32, No. 1, pp. 107–130.

7. V. G. Avakumovic. Uber die Eigenfunktionen auf geschlossenen Riemannschen Mannigfaltigkeiten, *Math. Z.*, 1956, Vol. 65, pp. 327–344.

8. K. I. Babenko. On the Asymptotic Behavior of the Eigenfunctions of Differential Operators, In: *Proceedings of the 3rd All-Union Mathematical Congress* (in Russian), Moscow, 1956.

9. H. Bateman and A. Erdélyi. Higher Transcendental Functions, Vol. 2, McGraw-Hill, New York, Toronto, London, 1953; Russian transl., Nauka, Moscow, 1974.

10. G. Bergendal. Convergence and Summability of Eigenfunction Expansions Connected with Elliptic Differential Operators, *Comm. Sem. Math de l'Universite de Lund*, 1959, Vol. 15, pp. 1–63.

11. S. Bochner. Summation of Multiple Fourier Series by Spherical Means, *Trans. Amer. Math. Soc.*, 1936, Vol. 40, pp. 175–207.

12. S. Bochner. Lectures on Fourier Integrals, No. 42, Princeton University Press, 1959; Russian transl., Fizmatgiz, Moscow, 1962.

13. T. Carleman. Propriétés Asymptotiques des Fonctions Fondamentales des Membranes Vibrantes. C. R. Sém. Congr. des Math. Scand. (Stockholm, 1934), Håkan Ohlsson, Lund, 1935, pp. 35–44.

14. T. Carleman. Über die Verteilung der Eigenwerte partieller Differential Gleichungen, *Ber. Sachs. Akad. Wiss.*, 1936, Vol. 86, pp. 119–132.

15. C. Chandrasekharan and S. Minakshisundaram. Typical Means, Oxford, 1952.

16. N. Dunford and J. T. Schwartz. Linear Operators. II. (Spectral Theory); Russian transl., Mir, Moscow, 1966.

17. K. O. Friedrichs. Spektraltheorie halbbeschränkter Operatoren, *Math. Ann.*, 1934, Vol. 109, pp. 465–489, 685–713; 1935, Vol. 110, pp. 777–779.

18. L. Gårding. The Eigenfunction Expansions Connected with Elliptic Operators, In: *12 Congr. Math. Scand. Lund* (1953), pp. 44–55 (see also "*Matematika*," Collection of Translated Papers (in Russian), 1957, 1:3, pp. 107–116).

19. L. Gårding. On the Asymptotic Properties of the Spectral Functions Belonging to a Self-Adjoint Semi-Bounded Extension of an Elliptic Differential Operator, *Kungl. Fysiorg. Sallsk. Lund Forh.*, 1954, Vol. 24, No. 21, pp. 1–18.

20. M. M. Gekhtman. On the Self-Adjoint Extensions of a Symmetric Operator with Totally Continuous Spectral Component, In: *Functional Analysis, Theory of Functions and Their Applications* (in Russian), 1974, Iss. 1, pp. 85–88, Makhachkala.

21. M. L. Goldman. On the Integral Representations in the Fourier Series of Differentiable Multivariable Functions (Cand. Sci. Dissertation), 1972, MIEM, Moscow.

22. M. L. Gorbachuk. Self-Adjoint Boundary-Value Problems for Differential Equations of Second Order with Unbounded Operator Coefficient, *Funkts. Anal. Ego Prilozhen.*, 1971, Vol. 5, No. 1, pp. 20–21.

23. G. H. Hardy, J. L. Littlewood and G. Polya. Inequalities: Cambridge, 1934; Russian transl., IL, Moscow, 1948.

24. L. Hormander. The Spectral Function of an Elliptic Operator, *Acta Math.*, 1968, Vol. 121, No. 3–4, pp. 193–218.

25. V. A. Il'in. The Fractional Kernels, *Mat. Sb.*, 1957, Vol. 41 (83), pp. 459–480.

26. V. A. Il'in. On the Convergence of Expansions in the Eigenfunctions of the Laplace Operator, *Usp. Mat. Nauk*, 1958, Vol. 13, No. 1, pp. 87–180.

27. V. A. Il'in. On the Uniform Convergence of the Eigenfunction Expansions within a Closed Domain, *Mat. Sb.*, 1958, Vol. 45 (87), No. 2, pp. 195–232.

28. V. A. Il'in. Sufficient Conditions for Expanding a Function into an Absolutely and Uniformly Convergent Series in Eigenfunctions, *Mat. Sb.*, 1958, Vol. 46 (88), No. 1, pp. 3–26.

29. V. A. Il'in. On the Eigenfunction Expansion of Functions of Singularities in a Conditionally Convergent Series, *Izv. Akad. Nauk SSSR, Ser. Mat.*, 1958, Vol. 22, No. 1, pp. 49–80.

30. V. A. Il'in. Problems in the Localization and Convergence of Fourier Series in the Fundamental Systems of Functions of the Laplace Operator, *Usp. Mat. Nauk*, 1968, Vol. 23, No. 2, pp. 61–120.

31. V. A. Il'in. The Generalized Localization Principle for the Riesz Means of a Fourier Series in an Arbitrary FSF of the Laplace Operator, *Differents. Uravn.*, 1970, Vol. 6, No. 7, pp. 1143–1158.

32. V. A. Il'in. The Generalized Localization Principle for the Riesz Means of an Arbitrary Self-Adjoint Nonnegative Extension of the Laplace Operator, *Differents. Uravn.*, 1970, Vol. 6, No. 7, pp. 1159–1169.

33. V. A. Il'in. The Conditions for Convergence of Spectral Decompositions Connected with the Self-Adjoint Extensions of Elliptic Operators, III. *Differents. Uravn.*, 1971, Vol. 7, No. 6, pp. 1036–1041.

34. V. A. Il'in. On the Equiconvergence of the Expansions in Eigenfunctions and in an N-fold Fourier Series, *Dokl. Akad. Nauk SSSR*, 1971, Vol. 198, No. 4, pp. 15–18.

35. V. A. Il'in. On the Riesz Equisummability of the Expansions in Eigenfunctions and in an N-fold Fourier Series, *Tr. Mat. Inst. im. V.A. Steklova*, 1972, Vol. 128, pp. 151–162.

36. V. A. Il'in. Convergence Conditions for the Spectral Decompositions Connected with the Self-Adjoint Extensions of Elliptic Operators, IV. *Differents. Uravn.*, 1973, Vol. 9, No. 1, pp. 49–73.

37. V. A. Il'in. Necessary and Sufficient Conditions for the Basis Property and Equiconvergence of Spectral Decompositions and a Trigonometric Series, I and II. *Differents. Uravn.*, 1980, Vol. 16, No. 5, pp. 771–794; Vol. 16, No. 6, pp. 980–1009.

38. V. A. Il'in. Necessary and Sufficient Conditions for the Basis Property in L_p and Equiconvergence of Spectral Decompositions, Exponent System Expansions, and a Trigonometric Series, *Dokl. Akad. Nauk. SSSR*, 1983, Vol. 273, No. 4, pp. 789–793.

39. V. A. Il'in. A Novel Method for Estimating the Spectral Functions of an Elliptic Operator, In: Spectral Theory for Problems in Mathematical Physics [in Russian], *Sb. Nauchn. Tr. MEI*, 1987, Vol. 141, pp. 5–16.

40. V. A. Il'in. Estimate for The Difference of the Riesz Means of Two Spectral Decompositions for a Function in the Class L_2, *Differents. Uravn.*, 1988, Vol. 24, No. 5, pp. 852–863.

41. V. A. Il'in and Sh. A. Alimov. Convergence Conditions for Spectral Decompositions Connected with the Self-Adjoint Extension of Elliptic Operators, I, II. *Differents. Uravn.*, 1971, Vol. 7, No. 4, pp. 670–710; Vol. 7, No. 5, pp. 851–882.

42. V. A. Il'in and Sh. A. Alimov. Convergence Conditions for Spectral Decompositions Connected with the Self-Adjoint Extension of Elliptic Operators, V. *Differents. Uravn.*, 1974, Vol. 10, No. 3, pp. 481–506.

43. V. A. Il'in and A. F. Filippov. On the Spectrum of a Self-Adjoint Extension of the Laplace Operator in a Bounded Domain (The Fundamental Systems of Functions with a Preassigned Subsequence of Fundamental Numbers), *Dokl. Akad. Nauk SSSR*, 1970, Vol. 191, No. 2, pp. 267–269.

44. V. A. Il'in and N. Yu. Kapustin. On the Convergence of Expansions Connected with Nonsemibounded Self-Adjoint Extensions of the Laplace Operator, In: Current Problems of Mathematical Physics and Computational Mathematics [in Russian], Nauka, Moscow, 1984, pp. 105–115.

45. V. A. Il'in and E. I. Moiseev. On the Spectral Decompositions Connected with an Arbitrary Nonnegative Extension of the General Self-Adjoint Elliptic Operator of Second Order, *Dokl. Akad. Nauk SSSR*, 1971, Vol. 191, No. 4, pp. 770–772.

46. V. A. Il'in and E. G. Poznyak. The Fundamentals of Mathematical Analysis (in Russian), Part II, Nauka, Moscow, 1980.

47. V. A. Il'in and I. A. Shishmarev. Fourier Series in the Fundamental Systems of Functions of a Polyharmonic Operator, *Dokl. Akad. Nauk SSSR*, 1969, Vol. 189, No. 4, pp. 707–709.

48. G. A. Il'yushina. On the Generalized Localization Principle of Spectral Decompositions Connected with the Beltrami Operator, *Differents. Uravn.*, 1971, Vol. 7, No. 10, pp. 1845–1853.

49. G. A. Il'yushina. On the Generalized Localization Principle of Spectral Decompositions Which are Connected with the Beltrami Operator Specified in an Arbitrary N-Dimensional Domain, *Differents. Uravn.*, Vol. 7, No. 11, pp. 2058–2065.

50. V. Ya. Ivrii. On the Second Term of a Spectral Asymptotics for the Laplace–Beltrami Operator on Manifolds with Boundary, *Funkts. Anal. Ego Prilozhen.*, 1980; Vol. 14, No. 2, pp. 25–30.

51. S. Kaczmarz and H. Steinhaus. Theory of Orthogonal Series (Russian transl.), Fizmatgiz, Moscow, 1958.

52. I. Kendzhaev. On the Absolute and Uniform Convergence of Fourier Series, *Dokl. Akad. Nauk Tajik SSR*, 1967, Vol. 10, No. 12, pp. 12–17.

53. I. Kendzhaev. On the Uniform and Absolute Convergence of Eigenfunction Expansions, *Dokl. Akad. Nauk Tajik SSR*, 1968, Vol. 11, No 1, pp. 10–15.

54. A. G. Kostyuchenko and B. S. Mityagin. On the Uniform Convergence of Spectral Decompositions, *Funkts. Anal. Ego Prolozhen.*, 1973, Vol. 7, No. 2, pp. 32–42.

55. M. A. Krasnosel'skii and E. I. Pustyl'nik. The Use of Fractional Powers in Studying Fourier Series in the Eigenfunctions of Differential Operators, *Dokl. Akad. Nauk SSSR*, 1958, Vol. 122, No. 6, pp. 978–981.

56. O. A. Ladyzhenskaya. On a Fourier Method for the Wave Equation, *Dokl. Akad. Nauk SSSR*, 1950, Vol. 75, No. 6, pp. 765–768.

57. B. M. Levitan. On the Eigenfunction Expansion of the Laplace Operator, *Mat. Sb.*, 1954, Vol. 35 (77), No. 2, pp. 267–316.

58. B. M. Levitan. On the Summation of Multiple Series and Fourier Integrals, *Dokl. Akad. Nauk SSSR*, 1955, Vol. 102, No. 6, pp. 1073–1076.

59. B. M. Levitan. On the Eigenfunction Expansion of a Self-Adjoint Partial Differential Equation, *Tr. Mosk. Mat. O-va*, 1956, Vol. 5, pp. 269–298.

60. C. Miranda. Partial Differential Equations of Elliptic Type (Russian transl.), IL, Moscow, 1957.

61. C. Mizohata. Sur le Proprietes Asymptotiques de Values Propres pour le Operateurs Elliptiques, *J. Math. Kyoto Univ.*, 1965, Vol. 4, No. 3, pp. 399–428.

62. E. I. Moiseev. The Mean-Value Formula for the Eigenfunctions of an Elliptic Self-Adjoint Operator, *Differents. Uravn.*, 1971, Vol. 7, No. 8, pp. 1490–1502.

63. E. I. Moiseev. The Uniform Convergence of Certain Expansions within a Closed Domain, *Dokl. Akad. Nauk SSSR*, 1977, Vol. 233, No. 6, pp. 1042–1045.

64. J. von Neumann. Allgemeine Eigenwerttheorie Hermitescher Funktionaloperatoren, *Math. Ann.*, 1925, Vol. 102, pp. 49–131.

65. S. M. Nikol'skii. Multivariable Function Approximation and Embedding Theorems (in Russian), Nauka, Moscow, 1969.

66. A. M. Olevskii. On the Continuation of a Sequence of Functions to a Complete Orthonormal System, *Mat. Zametki*, 1969, Vol. 6, No. 6, pp. 737–747.

67. A. M. Olevskii. Fourier Series with Respect to General Orthogonal Systems. Berlin, 1975.

68. P. K. Rashevskii. Riemannian Geometry and Tensor Analysis (in Russian), Nauka, Moscow, 1967.

69. Ya. Sh. Salimov. On the Riesz Means of the Spectral Function of a Non-Self-Adjoint Laplace Operator Involving Associated Functions of an Order in Excess of the Riesz Means Order, *Dokl. Akad. Nauk SSSR*, 1986, Vol. 289, No. 6, pp. 1311–1314.

70. Ya. Sh. Salimov. On the Riesz Means of the Biorthogonal Expansions in the Eigenfunctions and Associated Functions of Non-Self-Adjoint Extensions of the Laplace Operator, *Differents. Uravn.*, 1986, Vol. 22, No. 5, pp. 864–876.

71. R. T. Seeley. A Sharp Asymptotic Remainder Estimate for the Eigenvalues of the Laplacian in a Domain of R^3, *Adv. Math.*, 1978, Vol. 29, pp. 244–269.

72. Z. V. Shonia. The Conditions for Convergence of the Spectral Decompositions Connected with the Nonsemibounded Self-Adjoint Extensions of the Schroedinger Operator, *Differents. Uravn.*, 1986, Vol. 22, No. 3, pp. 483–495.

73. V. I. Smirnov. A Course in Higher Mathematics. Vol. 4. (in Russian) Gostekhizdat, Moscow, 1957.

74. S. L. Sobolev. Certain Applications of Functional Analysis in Mathematical Physics (in Russian), Izd. Sib. Otd. Akad. Nauk SSSR, Novosibirsk, 1962.

75. E. M. Stein. Localization and Summability of Multiple Fourier Series, *Acta Math.*, 1958, Vol. 100, No. 1–2, pp. 93–147.

76. E. C. Titchmarsh. Eigenfunction Expansions Associated with Second-Order Differential Equations. Part II. Oxford University Press, 1958; Russian transl., IL, Moscow, 1961.

77. G. N. Watson. Theory of Bessel Functions. Vol. 1 (Russian transl.) IL, Moscow, 1949.

78. H. Weyl. Das Asymptotische Verteilungsgezetz der Eigenwerte Linearer Partieller Differentialgleichungen (mit einer Anwendung auf die Theorie der Hohlraumstrahlung), *Math. Ann.*, 1912, Vol. 71, pp. 441–479.

79. A. Zygmund. Trigonometric Series. Vols. 1, 2. Cambridge University Press, 1959; Russian transl., Mir, Moscow, 1965.

References to the Epilogue

80. M. V. Keldysh. On the Completeness of the Eigenfunctions in Certain Classes of Non-Self-Adjoint Operators, *Usp. Mat. Nauk*, 1971, Vol. 26, No 4, pp. 15–41.

81. V. A. Il'in. Necessary and Sufficient Conditions for the Basis Property and Equiconvergence of Spectral Decompositions and a Trigonometric Series, I, II. *Differents. Uravn.*, 1980, Vol. 16, No. 5, pp. 771–794; Vol. 16, No. 6, pp. 980–1009.

82. V. A. Il'in. Necessary and Sufficient Conditions for the Basis Property in L_p and Equiconvergence of Spectral Decompositions, Exponent System Expansions, and a Trigonometric Series, *Dokl. Akad. Nauk SSSR*, 1983, Vol. 273, No. 4, pp. 789–793.

83. V. A. Il'in. On the Necessary Condition for the Convergence of a Trigonometric Series and the Spectral Decomposition of an Arbitrary Summable Function, *Differents. Uravn.*, 1985, Vol. 21, No. 3, pp. 371–379.

84. V. A. Il'in and E. I. Moiseev. On Systems Composed of the Subsets of Root Functions of Two Different Boundary-Value Problems, *Tr. Mat. Inst. im. V.A. Steklova*, 1991, Vol. 201, pp. 219–230.

85. V. A. Il'in. On the Existence of a Reduced System of the Eigenfunctions and Associated Functions in a Non-Self-Adjoint Ordinary Differential Operator, *Tr. Mat. Inst. im. V.A. Steklova*, 1976, Vol. 142, pp. 148–155.

86. V. A. Steklov. Sur les Expressions Asymptotiques de Certain Fonctions Definies par des Equations Differetielles Lineaires de Deuxieme Ordre, et Leur Applications au Probleme du Developpement d'une Fonction Arbitraire en Series Procedant Suivant les Dites Fonctions, *Kharkov. Soobshch. Mat. Ova.*, 1907–1909, Vol. 10, No. 2–6, pp. 97–199.

87. V. A. Il'in. On the Properties of the Reduced System of Eigenfunctions and Associated Functions of the Keldysh bundle for Ordinary Differential Operators, *Dokl. Akad. Nauk SSSR*, 1976, Vol. 130, No. 1, pp. 30–33.

88. V. A. Il'in. Equiconvergence with a Trigonometric Series of Expansions in Terms of Root Functions of a One-Dimensional Schrödinger Operator with a Complex Potential of the Class L_1. *Differents. Uravn.*, 1991, Vol. 27, No. 4, pp. 577–597.

89. V. A. Il'in. Component-wise Equiconvergence with a Trigonometric Series of Expansions in Eigenfunctions and Associated Functions of a Schrödinger Operator with a Nonhermite Potential Matrix of Summable Elements, *Differents. Uravn.*, 1991, Vol. 27, No. 11, pp. 1862–1879.

90. V. A. Il'in. On the Unconditional Basis Property of the Systems of Eigenfunctions and Associated Functions of a Differential Second-Order Operator within a Closed Interval, *Dokl. Akad. Nauk SSSR*, 1983, Vol. 273, No. 5, pp. 789–793.

91. V. A. Il'in. Necessary and Sufficient Conditions for the Riesz Basis Property of the Root Vectors of Discontinuous Second-Order Operators, *Differents. Uravn.*, 1986, Vol. 22, No. 12, pp. 2059–2071.

92. V. A. Il'in, K. V. Mal'kov and E. I. Moiseev. The Basis Property of the Root Functions of Non-Self-Adjoint Operators and Integrability of Nonlinear Evolution Equations Associated through Lax Representations, I, II. *Differents. Uravn.*, 1989, Vol. 25, No. 11, pp. 1956–1970; Vol. 25, No. 12, pp. 2133–2143.

93. V. A. Il'in. On the Riesz Basis Property of the Root Vector Function Systems of a Discontinuous Schroedinger Operator with Matrix Potential, *Dokl. Akad. Nauk SSSR*, 1990, Vol. 314, No. 1, pp. 59–62.

94. I. Ts. Gokhberg and M. G. Krein. An Introduction to the Theory of Non-Self-Adjoint Operators in Hilbert Space (in Russian), Moscow, 1965.

95. P. Lax. The Evolution Equation Integrals and Solitary Waves, *Commun. Pure Appl. Math.*, 1968, Vol. 21, p. 467; see also Russian transl. in the Collection "*Matematika*," 1969, Vol. 13, No. 5.

96. V. A. Il'in. On the Relations, Exact to within an Order of Magnitude, between the L_2-Norms of Eigenfunctions and the Associated Functions of an Elliptic Second-Order Operator, *Differents. Uravn.*, 1982, Vol. 18, No. 1, pp. 30–37.

97. V. A. Il'in. On the Absolute and Uniform Convergence of the Expansions in Eigenfunctions and Associated Functions of a Non-Self-Adjoint Elliptic Operator, *Dokl. Akad. Nauk SSSR*, 1984, Vol. 274, No. 1, pp. 19–22.

98. V. A. Il'in. The Form of Boundary Conditions and the Basis Property and Equiconvergence with a Trigonometric Series of Expansions in Terms of Root Functions for a Non-Self-Adjoint Differential Operator, *Differents. Uravn.*, 1994, Vol. 30, No. 9, pp. 1516–1629.

Index

Basis property in L_p 369
Bernstein inequality 92
Besov class of functions 90
Bessel differentiation 30
Bessel function 4
asymptotic formula 7
module estimation 15, 122
recurrence relations 47, 50
Bessel inequality 6
Bessel–MacDonald kernel 91
Bonnet mean-value theorem 60

Conditions for divergence in L_p of Fourier series in arbitrary FSF 79

Derivative
generalized 89
with respect to radius for functions of Nikol'skii class 148
Dirichlet kernel 210

Elliptic operator
fractional powers of self-adjoint extensions 275
self-adjoint 84, 254
Embedding theorems 93
Equiconvergence, component-wise 373
Estimate,
anti-a priori type 377
difference of the Riesz means
spectral functions in L_1 201
fractional kernels in L_1 215
two compactly supported spectral decompositions 219, 220
two spectral decompositions almost everywhere 227
Integral of the squared fundamental functions 99, 259
Lebesgue function of the Riesz means of spectral decompositions 111, 278, 297
local asymptotic 313
main asymptotic 342
Riesz means of a spectral function in L_1, lower-bound 110, 278, 297
remainder term of the Riesz means of a spectral function
in class L_2 174
in class L_∞ 180
spectral decomposition of fractional kernels
upper-bound in L_q 56
lower-bound in L_1 67
sum of the squares of fundamental functions 5, 13, 19
Exact conditions for
uniform convergence of multiple trigonometric series with spherical partial sums 352
eigenfunction convergence of the first three boundary-value problems for elliptic operator 356, 360
Extension, self-adjoint (semi-bounded)
elliptic operator of second order 84
elliptic operator of second order

with weight 254
Laplace operator 88

Finite function 85
Fourier coefficient 5, 361
Fourier image 87, 255
finite function of Nikol'skii class 135, 307
Fourier series in arbitrary FSF
divergence condition in L_p 79
localization condition 75
nonlocalization condition 76
uniform convergence condition 69
Fractional kernel
definition 28, 103
representation 29, 103
Friedrichs theorem 85
FSF, see Fundamental system of functions 2
Function $w_\nu(a, b)$ 257
Fundamental function
as element of FSF 3
as kernel for ordered spectral representation 87, 255
Fundamental system of functions (FSF) of Laplace operator
in domain 2
in subdomain 2

Gårding–Browder–Mautner theorem 87, 255
Gårding theorem 86
Generalized derivative 89
Generalized root functions of non-self-adjoint differential operator 368
Geodetical distance 257
Geodetical sphere 335
Green's function 29
integral operator 29

Helmholtz operator 29

Integral
fractional 250
of the squares of fundamental functions 99, 259
Isospectral problem 376

Korteweg-de Vries matrix equation 377

Laplace operator 2, 88
Lax representation 376
Lebesgue function of the Riesz means of spectral decompositions 110, 278
Local asymptotic estimate 313
Localization conditions
decompositions for elliptic operator of second order 333
Fourier series in arbitrary FSF 75
Riesz means of spectral decompositions 94–98
Riesz means of spectral decompositions almost everywhere 227
Localization principle, of spectral decompositions
classical 75, 94–103, 333
generalized 227
component-wise 373

MacDonald function 29
recurrence relation 31
Main inequality (4.2.14) 283
Mean-value formula for fundamental function 4, 98
Moiseev mean-value formula 257
Multiple trigonometric series with spherical sums 351
uniform convergence conditions 352

Neumann, von
theorem 85
Neyman function 4, 257
Nikol'skii class of functions 89
definition 89
derivatives with respect to radius 148
embedding theorem 93

Nikol'skii class of functions (*cont.*)
geodetical spherical means 335
norm 90, 92
spherical means 148
Nonnegative operator 85
Norm
in Besov class 90
in Nikol'skii class 90, 92
in Sobolev class 89
in Sobolev–Liouville class 91

Operation $\mathcal{D}F(r)$ 157, 228
Operation of averaging $S_{Ro}[F(r)]$ 182

Parseval equality 100, 13, 88, 255

Radial function 4, 99
Riesz basis 374
Riesz means, equisummability
in classical sense 199
in generalized sense 227
Riesz means of integrals 126
Riesz means of spectral decompositions
conditions for nonlocalization 94, 277
definition 86
equisummability in classical sense 199
equisummability in generalized sense 227
fractional kernels 129
localization in classical sense 95–96, 277
localization in generalized sense 227
uniform convergence 95–96
Riesz means of spectral function
remainder estimate in class L_2 174
remainder estimate in class L_∞ 179
Root functions
canonical system 367, 371
reduced system 371

Schroedinger
matrix-potential operator 372–374
Semi-bounded operator 85
Sobolev
class of functions 89
Sobolev–Liouville
class of functions 91
Space L_2
spectral ordered representation 87, 255
with weight 254
Spectral decomposition
arbitrary FSF 40
self-adjoint extension of elliptic operator 86, 255
Spectral function 40, 86
Spectrum, of arbitrary FSF 19
Stein
interpolation lemma 300
Symmetrical operator 85
System of root functions
canonical 367, 371
closed 368
minimal 368
reduced 371

Theorem(s)
embedding 93
equisummability of the Riesz means of spectral decompositions 199, 226
estimate of the remainder term of spectral function 41, 174, 179
negative type 109, 277
positive type 135, 303
resonance type 76, 133, 297

Uniform convergence, conditions for
multiple trigonometric series 352
expansions for arbitrary extensions of elliptic operators 303–305
eigenfunction expansions of the first three boundary-value problems for elliptic operator 356
Fourier series in an arbitrary FSF 69

Uniform convergence, conditions for (*cont.*)
Riesz means of spectral decompositions 95–98
Unit expansion 85

Zygmund–Hölder
class of functions 90